IFIP Advances in Information and Communication Technology 364

T0181045

IFIP – The International Federation for Information Processing

IFIP was founded in 1960 under the auspices of UNESCO, following the First World Computer Congress held in Paris the previous year. An umbrella organization for societies working in information processing, IFIP's aim is two-fold: to support information processing within ist member countries and to encourage technology transfer to developing nations. As ist mission statement clearly states,

> IFIP's mission is to be the leading, truly international, apolitical organization which encourages and assists in the development, exploitation and application of information technology for the bene t of all people.

IFIP is a non-profitmaking organization, run almost solely by 2500 volunteers. It operates through a number of technical committees, which organize events and publications. IFIP's events range from an international congress to local seminars, but the most important are:

- The IFIP World Computer Congress, held every second year;
- Open conferences;
- Working conferences.

The flagship event is the IFIP World Computer Congress, at which both invited and contributed papers are presented. Contributed papers are rigorously refereed and the rejection rate is high.

As with the Congress, participation in the open conferences is open to all and papers may be invited or submitted. Again, submitted papers are stringently refereed.

The working conferences are structured differently. They are usually run by a working group and attendance is small and by invitation only. Their purpose is to create an atmosphere conducive to innovation and development. Refereeing is less rigorous and papers are subjected to extensive group discussion.

Publications arising from IFIP events vary. The papers presented at the IFIP World Computer Congress and at open conferences are published as conference proceedings, while the results of the working conferences are often published as collections of selected and edited papers.

Any national society whose primary activity is in information may apply to become a full member of IFIP, although full membership is restricted to one society per country. Full members are entitled to vote at the annual General Assembly, National societies preferring a less committed involvement may apply for associate or corresponding membership. Associate members enjoy the same benefits as full members, but without voting rights. Corresponding members are not represented in IFIP bodies. Affiliated membership is open to non-national societies, and individual and honorary membership schemes are also offered.

Lazaros Iliadis   Ilias Maglogiannis
Harris Papadopoulos (Eds.)

Artificial Intelligence Applications and Innovations

12th INNS EANN-SIG International Conference, EANN 2011
and 7th IFIP WG 12.5 International Conference, AIAI 2011
Corfu, Greece, September 15-18, 2011
Proceedings , Part II

 Springer

Volume Editors

Lazaros Iliadis
Democritus University of Thrace
68200 N. Orestiada, Greece
E-mail: liliadis@fmenr.duth.gr

Ilias Maglogiannis
University of Central Greece
35100 Lamia, Greece
E-mail: imaglo@ucg.gr

Harris Papadopoulos
Frederick University
1036 Nicosia, Cyprus
E-mail: h.papadopoulos@frederick.ac.cy

ISSN 1868-4238   e-ISSN 1868-422X
ISBN 978-3-642-26965-3   ISBN 978-3-642-23960-1 (eBook)
DOI 10.1007/978-3-642-23960-1
Springer Heidelberg Dordrecht London New York

CR Subject Classification (1998): I.2, H.3, H.4, F.1, I.4, I.5

Typesetting: Camera-ready by author, data conversion by Scientific Publishing Services, Chennai, India

Printed on acid-free paper

Springer is part of Springer Science+Business Media (www.springer.com)

Preface

Artificial intelligence (AI) is a rapidly evolving area that offers sophisticated and advanced approaches capable of tackling complicated and challenging problems. Transferring human knowledge into analytical models and learning from data is a task that can be accomplished by soft computing methodologies. Artificial neural networks (ANN) and support vector machines are two cases of such modeling techniques that stand behind the idea of learning. The 2011 co-organization of the 12th Engineering Applications of Neural Networks (EANN) and of the 7th Artificial Intelligence Applications and Innovations (AIAI) conferences was a major technical event in the fields of soft computing and AI, respectively.

The first EANN was organized in Otaniemi, Finland, in 1995. It has had a continuous presence as a major European scientific event. Since 2009 it has been guided by a Steering Committee that belongs to the "EANN Special Interest Group" of the International Neural Network Society (INNS).

The 12th EANN 2011 was supported by the INNS and by the IEEE branch of Greece. Moreover, the 7th AIAI 2011 was supported and sponsored by the International Federation for Information Processing (IFIP).

The first AIAI was held in Toulouse, France, in 2004 and since then it has been held annually offering scientists the chance to present the achievements of AI applications in various fields. It is the official conference of the Working Group 12.5 "Artificial Intelligence Applications" of the IFIP Technical Committee 12, which is active in the field of AI. IFIP was founded in 1960 under the auspices of UNESCO, following the first World Computer Congress held in Paris the previous year.

It was the first time ever that these two well-established events were hosted under the same umbrella, on the beautiful Greek island of Corfu in the Ionian Sea and more specifically in the Department of Informatics of the Ionian University.

This volume contains the papers that were accepted to be presented orally at the 7th AIAI conference and the papers accepted for the First International Workshop on Computational Intelligence in Software Engineering (CISE) and the Artificial Intelligence Applications in Biomedicine (AIAB) workshops. The conference was held during September 15–18, 2011. The diverse nature of papers presented demonstrates the vitality of neural computing and related soft computing approaches and it also proves the very wide range of AI applications. On the other hand, this volume contains basic research papers, presenting variations and extensions of several approaches.

The response to the call for papers was more than satisfactory with 150 papers initially submitted. All papers passed through a peer-review process by at least two independent academic referees. Where needed a third referee was consulted to resolve any conflicts. In the EANN/AIAI 2011 event, 34% of the submitted manuscripts (totally 52) were accepted as full papers, whereas 21% were accepted as short ones and 45% (totally 67) of the submissions were rejected. The authors of accepted papers came from 27 countries all over Europe (e.g., Austria, Bulgaria, Cyprus, Czech Republic, Finland, France, Germany, Greece, Italy, Poland, Portugal, Slovakia, Slovenia, Spain, UK), America (e.g., Brazil, Canada, Chile, USA), Asia (e.g., China, India, Iran, Japan, Taiwan), Africa (e.g., Egypt, Tunisia) and Oceania (New Zealand). Three keynote speakers were invited and they gave lectures on timely aspects of AI and ANN.

1. Nikola Kasabov. Founding Director and Chief Scientist of the Knowledge Engineering and Discovery Research Institute (KEDRI), Auckland (www.kedri.info/). He holds a Chair of Knowledge Engineering at the School of Computing and Mathematical Sciences at Auckland University of Technology. He is a Fellow of the Royal Society of New Zealand, Fellow of the New Zealand Computer Society and a Senior Member of IEEE. He was Past President of the International Neural Network Society (INNS) and a Past President of the Asia Pacific Neural Network Assembly (APNNA). Title of the keynote presentation: "Evolving, Probabilistic Spiking Neural Network Reservoirs for Spatio- and Spectro-Temporal Data."
2. Tom Heskes. Professor of Artificial Intelligence and head of the Machine Learning Group at the Institute for Computing and Information Sciences, Radboud University Nijmegen, The Netherlands. He is Principal Investigator at the Donders Centre for Neuroscience and Director of the Institute for Computing and Information Sciences. Title of the keynote presentation: "Reading the Brain with Bayesian Machine Learning."
3. A.G. Cohn. Professor of Automated Reasoning, Director of Institute for Artificial Intelligence and Biological Systems, School of Computing, University of Leeds, UK. Tony Cohn holds a Personal Chair at the University of Leeds, where he is Professor of Automated Reasoning. He is presently Director of the Institute for Artificial Intelligence and Biological Systems. He leads a research group working on knowledge representation and reasoning with a particular focus on qualitative spatial/spatio-temporal reasoning, the best known being the well-cited region connection calculus (RCC). Title of the keynote presentation: "Learning about Activities and Objects from Video."

The EANN/AIAI conference consisted of the following main thematic sessions:

- AI in Finance, Management and Quality Assurance
- Computer Vision and Robotics
- Classification-Pattern Recognition
- Environmental and Earth Applications of AI
- Ethics of AI
- Evolutionary Algorithms—Optimization
- Feature Extraction-Minimization
- Fuzzy Systems
- Learning—Recurrent and RBF ANN
- Machine Learning and Fuzzy Control
- Medical Applications

- Multi-Layer ANN
- Novel Algorithms and Optimization
- Pattern Recognition-Constraints
- Support Vector Machines
- Web-Text Mining and Semantics

We would very much like to thank Hassan Kazemian (London Metropolitan University) and Pekka Kumpulainen (Tampere University of Technology, Finland) for their kind effort to organize successfully the Applications of Soft Computing to Telecommunications Workshop (ASCOTE).

Moreover, we would like to thank Efstratios F. Georgopoulos (TEI of Kalamata, Greece), Spiridon Likothanassis, Athanasios Tsakalidis and Seferina Mavroudi (University of Patras, Greece) as well as Grigorios Beligiannis (University of Western Greece) and Adam Adamopoulos (Democritus University of Thrace, Greece) for their contribution to the organization of the Computational Intelligence Applications in Bioinformatics (CIAB) Workshop.

We are grateful to Andreas Andreou (Cyprus University of Technology) and Harris Papadopoulos (Frederick University of Cyprus) for the organization of the Computational Intelligence in Software Engineering Workshop (CISE).

The Artificial Intelligence Applications in Biomedicine (AIAB) Workshop was organized successfully in the framework of the 12th EANN 2011 conference and we wish to thank Harris Papadopoulos, Efthyvoulos Kyriacou (Frederick University of Cyprus) Ilias Maglogiannis (University of Central Greece) and George Anastassopoulos (Democritus University of Thrace, Greece).

Finally, the Second Workshop on Informatics and Intelligent Systems Applications for Quality of Life information Services (2nd ISQLIS) was held successfully and we would like to thank Kostas Karatzas (Aristotle University of Thessaloniki, Greece) Lazaros Iliadis (Democritus University of Thrace, Greece) and Mihaela Oprea (University Petroleum-Gas of Ploiesti, Romania).

The accepted papers of all five workshops (after passing through a peer-review process by independent academic referees) were published in the Springer proceedings. They include timely applications and theoretical research on specific subjects. We hope that all of them will be well established in the future and that they will be repeated every year in the framework of these conferences.

We hope that these proceedings will be of major interest for scientists and researchers world wide and that they will stimulate further research in the domain of artificial neural networks and AI in general.

September 2011   Dominic Palmer Brown

Organization

Executive Committee

Conference Chair

Dominic Palmer Brown   London Metropolitan University, UK

Program Chair

Lazaros Iliadis	Democritus University of Thrace, Greece
Elias Maglogiannis	University of Central Greece, Greece
Harris Papadopoulos	Frederick University, Cyprus

Organizing Chair

Vassilis Chrissikopoulos	Ionian University, Greece
Yannis Manolopoulos	Aristotle University, Greece

Tutorials

Michel Verleysen	Universite catholique de Louvain, Belgium
Dominic Palmer-Brown	London Metropolitan University, UK
Chrisina Jayne	London Metropolitan University, UK
Vera Kurkova	Academy of Sciences of the Czech Republic, Czech Republic

Workshops

Hassan Kazemian	London Metropolitan University, UK
Pekka Kumpulainen	Tampere University of Technology, Finland
Kostas Karatzas	Aristotle University of Thessaloniki, Greece
Lazaros Iliadis	Democritus University of Thrace, Greece
Mihaela Oprea	University Petroleum-Gas of Ploiesti, Romania
Andreas Andreou	Cyprus University of Technology, Cyprus
Harris Papadopoulos	Frederick University, Cyprus
Spiridon Likothanassis	University of Patras, Greece
Efstratios Georgopoulos	Technological Educational Institute (T.E.I.) of Kalamata, Greece
Seferina Mavroudi	University of Patras, Greece
Grigorios Beligiannis	University of Western Greece, Greece

Adam Adamopoulos	University of Thrace, Greece	
Athanasios Tsakalidis	University of Patras, Greece	
Efthyvoulos Kyriacou	Frederick University, Cyprus	
Elias Maglogiannis	University of Central Greece, Greece	
George Anastassopoulos	Democritus University of Thrace, Greece	

Honorary Chairs

Tharam Dillon   Curtin University, Australia
Max Bramer   University of Portsmouth, UK

Referees

Aldanondo M.	Karpouzis Kostas	Senatore S.
Alexandridis G.	Kefalas P.	Shen Furao
Anagnostou K.	Kermanidis I.	Sideridis A.
Anastassopoulos G.	Kosmopoulos Dimitrios	Sioutas S.
Andreadis I.	Kosmopoulos D.	Sotiropoulos D.G.
Andreou A.	Kyriacou E.	Stafylopatis A.
Avlonitis M.	Lazaro J.Lopez	Tsadiras A.
Bankovic Z.	Likothanassis S.	Tsakalidis Athanasios
Bessis N.	Lorentzos Nikos	Tsapatsoulis N.
Boracchi G.	Malcangi M.	Tsevas S.
Caridakis George	Maragkoudakis M.	Vassileiades N.
Charalambous C.	Marcelloni F.	Verykios V.
Chatzioannou Aristotle	Margaritis K.	Vishwanathan Mohan
Constantinides A.	Mohammadian M.	Voulgaris Z.
Damoulas T.	Nouretdinov Ilia	Vouros G.
Doukas Charalampos	Olej Vladimir	Vouyioukas Demosthenis
Fox C.	Onaindia E.	Vovk Volodya
Gaggero M.	Papatheocharous E.	Wallace Manolis
Gammerman Alex	Plagianakos Vassilis	Wang Zidong
Georgopoulos E.	Portinale L.	Wyns B.
Hatzilygeroudis I.	Rao Vijan	Xrysikopoulos V.
Hunt S.	Roveri M.	Yialouris K.
Janssens G.	Ruggero Donida Labati	Zenker Bernd Ludwig
Kabzinski J.	Sakelariou I.	Zhiyuan Luo
Kameas A.	Schizas C.	

Sponsoring Institutions

The 12th EANN / 7th AIAI Joint Conferences were organized by IFIP (International Federation for Information Processing), INNS (International Neural Network Society), the Aristotle University of Thessaloniki, the Democritus University of Thrace and the Ionian University of Corfu.

Table of Contents – Part II

Fuzzy Systems

Learning and Novel Algorithms

Recurrent and Radial Basis Function ANN

Machine Learning

Generic Algorithms

Data Mining

Reinforcement Learning

Web Applications of ANN

Medical Applications of ANN and Ethics of AI

Environmental and Earth Applications of AI

Computational Intelligence in Software Engineering (CISE) Workshop

Artificial Intelligence Applications in Biomedicine (AIAB) Workshop

Table of Contents – Part I

Financial and Management Applications of AI

Fuzzy Systems

Support Vector Machines

Learning and Novel Algorithms

Reinforcement and Radial Basis Function ANN

Machine Learning

Evolutionary Genetic Algorithms - Optimization

Web Applications of ANN

Spiking ANN

Feature Extraction - Minimization

Medical Applications of AI

Environmental and Earth Applications of AI

Multi Layer ANN

Bioinformatics

The Applications of Soft Computing to Telecommunications (ASCOTE) Workshop

Computational Intelligence Applications in Bioinformatics (CIAB) Workshop

Informatics and Intelligent Systems Applications for Quality of Life information Services (ISQLIS) Workshop

Real Time Robot Policy Adaptation Based on Intelligent Algorithms

Genci Capi[1], Hideki Toda[1], and Shin-Ichiro Kaneko[2]

[1] Department of Electric and Electronic Eng.,
University of Toyama, Toyama, Japan
capi@eng.u-toyama.ac.jp
[2] Toyama National College of Technology

Abstract. In this paper we present a new method for robot real time policy adaptation by combining learning and evolution. The robot adapts the policy as the environment conditions change. In our method, we apply evolutionary computation to find the optimal relation between reinforcement learning parameters and robot performance. The proposed algorithm is evaluated in the simulated environment of the Cyber Rodent (CR) robot, where the robot has to increase its energy level by capturing the active battery packs. The CR robot lives in two environments with different settings that replace each other four times. Results show that evolution can generate an optimal relation between the robot performance and exploration-exploitation of reinforcement learning, enabling the robot to adapt online its strategy as the environment conditions change.

Keywords: Reinforcement learning, policy adaptation, evolutionary computation.

1 Introduction

Reinforcement learning (RL) ([1], [2]) is an efficient learning framework for autonomous robots, in which the robot learns how to behave, from interactions with the environment, without explicit environment models or teacher signals. Most RL applications, so far, have been constrained to stationary environments. However, in many real-world tasks, the environment is not fixed. Therefore, the robot must change its strategy based on the environment conditions. For small environment changes, Minato et al., (2000) has pointed out that current knowledge learned in a previous environment is partially applicable even after the environment has changed, if we only consider reaching the goal and thereby sacrifice optimality ([3]).

Efforts have also been made to move in more dynamic environments. Matsui et al. ([4]) proposed a method, which senses a changing environment by collecting failed instances and partially modifies the strategy for adapting to subsequent changes of the environment by reinforcement learning. Doya incorporated a noise term in policies, in order to promote exploration ([5]). The size of noise is reduced as the performance improves. However, this method can be applied when the value function is known for all the states.

L. Iliadis et al. (Eds.): EANN/AIAI 2011, Part II, IFIP AICT 364, pp. 1–10, 2011.

Previous approaches on combining learning and evolution ([6], [7], [8], [9]) reported that combination tends to provide earlier achievement of superior performance. Niv et al. have considered evolution of RL in uncertain environments ([10]). They solve near-optimal neuronal learning rules in order to allow simulated bees to respond rapidly to changes in reward contingencies. In our previous work, we considered evolution of metaparameters for faster convergence of reinforcement learning ([11], [12]). However, in all these approaches the robot learned the optimal strategy in stationary environment.

In difference from previous works, in this paper, we combine an actor-critic RL and evolution to develop robots able to adapt their strategy as the environment changes. The metaparameters, initial weight connection, number of hidden neurons of actor and critic networks, and the relation between the energy level and cooling factor are evolved by a real number Genetic Algorithm (GA). In order to test the effectiveness of the proposed algorithm, we considered a biologically inspired task for the CR robot ([13]). The robot must survive and increase its energy level by capturing the active battery packs distributed in the environment. The robot lives in two different environments, which substitute for each other four times during the robot's life. Therefore, the robot must adapt its strategy as the environment changes.

The performance of proposed method is compared with that of (a) RL and (b) evolution of neural controller. In the actor-critic RL, we used arbitrarily selected metaparameters, randomly initialized initial weight connections, and a linearly proportional relationship between cooling factor and energy level. The robot controlled by the evolved neural controller applies the same strategy throughout all its life, which was optimal only for one environment. The performance of the actor-critic RL was strongly related to the metaparameters, especially the relation between cooling factor and energy level. Combining learning and evolution gives the best performance overall. Because of optimized metaparameters and initial weight connections, the robot was able to exploit the environment from the beginning of its life. In addition, the robot switched between exploration and exploitation based on the optimized relation between the energy level and cooling factor.

2   Cyber Rodent Robot

In our simulations, we used the CR robot, which is a two-wheel-driven mobile robot, as shown in Fig. 1. The CR is 250 mm long and weights 1.7 kg. The CR is equipped with:

- Omni-directional C-MOS camera.
- IR range sensor.
- Seven IR proximity sensors.
- 3-axis acceleration sensor.
- 2-axis gyro sensor.
- Red, green and blue LED for visual signaling.
- Audio speaker and two microphones for acoustic communication.
- Infrared port to communicate with a nearby robot.
- Wireless LAN card and USB port to communicate with the host computer.

Five proximity sensors are positioned on the front of robot, one behind and one under the robot pointing downwards. The proximity sensor under the robot is used when the robot moves wheelie. The CR contains a Hitachi SH-4 CPU with 32 MB memory. The FPGA graphic processor is used for video capture and image processing at 30 Hz.

2.1 Environment

The CR robot has to survive and increase its energy level by capturing the active battery packs distributed in a rectangular environment of 2.5m x 3.5m (Fig. 2). The active battery packs have a red LED. After the charging time, the battery pack becomes inactive and its LED color changes to green and the battery becomes active again after the reactivation time. The CR robot is initially placed in a random position and orientation.

The robot lives in two different environments that alternatively substitute each-other four times. Based on environments settings, the robot must learn different policies in order to survive and increase its energy level. As shown in Fig. 2, the first and second environments have eight and two battery packs, respectively. In the first environment, the batteries have a long reactivation time. In addition, the energy consumed for 1m motion is low. Therefore, the best policy is to capture any visible battery pack (the nearest when there are more than one). When there is no visible active battery pack, the robot have to search in the environment. In the second environment, the reactivation time is short and the energy consumed during robot motion is increased. Therefore, the optimal policy is to wait until the previously captured battery pack becomes active again rather than searching for other active battery packs.

Fig. 1. CR robot and the battery pack

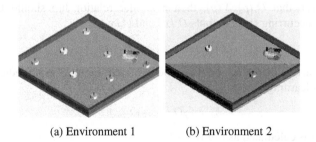

(a) Environment 1   (b) Environment 2

Fig. 2. Environments

3 Intelligent Algorithms

Consider the Cyber Rodent robot in an environment where at any given time t, the robot is able to choose an action. Also, at any given time t, the environment provides the robot with a reward r_t. Our implementation of the actor-critic has three parts: 1) an input layer of robot state; 2) a critic network that learns appropriate weights from the state to enable it to output information about the value of particular state; 3) an actor network that learns the appropriate weights from the state, which enable it to represent the action the robot should make in a particular state. Each time step, the robot selects one of the following actions: 1) Capture the battery pack; 2) Search for a battery pack; 3) Wait for a determined period of time. The wait behavior is interrupted if a battery becomes active. Both networks receive as input a constant bias input, the battery level and distance to the nearest active battery pack (both normalized between 0 and 1).

3.1 Critic

The standard approach is for the critic to attempt to learn the value function, $V(x)$, which is really an evaluation of the actions currently specified by the actor. The value function is usually defined as, for any state x, the discounted total future reward that is expected, on average, to accrue after being in state x and then following the actions currently specified by the actor. If x_t is the state at time t, we may define the value as:

$$V(x_t)=<r_t+ \gamma r_{t+1}+ \gamma^2 r_{t+2}+...>, \qquad (1)$$

where γ is a constant discounting factor, set such that $0<\gamma<1$ and $<\cdot>$ denotes the mean over all trials. $V(x)$ can actually suggest an improvement to the actions of the actor, since an action, which leads to a large increase in the value, is guaranteed to increase the battery level. Therefore, a good strategy for the actor is to try several actions for each state, with aim of choosing the action that involves the largest increase in value.

However, the value function is not given; the critic must learn it using TD learning, i.e., the weights must be adapted so that $O_c(x)=V(x)$. TD works by enforcing *consistency* between successive critic outputs. Specifically, the following relationship holds between successively occurring values, $V(x_t)$ and $V(x_{t+1})$:

$$V(x_t)=<r_t>+ \gamma V(x_{t+1}). \qquad (2)$$

If it were true that $O_c(p)=V(p)$, then a similar relationship should hold between successively occurring critic outputs $O_c(x_t)$ and $O_c(x_{t+1})$:

$$O_c(x_t)=<r_t>+ \gamma O_c(x_{t+1}). \qquad (3)$$

TD uses the actual difference between the two sides of eq. 5 as a prediction error, δ_t, which drives learning:

$$\delta_t=r_t+\gamma O_c(x_{t+1})-O_c(x_t). \qquad (4)$$

The TD error is calculated as follows:

$$\hat{r}[t+1] = \begin{cases} 0 \text{ if the start state} \\ r[t+1]+ \gamma^k v[t+1]-v[t] \text{ otherwise} \end{cases}, \qquad (5)$$

using the reward $r_{t+1} = (En\_level_{t+1} - En\_level_t)/50$. TD reduces the error by changing the weights.

3.2 Actor

The robot can select one of three actions and so the actor make use of three action cells, p_j, $j=1,2,3$. The captured behavior is pre-evolved (Capi et al. 2003) using the angle to the nearest battery pack as input of neural controller. When the search behavior is activated, the robot rotates 10 degrees clockwise. The robot does not move when the wait behavior becomes active. A winner-take-all rule prevents the actor from performing two actions at the same time.

The action is selected based on the softmax method as follows:

$$P(a, s_t) = \frac{e^{p_i(a,s)\beta}}{\sum_{i=1}^{3} (e^{p_i(a,s)\beta})}, \tag{6}$$

where β is the cooling factor.

Following the logic described above, the actor should try various actions at each state, with the aim of choosing the action, which produces the greatest increase in value. The stochastic action choice ensures that many different actions are tried at similar states. To choose the best action, a signal is required from the critic about the change in that result from taking an action. It turns out that an appropriate signal is the same prediction error, δ_t, used in the learning of the value function.

3.3 Combining Learning and Evolution

In our implementation, we used the actor-critic RL as explained previously. The metaparameters ρ_1, ρ_2, α_1, α_2, γ, the initial weight connections and the number of hidden neurons of both actor and critic networks are evolved by GA. In addition, GA optimizes the relation between the cooling factor β and energy level, by optimizing the values of β_0, en_1, β_1, en_2, β_2, and β_3, as shown in Fig. 3.

In order to force the evolution process to select individuals that live longer and have a higher energy level, the fitness is designed as follows:

$$fitness = \begin{cases} \dfrac{\sum_{i=1}^{4} En_i}{4} + \dfrac{CR_{max\_life}}{100} & \text{if the agent survives} , \\ En_{min} + \dfrac{CR_{life}}{100} & \text{if the agent dies} \end{cases} \tag{7}$$

where En_i is the energy level at the moment of each environment change and CR_{max_life} is maximum life time in seconds, En_{min} is the energy level when robot dies, CR_{life} is the time in seconds until the robot dies.

A real-value GA was employed in conjunction with the selection, mutation and crossover operators. Many experiments comparing real-value and binary GA show that real-value GA generates superior results in terms of the solution quality and CPU time.

3.4 Evolution of Neural Controller

We evolved the weight connections and number of hidden units of a feedforward neural network for the surviving behavior. The inputs of the neural network are the energy level and distance to the nearest battery pack. The weight connections have been fixed throughout the robot's life. Every time step one action is selected based on the selection probability, as follows:

$$P(a_i) = \frac{p'_i}{\sum_{i=1}^{3} p'_i} \tag{8}$$

where $P(a_i)$ is the probability of selecting action a_i, and p_i is the output of i-th neuron. The fitness value is calculated according to eq. 7.

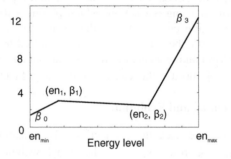

Fig. 3. Optimized cooling factor

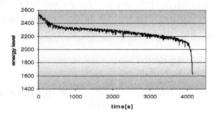

Fig. 4. Energy level during CR motion

4 Results

In order to determine the energy after each action, we recorded how the energy level changes by time, as the CR robot moves in the environment. The digital readings of energy level are shown in Fig. 4. Initially, the battery is fully recharged. The robot stops moving when the battery level goes under 1900. Because of the nonlinear relation, we used the virtual time to determine the energy level after each action, as shown in Table 2. Except for capturing the battery pack, all the other actions increased the virtual time. The maximum lifetime of the robot is 360 min and the environment changes every 90 minutes.

Table 1. Change in the virtual time for each action

Environment settings	Env. 1	Env. 2
Capturing the battery pack	-60s	-30s
Moving 1m distance	4s	15s
Searching	1s	1s
Waiting	5s	5s
Battery reactivation time	100s	10s

4.1 Evolution of Neural Controller

Initially 100 neural controllers were generated randomly. The performance of six different individuals selected from the first generation, is shown in Fig. 5. The robots A_1, A_2, and A_3 die shortly after they were placed in the environment. The robot A_4 reaches any visible active battery pack. When there is no visible active battery pack, the probability of selecting the wait action is higher than search action. Therefore, the robot performs better in the second environment. However, the search action is also selected, which resulted in a slow decrease of energy level in the first environment and some rapid energy decrease in second environment. Robots A_5 and A_6 apply a strategy, which is suitable for the first environment and the energy reduces rapidly in the second environment.

The best neural controller generated by evolution has 2 hidden units. Fig. 6 shows the energy level during the robot life. The robot applied the same strategy throughout its life, which is the optimal strategy for the first environment. When there is no visible active battery pack, mainly search action is generated. The probability of selecting the wait behavior was low, but sometimes it was generated which resulted in slower decrease of energy level in the second environment.

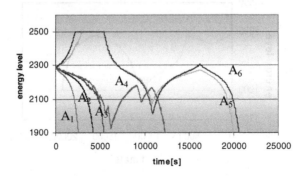

Fig. 5. Performance of different individuals from the first generation

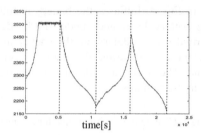

Fig. 6. Energy level of the best evolved neural controller

4.2 Actor-Critic RL

In this section, we compare the performance of the actor-critic RL using arbitrary selected metaparameters (Table 3) and random initial weight connections in the interval [-0.5 0.5]. The relation between β and energy level is considered linear where 1 and 10 correspond to empty and full battery level, respectively.

Fig. 6 shows the energy level during the time course of learning. At the beginning of robot life, the energy decreases because the initial weight connections are randomly generated. As the learning continues, the weight connections of both actor and critic networks are modified and energy level is increased. When the environment changes, the energy level decreases. Therefore, the robot starts to explore the environment. However, the robot was unable to survive in all four environments. The battery becomes empty after 20000 sec.

Table 3. Metaparameters used in actor-critic RL

RL parameters	Values
ρ_1	1
ρ_2	0.5
α_1	0.4
α_2	1
γ	0.9
β	Linear 1-10

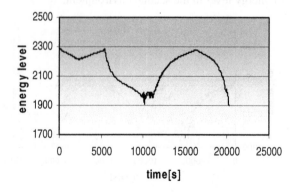

Fig. 7. Performance of actor-critic RL

4.3 Combining Learning and Evolution

In our simulations the population size is 100 and the evolution terminated after 20 generations. Each individual of the population has different values of metaparameters and initial weight connections. In the first generation, most of the individuals could not survive in four environments. Based on the fitness value, the individuals that survived longer have higher probability to continue in the next generation.

The searching interval and GA results are shown in Table 4. The actor and critic networks have 2 and 4 hidden neurons, respectively. The optimal relation between the energy level and cooling factor (β) shows that β is slightly increased when the energy level goes to minimum. The minimum value of β is for 59% of full battery level. When the energy level is higher than 72%, β becomes high.

Fig. 8 shows the energy level during the robot life, utilizing the evolved metaparameters, initial weight connections and optimized relation between cooling factor and energy level. At the beginning of robot life, due to large value of β and optimized initial weight connections, the robot starts exploiting the environment. Because the energy consumed for 1m motion is small, the best strategy is to capture any visible active battery pack or search otherwise.

When the environment changes, due to the large value of β, the robot follows the previous strategy. As the energy decreases, β gets lower. Therefore, the robot starts to explore the environment and to adapt its strategy to the new environment conditions. In the second environment, the reactivation time is very short and energy consumed for 1m motion is higher compared to the first environment. Therefore, the robot, instead of searching for an active battery pack, waits until the previous captured battery pack becomes active.

Table 4. Optimized metaparameters

Optimized parameters	Searching interval	Results
α_1	[0 1]	0.4584
α_2	[0 1]	0.4614
ρ_1	[0 1]	0.1722
ρ_2	[0 1]	0.4521
γ	[0 1]	0.6360

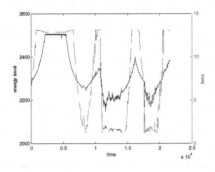

Fig. 8. Energy level and cooling factor during the robot life

5 Conclusion

In this paper, we considered combining learning and evolution in order to deal with non-stationary environments. The results of this paper can be summarized as follows:

- Metaparameters and initial weight connections optimized by GA helped the robot to adapt much faster during the first stage of life.

- Based on the relation between the energy level and cooling factor, the robot was able to adapt its strategy as the environment changed.

- The robot controlled by an evolved neural controller applied always the same strategy, which was the optimal only in one of the environments.

The performance of actor-critic RL was strongly related to the values of metaparameters, especially the relation between cooling factor and energy level.

References

1. Sutton, R.S., Barto, A.G.: Reinforcement learning. MIT Press, Cambridge (1998)
2. Kaelbling, L.P., Littman, M.L., Moore, A.W.: Reinforcement learning: A survey. Journal of Artificial Intelligence Research 4, 237–285 (1996)
3. Minato, T., Asada, M.: Environmental change adaptation for mobile robot navigation. Journal of Robotics Society of Japan 18(5), 706–712 (2000)
4. Matsui, T., Inuzuka, N., Seki, H.: Adapting to subsequent changes of environment by learning policy preconditions. Int. Journal of Computer and Information Science 3(1), 49–58 (2002)
5. Doya, K.: Reinforcement learning in continuous time and space. Neural Computation 12, 219–245 (2000)
6. Belew, R.K., McInerney, J., Schraudolph, N.N.: Evolving networks: using the genetic algorithm with connectionist learning. In: Langton, C.G., et al. (eds.) Artificial Life II, pp. 511–547. Addison Wesley, Reading (1990)
7. Unemi, T.: Evolutionary differentiation of learning abilities - a case study on optimizing parameter values in Q-learning by a genetic algorithm. In: Brooks, R.A., Maes, P. (eds.) Artificial Life IV - Proceedings of the Fourth International Workshop on the Synthesis and Simulation of Living Systems, pp. 331–336. MIT Press, Cambridge (1994)
8. French, R.M., Messinger, A.: Genes, phenes and the baldwin effect: learning and evolution in a simulated population. In: Proc. of Forth Int. Workshop on the Synthesis and Simulation of Living Systems, pp. 277–282 (1977)
9. Nolfi, S., Parisi, D.: Learning to adapt to changing environments in evolving neural networks. Adaptive Behavior 5(1), 75–98 (1996)
10. Niv, Y., Joel, D., Meilijson, I., Ruppin, E.: Evolution of reinforcement learning in uncertain environments: A simple explanation for complex foraging behaviors. Adaptive Behavior 10(1), 5–24 (2002)
11. Capi, G., Doya, K.: Application of evolutionary computation for efficient reinforcement learning. Applied Artificial Intelligence 20(1), 1–20 (2005)
12. Eriksson, A., Capi, G., Doya, K.: Evolution of metaparameters in reinforcement learning. In: IROS 2003 (2003)
13. Capi, G.: Multiobjective Evolution of Neural Controllers and Task Complexity. IEEE Transactions on Robotics 23(6), 1225–1234 (2007)

A Model and Simulation of Early-Stage Vision as a Developmental Sensorimotor Process

Olivier L. Georgeon[1], Mark A. Cohen[2], and Amélie V. Cordier[1]

[1] Université de Lyon, CNRS
Université Lyon 1, LIRIS, UMR5205, F-69622, France
[2] Lock Haven University, PA, USA
{olivier.georgeon,amelie.cordier}@liris.cnrs.fr, mcohen@lhup.edu

Abstract. Theories of embodied cognition and active vision suggest that perception is constructed through interaction and becomes meaningful because it is grounded in the agent's activity. We developed a model to illustrate and implement these views. Following its intrinsic motivation, the agent autonomously learns to coordinate its motor actions with the information received from its sensory system. Besides illustrating theories of active vision, this model suggests new ways to implement vision and intrinsic motivation in artificial systems. Specifically, we coupled an intrinsically motivated schema mechanism with a visual system. To connect vision with sequences, we made the visual system react to movements in the visual field rather than merely transmitting static patterns.

Keywords: Cognitive development, Intrinsic Motivation, Artificial Intelligence, Cognitive Science, Intelligent agents, Machine learning, Computer simulation.

1 Introduction

We address the question of how autonomous agents can learn *sensorimotor contingencies*—contingencies between the agent's motor actions and the signal received through the sensors. We propose a model that learns such contingencies in rudimentary settings. The agent has primitive possibilities of interaction in a two-dimensional grid, and distal sensors that reflect some remote properties of the grid. The learning process is driven by intrinsic motivations hard coded in the agent, and results in the agent gradually improving its capacity to exploit distal sensory information to orient itself towards targets within the environment.

The idea that visual perception is actively constructed through interaction was proposed by theories of active vision [e.g., 1]. Specifically, O'Regan and Noë [2] proposed the *sensorimotor hypothesis* of vision. They used the metaphor of a submarine controlled from the surface by engineers, but with connections that have been mixed up by some villainous marine monster. They argue that the engineers would have to learn the *contingencies* between the commands they send and the signals they receive. O'Regan and Noë's sensorimotor hypothesis of perception posits that making sense of perception precisely consists of knowing these contingencies.

L. Iliadis et al. (Eds.): EANN/AIAI 2011, Part II, IFIP AICT 364, pp. 11–16, 2011.

To implement these views, we rely on our previous work regarding intrinsically-motivated hierarchical sequence learning [3]. In this previous work, we implemented an original algorithm that learned regularities of interaction. This algorithm was inspired both by constructivist schema mechanisms (e.g. [4, 5]) and by principles of intrinsic motivation [6, 7]. The algorithm implements an innovative way to associate these two notions by incorporating intrinsic motivation into the schema mechanism, with schemas representing hierarchical sequences of interaction. Intrinsic motivation makes it possible to address the scalability issues of traditional schema mechanisms by driving the selection of schemas.

We demonstrated that an agent could use this algorithm to learn sequential contingencies between *touch* interactions and *move* interactions to avoid bumping into obstacles. In this paper, we report an extension to this algorithm that allows the agent to learn contingencies when perception is not a direct feedback from motion. For example, unlike *touch* in our previous agent, vision does not directly result from motion. Yet, because our previous algorithm succeeded in learning "touch/motor" contingencies, we expect it to prove useful for learning "visio/motor" less direct contingencies. Specifically, we envision coupling the algorithm with a complementary sensory mechanism as suggested by theories of *dual process* [e.g., 8].

More broadly, this work seeks to model and simulate *ab-nihilo* autonomous learning, sometimes referred to as *bootstrapping cognition* [9]. We relate this developmental approach to Piaget's [10] notion of an *early stage* in human's ontological development (pre-symbolic). For this work, though, this early-stage notion can also fit within the framework of phylogenetic evolution of animal cognition, as discussed for example by Sun [11].

2   The Model

In this model, the agent has six primitive behaviors. Each primitive behavior consists of the association of primitive actuators with binary feedback. These six primitive interaction patterns are listed in the first six lines of Table 1. Similar to our previous work [3], the binary feedback corresponds to a proximal sense that can be thought of as *touch*. If the agent tries to move forward, he can either succeed and touch no wall,

Table 1. Primitive actuators and sensors

Symbols		Actuators	Sensors	Description	Intrinsic satisfaction	
^	(^)	Turn left	True	Turn 90° left toward adjacent empty square	0	(indifferent)
	[^]		False	Turn 90° left toward adjacent wall	-5	(dislike)
>	(>)	Forward	True	Move forward	0	(indifferent)
	[>]		False	Bump wall	-8	(dislike)
v	(v)	Turn right	True	Turn 90° right toward adjacent empty square	0	(indifferent)
	[v]		False	Turn 90° right toward adjacent wall	-5	(dislike)
*			Appear	Target appears in distal sensor field	15	(love)
+			Closer	Target approaches in distal sensor field	10	(enjoy)
x			Reached	Target reached according to distal sensor	15	(love)
o			Disappear	Target disappears from distal sensor field	-15	(hate)

or fail and bump into a wall. Succeeding in moving forward is associated with the intrinsic satisfaction of 0 (indifference) while bumping is associated with -5 (dislike). When the agent turns, he receives *tactile* information about the adjacent square that he turned towards; touched a wall (-8, dislike), or not touched (0, indifferent).

In addition, we have now implemented the rudimentary distal sensory system depicted in Figure 1a. This system consists of two *eyes* that detect the blue color in the environment (*target* in Figure 1a). Each eye has a *visual field* that covers 90° of the agent's surrounding environment. The two visual fields overlap in the line straight in front of the agent, including the agent's location. Each eye generates a single integer value that indicates the amount of blue color detected, also reflecting the distance to a blue square if there is only one. This distal sensory system can thus be seen as a rudimentary monochromic visual system with a resolution of two pixels.

We want the visual system to forward visual information to the agent's situational representation in working memory to inform the selection of behavior. Inspired by the dual process argument, we first considered an iconic approach that would incorporate the two-pixel icon provided by the visual system as an element of context in our existing schemas. This approach proved inefficient because it added too many random combinations in the schema mechanism and the agent was unable to learn the contingency between the perceived icons and the actions. Moreover, the combinatorial growth would be prohibitive with wider icons.

To better support contingency learning, we looked for inspiration from studies of biological primitive organisms. We found useful insights from the limulus (horseshoe crab), an archaic arthropod whose visual system has been extensively studied [12, 13]. Specifically, we retained the two following principles:

1) Sensibility to movement: the signal sent to the brain does not reflect static shape recognition but rather reflects changes in the visual field. A horseshoe crab's *"eye is highly sensitive to images of crab-size objects moving within the animal's visual range at about the speed of a horseshoe crab (15 cm/s)"* [13, p. 172].

2) Visio-spatial behavioral proclivity: male horseshoe crabs move toward females when they see them with their compound eyes, whereas females move away from other females.

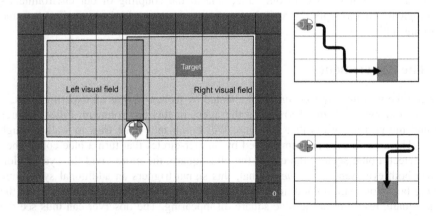

Fig. 1. a) The agent's vision of its environment (left). **b)** The diagonal strategy (top-right). **c)** The tangential strategy (bottom-right).

Similar to our agent, horseshoe crabs' eyes are fixed to their body and have poor resolution (roughly 40*25 pixels).

From these insights, we modeled the visual system so that it updated the schema mechanism only when a change occurred in the visual field. We identified four different signals that each eye would generate: *appear*, *closer*, *reached*, and *disappear*. Additionally, to generate a visio-spatial behavioral proclivity, the schema mechanism receives an additional intrinsic satisfaction associated with each of these signals. These four signals are listed with their satisfaction values in the last four lines of Table 1. For example, an eye sends the signal *closer* when the amount of blue color has increased in this eye's visual field over the last interaction cycle, meaning the square has gotten closer (with this regard, our agent's visual acuity is more than two pixels because the agent can detect the enlargement of the target's span in the visual field). We associate the *closer* signal with a positive inborn satisfaction value (10) to generate the proclivity to move toward blue squares.

With these settings (as reported in Table 1), we expect our agent to learn to coordinate its actions with its perception and orient itself toward the blue square. We must note that nothing tells the agent *a priori* that moving would, in some contexts, get it closer, or that turning would shift the blue color in the visual field. These are the kind of contingencies that the agent will have to learn through experience.

After an initial learning phase, we expect to see different behaviors emerge. One possible behavior is the *diagonal strategy* depicted in Figure 1b. This behavior consists of alternatively moving forward and turning toward the blue square until the agent becomes aligned with the blue square. At this point, the agent will continue to move forward.

Another possible behavior is the *tangential strategy* depicted in Figure 1c. The tangential strategy consists of approaching the blue square in a straight line. The trick with the tangential strategy is that the agent cannot accurately predict when he should turn toward the blue square before he passed it. The tangential strategy thus consists of moving on a straight line until the blue square disappears from the visual field, then returning one step backward, and then turning toward the blue square.

Of course, such specific strategies have probably little to do with real horseshoe crabs. These strategies would only arise due to the coupling of our environment's topological structure with our agent's sensorimotor system, intrinsic motivations, and learning skills.

3   The Experiment

We use the *vacuum environment* implemented by Cohen [14] as an experimental test bed for our agent. Figure 1.a) shows the agent in this environment. Filled squares around the grid are walls that the agent will bump into if he tries to walk through them. The agent's eyes are represented by quarter-circles that turn a blue color when they detect a blue square; the closer the blue square, the more vivid the eye's color. When both eyes send a *reached* signal, this signal triggers an additional systematic *eating* behavior (with no additional satisfaction value) that makes the agent "eat" the blue square, resulting in the blue square disappearing. The observer can thus see the agent's behavior as a quest for food. The observer can click on the grid to insert a new blue square when the agent has eaten the previous one.

We provide online videos of different runs of the agent[1]. At the beginning, these videos show the agent acting frantically because it has not yet learned the contingency between its actions and its perceptions. The agent picks random behaviors when it has no knowledge of what to do in a specific context. It learns to categorize contexts in terms of possibilities of behavior, in parallel with learning interesting composite behaviors. After the initial pseudo random activity, the videos show the agent more often orienting itself toward the blue square. After eating one or two blue squares, the agent starts to stick to a specific strategy. Our website shows example videos where the agent has learned the diagonal strategy[2] and where it has learned the tangential strategy[3]. The website also provides a detailed analysis of activity traces to discuss the learning process[4]. An interactive demonstration is also available[5] where the visitor can interact with the agent by adding food on the grid (this online demonstration is based on a slightly different configuration than the experiment reported here; in particular, the agent is in a continuous world rather than a grid, but the underlying algorithm remains the same).

Different runs show that the agent always learns a strategy within the first hundred steps, and that the most frequently found strategy is the diagonal strategy, with the settings defined in Table 1. The experiment therefore demonstrates that the agent always succeeds in learning sensorimotor contingencies.

The learning performance varies with the initial settings and the environmental conditions during training. Our goal was not to optimize the learning performance but to qualitatively demonstrate the nature of this developmental learning mechanism. In particular, the agent does not encode strategies or task procedures defined by the programmer, but rather autonomously constructs a strategy, as opposed to traditional cognitive models [15]. Our model is also consistent with studies of horseshoe crabs that show male horseshoe crabs can orient themselves toward static females because their visual system reacts to female-like objects that appear to be moving (relatively to their own speed) in a uniform background (a sandy shallow ocean bottom or beach) [12].

4 Conclusion

This work demonstrates a technique for implementing vision as an embodied process in an intrinsically motivated artificial agent. In this technique, the visual system does not send static images to the central system, but rather sends signals denoting change in the visual field. Notably, this technique allows the agent to see static objects, because changes in the visual field can result from the agent's own movements. This work opens the way to more complex models where the eye's resolution will be increased and where the agent will have the capacity to move its eyes independently from its body. Such developments inform our understanding of visual systems in

---

[1] http://liris.cnrs.fr/ideal/
[2] http://e-ernest.blogspot.com/2011/01/
ernest-82-can-find-his-food.html
[3] http://e-ernest.blogspot.com/2011/01/tengential-strategy.html
[4] http://e-ernest.blogspot.com/2011/01/
tangential-strategy-details.html
[5] http://liris.cnrs.fr/ideal/demo/ernest83/Seca.html

natural organisms and suggest new techniques to implement vision in intrinsically motivated robots.

Acknowledgments. This work was supported by the Agence Nationale de la Recherche (ANR) contract RPDOC-2010-IDEAL. We gratefully thank Pr. Alain Mille and Jonathan H. Morgan for their useful comments, and Olivier Voisin for his implementation of the online demonstrations.

References

1. Findlay, J., Gilchrist, I.: Active Vision: The Psychology of Looking and Seeing. Oxford University Press, USA (2003)
2. O'Regan, J.K., Noë, A.: A sensorimotor account of vision and visual consciousness. Behavioral and Brain Sciences 24, 939–1031 (2001)
3. Georgeon, O.L., Morgan, J.H., Ritter, F.E.: An Algorithm for Self-Motivated Hierarchical Sequence Learning. In: International Conference on Cognitive Modeling, Philadelphia, PA, pp. 73–78 (2010)
4. Drescher, G.L.: Made-up minds, a constructivist approach to artificial intelligence. MIT Press, Cambridge (1991)
5. Arkin, R.: Motor schema-based mobile robot navigation. The International Journal of Robotics Research 8, 92–112 (1987)
6. Blank, D.S., Kumar, D., Meeden, L., Marshall, J.: Bringing up robot: Fundamental mechanisms for creating a self-motivated, self-organizing architecture. Cybernetics and Systems 32 (2005)
7. Oudeyer, P.-Y., Kaplan, F.: Intrinsic motivation systems for autonomous mental development. IEEE Transactions on Evolutionary Computation 11, 265–286 (2007)
8. Norman, J.: Two visual systems and two theories of perception: An attempt to reconcile the constructivist and ecological approaches. Behavioral and Brain Sciences 25, 73–144 (2002)
9. Dennett, D.: Brainchildren. Essays on designing minds. Penguin Books (1998)
10. Piaget, J.: The construction of reality in the child. Basic Books, New York (1937)
11. Sun, R.: Desiderata for cognitive architectures. Philosophical Psychology 17, 341–373 (2004)
12. Shuster, C., Barlow, R.B., Brockmann, J.: The american horseshoe crab. Harvard University Press, Harvard (2004)
13. Barlow, R.B., Hitt, J.M., Dodge, F.A.: Limulus vision in the marine environment. Biological Bulletin 200, 169–176 (2001)
14. Cohen, M.A.: Teaching agent programming using custom environments and Jess. AISB Quarterly 120, 4 (2005)
15. Newell, A.: Unified Theories of Cognition. Harvard University Press, Cambridge (1990)

Enhanced Object Recognition in Cortex-Like Machine Vision

Aristeidis Tsitiridis[1], Peter W.T. Yuen[1], Izzati Ibrahim[1], Umar Soori[1], Tong Chen[1], Kan Hong[1], Zhengjie Wang[2], David James[1], and Mark Richardson[1]

[1] Cranfield University,
Department of Informatics and Systems Engineering,
Defence College of Management and Technology,
Shrivenham, Swindon, SN6 8LA, United Kingdom
[2] Electrical Engineering Dept., Beijing Institute of Tech, Beijing, P.R. China

Abstract. This paper reports an extension of the previous MIT and Caltech's cortex-like machine vision models of Graph-Based Visual Saliency (GBVS) and Feature Hierarchy Library (FHLIB), to remedy some of the undesirable drawbacks in these early models which improve object recognition efficiency. Enhancements in three areas, a) extraction of features from the most salient region of interest (ROI) and their rearrangement in a ranked manner, rather than random extraction over the whole image as in the previous models, b) exploitation of larger patches in the C1 and S2 layers to improve spatial resolutions, c) a more versatile template matching mechanism without the need of 'pre-storing' physical locations of features as in previous models, have been the main contributions of the present work. The improved model is validated using 3 different types of datasets which shows an average of ~7% better recognition accuracy over the original FHLIB model.

Keywords: Computer vision, Human vision models, Generic Object recognition, Machine vision, Biological-like vision algorithms.

1 Introduction

After millions of years of evolution visual perception in primates is capable of recognising objects independent of their sizes, positions, orientations, illumination conditions and space projections. Cortex-like machine vision [1] [2] [3] [4] [5] [6], attempts to process image information in a similar manner to that of biological visual perception. This work is different from other studies on human vision models such as [] [7] [8], in which the features are extracted with a statistical distance based descriptor methodology rather than a biologically inspired saliency approach [9]. In Itti's recent biological visual research [10] [11], which utilised a 44-class data set for the scene classification work, the authors reported a slightly inferior performance of the biological inspired C2 feature than the statistical SIFT. However, it is difficult to draw conclusions based on a single dataset and more work is needed to confirm their results. Other models such as Graph-Based Visual Saliency (GBVS) [12] and Feature Hierarchy Library (FHLib) [13], have implemented biological vision in hierarchical

L. Iliadis et al. (Eds.): EANN/AIAI 2011, Part II, IFIP AICT 364, pp. 17–26, 2011.

layers of processing channels similar to the ventral and dorsal streams of the visual cortex [14]. In these models the dorsal stream process has been implemented by means of a saliency algorithm to locate the positions of the most "prominent" regions of interest for visual attention. To mimic the simple and complex cell operations in the primary visual cortex for object recognitions, the cortex ventral stream process has been commonly presented in alternating layers of simple (S-layers) and complex (C-layers) operations. In the centre of these cortex-like models is the centre-surround operation over different spatial scales (so called image pyramids) of the image in both the dorsal-like and ventral-like processing. Although previous models have achieved rather impressive results using the Caltech 101 data as the test set [13], additional improvements are needed. Firstly, the previous work [13] considered datasets where all objects in the images are in the centre of the scene, like that in the Caltech 101 dataset. Secondly, FHLIB has been testified using a single dataset and there is real need to extend the tests using different datasets. Thirdly, the existing model extracts features for the template library in a random manner, which may reduce performance due to the inclusion of non-prominent and/or repeated features. Fourthly, the template matching process itself in FHLIB utilises a pre-stored location of the templates from the training data and then "searches" around the vicinity of this location to perform matching. This "blind" searching mechanism is neither efficient nor adaptive.

This paper is largely based on MIT and Caltech's work and it addresses the above four points. All enhancements are implemented within the cortex-like FHLIB framework. All codes utilised in this work have been implemented in MATLAB and results are compared with that of the "original" FHLIB algorithm which has been re-coded in this work, according to paper [13] and it is referred as MATLAB-FHLIB [MFHLIB]. One main contribution in this work is the incorporation of saliency within the template feature extraction process and this is termed as saliency FHLIB [SFHLIB]. Other enhancements have been the substitution of the computationally expensive Gabor filters for multiple orientations with a single circular Gabor filter, the improvement of the feature representation using larger patches as well as the addition of an extra layer to refine the feature library thereby eliminating redundant patches while at the same time ranking features in the order of significance. The classification in the present paper is achieved through a Support Vector Machine (SVM) classifier using the features extracted in a cortex-like manner similar to the previous work.

2 Cortex-Like Vision Algorithms

2.1 Graph-Based Visual Saliency (GBVS)

In GBVS the saliency of intrinsic features is obtained without any intentional influence while incorporating a very important characteristic of biological vision, the centre-surround operation. Initially for a digital RGB input image, the algorithm extracts the fundamental features i.e. average intensity over RGB bands, double opponency colour and Gabor filter orientations, and activation maps are formed by using image pyramids across different scales under centre surround operations. The final saliency map that highlights the most prominent regions of interest (ROI) in the image is then constructed from the normalised Markov chains of these activation maps. For more information on GBVS please refer to [12].

2.2 Feature Hierarchy Library (FHLIB)

FHLIB is an unsupervised generic object recognition model which consists of five layers. Their operations follow the early discoveries of simple and complex cells in the primary visual cortex [15] and alternate in order to simulate the unsupervised behaviour of the ventral stream as it propagates visual data up to higher cortical layers. FHLIB's layers are the input image layer, Gabor filter (S1) layer, Local invariance (C1) layer, Intermediate feature (S2) layer, Global invariance (C2) layer. S2 patches are stored in a common featurebook during the training phase and used as templates to be matched against testing images. The stored C2 vectors from the training phase can be compared against the C2 vectors of a given test image, for example by means of a linear classifier such as a Support Vector Machine (SVM). In addition, FHLIB introduced further modifications with respect to previous biologically inspired vision models to improve performance such as the sparsification of S2 units, inhibition of S1/C1 units and limitation of position/scale invariance in S2 units. For more information on FHLIB please refer to [13].

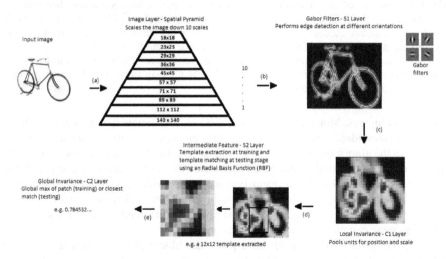

Fig. 1. FHLIB's architecture by forming a pyramid of various resolutions of the image, followed by tuning the Gabor features in the S layers and max-pooling across the adjacent C layers in the pyramid, then brings spatial information down to feature vectors for classification.

3 The Cranfield University Algorithm (SFHLIB)

3.1 Feature Extractions from a Salient Region Of Interest (ROI)

Unless there is a task in which even the most refined features are required to distinguish subtle differences or similarities between objects (often of the same category) then retaining all visual information is computationally expensive and unnecessary. Currently in FHLIB, there is no specific pattern by which features are extracted and the selection process of both feature size and locations occurs randomly

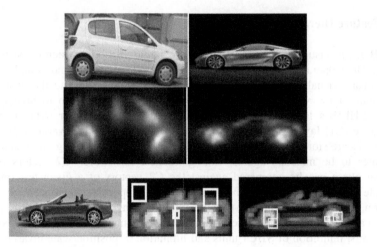

Fig. 2. Two images of cars. Top row shows the original images and second row their saliency maps using GBVS (12 Gabor angles). Highest attention accumulates on the areas of the wheels which is a common saliency feature and it is evident that saliency can effectively ignore background information. Third row shows the effect of accurate feature extraction via salience in a C1 layer map. Rectangular boxes illustrate the feature templates of varying sizes. Extraction occurs in FHLIB (centre) "blindly" while in C1 map from SFHLIB (right) patches are extracted from the salient ROI.

across input images. Moreover, it becomes difficult to estimate the number of features or feature sizes ideally required. Solving this problem by introducing a geometric memory in the algorithm (section 2.2) i.e. storing the location coordinates from which an S2 template was found so that a respective area in a testing image is compared, led to the conclusion that such a system becomes specialised in recognising objects in similar locations [16]. This however is impractical for real-world situations since objects may appear at other locations or may become differently orientated and so the algorithm must generically overcome this problem.

By applying the GBVS model on a particular object points to salient areas and evaluates an activation map according to priority of attention. For objects of the same category the most prominent areas are nearly the same and thus condensation of structured objects can be achieved (figure 2). We use the orientation feature only and rank salient areas of a certain circumference around the highest values. These areas effectively represent the local features which can be used along with more global shape areas (i.e. larger types of features extracted freely from any point in the image) and can be combined for the recognition stage.

3.2 Higher Resolution, Patch Sizes and Circular Gabor Filters

Salient areas can be very specific to small regions of an image. At low resolutions spatial information is also low and therefore extractions yield to incoherent representations of the object. To overcome this problem and to improve spatial representation, the resolution of images has to be increased. At the same time, in order

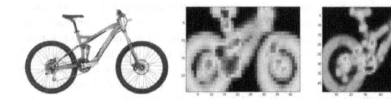

Fig. 3. An example of an original image (left) which at a lower resolution (input image 140 pixels) at the C1 layer (middle, first –finest scale) has retained little of the object's structure. At a higher resolution (input image 240 pixels) the C1 layer shows a more detailed representation (right).

to maintain the spectrum of patch sizes required to store suitable features the patch sizes have to be enlarged. To tackle these issues, we increase the size of the short edge of an image to 240 pixels (thus preserving the aspect ratio) and include S2 feature patches of size 20x20 and 24x24.

The use of Gabor filter banks in object recognition simulates the tuning of V1 simple cell at different orientations (θ) well and highlights their role in bottom-up mechanisms of the brain. However, constructing S1 responses for different orientations requires the creation of an equal amount of Gabor pyramids for each orientation which is computationally expensive and time consuming as the number of orientations increases to improve an object's description. To eliminate this, we generalise the S1 responses by varying the sinusoid across all orientations which then becomes circular symmetric [17]. Using this single circular Gabor filter, one S1 pyramid is obtained and at the same time FHLib's sparsifying step over orientations (section 2.2) become redundant and are removed. The circular Gabor filter is given below:

$$G(x, \ y) = \exp\left(-\frac{(X^2 + Y^2)}{2\sigma^2}\right) \exp\left(2\pi\sqrt{X^2 + Y^2}\right) \tag{1}$$

Note that in equation 1, θ and γ are no longer parameters for this equation and this equation now only depends on σ and λ.

3.3 Adding S3 and C3 Layers

At the object recognition part, when training template patches are extracted randomly from salient ROI, it is inevitable to have patches extracted more than once from the same location and scale, especially as the required total number of training patches is increased. Furthermore, there is no refinement mechanism currently in FHLIB that evaluates the extracted patches' performance and as such the algorithm may store patches that do not explicitly and accurately represent each class. In FHLIB, a refinement was made at the classification stage [13], however it is a post-processing, non-biological and time consuming remedy.

We address both aforementioned issues by introducing two more layers, namely S3 and C3. In the S3 layer, all patches of a particular class are directly grouped together from the S2 featurebook (section 2.2) and are organised according to their extraction sequence. The algorithm continues by extracting the training C2 vectors (as in FHLIB) which are again grouped so that the responses of every patch from each class across all images now exist together. By examining the C2 responses of each patch for every class on objects of the same class, e.g. if the class was 'bikes' and a patch was extracted from one of its images then C2 responses for this patch from all images portraying bikes are grouped together. Patches that have yielded identical C2 responses (in practice C2 vectors are float numbers and identical responses can only be obtained from identical patches) are dropped and only one unique patch is retained therefore eliminating co-occurrences. We remember the origin of the retained C2 vectors and refine the S3 featurebook accordingly.

Additionally, the performance of each patch can be measured for every class against objects of the same category to deduce to sampled patches that best describe that class. By summing the C2 responses for every patch we rank the S3 patches from high to low (high showing patches that are most commonly found for a particular object, low showing less generalisation and thus uncommon patches that do not exist across all images). At this point, a percentage number is introduced i.e. the amount of patches to be retained and for example, setting it to a certain value means that the featurebook is reduced by a percentage and the patches retained maximally express the trained classes. The final version of the significantly reduced S3 featurebook refined from co-occurences and uncommon patches, is used over the training images to create C3 vectors which in turn are used to train the SVM classifier. Similarly, at the testing phase, the stored S3 featurebook is used over the testing images and their C3 responses are compared against the trained C3 vectors.

4   Experiments

4.1   Image Datasets

Three image datasets were used, the Cranfield University Uncluttered Dataset (CUUD), the Cranfield University Cluttered Dataset (CUCD) and the Caltech 101. CUUD consists of four categories of vehicle images that were collected from the internet and are namely airplanes, bikes, cars and tanks. Each image contains only a particular vehicle without any clutter or obscurances (figure 3a). The images are of varying aspect ratios and their resolutions are always higher than a minimum of 240 pixels for their shortest edge. Objects are in varying directions and portray some variation in spatial position. Naturally, we have separated the dataset into different training and testing images.

CUCD has also been partly assembled from the internet and in part from our own image database. All images in the dataset contain background clutter and belong to four categories background, bikes, cars, and people. The background category shows a great variability of information, i.e. buildings, roads, trees etc. The people's category is the only category of non-rigid objects and therefore within this category pose and position vary significantly. Another difference with respect to CUUD is that in an image there may be more than one object (of the same category) present. Similarly to CUUD, the images are of varying aspect ratios and their resolutions are always higher than a minimum of 240 pixels for their shortest edge (figure 3b).

Fig. 4. (a) Four example images (airplanes, bikes, cars, tanks) from the CUUD vehicle classes. No background clutter is present and images contain only one object for classification, **(b)** four example images (background, bikes, cars, people) from the CUCD classes. Background clutter is present and images in some cases contain more than one object in each image.

Fig. 5. Some examples of classes from the 101 Caltech dataset

The multiclass image dataset Caltech 101 [18] consists of 101 different object categories including one for backgrounds. A total of 9197 images on various spatial poses include unobstructed objects mostly centred in the foreground with both cluttered and uncluttered background environments. All images have been taken at different aspect ratios and are always higher than a minimum of 200 pixels for their shortest edge.

4.2   Experiments Setup

In this work, we directly use GBVS MATLAB code with some modifications while all code regarding the recognition part of the algorithm has been inspired from [13] but otherwise created from the authors .

The algorithm is first tested with FHLIB-like parameterisation, 140 pixels for the images' short edge and 4 different size patches (4x4, 8x8, 12x12, 16x16), 11x11 Gabor filter banks while a sliding window approach was used to extract the maximum C2 responses across the entire image. At this point, the Gabor filters consist of 12 banks i.e. one per orientation angle. Subsequently, we enhance this algorithm with our improvements gradually by introducing a higher resolution for each image (240 pixels, short edge) and adding two more patch sizes 20x20 and 24x24. We then attach our feature extraction method using saliency and also substitute the 12 Gabor filters with one circular Gabor filter. Finally, the S3 and C3 layers are in turn embedded.

Efficient and fast biological-like detection and object recognition requires parallel execution. For our experiments, we concentrate on the results and use a sequential approach in order to prepare the saliency maps of both training and testing images of our dataset beforehand.

Each saliency map from GBVS matches the size of the original image that is later used in object recognition exactly i.e. 240 pixels for the shortest edge, and the only feature used is orientation at 12 Gabor angles spanning from 0 to π. For all experiments during training, we choose an abundant number of features (10000) to avoid underrepresentation of objects and Gabor filter parameters γ, σ and λ are all fixed according to [13]. For datasets CUUD and CUCD, 50 different images for each class were chosen for training and another 50 per class for training. For the Caltech dataset, we use 15 per class for training and 15 per class for testing. Classification accuracies are obtained as an average of 3 independent runs for all experiments. Finally, all classification experiments were conducted using an SVM classifier using one-against-one decomposition.

5   Results – Discussion

MFHLIB is the foundation upon which all improvements of this work are established. It extracts the maximum responses by using a sliding window approach which is overall inefficient, time-consuming and as table 1 shows the least accurate. Salience FHLIB (SFHLIB) on other hand illustrates higher performance and robust behaviour.

Table 1. Average percentage classification accuracies over 3 independent runs for the three datasets. Note that descending order algorithms in the left column include the enhancements of the previous algorithms. All results typically vary at ±1.5% (see our discussion for more detail).

Method	Dataset CUUD – Classification Accuracy (%)	Dataset CUCD– Classification Accuracy (%)	Dataset Caltech– Classification Accuracy (%)
MFHLIB	80	70.6	18.75
SFHLIB + Circular Gabor	85	80.4	22.4
SFHLIB + S3/C3 Layers (60% features)	86	76.6	19
SFHLIB + S3/C3 Layers (100% features)	81	80.4	21.4

From table 1, the results portray for all enhancements a gradual improvement over both the accuracy itself and time. CUUD being uncluttered, presents minimal difficulty for an algorithm and classification accuracies were overall the highest. Under this dataset, a 6% percentage improvement was observed between MFHLIB and SFHLIB variants (excluding SFHLIB with 100% features).

A higher difference between the MFHLIB and our enhancements was noticed in CUCD. In this dataset even though the number of classes remains the same, the added background information and more complicated poses, affect the performance of all algorithms, particularly in MFHLIB. As a first step by increasing the resolution and tampering the number and size of patches has increased performance by 6% and a total of 10% better performance was achieved by using SFHLIB with circular Gabor filters. A drop of nearly 10% for MFHLIB between CUUD and CUCD signifies its inefficiency as a dataset becomes more realistic. A decrease in performance (4.5%) can be also observed for SFHLIB though it is almost half compared with MFHLIB.

Experiments with the benchmark Caltech 101 dataset have revealed a decrease in performance with respect to the other two datasets which was primarily caused by the large number of classes and different setup. However, within this set of experiments an incremental difference between FHLIB and SFHLIB is apparent.

Classification accuracies for S3/C3 layers show that although for the cluttered datasets an improvement can be claimed the trend is not followed in CUUD. A major difference between previous variants of the code is that the number of features required to achieve this performance was lower and thus computationally cheaper. Having selected a fixed number of features (10000) for the library, by running the S3/C3 on the CUUD, reductions of an average of 15% were observed for a 100% of the features used. Similarly for the CUCD, the average percentage of identical feature discards reached 22% and for the Caltech dataset 10%. The difference of this percentage between the three datasets can be explained by the larger number of images used in the Caltech data. The same total number of features corresponds to fewer features per image thus reducing the probability of identical patches extracted randomly across salient regions. Discarding identical features improves time (by approximately the same percentage) and computational requirements.

6 Conclusions

Following the basic cortex-like machine vision models of FHLIB and GBVS, the contribution here has been to enhance object recognition performances in these early models by incorporating the visual saliency into the ventral stream process. The SFHLIB version being a fusion of salience and recognition, has achieved ~6% classification accuracy for the CUUD, 10% for the CUCD and 4.5% for the Caltech dataset better than that of the MFHLIB model. The present work has also highlighted the need of an efficient feature extraction method from the dataset and further alterations on the mechanism of the algorithm revealed the significance of refining the extracted features to improve the integrity of the feature library itself. It has been also found that the computational time of the proposed SFHLIB is faster by a significant percentage than the MFHLIB.

It is planned to use more extensive datasets to verify the newly developed SFHLIB algorithm against its portability and adaptability. Moreover, it is planned to employ a pulsed (spiking) neural network to replace the SVM classifier for object classification in the near future.

Acknowledgements. The authors thank Drs C Lewis, R Bower & R Botton of the CPNI. AT thanks EPSRC for the provision of DTA grant and TC & KH thank the DCMT internal funding of their studentships.

References

1. Treisman, A., Gelade, G.: A feature-integration theory of attention. Cognitive Psychology 12(1), 97–136 (1980)
2. Itti, L., Koch, C., Niebur, E.: A model of saliency-based visual attention for rapid scene analysis. IEEE Transactions on Pattern Analysis and Machine Intelligence 20(11) (1998)
3. Itti, L.: Visual Attention. In: The Handbook of Brain Theory and Neural Networks, pp. 1196–1201. MIT Press, Cambridge (2003)
4. Riesenhuber, M., Poggio, T.: Hierarchical models of object recognition in cortex. Nature Neuroscience 2(11), 1019–1025 (1999)
5. Riesenhuber, M., Poggio, T.: Models of object recognition. Nature Review (2000)
6. Serre, T., Wolf, L., Poggio, T.: Object recognition with features inspired by visual cortex. In: CVPR (2005)
7. Fukushima, K., Miyake, S., Ito, T.: Neocognitron: a neural network model for a mechanism of visual pattern recognition. IEEE Transactions on Systems, Man, and Cybernetics SMC-13(3), 826–834 (1983)
8. Wysoski, S., Benuskova, L., Kasabov, N.: Fast and adaptive network of spiking neurons for multi-view and pattern recognition, pp. 2563–2575 (2008)
9. Zhang, W., Deng, H., Diettrich, G., Mortensen, N.: A Hierarchical Object Recognition System Based on Multi-scale Principal Curvature Regions. In: 18th International Conference on Pattern Recognition, ICPR 2006 (2006)
10. Elazary, L., Itti, I.: A Bayesian model for efficient visual search and recognition, pp. 1338–1352 (2010)
11. Borji, A., Itti, L.: Scene Classification with a Sparse Set of Salient Regions. In: IEEE International Conference on Robotics and Automation (ICRA), Shanghai, China (February 2011)
12. Harel, J., Koch, C., Perona, P.: Graph-Based Visual Saliency. MIT Press, Cambridge (2007)
13. Mutch, J., Lowe, D.: Object class recognition and localisation using sparse features with limited receptive fields. International Journal of Computer Vision 80(1), 45–57 (2008)
14. Ungerleider, L., Mishkin, M.: Two cortical visual systems. MIT Press, Cambridge (1982)
15. Hubel, D., Wiesel, T.: Receptive fields and functional architecture of monkey striate cortex. Journal of Physiology 195 (1967)
16. Tsitiridis, A., Yuen, P., Hong, K., Chen, T., Ibrahim, I., Jackman, J., James, D., Richardson, M.: An improved cortex-like neuromorphic system for target recognitions. In: Remote Sensing SPIE Europe, Toulouse (2010)
17. Zhang, J., Tan, T., Ma, L.: Invariant Texture Segmentation Via Circular Gabor Filters. In: 16th International Conference on Pattern Recognition, ICPR 2002 (2002)
18. Fei-Fei, L., Fergus, R., Perona, P.: Learning generative models from few training examples: an incremental bayesian approach tested on 101 object cagories. In: CVPR Workshop on Generative-Model Based Vision (2004)
19. Serre, T., Wolf, L., Bilecshi, S., Riesenhuber, M., Poggio, T.: Robust Object Recognition with Cortex-like Mechanisms. IEEE Transactions on Pattern Analysis and Machine Intelligence 29(3), 411–425 (2007)

A New Discernibility Metric and Its Application on Pattern Classification and Feature Evaluation

Zacharias Voulgaris

Independent Contractor, 7750 Roswell Rd. #3D,
Atlanta, Georgia 30350, USA
research@voulgaris.tk

Abstract. A novel evaluation metric is introduced, based on the Discernibility concept. This metric, the Distance-based Index of Discernibility (DID) aims to provide an accurate and fast mapping of the classification performance of a feature or a dataset. DID has been successfully implemented in a program which has been applied to a number of datasets, a few artificial features and a typical benchmark dataset. The results appear to be quite promising, verifying the initial hypothesis.

Keywords: Discernibility, Feature Evaluation, Classification Performance, Dataset Evaluation, Information Content, Classification, Pattern Recognition.

1 Introduction

The potential of accurate classification of a feature or a set of features has been a topic of interest in the field of pattern classification. Especially in cases where the classification process is a time-consuming or generally impractical process, knowing beforehand how well a classifier will perform on that data can be a very useful insight. The concept of Discernibility aims at exactly that [1], through its metrics. Yet it was only with the creation of the latest index that this insight can be yielded in a very efficient way, making it a viable alternative for feature evaluation among other applications. For this purpose, a number of artificial features, of different class overlap levels were created. These features, along with a typical machine learning benchmark dataset [2] were applied to four different classifiers as well as the proposed metric.

The rest of the paper is structured as follows. In Section 2, a review of the relevant literature is conducted. This is followed by description of the methodology of the introduced metric (Section 3). In Section 4, the experiments related to the aforementioned datasets are described and a discussion of the results is presented. This is followed by the conclusions along with future avenues of research based on this work (Section 5).

2 Literature Review

The concept of Discernibility was formally introduced in previous work of the author [1]. However, this notion has been used even before that, since the idea of class

L. Iliadis et al. (Eds.): EANN/AIAI 2011, Part II, IFIP AICT 364, pp. 27–35, 2011.
© IFIP International Federation for Information Processing 2011

distinguishability has been present in the field of clustering and pattern classification for a while.

In particular, a metric called *Silhouette Width* [3] has been a popular choice for measuring this, in the context of clustering. This measure makes use of inter- and intra-class distances, though it only considers the inter-class distance of the closest class. It has been shown in [1, 4] that it is outperformed by the Spherical and the Harmonic Indexes of Discernibility, which were developed for this particular task. The latter have been tested in a variety of pattern recognition problems with success.

Another metric that performs this task is the *Fisher Discriminant Ratio* [5] which makes use of statistical analysis to evaluate the class overlap. The downside of this measure is that, because of its nature, it only works for one-dimensional data (a single feature). Also, while the other metrics yield a value in a bound interval ([-1, 1] for SW and [0, 1] for SID and HID), FDR may yield any positive value, making its output sometimes difficult to incorporate in larger frameworks, or to compare with other metrics. This shortcoming is addressed in the metric introduced in this paper.

Alternative metrics for this task have been proposed in [6], although they share the same drawback as FDR, as they were designed to evaluate a single feature at a time. Although most of them concentrate on measuring the trend of the features in relation to the class vector (something quite significant for Fault Diagnosis applications), one of them focuses on class distinguishability. This metric, the *Assumption Free Discriminant Ratio*, is basically another version of FDR with the difference that it does not assume any distribution for the data at hand, an approach that is shared by SW, SID and HID as well.

Another metric is that of the Kernel Class Separability method [7], which has application in feature selection [8]. However, this metric this metric is very heavy in terms of parameters, which although they can be fine-tuned, they make this technique quite cumbersome and impractical for real-world problems. In addition, this metric's use in feature selection was tested only using SVMs, a powerful classifier type but a single classifier type nevertheless. Therefore, this metric cannot be considered as a viable alternative for the class separability measurement task, at least not until it is further refined and optimized.

An interesting alternative is presented in [9] where a statistical analysis is performed to evaluate the class separability potential of various features. This is very similar to the FDR technique, though more analytical and therefore computationally expensive. This method has the inherent weakness of the statistical approach, namely its limitation to a single-dimensional dataset, rendering it ideal for feature evaluation but inept for anything more complex.

All of the aforementioned metrics, with the exception of FDR, are to some extent computationally expensive when it comes to larger datasets, a drawback that is addressed by the proposed metric, as it will become evident later on.

3 Methodology

The idea of the metric introduced in this work is to provide a measure of a dataset's Discernibility using inter-class and intra-class distances for each class, for each class pair combination. This is why the metric is called Distance-based Index of Discernibility (DID). Contrary to the Spherical Index of Discernibility, it does not use hyper-spheres, therefore cutting down the computational cost significantly. In addition, it provides only an overall estimate of the Discernibility of the dataset, avoiding the individual Discernibility scores of the comprising data points. This gives it an edge in the computational cost towards both SID and HID, which base their Discernibility estimate on the individual discernibilities of the patters of the dataset.

DID also has the option of using only a sample of the data points, instead of taking the whole dataset. This allows it to tackle datasets with forbiddingly large sizes, without much loss of accuracy in the Discernibility estimation (as it will be seen in Section 4). If the sample size is omitted in the inputs, the whole dataset is used.

Moreover, the DID metric provides the inter-class Discernibility for each class pair, something that, to the best of the author's knowledge, no other similar measure yields as an output. This is particularly useful as it provides insight to the dataset structure, something essential in datasets which due to their dimensional complexity cannot be viewed graphically.

DID's Discernibility estimate is computed by averaging the various inter-class Discernibility scores. The latter are calculated as follows. First the centers of the various classes are found, by averaging all the data points of each class. For example, in a 2-D feature space, if in class A there are 3 points (x_1, y_1), (x_2, y_2) and (x_3, y_3), the center of class A would be $(x_A, y_A) = ((x_1+x_2+x_3)/3, (y_1+y_2+y_3)/3)$.

Then the radius of each class is then calculated, by taking the largest distance from the center of the class to the various class points. In the previous example, if point 2 is the farthest from the center (x_A, y_A), then the radius of class A would be $R_A = $ sqrt($(x_2-x_A)^2 + (y_2-y_A)^2$).

Afterwards, a small number is added to all the radii to ensure that there are all non-zero. Then, the sample is divided according to the class proportions and the corresponding amount of data points from each class are selected randomly, for each class. Following that, the distances of these points to each class center are calculated (using the Euclidean distance formula). The ratio of each distance of a pattern of a class i over the radius of the class j is then computed and adjusted so that it does not surpass the value of 1.

Following that, these ratios are added up for each class pair and then divided by the number of patterns used in the calculations. So if due to the class distributions of the dataset the sample used comprise of n patterns in class i and m patterns in class j, the inter-class discernibility of classes i and j (ICD_{ij}) would be:

$$ICD_{ij} = \frac{\sum_{i=1}^{n} \min(1, dist_{ij} / R_i) + \sum_{j=1}^{m} \min(1, dist_{ij} / R_j)}{n+m} . \tag{1}$$

where ICD_{ij}: inter-class disc. of classes i and j
         $dist_{ij}$: distance between patterns i and j
         R_i: radius of hypersphere of class i
         R_j: radius of hypersphere of class j
         n: number of patterns in sample of class i and
         m: number of patterns in sample of class j

Note that the above measure is calculated only for different classes ($i \neq j$), for all possible combinations. Moreover, as one would expect, $\text{ICD}_{ij} = \text{ICD}_{ji}$. So, in a dataset having 4 classes, 6 class combinations will be taken, yielding 6 inter-class distances. As mentioned previously, once the inter-class discernibility scores for each class pair have been calculated, their average yields the overall discernibility score for the whole dataset. So, for a 3-class dataset, DID would be equal to ($\text{ICD}_{12} + \text{ICD}_{13} + \text{ICD}_{23}$) / 3.

4 Experiments and Results

4.1 Data Description

The data used for this research are a combination of 5 artificial datasets, generated by simple Gaussian distributions, and a benchmark dataset from the UCI machine-learning repository, titled *balance*.

The artificial datasets were designed to manifest five distinct class overlap levels and are single-dimensional (in essence they are features of various quality levels). These datasets comprised of 3000 data points, divided evenly among three classes, which followed a Gaussian distributions with $\sigma = 1$ and various μ's. A typical such dataset is feature3, which exhibits a moderate class overlap, as seen in Fig. 1.

The *balance* dataset was created as part of a cognitive research [11] and comprises of 3 classes as well. It has 625 points of 4 numeric attributes each. The class distribution is unbalanced (288 points in class 1, 49 points in class 2, and 288 points in class 3). The attributes (which are all integers from 1 to 5) describe the following measures: Left-Weight, Left-Distance, Right-Weight and Right-Distance, of a balance scale tip. This dataset has been used mainly for classification research and has no missing values.

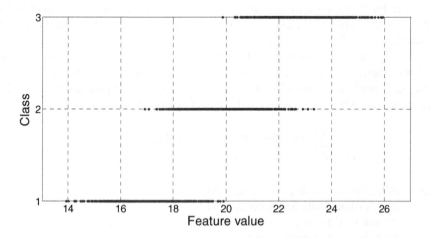

Fig. 1. Mapping of a typical artificial dataset (feature3)

4.2 Experiment Setup

A number of experiment rounds over four quite diverse classifiers were carried out. Each round comprised of a 10-fold cross-validation scheme, to ensure an unbiased partitioning of the training and testing sets. The classifiers used were the classic k Nearest Neighbor (using k = 5, which is a popular value for the number of neighbors parameter), the ANFIS neuro-fuzzy system (30 epochs for training), the Linear Discriminant Analysis statistical classifier (LDA) and the C4.5 decision tree. These classifiers were selected because they cover a wide spectrum of classification techniques. Also, in order to ensure more robust results, the number of experiment rounds was set to 30. Most of these classifiers, along with a few others, are thoroughly described in [10].

These experiments were conducted for each dataset and afterwards, the Accuracy Rates of the four classifiers were averaged. The end result was an Accuracy Rate for each dataset, reflecting in a way the classification potential of that data. In addition, each dataset was evaluated using the proposed metric, as well as a few other representative measures: SID, HID, FDR and AFDR. Note that the last two metrics were not applied on the balance dataset as they are limited to one-dimensional data. Also, all of the aforementioned measures were applied on the whole datasets, although their outputs are not significantly different when applied on the training sets alone (which constituted 90% of the whole datasets, for each classification experiment).

Another set of experiments was conducted in order to perform a sensitivity analysis of the proposed metric and the sample size used. These experiments constituted of 100 rounds and two of the aforementioned datasets were used.

An additional set of experiments was carried out, this time using only the Discernibility metrics, in order to establish a comparison in terms of computational complexity. These experiments comprised of 100 rounds and all of the aforementioned datasets were used.

4.3 Evaluation Criteria

The various outputs of the experiments were evaluated using three evaluation criteria, one for each set of experiments. The relationship between a Discernibility metric with the (average) Accuracy Rate is assessed using the Pearson Correlation (over the six datasets used). For the sensitivity analysis experiments the relative error (in relation to the Discernibility score of the whole dataset) was employed. As for the computational complexity series of experiments, the CPU time measure was used.

4.4 Results

The experiments described previously yielded some interesting results that appear to validate the initial aim of acquiring a reliable insight of a dataset's classification potential, in a way that is computationally inexpensive.

As it can observed from the results of the Accuracy Rates experiments (Table 1), the DID metric appears to follow closely the average Accuracy Rate, for the six datasets used. This close relationship can be more clearly viewed in Figure 2. The correlation coefficient was calculated to be an impressive 99.8%, verifying statistically the above observation.

Table 1. Experimental results for examining the relationship between classification accuracy and DID scores. The accuracy rate is averaged over all four classifiers used and over all thirty rounds.

Dataset	Mean Accuracy Rate	DID score
Feature1	0.9998	1.0000
Feature2	0.9887	0.9921
Feature3	0.8970	0.8909
Feature4	0.5354	0.4265
Feature5	0.3361	0.2468
Balance	0.8116	0.7599

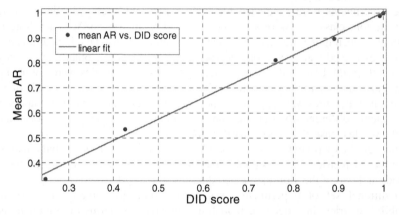

Fig. 2. Relationship between mean Accuracy Rate and DID scores, for all of the datasets tested. It is clear that a vivid (linear) correlation exists.

In order to ensure that the proposed metric is a viable alternative to the existing measures that opt to accomplish the same task, a comparison was made among them. Since a couple of these metrics cannot be applied in multi-dimensional data, two sets of comparisons were made, one using all datasets and one using only the single-dimensional ones (the artificial features created for this research). The results can be viewed in Table 2.

Table 2. Comparison of DID with other Discernibility metrics, based on the (Pearson) correlation with the classification Accuracy Rate, using all 6 datasets and the 5 single-dimensional datasets respectively.

Discernibility Metric	Correlation w. Mean Accuracy Rate	
	All datasets	Only 1-dim datasets
SID	0.997	0.998
HID	0.976	0.999
DID	0.998	0.998
FDR	–	0.946
AFDR	–	0.971

As the proposed metric has the option of using a sample of the data points in the dataset it is applied on, it is worthwhile investigating how the size of the sample influences the metric's output. This was done in the second set of experiments, which involved two datasets, the balance one and one of the features (feature3). The output of the metric when it is applied using the whole dataset is taken to be the correct Discernibility score, with which all the other outputs are compared (Table 3).

Table 3. Sensitivity analysis of DID scores, over different sample sizes, for two of the datasets used in the classification experiments. The DID scores were calculated over 100 runs. The original DID scores for the two datasets were 0.7599 and 0.8909 respectively.

Dataset	Sample size	Mean	St. Dev.	Rel. Error (%)
Balance	50%	0.7593	0.0060	0.0734
	25%	0.7605	0.0104	0.0790
	12.5%	0.7611	0.0153	0.1535
	5%	0.7558	0.0211	0.5457
Feature3	50%	0.8911	0.0018	0.0197
	25%	0.8911	0.0029	0.0197
	12.5%	0.8911	0.0051	0.0197
	5%	0.8911	0.0076	0.0197

In the third set of experiments the computational cost of the proposed metric, in comparison with the other metrics, is examined. The results of these experiments can be best viewed graphically, as seen in Figure 3 below.

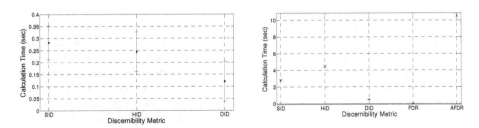

Fig. 3. Computational cost analysis. The calculation time of the Discernibility scores for the various metrics are shown (in sec). Two datasets were used for 100 runs   (left = *balance*, right = *feature3*). The error bars depict the 95% confidence intervals of the calculation time.

4.5 Discussion

It is clear from the aforementioned results that the proposed metric maps quite accurately the (average) Accuracy Rate. This in essence makes it a reliable predictor of a classifier's performance, something can translates into a significant advantage in applications where the classification cost is relatively high. Also, it enables the user to have a better understanding of the dataset before the actual classification, something that may help him/her make a better decision regarding the classifier used.

The results of the second series of experiments dictate that the proposed metric is at least as good the other ones, in mapping the Accuracy Rate of the classifiers. Also, in appears to be somewhat better in that aspect, when compared to FDR which has been extensively used in the past for the same purpose.

The results of the sensitivity analysis are quite interesting as they show that the metric's output is not greatly affected by the sample size. As one would expect, the output varies more as the sample becomes smaller, yet the relative error remains quite low (<1%). even at samples of only 1/20th of the original dataset. It is noteworthy that in the case of feature3, where the number of data points is relatively high, the metric's output is quite stable and close to the correct value, even though it varies a bit, as the sample gets smaller.

The computational cost experiments verified the original hypothesis that the proposed metric is an efficient alternative to the other Discernibility measures. When tested on a multi-dimensional dataset against SID and HID, it is clear that it is generally faster. Also, on a single-dimensional data with more data points, the advantage over these two measures is even more evident. It is still not as fast as FDR, but it is significantly faster than AFDR, which is in general a more robust metric than FDR.

Further analysis, using the ROC evalutation criterion for example, could have been performed. However, it is evident from the analysis so far that the proposed metric is adequate regarding the task it undertakes. Besides, a more extensive analysis is beyond the scope of this paper and can be part of a future publication based on further research on the subject.

5   Conclusions and Future Work

From the research conducted it can be concluded that the proposed method is a robust Discernibility metric, yielding a very high correlation with the average Accuracy Rate over the datasets used in the experiments of this research. Apparently it is not as fast as FDR, yet DID provides a better performance that this metric plus it is applicable on multi-dimensional data as well. Also, it has the option of using a sample of the dataset, without deviating much in its output, even for quite small sample sizes.

Future work on this topic will include more extensive testing of the method, in a larger variety of datasets, as well as use of it in other classification-related applications. Also, ways of making it applicable on the data point level will be investigated, so that it can yield Discernibility scores for the individual patterns of the dataset it is applied on. Finally, ways of making use of the inter-class Discernibility assessments of the various class pairs of a dataset will be also explored.

References

1. Voulgaris, Z.N.: Discernibility Concept for Classification Problems. Ph. D. thesis, the University of London (2009)
2. UCI repository,
   http://archive.ics.uci.edu/ml/datasets.html
   (last accessed: January 2011)

3. Kaufman, L., Rousseeuw, P.J.: Finding groups in data, pp. 83–85. Wiley Interscience Publications, New York (1990)
4. Voulgaris, Z., Mirkin, B.: Choosing a Discernibility Measure for Reject-Option of Individual and Multiple Classifiers. International Journal of General Systems (2010) (accepted and pending publication)
5. Fisher, R.A.: The use of Multiple Measurements in Taxonometric Problems and. Eugenics 7, 179–186 (1936)
6. Voulgaris, Z., Sconyers, C.: A Novel Feature Evaluation Methodology for Fault Diagnosis. In: Proceedings of World Congress on Engineering & Computer Science 2010, USA, vol. 1, pp. 31–34 (October 2010)
7. Wang, L., Chan, K.L.: Learning Kernel Parameters by using Class Separability Measure. In: 6th Kernel Machines Workshop (2002)
8. Wang, L.: Feature Selection with Kernel Class Separability. IEEE Transactions on Pattern Analysis and Machine Intelligence 30(9), 1534–1546 (2008), doi:10.1109/TPAMI.2007.70799
9. Cantú-Paz, E.: Feature Subset Selection, Class Separability, and Genetic Algorithms. In: Deb, K., et al. (eds.) GECCO 2004. LNCS, vol. 3102, pp. 959–970. Springer, Heidelberg (2004), doi:10.1007/978-3-540-24854-5_96
10. Duda, R.O., Stork, D.G., Hart, P.E.: Pattern Classification, 2nd edn. John Wiley & Sons, Chichester (2001)
11. Siegler, R.S.: Three Aspects of Cognitive Development. Cognitive Psychology 8, 481–520 (1976)

Time Variations of Association Rules in Market Basket Analysis

Vasileios Papavasileiou and Athanasios Tsadiras

Department of Economics, Aristotle University of Thessaloniki
GR-54124 Thessaloniki, Greece
{vapapava,tsadiras}@econ.auth.gr

Abstract. This article introduces the concept of the variability of association rules of products through the estimate of a new indicator called overall variability of association rules (OCVR). The proposed indicator applied to super market chain products, tries to highlight product market baskets, with great variability in consumer behavior. Parameter of the variability of association rules in connection with changes in the purchasing habit during the course of time, can contribute further to the efficient market basket analysis and appropriate marketing strategies to promote sales. These strategies may include changing the location of the products on the shelf, the redefinition of the discount or even policy or even the successful of recommendation systems.

Keywords: Market Basket Analysis, Association Rules, Data Mining, Marketing Strategies, Recommendation Systems.

1 Introduction

1.1 Market Basket Analysis

The consumer behavior data collection, for a super market chain, via an appropriate strategy, may lead the company to significant economies of scale. For the same reason, the extension of the use of loyalty cards, constitutes a primary goal for the administration of a super market chain. In particular, the market basket analysis [1] can bring out the combinations of products that are susceptible of marketing strategies and may lead to better financial results. With market basket analysis [2], the administration of a super market can understand the behavior and purchasing habits of customers, through combinations of product market [3], which is repeated in large or small degree, as a habit. These combinations are called products association rules [4,5,6,7] and are the result of market basket analysis procedure.

The simplest form of an association rule, shows two products and has the form X > Y, which means that the purchase of product X leads to the purchase of product Y. For rules evaluation there are used three main indicators [8], which is the degree of confidence, the degree of support and the degree of lift. The degree of confidence expresses the possibility of realization of rule X > Y in the set of transactions involving the purchase of the product X. Respectively, the degree of support expresses the possibility of realization of rule X > Y in the set of all transactions.

L. Iliadis et al. (Eds.): EANN/AIAI 2011, Part II, IFIP AICT 364, pp. 36–44, 2011.

Finally, lift Indicates how much better the rule is at predicting the "result" or "consequent" as compared to having no rule at all, or how much better the rule does rather than just guessing.

1.2 Dataset

The data of this survey are collected from a Greek known super market chain and are related to annual customer transactions [9,10] for the period from 01/09/2008 to 31/08/2009. Data are sourced from the information system of the company and come exclusively from the customer card transactions through loyalty cards. Every transaction is a record of the purchase of specific products carried out by the customer who makes use of the card at the time of purchase.

As far as data distribution is concerned there has been observed the long tail effect, which belongs to the broader category of low power distributions that appear quite often in nature. The long tail effect has assumed its present connotation and dynamics by Chris Anderson [11]. Figure 1 below shows the curve of long tail effect.

This term often refers to data products purchase in supermarkets describing their distribution as a long tail in which a small number of products is purchased more frequently whereas a large one is purchased less frequently. This phenomeno creates data sparsity problem and worsens even more their elaboration. For this survey from the total number of transactions there were selected, those that included at least the purchase of 8 products. That is how, to some extent, there was tried to resolve the data sparsity problem and prevent the removal of any market basket product, which may present a risk of information loss.

During the stage of preprocessing [12,13,14], the products were grouped into categories and subcategories, depending on their utility and the classification in supermarket shelves. The number of codes of the products sold by the chain is approximately 95.000 and with the categorization applied there was diminished in 247 product subcategories.

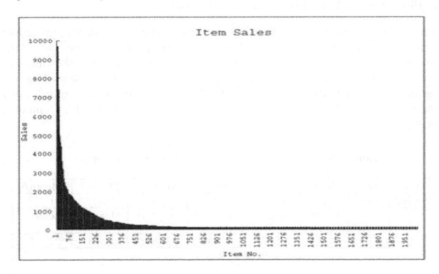

Fig. 1. Curve of Long Tail Effect

Table 1. Number of Transactions per Year and per Month

Time	Year	Month 1	Month 2	Month 3	Month 4	Month 5	Month 6
Number	183.827	15.603	16.027	14.621	18.141	15.323	15.982
		Month 7	Month 8	Month 9	Month 10	Month11	Month 12
		16.189	15.642	15.907	14.624	13.189	12.579

Occasionally, there have been proposed various algorithms for efficient market basket analysis. The most widely established, is the a priori algorithm, delivered for the first time by Rakesh Agrawal and Ramakrishnan Srikant [15]. For the extraction of data there was used the free software Tanagra [16]. This specific data mining tool is specialized in market basket analysis and is applied in the a priori algorithm. For compatibility with the program, the sets of data were formatted in binary tables [17], where each row is a separate transaction and each column represents one of the 247 sub-categories of products.

Each product of a transaction corresponds to 1, when purchased and to 0 when it isn't. Hence, there were formed 12 binary transaction tables respectively to the 12 months of analysis and 1 binary transaction table for the annual purchase data. The market basket analysis was applied to each of the 12 subsets separately, in a subcategory level of products and the results were studied in a association with time. In table 1 there is the number of transactions elaborated, both for the whole year and the 12 sub-totals. With the indication Month 1 we refer to the total number of all purchasing data for September 2008, while for Month 12, it is implied the total number of purchasing data for August 2009 respectively.

2 Time Variations of Association Rules

2.1 Discoveries of Association Rules from Dataset

For all of the data and for each of the subsets there has been activated the procedure of market basket analysis and recorded the results of the three indicators, confidence, support and lift. For every application of a market basket analysis process, the minimum level of confidence, support and lift established in the program, is 0.25, 0.01 and 1.1 respectively. The results of the number of rules that came out, are registered in table 2 and correspond to the level of 247 product subcategories.

Table 2. Number of Annual and Monthly Association Data Rules

Time	Year	Month 1	Month 2	Month 3	Month 4	Month 5	Month 6
Number	1.803	2.262	2.221	2.537	2.429	2.440	1.929
		Month 7	Month 8	Month 9	Month 10	Month11	Month 12
		1.619	1.875	2.149	1.696	1.651	2.215

2.2 Time Development of Association Rules

The 1803 rules, at the level of 247 subcategories of products, resulting from the extraction process of overall annual data were used as a base for additional rules analysis at monthly level. For each of the 1803 rules, there have been recorded the values for the three indicators, lift, support and confidence during the 12 months of the year. Thus was created the table 3, with the monthly development of association rules measurement indicators. In table 3, L indicates lift, S is for the value of support and C denotes the value of confidence.

In an attempt to show the price development of the degree of confidence in months 1, 2 and 12 it comes out the graph of time change of the degree of confidence of association rules.

Rules 5, 6 and 7 present a significant slope owing to the strong variability of the degree of customer confidence in September 2008, October 2008 and August 2009. It is also noted that when the confidence of customers is increased for the market basket of rule 5, the same period the confidence of customers for the market basket of rule 6 is reduced. The same but to a lesser extent seems to be the case with rules 5 and 7. In such cases the market baskets between them seem to work as substitute.

Table 3. Monthly Development of Association Rules Measurement Indicators

A/A	Rules	Month 1 (Sept. 2008)			Month 2 (Oct. 2008)		
		L	S %	C %	L	S %	C %
1	Yoghurt – Bread > Fruits	1.83	1.27	86.84	1.72	1.49	81.02
2	Toast- Cereals> Fresh Milk	1.20	2.26	71.89	1.23	2.52	73.81
3	Fruits – Sweet Biscuits > Vegetables	1.73	2.63	64.70	1.79	2.44	61.77
4	Fruits – Pasta > Fresh Milk - Vegetables	1.83	1.87	40.70	1.84	1.69	36.44
5	Fresh Milk – Beers > Cola	1.86	1.27	40.60	11.2	1.12	66.42
6	Cola – Salad (Not Packed, Packed) > Vegetables	12.8	1.02	79.50	1.14	1.41	68.28
7	Chicken Not Packed> Fruits	14.1	1.57	53.49	1.17	1.01	30.55
.	....	...	...	...	...	...	....
.	....	...	...	...	...	...	...
1803	Fresh Milk – Beef Not Packed> Bread	1.63	1.28	33.67	1.73	1.18	34.36

		......			Month 12 (Aug. 2009)		
		L	S%	C%	L	S%	C%
		...	...	...	1.85	1.26	85.95
		...	...	...	1.38	2.58	79.41
		...	...	....	1.56	2.70	62.62
		...	...	....	1.61	1.53	36.42
		...	...	...	1.69	1.71	37.79
		...	...	...	15.2	1.19	99.34
		...	...	....	1.16	1.26	32.78
		...	...	....	...	...	....
		...	...	....	...	...	....
		...	...	....	1.58	1.34	31.12

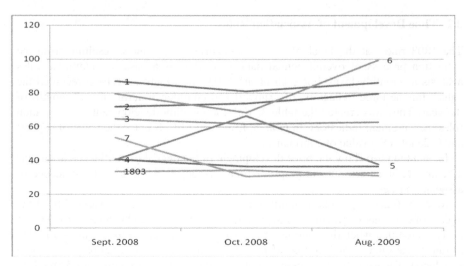

Fig. 2. Graph of the Time Change Confidence Degree of the Association Rules

As shown by the results in table 3, the values of association rules indicators lift, confidence, support, of super market chain products, present monthly fluctuations. For example, the degree of confidence in the rule 1 appears with percentage 86.84% in September and is reduced to 81.02% in the month of October and is increased to 85.95% the month of August. The option to purchase fruit after the selection of bread and yoghurt in the basket market seems to be less random in September compared to October and more random than that in August. Also on rules 5, 6 and 7, for the first three months under consideration, there are significant changes in levels of confidence in connection with the other rules. In particular, for rule 6 the changes are made to a high degree of confidence.

These monthly fluctuations in prices of lift and confidence are the subject of this research study. The research of rules association products should not be concluded solely on the assessment of the three indicators, without taking account of changes in these prices during the year. Through the assessment of variation in the degree of confidence and lift of a rule, one can draw useful conclusions about the marketing. The concept of variability has occupied the research of the export association rules geographical and temporal data within the study mainly climatic [18] and weather [19] phenomena. This article attempts to highlight the term of variability in critical factor in analyzing purchasing data of super markets.

3 Define of OCVR

For the evaluation of changes in the lift and confidence values of association rules, there has been calculated the statistical indicator of relative variability or standard deviation [20]. The indicator of the relative standard deviation, was chosen as the most effective for the comparison of the variability of the rules, as it is the most appropriate to compare the observations expressed in the same units but whose arithmetic average differ significantly. The variability index CV, is defined as the ratio of the standard deviation s by the mean value $\overline{X}$.

$$CV = \frac{s}{\overline{X}} \ . \tag{1}$$

For each of the 1803 association rules during the 12 months there has been calculated the index variability lift (CVL) and the index variability confidence (CVC).

In order to aggregate both CVL and CVC indexes and combine these indexes into a single index, we introduced a new indicator the one of the overall variability of association rules (OCVR), as a result of the average variability of lift and confidence.

$$OCVR = \frac{CVL + CVC}{2} \ . \tag{2}$$

4 Experimental Study

Table 4 lists the results of price variability index lift (CVL), the index variability confidence (CVC) and the index of overall variance (OCVR) specific association rules.

Table 4. Overall Variability Association Rules

A/A	Rules	CVL %	CVC %	OCVR %
1	Yoghurt – Bread> Fruits	4.19	4.46	4.33
2	Toast- Cereal > Fresh Milk	4.05	2.86	3.46
3	Fruits – Sweet Biscuits > Vegetables	5.40	4.27	4.83
4	Fruits – Pasta > Fresh Milk - Vegetables	5.11	5.66	5.39
.	…………………………	…	…	…
.	…………………………	…	…	…
.	…………………………			
1803	Fresh Milk – Beef not Packed > Bread	11.10	24.80	17.95

We note that the indicator of overall variability of rule 1 is 4.33% and the corresponding value of confidence for the three months that were presented in table 3, ranges from 81.02% to 86.84%. The indicator of overall variability of rule 2 is 3.46 and the corresponding value of the confidence ranges from 71.89% to 79.41%. This could mean that the rule 1 which is realized by a greater degree of consumer confidence than rule 2, can be changed more easily compared to rule 2, which seems to be more compact and inflexible. The indicator of overall variability of rule 3 is 4.83 and the corresponding value of the confidence ranges from 61.77% to 64.70%. This could mean that the rule 3 which is realized with a lesser extent consumer confidence compared to rule 2, can be changed more easily than rule 2. Rule 3 compared to rule 2 could be perceived as a more sensitive and flexible.

Figure 3 below shows the curve of the overall variance ratio (OCVR) for all of 1803 association rules for products of super market, for the period from 01.09.2008 to 31.08.2009.

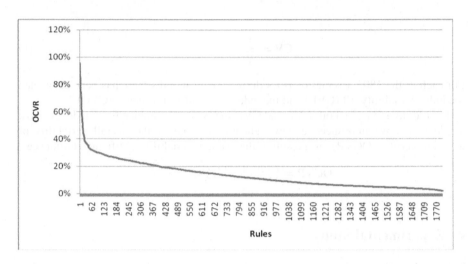

Fig. 3. Overall Variability Curve of Association Rules

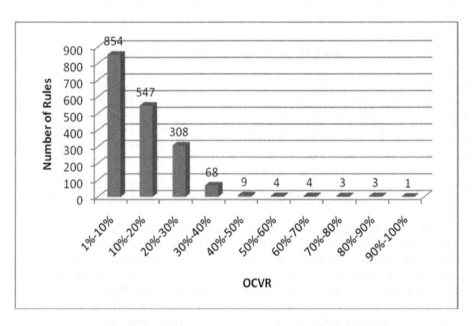

Fig. 4. Histogram of Overall Variability of Association Rules

We can observe that the form of the distribution of values of OCVR follows the long tail distribution. More specifically, a small number of rules presents a high overall variability. In Figure 4 is shown the histogram of the values of the overall variability. We observe that the majority of the overall variability index values range from 1% of space – 30% while there are very few rules between values 30%-100%.

5 Conclusions

The OCVR allows for the evolution and understanding of the association rules. There might be a particular interest in rules that receive high values of OCVR index. These rules seem to vary systematically and with intense degree, which means that the purchasing behavior of consumers in this particular market basket is not so stable and predefined. However, this does not mean that these rules do not have value for further analysis. On the contrary, this research sustains that association rules products with high OCVR values, are more susceptible of marketing strategies, as the determination and the dedication of the consumer in selecting the specific market basket presents significant variances. At the same time there should be observed the degree of confidence of rules with high OCVR and initially to control those who display a high degree of confidence. By taking action with appropriate marketing strategies it is more likely to change the level of confidence of those rules into a higher level, which already seems to be accomplished randomly and occasionally during the year. To change the location of the products on the shelf, to define the discount policy but also to form recommendation systems, the analysis of OCVR index values can be particularly useful. The evolution of consumer behavior in the course of time can reveal the combinations of products and, in particular, those products that are more suitable for promotion sales strategies.

6 Summary and Future Work

This research was conducted in Greek market product data and known super market chain, collected from customer purchases through loyalty cards. During the process of market basket analysis there were created 12 subsets of data corresponding to the data collected each month of the year. For each subset there were calculated lift and confidence indicators of association rules and at the same time a new indicator was co-estimated, the one of overall variability (OCVR). The OCVR results from the average of the values of the variability of lift and confidence indicators and describes degree of the overall variability of association rules. Each marketing strategy must be implemented to streamline the consumer behavior and to increase customer confidence. For this reason you must first identify application rules with high values of OCVR and high degree of confidence and attempt with the appropriate marketing strategy to reduce the value of OCVR while increasing the confidence of its customers.

In a future research it would be very interesting to establish an indicator which incorporates and includes the above information. It would also be quite interesting in the evaluation of OCVR to find weight factors for the representation of the variability of the degree of confidence and the degree of lift in the composition of the index. The existence of possible relations of substitution or even complementation between the association rules it may also constitute a subject for further research. Finally, it would be interesting to examine the existence of annual periodicity of the indexes, using additional data of more than one year.

Acknowledgments. We would like to thank Super Markets "Afroditi" for providing data and for their help.

References

1. Chen, Y.L., Tang, K., Shen, R.J., Hu, Y.H.: Market Basket Analysis in a Multiple Store Environment. Decision Support Systems 40, 339–354 (2005)
2. Shaw, M.J., Subramaniam, C., Tan, G.W., Welge, M.E.: Knowledge Management and Data Mining for Marketing. Decision Support Systems 31, 127–137 (2001)
3. Lawrence, R.D., Almasi, G.S., Kotlyar, V., Viveros, M.S., Duri, S.S.: Personalization of Supermarket Product Recommendations. Data Mining and Knowledge Discovery 5, 11–32 (2001)
4. Margaret, H.D.: Data Mining Introductory and Advanced Topics. Prentice Hall, Englewood Cliffs (2002)
5. Agrawal, R., Imielinski, T.: Mining Association Rules Between Sets of Items in Large Databases. In: Proc. of the ACM SIGMOD International Conference on Management of Data, Washington, D.C. (1993)
6. Ale, J.M., Rossi, G.H.: An Approach to Discovering Temporal Association Rules. In: Proc. of the 2000 ACM Symposium on Applied Computing, Como, Italy (2000)
7. Bayardo, R.J., Agrawal, R.: Mining the Most Interesting Rules. In: Proc. of the 5th ACM SIGKDD International Conference on Knowledge Discovery and Data Mining, San Diego, CA, USA (1999)
8. Wikipedia: The Free Encyclopedia, `http://en.wikipedia.org/wiki/Association_rule_learning`
9. Bhanu, D., Balasubramanie, P.: Predictive Modeling of Inter-Transaction Association Rules-A Business Perspective. International Journal of Computer Science & Applications 5, 57–69 (2008)
10. Brin, S., Motwani, R., Ulman, J.D., Tsur, S.: Dynamic Itemset Counting and Implication Rules for Market Basket Data. In: Proc. of the 1997 ACM-SIGMOD Conference on Management of Data, Arizona, USA (1997)
11. Anderson, C.: The Long Tail: Why The Future of Business Is Selling Less of More. Hyperion (2006)
12. Roiger, R.J., Geatz, M.W.: Data Mining: A Tutorial-Based Premier. Pearson Education, London (2003)
13. Bose, I., Mahapatra, R.K.: Business Data Mining – A Machine Learning Perspective. Information and Management 39, 211–225 (2001)
14. Chen, M.S., Han, J., Yu, P.S.: Data Mining: An Overview from a Database Perspective. IEEE Transactions on Knowledge and Data Engineering 8, 866–883 (1996)
15. Agrawal, R., Srikant, R.: Fast Algorithms for Mining Association Rules. In: Proc. of the 20th Int'l Conference on Very Large Databases, Santiago, Chile (1994)
16. TANAGRA, a free DATA MINING software for academic and research purposes, `http://eric.univ-lyon2.fr/~ricco/tanagra/en/tanagra.html`
17. Mild, A., Reutterer, T.: Collaborative Filtering Methods For Binary Market Basket Data Analysis. In: Liu, J., Yuen, P.C., Li, C.-H., Ng, J., Ishida, T. (eds.) AMT 2001. LNCS, vol. 2252, pp. 302–313. Springer, Heidelberg (2002)
18. Hong, S., Xinyan, Z., Shangping, D.: Mining Association Rules in Geographical Spatio-Temporal Data. The International Archives of the Photogrammetry, Remote Sensing and Spatial Information Sciences 196, 225–228 (2008)
19. Yo-Ping, H., Li-Jen, K., Sandnes, F.E.: Using Minimum Bounding Cube to Discover Valuable Salinity/Temperature Patterns from Ocean Science Data. In: Proc. of the 2006 IEEE International Conference on Systems, Man and Cybernetics, vol. 6, pp. 478–483 (2006)
20. Papadimitriou, J.: Statistical: Statistical Inference. Observer, Athens (1989)

A Software Platform for Evolutionary Computation with Pluggable Parallelism and Quality Assurance

Pedro Evangelista[1,2], Jorge Pinho[1], Emanuel Gonçalves[1],
Paulo Maia[1,2], João Luis Sobral[1], and Miguel Rocha[1]

[1] Department of Informatics / CCTC - University of Minho
{jls,mrocha}@di.uminho.pt
[2] IBB - Institute for Biotechnology and Bioengineering
Centre of Biological Engineering - University of Minho
Campus de Gualtar, 4710-057 Braga - Portugal
{ptiago,paulo.maia}@deb.uminho.pt

Abstract. This paper proposes the Java Evolutionary Computation Library (JECoLi), an adaptable, flexible, extensible and reliable software framework implementing metaheuristic optimization algorithms, using the Java programming language. JECoLi aims to offer a solution suited for the integration of Evolutionary Computation (EC)-based approaches in larger applications, and for the rapid and efficient benchmarking of EC algorithms in specific problems. Its main contributions are (i) the implementation of pluggable parallelization modules, independent from the EC algorithms, allowing the programs to adapt to the available hardware resources in a transparent way, without changing the base code; (ii) a flexible platform for software quality assurance that allows creating tests for the implemented features and for user-defined extensions. The library is freely available as an open-source project.

Keywords: Evolutionary Computation, Open-source software, Parallel Evolutionary Algorithms, Software Quality.

1 Introduction

The field of EC has been rapidly growing in the last decades, addressing complex optimization problems in many scientific and technological areas. Also, there has been the proposal of several software platforms with varying sets of functionalities, implemented in distinct programming languages and systems. Although the authors acknowledge the existence of interesting platforms implementing EC approaches, none of the ones tested was able to simultaneously respond to the set of requirements underlying our needs, namely: (i) robustness, modularity and quality assurance in the process of developing EC based components for other systems and applications; (ii) efficiency in allowing the rapid benchmarking of distinct approaches in specific optimization tasks, mainly when taking advantage of the currently available parallel hardware.

L. Iliadis et al. (Eds.): EANN/AIAI 2011, Part II, IFIP AICT 364, pp. 45–50, 2011.

This paper introduces the Java Evolutionary Computation Library (JECoLi), a reliable, adaptable, flexible, extensible and modular software framework that allows the implementation of metaheuristic optimization algorithms, using the Java programming language. The main contributions are its built-in capabilities to take full advantage of multicore, cluster and grid environments and the inclusion of a quality assurance platform, an added value in validating the software.

In this work, we show how independent parallelism models and platform mappings can be attached to the basic framework to take advantage of new computing platforms. Our objective is to support parallel processing allowing the user to specify the type of parallelization needed. The parallelization models are independent of the framework, and the user can specify the parallel execution model to meet specific computational methods and resources, taking advantage of the processing power of each target platform.

On the other hand, the quality assurance framework developed aims at providing a flexible way for both the developers of JECoLi and JECoLi-based applications to be able to validate the available features such as algorithms, operators, parallelization features, etc. This framework should allow to conduct software tests with minimal programming.

2   Main Functionalities

JECoLi already includes a large set of EC methods: general purpose Evolutionary/ Genetic Algorithms, Simulated Annealing (SA), Differential Evolution, Genetic Programming and Linear GP, Grammatical Evolution, Cellular Automata GAs and Multi-objective Evolutionary Algorithms (NSGA II, SPEA2). Among the parameters that can be configured, we find the encoding scheme, the reproduction and selection operators, and the termination criteria.

In JECoLi, solutions can be encoded using different representations. A general purpose representation using linear chromosomes is available including binary, integer or real valued genes. Also, individuals can be encoded as permutations, sets and trees. Numerous reproduction operators are provided for each of these representations. It is also possible to add new user-defined representations and/ or reproduction operators.

A loose coupling is provided between optimization algorithms and problems allowing the easy integration with other software. New problems are added by the definition of a single class specifying the evaluation function. An adequate support is also given to the development of hybrid approaches, such as local optimization enriched Evolutionary Algorithms (EAs) (e.g. Memetic Algorithms), hybrid crossover operators and initial population enrichment with problem-specific heuristics.

The framework is developed in Java, being portable to the major systems. The project is open source, released under the GPL license. Extensive documentation (howto's, examples and a complete API), binaries and source code releases are available in the Wiki-based project's web site `http://darwin.di.uminho.pt/ jecoli`.

3   Overall Architecture and Development Principles

The JECoLi architecture was built to be an adaptable, flexible, extensible, reliable and modular software platform. In Figure 1 a simplified class diagram is shown, containing the main entities. The *Algorithm* abstraction represents the optimization method. The information common to all algorithms was abstracted in the *AbstractConfiguration* class. The algorithm dependent configuration details are stored in the concrete implementation. When an algorithm starts executing, it checks all setup components, executes a specific initialization routine, and finally executes a series of iterations. During the execution, each algorithm holds an internal state that stores the current solution set, results and statistics collected during the previous iterations. Several alternative termination criteria can be defined (e.g. number of generations, function evaluations, CPU time).

Classes representing individuals implement interface *ISolution*, including the genome (with a specific representation) and the fitness value(s). Implementation of specific representations follow the *IRepresentation* interface. The solution factory design pattern was used to allow building new solutions and copying existing ones. These factories are a centralized location for restrictions applied to a specific genome. Populations are implemented by the *SolutionSet* that implements lists of solutions with enhanced functionalities.

An evaluation function is responsible for decoding an individual to a particular problem domain and providing for fitness evaluation. This makes the connection between the problem and the algorithm domains. One of the major aims was the extensibility of the platform, implemented by defining contracts (using interfaces) for specific components enabling the addition of new algorithms, representations, operators, termination criteria, etc.

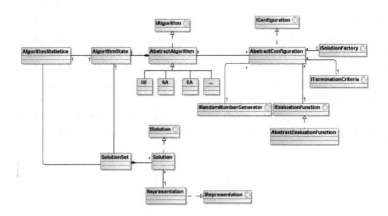

Fig. 1. Architecture of the JECoLi: main classes

4   Parallelization of the Library

The parallelization strategy is based on three key characteristics:

– Non-invasive - parallelization should have minimal impact on the original code; using/ developing functionalities in the base JECoLi does not require the developer to have knowledge about the parallelism models/mappings;
– Localized - parallelization code should be localized in well defined modules;
– Pluggable - the original code can execute when parallelism modules are added/ removed, i.e. parallelism modules are included on request.

Thus, the parallelization of the library consists of (i) the base JECoLi; (ii) a set of non-invasive and pluggable modules that adapt the base framework to support parallel execution and a set of mappings that adapt the supported models of parallel execution to each target platform; (iii) a tool that composes the base framework with modules for parallel execution and platform mappings, according to the user requests. Two distinct conceptual models for the parallelization of EAs are implemented: (i) the island model; and (ii) the parallel solution evaluation. These models do not imply any specific parallel environment and can be executed even in sequential architectures.

The parallelization of JECoLi is based on a three-layer architecture depicted in Figure 2. The features were made based on layers: the first layer is the original JECoLi; the second contains the parallelism models to use, and the last one introduces the mapping of the parallel behavior to the execution platform. The parallelism model layer contains the parallel models presented above: island model and parallel execution of the evaluation. Three distinct modules were implemented in this layer: *Parallel Eval*, *BuildIslandModel* and *Abstract Migration*, implementing the parallelization features that encapsulate platform independent behaviour. The first implements the model related to the parallel execution of the evaluation of solutions, while the second and third are related to the implementation of the island model. The mappings layer contains all the specific code to be loaded regarding a specific parallel platform. It encapsulates the mapping of the parallelism models to meet specific target parallel environments: single machine with multicore processor, clusters and grid environments.

The implementation of the modules resorted to the use of Aspect Oriented Programming (AOP) [2], a programming technique used to encapsulate crosscutting phenomena into specific modules of the programs. Technical details of the full implementation of the modules are provided in the web site documentation.

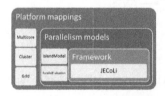

Fig. 2. Architecture of the Parallelization of JECoLi

5   Quality Assurance Framework

A generic test framework was engineered in order to validate JECoLi's function-alities. The main aim is to verify the correctness of the distinct components. The tests treat each component of the library as a black-box entity with a predefined set of inputs and a desired output. These are defined in order to cover all the possible outcomes of an entity for a certain input even if the library component only returns a partial result set. If the outcome of an entity is contained within the full result set the test is considered valid.

A domain specific language (DSL) with an LL(1) grammar was conceived to create *JUnit* tests [1] for the distinct components (algorithms, operators and parallelization methods). This language captures the tests configuration skeleton. Examples of configurations of specific tests are provided in the project web site.

Also, a test framework was developed encompassing a class hierarchy that is given in the web site documentation. The implementation of the abstract classes in the lower level of the class hierarchy only needs to detail component configuration information. These structures allow to capture seamlessly the test structure and to implement it in a domain specific language without the need to program in Java to add new tests for existing components.

6   Applications

The library has already been used in several research projects, described in the web site. Here, we highlight a few of these examples that have led to the development of software applications:

– Metabolic engineering: the aim is microbial strain optimization, using *in silico* simulations based on genome-scale metabolic models. EAs and SA using a set-based representation have been used [5] and implemented as part of the OptFlux software platform (www.optflux.org) [4], that represents a reference open-source software platform.
– Fermentation optimization: aims at the numerical optimization of feeding profiles in fed-batch fermentation processes in bioreactors [3]. EAs with a real value representations and several variants of DE were tested and eval-uated. The implementation of these approaches was included in a software application named OptFerm (darwin.di.uminho.pt/optferm).
– Network traffic engineering: the aim is to implement methods to improve the weights of intra-domain routing protocols [6]. Two approaches were used, both using an integer representation: single objective EAs with linear weight-ing objective functions and multiobjective EAs. These methods have been incorporated into an application - NetOpt (darwin.di.uminho.pt/netopt).

7   Conclusions

This paper described the framework JECoLi, a Java-based platform for the im-plementation of metaheuristic methods. The main strengths of this platform rely on its efficiency, portability, flexibility, modularity and extensibility which make it ideal to support larger applications that need to include an optimization engine based on EC methods. The main focus of this library has been to support this embedding capabilities also by making available a quality assurance framework, including a domain specific language for building unit tests, that allows the full coverage of the implemented features.

In terms of computational efficiency, one of the major concerns has been the development of modules to support parallelism. The parallelization of JECoLi was pursued in a pluggable, non-invasive and localized way, thus allowing the original library to evolve with new features being added and also to allow the automatic adaptation of the code to distinct scenarios in terms of hardware resources (multicore, cluster, grid).

Most of the future work includes the improvement of the existing functionalities. Also, we aim to develop new capabilities including new algorithms, representations or selection/reproduction operators.

Acknowledgments. This work is supported by project PTDC/EIA-EIA/ 115176/2009, funded by Portuguese FCT and Programa COMPETE.

References

1. Cheon, Y., Leavens, G.: A simple and practical approach to unit testing: The JML and JUnit way. In: Deng, T. (ed.) ECOOP 2002. LNCS, vol. 2374, pp. 231–1901. Springer, Heidelberg (2002)
2. Kiczales, G.J., Lamping, J.O., Lopes, C.V., Hugunin, J.J., Hilsdale, E.A., Boyapati, C.: Aspect-oriented programming, October 15, US Patent 6,467,086 (2002)
3. Mendes, R., Rocha, I., Ferreira, E., Rocha, M.: A comparison of algorithms for the optimization of fermentation processes. In: 2006 IEEE Congress on Evolutionary Computation, Vancouver, BC, Canada, pp. 7371–7378 (July 2006)
4. Rocha, I., Maia, P., Evangelista, P., Vilaça, P., Soares, S., Pinto, J.P., Nielsen, J., Patil, K.R., Ferreira, E.C., Rocha, M.: Optflux: an open-source software platform for in silico metabolic engineering. BMC Systems Biology 4(45) (2010)
5. Rocha, M., Maia, P., Mendes, R., Ferreira, E.C., Patil, K., Nielsen, J., Rocha, I.: Natural computation meta-heuristics for the in silico optimization of microbial strains. BMC Bioinformatics 9(499) (2008)
6. Sousa, P., Rocha, M., Rio, M., Cortez, P.: Efficient OSPF Weight Allocation for Intra-domain QoS Optimization. In: Parr, G., Malone, D., Ó Foghlú, M. (eds.) IPOM 2006. LNCS, vol. 4268, pp. 37–48. Springer, Heidelberg (2006)

Financial Assessment of London Plan Policy 4A.2 by Probabilistic Inference and Influence Diagrams

Amin Hosseinian-Far, Elias Pimenidis,
Hamid Jahankhani, and D.C. Wijeyesekera

School of Computing, IT and Engineering,
University of East London
{Amin,E.Pimenidis,Hamid.Jahankhani,Chitral}@uel.ac.uk

Abstract. London Plan is the London mayor's Spatial Development Strategy. This strategic long-term plan comprises of proposals for different aspects of change within the London boundary. Furthermore, the proposals include chapters outlining adjustments in each facet. Policy 4A.2 reflects the Climate Change Mitigation scheme. Some consultations and research works have been performed to this point, but an extensive cost assessment has not been done. This paper reflects a financial assessment by means of Influence Diagrams based upon the London Plan policies 4A.X.

Keywords: London Plan, Climate Change Mitigation, Influence Diagrams, Probabilistic Inference, Analytica, Sustainability.

1 Introduction

There are various definitions for Sustainable City which is also known as eco-city. They generally denote to a city which tries to have the minimum energy, and food input and has the minimum CO_2 emission, energy waste, general waste, methane and water pollution. The term was first introduced by Richard Register in 1987 as "*An Ecocity is an ecologically healthy city*" [1]. There are different plans for attaining the status of a sustainable city; they even vary from town to town and city to city. The United Kingdom is among the contributors to ecocity plans.

The aim of this paper is to demonstrate how influence diagrams can be used to do knowledge representation for a strategic, uncertain scenario. The case study selected is the London Plan Policy 4A.2. Section 1 discusses the key features of the London Plan relevant to this work; Section 2 outlines Influence Diagrams and the advantages of using them for this scenario. Finally the introduction to ID implementation is considered.

2 London Plan

The London Plan is the Mayor's Spatial Development Strategy, which covers different aspects of London's development; and a draft version of the plan has been out for public consultation since 2006 [2]. The section of the plan that this research

L. Iliadis et al. (Eds.): EANN/AIAI 2011, Part II, IFIP AICT 364, pp. 51–60, 2011.

focuses on is that of tackling the climate change and addressing sustainability analysis for London as a city. It is a strategic plan using a planning system and a set of aims and objectives relating to Energy Strategy and the control of climate change. The policies in the London Plan outline the proposals and schemes from London Transport strategic arrangement to the city Design. Climate Change is a sub division of the overall proposal.

2.1 Climate Change Policy Area

Climate Change as a sub division of the London Plan includes policies concerning different aspects of climate change from Air, Noise, Tackling Climate Change and even to Hazardous Substances. The Tackling Climate Change is the main concern of this research study. The policies involved are 4A.1 Tacking Climate Change, 4A.2 Mitigating Climate Change, and finally 4A.3 Sustainable Design and Construction. In spite of the categorization, these policies have overlapping areas between each other. The basis for the model which will be described in this work is Policy 4A.2, but some statistics and proposals will be borrowed from 4A.1 and 4A.3.

2.2 Tackling Climate Change

The main focus of this policy is on 60 % carbon dioxide reduction in London by 2050 relatively to the 1990 base. Comparatively, 15%, 20%, 25%, and 30% reduction targets have been set for 2010, 2015, 2020, and 2025 [2]. According to the plan draft new Building Development regulations are needed to be introduced; not only the new built properties should be regulated, but also the existing houses and properties should adapt to the criteria accordingly. According to Day, Ogumka, & Jones 20% out of the 60% reduction can be gained through the use of renewable energies [3]. Foreman Roberts research debates that the 20% out of 60% solely by renewables is not achievable [4]. Despite of all discussions, the cost assessment will be performed by only considering the policies published by the Greater London Authority (GLA). There are 147 case studies by further plan analysis, as approved by the London mayor. One of the main concerns of the plan is that of the use of Combined Heat and Power (CHP). CHP utilization would the most cost effective approach in the London Plan [5].

The other two policies focus on more thorough details of the London Climate Change route. According to research conducted by London South Bank University the impact assessment of the approach shown in Figure 1 should prove suitable to assess the general perspective of the London Plan climate change policies:

Fig. 1. Climate Change policy in three main phase [6]

According to the same research, the level of CO2 emission savings based on Energy Efficiency is higher than what could be expected by the use of renewable energy, as shown in table 1 below.

Table 1. Energy Efficiency and Renewables savings in the London Plan

	Savings to date from 113 energy statements		Savings to date – scaled for 350 developments
	tonnes CO_2/y	%	tonnes CO_2/y
EE	111,492	21.3	345,329
RE	24,039	5.8	74,457
Total	135,528	25.8	419,777

One of the debates relating to the London Plan (LP) is that of the targets set by the government. The government has set 60% CO2 reduction by 2050; below with the specific figure of 60% being quite unclear as to what the government envisages. Fig. 2 illustrates the most predominant understanding and it this viewpoint that the work presented here adopts:

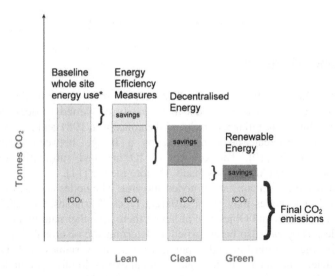

Fig. 2. Calculation of energy/carbon dioxide savings [7]

Energy Efficiency is the first phase according to the plan and also based on the LSBU research data is shown as having the highest reduction. The use of Combined Heat and Power (CHP) which is also called cogeneration refers to generation of electricity as well as heating the development. Renewable energy systems do not play a major role in the reduction process, but the introduction of more efficient renewables could yield an advantage. The application of the plan on 260 approved developments has revealed the following results:

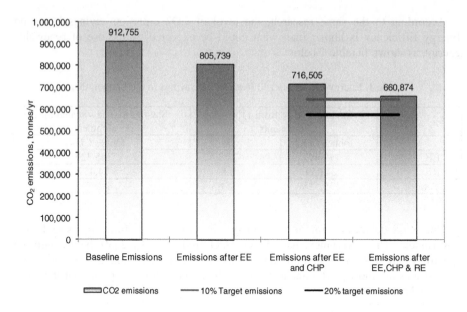

Fig. 3. Savings and approaches [6]

3 Influence Diagrams

An Influence Diagram (ID) is a network visualization for probabilistic models [8]. It also includes decision analysis and can be easily converted to decision trees [9]. Howard and Matheson initially introduced the concept in 1981 and according to them Influence Diagrams can be considered as a bridge between qualitative and quantitative analysis [10]. The concept of ID has had substantial influence on Artificial Intelligence (AI) and has made AI more mature [11].

An influence diagram consists of nodes and arcs. The nodes and arcs form a visual network [12]. The nodes may be of different types: Variables, Constants, Decision Nodes, Chance Node, and Objective; although there are other naming conventions for the ID nodes. They can also be classified as deterministic or probabilistic nodes. The Arcs specify the relationship and dependencies of the variables [13]. Definition of each node determines an associated mathematical expression for each node.

Influence Diagrams are a proper tool for decision making and analyzing a scenario under uncertainty [14]. Applying ID to London Plan policy analysis is an appropriate approach as the London Plan involves some uncertainty points. Furthermore Influence Diagrams can be furthermore broken down and illustrate more details. In complex situations such as the London Plan, the boundaries and level of details are in the hands of the analyst and can even be further refined later.

A Well-Formed Influence Diagram (WFID) is an ID which is not only well visualized, but also includes well-defined mathematical expressions for the nodes [15]. In addition WFIDs can be evaluated using reversal and removal approaches. This can be used for complex inferential and probabilistic scenarios [16].

In terms of categorization and classification of the subject, Influence Diagrams are classified under 'Bayesian Updating, Certainty Theory and Fuzzy Logic' category (Fig. 4).

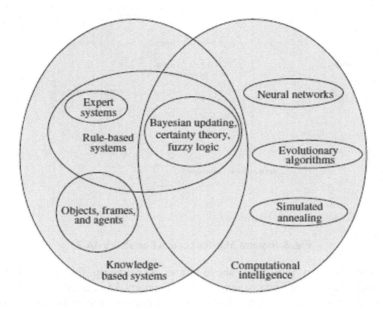

Fig. 4. Artificial Intelligence Classification [17]

4 Implementation

4.1 Indicators

A set of indicators for the London Plan monitoring phase is introduced, but the list does not cover the 4A.X policies. Therefore the indicators and variables used in the model are derived from the LP qualitative and quantitative statements within the boundary of this study.

4.2 Model Creation

The Integrated Development Environment (IDE) used for this research is called Analytica, many academic and industrial users of the software review Analytica as one of the most powerful applications for modeling using Influence Diagrams; even though other modeling environments can be used.

There are various recommendations for the starting point of influence diagrams. A systemic approach is to create the systems map first, and based on the systems map the influence diagram can be created. By using the systems map, the areas of the analysis and boundaries of the system under investigation will be determined [18]. In addition the influential parts of the research would be easier to find. A tentative and general systems map for this research is outlined in Fig. 5:

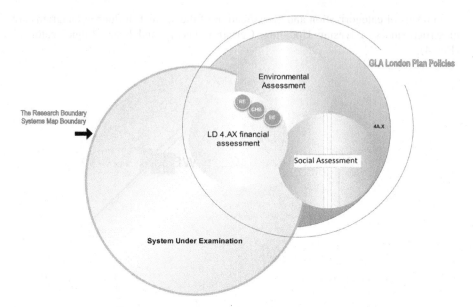

Fig. 5. Systems Map for London Plan Policy 4A.2

The systems map illustrated above draws a boundary for the system. The Environmental and Social assessments of the LD policy 4A.X overlap with these two domains. The economic or financial assessment domain is fully within the boundary. The use of renewable energies, energy efficiency procedures, and also less use of energy here outlined as CHB are the sub components which are located in the overlapping area of the three domains of sustainability. Based on this diagram a general visual representation of policy4A.2 analysis is demonstrated using a Probabilistic Network in Fig. 6:

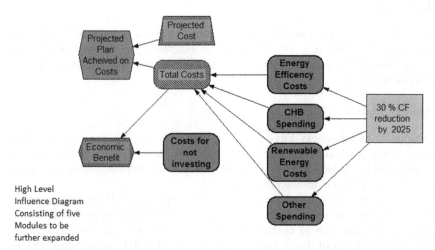

Fig. 6. High level Influence Diagram for London Plan Policy 4.A.2

The above general model consists of five modules which will be further detailed in the context of this work. The decision node or policy node is '30% Carbon Footprint (CF) reduction by 2025'. This is one of the nodes that do not include a mathematical expression while it only demonstrates an information passing process, rather than actual calculability.

The above decision node includes:

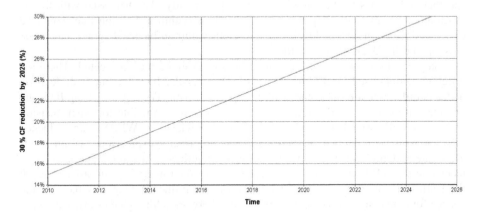

Fig. 7. Decision node: 30% CF Reduction by 2025 considering the 1990's base

The linearity of the output is due to the input being the contents of a table of data rather than a non linear equation. The probability distribution of the cost for not investing is as shown below:

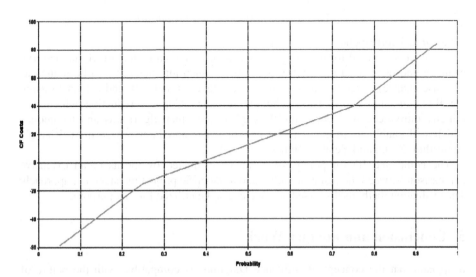

Fig. 8. Cost of Carbon Footprint by 2030, a tentative output illustrating different possible degrees of membership and the CF cost

If Projected_cost>Total_Costs, Then True, Else False

The above influence diagram and its modules and sub diagrams are to be created after this initial generalized model. But the reasons for choosing the influence diagrams for this type of policy modeling are due to its advantages. Influence diagrams have many advantages over decision trees. In decision trees adding nodes and generally expanding the model with more details requires to add nodes as child within the tree and that adds nodes to the whole diagram exponentially. Another issue with that is articulating conditional independence relationships within a decision tree would be easier said than done. Some levels of expressions within influence diagrams cannot be developed using decision trees [10].

A key question here would be that of how to evaluate the ID created. There are different naming used for different types of ID. A regular ID is the one with decision nodes in order, one after the other within a specified path. The not-forgetting ID is the one that information for previous decision nodes is also available for other decision nodes. A well-formed influence diagram or abbreviated as WFID is an influence diagram with well defined mathematical and probability expressions within itself. There are different ways in order to evaluate an influence diagram. The evaluation technique for WFID would be the reversal and removal operations in order to reach the starting node. For regular and other forms of influence diagrams there is the chance for conversion of the ID into a Bayesian network, although these two are quite the same. Converting an ID into a Bayesian Network is a straightforward task, as both have many similarities. All the nodes within the ID should be converted into chance nodes. The reason for that is in a Bayesian network the nodes are representing a probability distribution and no exact node should be defined. For conversion of the decision nodes into probability nodes or chance nodes, the newly defined chance node should have an even distribution:

$$\forall \langle di \in dom(di), P(\langle di / \pi di) = 1 // dom(di)/$$

Where $di \in D$ corresponds to decisions to be made.

It is once simulated for a probability less than 50% (can be tentatively named true chance), and once for a probability over 50% (which alternatively can be named as the false chance). Then it can be simulated as Bayesian network and evaluated using the Bayesian algorithms. This technique is called Coppers reduction which was initially introduced by Cooper in 1988. There are other algorithms and techniques which will enable us to evaluate ID e.g. Shachter and Peot's Algorithm, Zhang's Algorithm, Xiang and Ye algorithm and etc [19].

The completed ID for policy 4A.2 can be evaluated by means of an extensive comparison between the result of the ID and a scientific project management approach. The evaluation methods discussed above can also be utilized for that purpose.

5 Conclusion and Further Work

It appears that the concept of Influence Diagrams is compatible with the nature of decision making challenges paused by the London Plan and in particular the financial viability of the use of Renewable Energy Systems to partly meet LP's targets.

Completion of the above ID would be a proper means for evaluating the financial domain of the London Plan 4A.x policy. Apart from completing the ID, evaluation of the ID is essential.

Policy making for tackling climate change, energy crisis and a probable catastrophe has been urged by many governments. Some have started it already using various technical and scientific research studies. However, some policies are still at political level and there is no proper research work to allow for the validation of such plans.

According to the United Nations MDG indicators datasets, China and United States are the most CO2 polluters. Political and cultural similarities between the UK and the USA confirm that the next step for the use this climate change policy modelling approach would be the US cities with the highest CO2 emission [20].

Similar policies can be applied to them. Finally, use of other renewable energy systems would not cause any amendments to the general structure of the research using Influence Diagrams.

References

1. Register, R.: Ecocity Berkeley, Building Cities for a Healthy Future. North Atlantic Books, US (1987)
2. Friends of Earth: The Mayor's London Plan Briefing Report. Friends of Earth (FEO) Website (2006)
3. Day, A.R., Ogumka, P., Jones, P.G., Dunsdon, A.: The Use of the Planning System to Encourage Low Carbon Energy Technologies in Buildings. Renewable Energy 34(9), 2016–2021 (2009)
4. Roberts, F.: Renewables and the London Plan. British Council for Offices (March 2007)
5. Cibse: Small-scale combined heat and power for building. Guide 176 in Cibse website (2011)
6. Day, T.: Renewable Energy in the City. In: Advances in Cumputing and Technology Conference, London (2011)
7. Greater London Authority: The London Plan, Spatial Development Strategy for Greater London, Consolidated with Alterations since 2004. London Mayor's website (February 2008)
8. Shachter, R.D.: Probabilistic Inference and Influence Diagrams. Stanford University, Stanford (1987)
9. Ronald, A.H., James, E.M.: Influence Diagrams. Decision Analysis 2(3), 127–143 (2005)
10. Howard, R.A., Matheson, J.E.: Influence diagrams (1981). In: Howard, R.A., Matheson, J.E. (eds.) Readings on the Principles and Applications of Decision Analysis, vol. 2. Strategic Decisions Group, Menlo Park (1984)
11. Boutilier, C.: The Influence of Influence Diagrams on Artificial Intelligence. Decision Analysis 2(2), 229–231 (2005)
12. Shachter, R.D.: Model Building with Belief Networks and Influence Diagram. In: Edwards, W., Ralph, F., Miles, J., Winterfeldt, D.V. (eds.) Advances in Decision Analysis: From Foundations to Applications, pp. 177–201. Cambridge University Press, Cambridge (2007)
13. Influence Diagrams,
    http://www.lumina.com/technology/influence-diagrams/

14. Liao, W., Ji, W.: Efficient Non-myopic Value-of-information Computation for Influence Diagrams. International Journal of Approximate Reasoning 49(2), 436–450 (2008)
15. Sage, A.P.: Systems Engineering. Wiley Series in System Engineering (1992)
16. Shachter, R.: Evaluating Influence Diagrams. Operations Research 34(6), 871–882 (1986)
17. Hopgood, A.A.: Intelligent Systems for Engineers and Scientists, 2nd edn. CRC Press, New York (2001)
18. Open University: Influence Diagrams,
    http://openlearn.open.ac.uk/mod/oucontent/
    view.php?id=397869&section=1.9.4
19. Crowley, M.: Evaluating Influence Diagrams: Where we've been and where we're going. Unpublished Literatrue review (2004)
20. Daily Green: Top 20 CO2-Polluting U.S. Cities and Suburbs (2011), retrieved April 2011, from Daily Green: http://www.thedailygreen.com/environmental-news/latest/carbon-emissions-47041804

Disruption Management Optimization for Military Logistics

Ayda Kaddoussi[1], Nesrine Zoghlami[1], Hayfa Zgaya[2],
Slim Hammadi[1], and Francis Bretaudeau[3]

[1] Ecole Centrale de Lille, Cité Scientifique –BP 48, 59651,
Villeneuve d'Ascq, France
[2] ILIS, 42, rue Ambroise Paré, 59120 – LOOS France
[3] Logistics Department EADS DS Systems, Saint Quentin en Yvelines, France
{ayda.kaddoussi,nesrine.zoghlami,slim.hammadi}@ec-lille.fr,
hayfa.zgaya@univ-lille2.fr,
francis.bretaudeau@eads.com

Abstract. To ensure long-term competitiveness, companies try to maintain a high level of agility, flexibility and responsiveness. In many domains, hierarchical SCs are considered as dynamic systems that deal with many perturbations. In this paper, we handle a specific type of supply chain: a Crisis Management Supply Chain (CMSC). Supply during peacetime can be managed by proactive logistics plans and classic supply chain management techniques to guaranty the availability of required needs. However, in case of perturbations (time of war, natural disasters...) the need for support increases dramatically and logistics plans need to be adjusted rapidly. Subjective variables like risk, uncertainty and vulnerability will be used in conjunction with objective variables such as inventory levels, delivery times and financial loss to determine preferred courses of action.

Keywords: Logistics, supply chain, risk management, multi-agent.

1 Introduction

There is an urgent need for assuring the defense and support mission in time of crisis with a reasonable financial cost. The establishment of an efficient risk assessment that identifies and mitigates deficiencies that could impact mission success must be developed. In the military, decisions are the means by which the commander or decision maker translates their vision of the end state into actions [1]. Hence, there is a need to have the relevant facts and information to make the more adequate and prudent decisions. Experts in all specialties, including logistics, intelligence, medical, etc. deliberate to develop different possible courses of actions and scenarios, and determine what the most likely solution based on their judgment and the feedback from the risk management module and the CMSC in general.

In this paper, a multi-agent based system is proposed to manage the risk induced by the complexity of the CMSC environment. The reminder of the paper is organized as follows: in the following section, definitions about risk and uncertainty are presented, and the usefulness of a multi-agent based framework for SCs risk

L. Iliadis et al. (Eds.): EANN/AIAI 2011, Part II, IFIP AICT 364, pp. 61–66, 2011.

management is discussed. The third section presents a general description of our context and problematic. Then the disruption management process is detailed in section 4. Conclusion and possible future works are addressed in last section.

2 Literature Review

Risk can be described as the product of the probability of an event occurring and its consequences, in other words, risk is about uncertainty and its impact [2].Uncertainty is the general property of non-deterministic environment. It generates events that disrupt the pre-established behavior of any complex system such as Supply Chains (SCs). Uncertainty raises risk which can be identified, analyzed, controlled and regulated.

There are different types of disturbances that can be divided to purposeful disturbances (such thefts, terrorism, piracy and financial misdeeds) and non-purposeful disturbances (such demand fluctuations and bullwhip-effect). The deviations from the expected outcome may affect operations, processes, plans, goals or strategies and would manifest as under achievement of performance, poor reliability, time delay or financial loss. Therefore corrective measures need to be taken to adjust the SC.

Different methods are used for risk management in supply chain. Table 1 summarizes the different methods.

Table 1. Risk management methods

Retrospective approaches	
Methods	*Description*
Pareto chart	Used during the analysis phase of reported undesirable events, it prioritizes the relative importance of different events and ranks them by decreasing frequency.
Ishikawa diagram	It is a causal diagram that shows the causes of a certain event. Each cause or reason for imperfection is a source of variation. Causes are usually grouped into major categories to identify these sources of variation. The categories typically include People, Methods, Machines, Materials and Environment.
Predictive approaches	
Methods	*Description*
FMECA (Failure Mode, Effects, and Criticality Analysis)	It is a procedure for analysis of potential failure modes within a system by classification of the severity and likelihood of the failures. A successful FMECA activity helps to identify potential failure modes based on past experience with similar products or processes, enabling the team to design those failures out of the system with the minimum of effort and resource expenditure, by reducing development time and costs.
HACCP (Hazard Analysis Critical Control Point)	Aims to assess potential hazards on a given process, and identify appropriate preventive measures and required monitoring system. Its application is mostly in catering and food industry.
HAZOP (HAZard and OPerability study)	It was initially developed to analyze chemical process systems, but has later been extended to other types of systems and also to complex operations and to software systems, to identify and evaluate problems that may represent risks to personnel or equipment.

There is an urgent need for SCs planners to have specific strategies to manage these disruptions. Multi-agent technology is very well suited as a technological platform for supporting SC concepts through modeling the different heterogeneous and linked entities. In computer science, an agent can be defined as a software entity, which is autonomous to accomplish its designed goals through the axiom of communication and coordination with other agents [4]. These agents interact and cooperate in order to solve problems beyond their individual knowledge or expertise, and to promote a higher performance for the entire system [5]. Through their learning capability, MAS can demonstrate efficiently the proactive and autonomous behavior of the actors to mitigate disruptions and rectify the SC functioning in real time [6-7]. Agent based technology has also been used for the management of disruptions within a supply chain in some studies. Kimbrough et al. [8], for example, use it for the reduction of the bullwhip effect through modeling a SC with agents. Bansal et al.[9] provide a generalized collaborative framework for risk management oriented to refinery SCs.

3   Problem Description and Objectives

The presented work proposes to support the logistics planners in dealing with events that may disrupt the CMSC and get the plan deviate from its intended course. In the case of distributed systems such SCs, this implies a design that unsure an efficient disturbance handling and relevant deployment of individual recovery behaviors. The aim of our approach is to help the different actors of the CMSC to improve their performance and to minimize the impact of disruptions on the whole SC. Our objective are to determine how to deploy the military units and its associated resources and equipment, including personnel, vehicles, and aircraft; and sustain them throughout the operation, by providing supplies (food, water, fuel, etc.) and meeting other needs, such as medical support. Military logistics is a complex process, involving collaboration and coordination among many organisational and informational entities, such as supply, transport and troupes, which are geographically distributed and contain complex information. The Optimisation Based on Agents' Communication (OBAC) is a demonstrator aimed at developing a support system for military logistic that automates the functioning of the military supply chain under crisis. OBAC is prototyped by our LAGIS[1] research team in collaboration with the Logistics department of EADS, and is investigating, developing and demonstrating technologies to make a fundamental improvement in logistics planning. Agent-based technology has already been applied to different areas in military logistics.

4   Disruption Management Process

The risk management module is composed of a set of agents called *Disruption Manager Agent (DMA)*. It is constituted of a *watch_agent* responsible for providing

---

[1] LAGIS: Laboratoire d'Automatique, Génie Informatique et Signal.

monitoring information. It has the ability to trigger an alarm when a disruption event appears. For example, it provides supervision of the delivery process, by monitoring the actual delivery time and comparing it to the planned one. A second agent involved in the DMA is the *proposer_rectifying_agent* responsible for the suggestion of corrective actions for the emerging risks, based on the alert triggered by the *watch_agent*. This initiates a process that intents to eliminate or reduce the prominent risk. Other agents will intervene, and will help to facilitate exchange of information and resources, as *Zone Agent (ZA:* each zone of our CMSC is represented by an agent), *GUI Agent (Interface agent), Weather Agent (WA), Need Estimating Agent (NEA), Posts Coordinator Agent (PCA) and Consumption Agent (CA).*

4.1 Risk Identification Phase

To identify risks, we use quantitative models, and for each activity, risk sources are listed and described. This stage is based on the monitoring of what we called performance indicators (PIs) related to the performance of all actors in the CMSC. The level of in-stock inventory, the amount of resources consumed and delivery dates are some of the PIs that can be used to identify abnormal situations that may engender risks and deficiencies on the CMSC. Actual values of the PIs are monitored by the *watch_agent* which compares them to predefined values. In case of a significant violation, *watch_agent* triggers an alarm characterizing the type of violation detected.

4.2 Risk Assessment Phase

This phase analyses the impact of the risk identified and provides information about the parameters to be corrected during the decision and selection of the strategy to adopt. To do that, the *proposer_rectifier_agent*, using the FMEC analysis and based on a risk rating mechanism, estimates the probabilities of occurrence of the event (very unlikely, improbable event, moderate event, probable event, very probable), the risk impact (no impact, minor impact, medium impact or serious impact), and probability of risk non-detectability (very low, low, moderate, high, very high). This assessment is generally based on historical data, experience and advice from experts. The monitoring of the PIs by the *watch-agent* is maintained in order to provide updates to the *DMA*. During this step, the success of past decisions applied to the same risk is taken into account for evaluation.

4.3 Decision and Selection of Corrective Actions Phase

During this phase, the *proposer_rectifier_agent* proposes the corrective action to perform for the identified risk. The selected action or strategy is transferred to the agent concerned by the emerging risk to execute it. An example of the behavior of *ZA* is illustrated by the diagram of activity of figure 1.

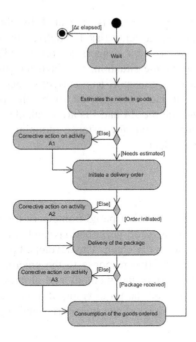

Fig. 1. Behavior of the ZA

Activity	Risk	Corrective Action
A1: Need estimation	Inaccurate estimation	Prevent variation with safety stock.
A2: Order delivery	Error in the command.	Change the mean of transportation.
A3: Order reception	Delay in reception.	Initiate an emergency order.

4.4 Evaluation Phase

In the final step of the risk management process, the global supply chain performance is evaluated by calculating the *avoided cost*. This cost is composed of the amount of financial loss that the disruption could have engendered if it had becomes a reality and the investment cost to mitigate the risk. The aim is to get the CMSC efficiency at a higher level or to maintain it at the same level.

This approach is based on a simulation tool which enables the implementation of different models for the planning activities of each actor of the CMSC and takes into account interactions between them.

5 Validation

We are developing our system, with JADE platform (Java Agent Development platform) [10]. It is a middleware which permits a flexible implementation of multi-agents systems and offers an efficient transport of ACL (Agent Communication Language) messages for agents' communication. To demonstrate the reliability and the applicability of our framework, we can assume a scenario in which the actual in-stock inventory is lower than what is indicated by the warehouse manager interface. The situation suggests a delay risk for delivering a specific order to one ZA, and as a result the *watch-agent* triggers a corresponding alarm. This alarm initiates the process to remedy the situation, in which the *proposer_rectifier_agent* proposes to change the mean of transport as a corrective action. This action is then evaluated and executed if it has been proved to be the optimal one. Else, the *proposer_rectifier_agent* notifie the ZA and PIs are recalculated in the perspective of performing another corrective action (for instance rationing the consumption on the ZA until the reception of the order).

6 Conclusion

This paper has presented a general agent-based framework for disruption management aiming at minimizing the effects of perturbations and uncertainties on a highly distributed crisis management supply chain. The developed process generates courses of actions to have the most effective response to the emergence of a risk. As this work is based on validated models and methods and already existent software tool (OBAC), the next steps consist on the prototyping of a warehouse managing system based on RFID tags in order to track down inventories and so insure better reactivity in a distributed SC.

References

1. Field Manual - FM No. 101-5, Staff organization and operations contents. Headquarters, Department of the army, Washington, DC, May 31 (1997)
2. Hull, K.: Risk analyses techniques in defense procurement. In: The IEEE Colloquium on Risk Analysis Methods and Tools, London, June 3, pp. 3/1–317 (1992)
3. Giannakis, M., Louis, M.: A multi-agent based framework for supply chain risk management. Journal of Purchasing & Supply Management 17, 23–31 (2011)
4. Stone, P., Veloso, M.: Multiagent systems: A survey from a machine learning perspective. Autonomous Robots 8(3), 345–383 (2000)
5. Kwon, O., Im, G.P., Lee, K.C.: MACE-SCM: a multi-agent and case-based reasoning collaboration mechanism for supply chain management under supply and demand uncertainties. Expert Systems with Applications 33, 690–705 (2007)
6. Lu, L., Wang, G.: A study on multi-agent supply chain framework based on network economy. Computers and Industrial Engineering 54(2), 288–300 (2007)
7. Kimbrough, S.O., Wu, D.J., Zhong, F.: Computers play the beer game: can artificial agents manage supply chains? Decision Support Systems 33(3), 323–333 (2002)
8. Bansal, M., Adhitya, A., Srinivasan, R., Karimi, I.A.: An online decision support framework for managing abnormal supply chain events. Computer-aided Chemical Engineering 20, 985–990 (2005)
9. Java Agent DEvelopment framework, http://jade.titlab.com/doc

Using a Combined Intuitionistic Fuzzy Set-TOPSIS Method for Evaluating Project and Portfolio Management Information Systems

Vassilis C. Gerogiannis[1,2], Panos Fitsilis[1,2], and Achilles D. Kameas[2]

[1] Project Management Department, Technological Education Institute of Larissa, 41110, Larissa, Hellas
{gerogian,fitsilis}@teilar.gr
[2] Hellenic Open University, 167 R. Feraiou & Tsamadou str., 26222, Patras, Hellas
kameas@eap.gr

Abstract. Contemporary Project and Portfolio Management Information Systems (PPMIS) have embarked from single-user, single-project management systems to web-based, collaborative, multi-project, multi-functional information systems which offer organization-wide management support. The variety of offered functionalities along with the variation among each organization needs and the plethora of PPMIS available in the market, make the selection of a proper PPMIS a difficult, multi-criteria decision problem. The problem complexity is further augmented since the multi stakeholders involved cannot often rate precisely their preferences and the performance of candidate PPMIS on them. To meet these challenges, this paper presents a PPMIS selection/evaluation approach that combines TOPSIS (Technique for Order Preference by Similarity to Ideal Solution) with intuitionistic fuzzy group decision making. The approach considers the vagueness of evaluators' assessments when comparing PPMIS and the uncertainty of users to judge their needs.

Keywords: Project and Portfolio Management Information Systems, Multi-Criteria Decision Making, Group Decision Making, TOPSIS, Intuitionistic Fuzzy Sets.

1 Introduction

Research studies [1] present that increasing organizational requirements for the management of the entire life-cycle of complex projects, programs and portfolios motivate the further exploitation of Project and Portfolio Management Information System (PPMIS) from enterprises of any type and size. PPMIS have embarked from stand-alone, single-user, single-project management systems to multi-user, multi-functional, collaborative, web-based and enterprise-wide software tools which offer integrated project, program and portfolio management solutions, not limited to scope, budget and time management/control. Modern PPMIS support, through a range of features, processes in all knowledge areas of the "Project Management Body of Knowledge" [2], by covering an expansive view of the "integration management"

L. Iliadis et al. (Eds.): EANN/AIAI 2011, Part II, IFIP AICT 364, pp. 67–81, 2011.

area that includes alignment and control of multi-project programs and portfolios. The market of PPMIS is rapidly growing and includes many commercial software products offering a number of functionalities such as time, resource and cost management, reporting features and support for change, risk, communication, contract and stakeholder management. Interested readers are referred to [3] where detailed information is given for 24 commercial leading PPMIS.

This variety of offered functionalities along with the variation among each organization needs and the plethora of PPMIS in the market, make their evaluation a complicate multi-criteria decision problem. The problem is often approached in practice by ad hoc procedures based only on personal preferences of users or any marketing information available [3], [4]. Such an approach may lead to a final selection that does not reflect the organization needs or, even worse, to an unsuitable PPMIS. Therefore, the use of a robust method from the multi-criteria decision making (MCDM) domain can be useful to support PPMIS selection. Review studies [5], [6], [7] reveal that the Analytic Hierarchy Process method (AHP) and its extensions have been widely and successfully used in evaluating several types of software packages (e.g., MRP/ERP systems, simulation software, CAD systems and knowledge management systems).

Although AHP presents wide applicability in evaluating various software products, little work has been done in the field of evaluating PPMIS. For example, in [8] the authors admit that their work is rather indicative with main objective to expose a representative case for illustrating the PPMIS selection process and not to create a definitive set of criteria that should be taken into account in practice. This lack of applicability of AHP in the PPMIS selection problem domain can be attributed to the fact that, despite its advantages, the method main limitation is the large number of pairwise comparisons required. The time needed for comparisons increases geometrically with the increase of criteria and alternatives involved, making AHP application practically prohibitive for complicate decisions, such as the selection of a PPMIS. As a response to this problem, in the recent past [9], we presented an approach for evaluating alternative PPMIS that combines group-based AHP with a simple scoring model. This group-AHP scoring model, although practical and easy to use, does not consider the vagueness or even the unawareness of users, when they evaluate their preferences from a PPMIS by rating their requirements. Also the approach does not take into account the uncertainty of experts, when they judge the performance of alternative PPMIS on the selected criteria, expressed as user requirements.

These ambiguities in evaluation of software systems and the incomplete available information expose the need to adopt a fuzzy-based approach [10]. Fuzzy-based methods provide the intuitive advantage to utilize, instead of crisp values, linguistic terms to evaluate performance of the alternatives and criteria weights. A fuzzy-based approach can be even more beneficial when it is combined with other decision making techniques. For example, fuzzy AHP [11] is proposed to handle the inherent imprecision in the pairwise comparison process, while fuzzy TOPSIS (Technique for Order Preference by Similarity to Ideal Solution) [12] can be used to jointly consider both positive (benefit/functional oriented) and negative (cost/effort oriented) selection criteria. Fuzzy-based MCDM techniques have been used to select various types of software products (see, for example, [10], [13], [14]), but in the relevant literature there is lack of a structured fuzzy-based approach for the selection of PPMIS under uncertain knowledge.

The main objective of this paper is to present such an approach that involves both users and evaluators (decision makers) in the decision making process and tries to exploit the interest/expertise of each one in order to strengthen the final evaluation results. This is achieved by aggregating all weights of criteria (requirements) and all ratings of performance of the alternative systems, as they are expressed, by individual stakeholders, in linguistic terms. The approach is based on intuitionistic fuzzy sets, an extension of fuzzy sets proposed by Atanassov [15], which has been applied in a variety of decision making problems [16], [17], [18], [19], [20]. An intuitionistic fuzzy set includes the membership and the non-membership function of an element to a set as well as a third function that is called the hesitation degree. This third function is useful to express the lack of knowledge and the hesitancy concerning both membership and non-membership of an element to a set. Expression of hesitation is particularly helpful for both decision makers and users when they select a software product for an organization such as, in our case, a PPMIS. On one hand, decision makers often cannot have a full knowledge upon all functionalities included in the newest version of each candidate system. Thus, they base their ratings only on experience from using previous system versions as well by referencing to system assessments which can be found in products survey reports. On the other hand, users are often unfamiliar with how a PPMIS can support project management processes tasks and, therefore, cannot precisely express which tasks require more to be supported by a PPMIS. The presented approach mainly utilized the method proposed in [18] which combines intuitionistic fuzzy sets with TOPSIS for supporting the supplier selection problem. The advantage of this combination in case of PPMIS evaluation is that we can distinguish between benefit criteria (e.g., functionalities/tasks supported by the PPMIS) and cost criteria (e.g., effort for system customisation and price for ownership). The PPMIS that is closest to the positive to ideal solution and most far from the negative ideal solution could be probably the most appropriate PPMIS to cover the organization needs.

The rest of the paper is organized as follows. In section 2, we present the aspects of the PPMIS evaluation problem and we justify how, in our case study, the PPMIS selection criteria were determined. In section 3, we briefly discuss the basic concepts of intuitionistic fuzzy sets. Section 4 presents the detailed description of the approach and Section 5 presents the conclusions and future work.

2 Evaluation of PPMIS and Selection Criteria

Empirical studies [21] demonstrate that a number of project managers indicate a strong impact of PPMIS upon successful implementation of their projects, while others do not. The PPMIS selection process is usually supported by referencing to market surveys [3], [4], [22] or by considering the users' perceptions and satisfaction from a PPMIS usage [23]. Detailed assistance in evaluating PPMIS is provided by evaluation frameworks which propose to consider an extensive list of system characteristics. These characteristics can be either functional or process oriented selection criteria. NASA, for example, in the past has convened a working group to evaluate alternative PPMIS for NASA's departments, upon a number of functional requirements. In the group's report [24] thirteen clusters of functional requirements are identified. Each cluster further includes a set of functional features and, in total, 104 functional criteria are identified to be evaluated. This vast number of criteria prevents decision makers from using a typical hierarchical MCDM approach like AHP.

As far as process oriented evaluation is concerned, evaluators may utilize as reference the set of criteria offered by a conceptual software architecture for PPMIS, like, for example, is the M-Model [25]. The M-Model was used in [3] to evaluate commercial PPMIS according to project phases/tasks and corresponding required functionality (Table 1). Each PPMIS was evaluated according to the extent that it offers the required functionality and the overall support for the corresponding project phase/task was specified with a "4-stars" score.

Table 1. Evaluation criteria (source: [3])

Phase/Task	Required Functionality
1. Idea Generation / Lead Management (IGLM)	Creativity Techniques, Idea / Project Classification, Lead Management (Mgmt.), Project Status / Project Process Mgmt.
2. Idea Evaluation (IE)	Estimation of Effort, Resource Needs Specification, Risk Estimation, Profitability Analysis, Project Budgeting, Offer Mgmt.
3. Portfolio Planning (PP1)	Organizational Budgeting, Project Assessment, Project Portfolio Optimization, Project Portfolio Configuration
4. Program Planning (PP2)	Project Templates, Resource Master Data, Resource Assignment Workflow, Resource Allocation
5. Project Planning (PP3)	Work Breakdown Structure Planning, Scope/ Product Planning, Network Planning, Scheduling, Resource Leveling, Risk Planning, Cost Planning
6. Project Controlling (PC1)	Change Request Mgmt., (Travel) Expense Mgmt., Timesheet, Cost Controlling, Meeting Support
7. Program Controlling (PC2)	Status Reporting, Deviation / Earned Value Analysis, Quality Controlling, Versioning, Milestone Controlling
8. Portfolio Controlling (PC3)	Performance Measurement, Dashboard, Organizational Budget Controlling
9. Program Termination (PT1)	Knowledge Portal, Competence Database / Yellow Pages, Project Archiving, Searching
10. Project Termination (PT2)	Invoicing, Document Mgmt., Supplier & Claim Mgmt.
11. Administration/Configuration (AC)	Workflow Mgmt., Access Control, Report Development, Form Development, User-Defined Data Structures, MS Office Project Interface, Application Programming Interface, Offline Usage

3   Basic Concepts of Intuitionistic Fuzzy Sets

Before proceeding to describe how the PPMIS selection problem was tackled, we briefly introduce some necessary introductory concepts of intuitionistic fuzzy sets (IFS). An IFS A in a finite set X can be defined as [15]:

$$A = \{< x, \mu_A(x), v_A(x) >| x \in X\}$$

where $\mu_A : X \rightarrow [0,1]$, $v_A : X \rightarrow [0,1]$ and $0 \leq \mu_A(x) + v_A(x) \leq 1$ $\forall x \in X$. $\mu_A(x)$ and $v_A(x)$ denote respectively the degree of membership and non-membership of x to A. For each IFS A in X, $\pi_A(x) = 1 - \mu_A(x) - v_A(x)$ is called the hesitation degree of whether x belongs to A. If the hesitation degree is small then knowledge whether x belongs to A is more certain, while if it is great then knowledge on that is more uncertain. Thus, an ordinary fuzzy set can be written as:

$$\{< x, \mu_A(x), 1 - \mu_A(x) >| x \in X\}$$

In the approach we will use linguistic terms to express: i) the importance of decision stakeholders (users/decision makers), ii) judgements of decision makers on the performance of each PPMIS and iii) perceptions of users on the importance of each selection criterion. These linguistic terms can be transformed into intuitionistic fuzzy numbers (IFNs) in the form of $[\mu(x), v(x)]$. For example, an IFN $[0.50, 0.45]$ represents membership $\mu = 0.5$, non-membership $v = 0.45$ and hesitation degree $\pi = 0.05$. In the approach we will also use addition and multiplication operators of IFNs. Let $a1 = (\mu_{a1}, v_{a1})$ and $a2 = (\mu_{a2}, v_{a2})$ be two IFNs. Then these operators can be defined as follows [15], [27], [28]:

$$a1 \oplus a2 = (\mu_{a1} + \mu_{a2} - \mu_{a1} \cdot \mu_{a2}, v_{a1} \cdot v_{a2})$$

$$a1 \otimes a2 = (\mu_{a1} \cdot \mu_{a2}, v_{a1} + v_{a2} - v_{a1} \cdot v_{a2})$$

$$\lambda \cdot a1 = (1 - (1 - \mu_{a1})^{\lambda}, v_{a1}^{\lambda}), \quad \lambda > 0$$

4   Evaluation of PPMIS with Intuitionistic Fuzzy Sets and TOPSIS

In this section we will describe how an intuitionistic fuzzy MCDM method was applied with the overall goal to select the most appropriate PPMIS system for the Hellenic Open University (HOU) (www.eap.gr). HOU is a university that undertakes various types of national and international R&D projects and programs, particularly in the field of lifelong adult education. The university does not maintain an integrated project/portfolio management infrastructure. In order to increase project management effectiveness and productivity, the management of HOU has decided to investigate the adoption of a collaborative PPMIS. The Department of Project Management (DPM) at the Technological Education Institute of Larissa in Greece was appointed to act as an expert and aid this decision making process. Three experts D_1, D_2 and D_3 (decision makers) from DPM, with an average of seven years teaching/professional experience in using PPMIS, were involved in this process, aiming to identify HOU requirements from a PPMIS and to select an appropriate system that will cover these requirements. Three project officers/managers U_1, U_2 and U_3 (users) from the HOU site were also involved in the decision making. These persons have high expertise in contract management, multi-project coordination and planning of R&D projects and portfolios, but they present low experience in systematically using PPMIS. It should be mentioned here that the presented approach mainly utilized the method presented in [18] to handle a hypothetical supplier selection problem with five alternative suppliers and four selection criteria. In the current paper we validate the method in an actual context and we show its applicability with an extensive set of selection criteria. In addition, we show how sensitivity analysis can be applied to evaluate the influence of criteria weights on the final selection results.

The application of the approach for selecting an appropriate PPMIS for the case organization has been conducted in eight steps presented as follows.

Step 1: Determine the weight of importance of decision makers and users. In this first step, the expertise of both decision makers and users was analysed by specifying corresponding weights. In a joint meeting, the three decision makers D_1, D_2, D_3 agreed to qualify their experience in using PPMIS as "Master", "Proficient" and "Expert", respectively. The three users U_1, U_2, U_3 also agreed that their level of expertise in managing large projects can be characterized as "Master", "Proficient" and "Expert", respectively. These linguistic terms were assigned to IFNs by using the relationships presented in Table 2 between values in column 1 and values in column 3. In general, if there are l stakeholders in the decision process, each one with a level of expertise rated equal to the IFN $[\mu_k, v_k, \pi_k]$, the weight of importance of k stakeholder can be calculated as [18]:

$$\lambda_k = \frac{(\mu_k + \pi_k(\frac{\mu_k}{\mu_k + v_k}))}{\sum_{k=1}^{l}(\mu_k + \pi_k(\frac{\mu_k}{\mu_k + v_k}))} \tag{1}$$

where $\lambda_k \in [0,1]$ and $\sum_{k=1}^{l}\lambda_k = 1$.

Table 2. Linguistic terms for the importance of stakeholders and criteria

Level of Stakeholder Expertise (1)	Importance of Selection Criteria (2)	IFN (3)
Master	Very Important (VI)	[0.90,0.10]
Expert	Important (I)	[0.75,0.20]
Proficient	Medium (M)	[0.50,0.45]
Practitioner	Unimportant (U)	[0.35,0.60]
Beginner	Very Unimportant (VU)	[0.10,0.90]

According to eq. (1), the weights of decision makers were calculated as follows: $\lambda_{D1} = 0.406$, $\lambda_{D2} = 0.238$, $\lambda_{D3} = 0.356$. Since users were assigned to the same linguistic values, their weights were respectively the same: $\lambda_{U1} = 0.406$, $\lambda_{U2} = 0.238$, $\lambda_{U3} = 0.356$.

Step 2: Determine the level of support provided by each alternative PPMIS. Though there is a large number of available PPMIS, decision makers were queried to express their general opinion on ten commercial PPMIS which in market survey results [4] are characterised as leaders and challengers in this segment of enterprise software market. Five from these systems were excluded for two reasons: because they do not have presence in the national market and because decision makers were persuaded that their usage was inappropriate for the specific case, mainly due to lack of technical support and non-availability of training services. This first-level screening resulted in a list of five powerful PPMIS with strong presence (i.e., technical/training support) in the national market. For confidentiality reasons and aiming at avoiding the promotion of any software package, we will refer to these PPMIS as A_1, A_2, A_3, A_4 and A_5.

Table 3. Linguistic terms for rating the performance of PPMIS

Level of Performance/Support	IFN
Extremely high (EH)	[1.00,0.00]
Very very high (VVH)	[0.90,0.10]
Very high (VH)	[0.80,0.10]
High (H)	[0.70,0.20]
Medium high (MH)	[0.60,0.30]
Medium (M)	[0.50,0.40]
Medium low (ML)	[0.40,0.50]
Low (L)	[0.25,0.60]
Very low (VL)	[0.10,0.75]
Very very low (VVL)	[0.10,0.90]

In order to evaluate the candidate PPMIS in a manageable and reliable way, decision makers rated the performance of each system with respect to the criteria previously identified. Each decision maker was asked to carefully rate the support provided by each system on each of the 11 criteria (project phases/tasks) presented in Table 1. In addition to these 11 "positive" (benefit oriented) criteria, two "negative" (cost oriented criteria) was decided to be included in the list. These are the total price for purchasing/ownership (PO) and the effort required to customise/configure the PPMIS (CC). Thus, 13 criteria in total were adopted. All decision makers provided a short written justification for every rating they gave in linguistic terms. For ratings they used the linguistic terms presented in Table 3. Decision makers were also asked to cross-check their marks, according to the corresponding "4-stars" scores, as they are listed for each tool in [3]. Due to space limits, Table 4 presents ratings given by the three decision makers to the five PPMIS for the first three of the 13 criteria.

Based on these ratings and the weights of decision makers, the aggregated intuitionistic fuzzy decision matrix (AIFDM) was calculated by applying the intuitionistic fuzzy weighted averaging (IFWA) operator [28]. The basic steps of the IFWA operator are that it first weights all given IFNs by a normalized weight vector, and then aggregates these weighted IFNs by addition. Each result derived by using the IFWA operator is an IFN. If $A = \{A_1, A_2, ..., A_m\}$ is the set of alternatives and $X = \{X_1, X_2, ..., X_n\}$ is the set of criteria, then AIFDM R is an m x n matrix with elements IFNs in the form of $r_{ij} = [\mu_{A_i}(x_j), v_{A_i}(x_j), \pi_{A_i}(x_j)]$, where $i = 1,2,...,m$ and $j = 1,2,...,n$. By considering weights λ_k ($k = 1,2,...,l$) of l decision makers, elements r_{ij} of the AIFDM can be calculated using IFWA as follows:

$$r_{ij} = IFWA_\lambda(r_{ij}^{(1)}, r_{ij}^{(2)}, ..., r_{ij}^{(l)}) = \lambda_1 r_{ij}^{(1)} \oplus \lambda_2 r_{ij}^{(2)} \oplus \lambda_3 r_{ij}^{(3)} \oplus ... \oplus \lambda_l r_{ij}^{(l)}$$

$$= \left[1 - \prod_{k=1}^{l}(1-\mu_{ij}^{(k)})^{\lambda_k}, \prod_{k=1}^{l}(v_{ij}^{(k)})^{\lambda_k}, \prod_{k=1}^{l}(1-\mu_{ij}^{(k)})^{\lambda_k} - \prod_{k=1}^{l}(v_{ij}^{(k)})^{\lambda_k} \right] \quad (2)$$

Table 4. The ratings of the alternative PPMIS (excerpt)

Criteria	Decision Makers	PPMIS				
		A_1	A_2	A_3	A_4	A_5
IGLM	D_1	VH	VH	H	MH	H
	D_2	H	VH	MH	H	H
	D_3	H	H	H	H	MH
IE	D_1	H	M	VH	M	M
	D_2	MH	M	H	H	H
	D_3	M	MH	H	MH	H
PP1	D_1	MH	H	VVH	VH	VH
	D_2	MH	MH	VH	MH	VH
	D_3	MH	MH	H	H	VH

Table 5. Aggregated intuitionistic fuzzy decision matrix (excerpt)

	A_1	A_2	A_2	A_4	A_5
IGLM	0.746	**0.769**	0.679	0.663	0.668
	0.151	**0.128**	0.220	0.236	0.231
	0.104	**0.103**	0.101	0.101	0.101
IE	0.615	0.538	0.746	0.591	0.631
	0.282	0.361	0.151	0.306	0.265
	0.103	0.101	0.104	0.103	0.104
PP1	**0.600**	0.644	0.826	0.728	0.800
	0.300	0.254	0.128	0.166	0.100
	0.100	0.101	0.046	0.106	0.100

An excerpt of the AIFDM for the case problem is shown in Table 5. The matrix IFNs were calculated by substituting in eq. (2) the weights of the three ($l = 3$) decision makers ($\lambda_{D1} = 0.406$, $\lambda_{D2} = 0.238$, $\lambda_{D3} = 0.356$) and the IFNs ($\mu_{ij}^{(k)}, v_{ij}^{(k)}, \pi_{ij}^{(k)}$) produced by using the relationships of Table 3 (i.e., these IFNs correspond to ratings given by the k decision maker on each system A_i ($i = 1,2,...,5$) with respect to each criterion j ($j = 1,2,...,13$)). For example, in Table 5, the IFN [0.769, 0.128, 0.103], shown in bold, is the aggregated score of PPMIS A_2 on criterion IGLM (Idea Generation/Lead Mgmt.), while the IFN [0.600, 0.300, 0.100], also shown in bold, is the aggregated score of PPMIS A_1 on criterion PP1 (Portfolio Planning).

Step 3: Determine the weights of the selection criteria. To analyse users' requirements from a PPMIS we disseminated to the three users/members of HOU a structured questionnaire, asking them to evaluate the 13 selection criteria and express their perceptions on the relative importance of each one criterion with respect to the overall performance and benefits provided from a candidate PPMIS. Each of the 3 users was requested to answer 13 questions by denoting a grade for the importance of each criterion in a linguistic term, as it is shown in column 2 of Table 2. Opinions of users U_1, U_2 and U_3 on the importance of the criteria are presented in columns of Table 6 entitled with the label "Users". These opinions are assigned to corresponding IFNs by using the relationships between values in column 2 and values in column 3 of Table 2.

Table 6. Importance values and weights of the criteria

Criteria	Users			Weights		
	U_1	U_2	U_3	μ	ν	π
IGLM	VI	I	M	0.779	0.201	0.019
IE	M	VI	I	0.734	0.236	0.031
PP1	M	VI	VI	0.808	0.184	0.008
PP2	VI	VI	VI	0.900	0.100	0.000
PP3	I	VI	VI	0.855	0.133	0.013
PC1	M	VI	VI	0.808	0.184	0.008
PC2	M	VI	I	0.734	0.236	0.031
PC3	M	M	VI	0.718	0.263	0.018
PT1	I	VI	VI	0.855	0.133	0.013
PT2	VI	M	I	0.797	0.183	0.020
AC	VI	I	I	0.828	0.151	0.021
PO	VI	VI	M	0.823	0.171	0.007
CC	I	M	VI	0.787	0.189	0.023

The IFWA operator was also used to calculate the weights of criteria by aggregating the opinions of the users. Let $w_j^{(k)} = (\mu_j^{(k)}, v_j^{(k)}, \pi_j^{(k)})$ be the IFN assigned to criterion j ($j = 1,2,...,n$) by the k user ($k = 1,2,...,l$). Then the weight of j can be calculated as follows:

$$w_j = IFWA_\lambda(w_j^{(1)}, w_j^{(2)},...,w_j^{(l)}) = \lambda_1 w_j^{(1)} \oplus \lambda_2 w_j^{(2)} \oplus \lambda_3 w_j^{(3)} \oplus ... \oplus \lambda_l w_j^{(l)}$$

$$= \left[1 - \prod_{k=1}^{l}(1-\mu_j^{(k)})^{\lambda_k}, \prod_{k=1}^{l}(v_j^{(k)})^{\lambda_k}, \prod_{k=1}^{l}(1-\mu_j^{(k)})^{\lambda_k} - \prod_{k=1}^{l}(v_j^{(k)})^{\lambda_k} \right] \quad (3)$$

Thus, a vector of criteria weights is obtained $W = [w_1, w_2,...,w_j]$, where each weight w_j is an IFN in the form $[\mu_j, v_j, \pi_j]$ ($j = 1,2,...,n$). In the case problem, substituting in eq. (3) the weights of three users ($\lambda_{U1} = 0.406$, $\lambda_{U2} = 0.238$, $\lambda_{U3} = 0.356$) and using IFNs which correspond to linguistic values of Table 6 yielded the criteria weights shown in the columns of the same table entitled with the label "Weights".

Step 4: Compose the aggregated weighted intuitionistic fuzzy decision matrix. In this step, the aggregated weighted intuitionistic fuzzy decision (AWIFDM) matrix R' is composed by considering the aggregated intuitionistic fuzzy decision matrix (i.e., table R produced in step 2) and the vector of the criteria weights (i.e., table W produced in step 3). Step 4 is necessary to synthesize the ratings of both decision makers and users. In particular, the elements of the AWIFDM can be calculated by using the multiplication operator of IFSs as follows:

$$R \otimes W = \{< x, \mu_{A_i}(x) \cdot \mu_W(x), v_{A_i}(x) + v_W(x) - v_{A_i}(x) \cdot v_W(x) >| x \in X \} \qquad (4)$$

R' is an m x n matrix composed with elements IFNs in the form of $r'_{ij} = [\mu_{A_iW}(x_j), v_{A_iW}(x_j), \pi_{A_iW}(x_j)]$, where $\mu_{A_iW}(x_j)$ and $v_{A_iW}(x_j)$ are values derived by eq. (4) and $\pi_{A_iW}(x) = 1 - v_{A_i}(x) - v_W(x) - \mu_{A_i}(x) \cdot \mu_W(x) + v_{A_i}(x) \cdot v_W(x)$.

In the case problem, substituting in eq. (4) the IFNs of Table 5 (table R) and IFNs of Table 6 (table W) yielded the IFNs of the AWIFDM (table R'), an excerpt from which is presented in Table 7. For example, in Table 7, the IFN [0.599, 0.304, 0.097], shown in bold, is the aggregated weighted score of PPMIS A_2 on criterion IGLM (Idea Generation/Lead Mgmt.), while the IFN [0.485, 0.429, 0.086], also shown in bold, is the aggregated score of PPMIS A_1 on criterion PP1 (Portfolio Planning).

Step 5: Compute the intuitionistic fuzzy positive ideal solution and the intuitionistic fuzzy negative ideal solution. In order to apply TOPSIS, the intuitionistic fuzzy positive ideal solution (IFPIS) A^* and the intuitionistic fuzzy negative ideal solution (IFNIS) A^- have to be determined. Both solutions are vectors of IFN elements and they are derived from the AWIFDM matrix as follows. Let B and C be benefit and cost criteria, respectively. Then A^* and A^- are equal to:

Table 7. Aggregated weighted intuitionistic fuzzy decision matrix (excerpt)

	A_1	A_2	A_3	A_4	A_5
IGLM	0.581	**0.599**	0.529	0.517	0.520
	0.322	**0.304**	0.377	0.390	0.386
	0.097	**0.097**	0.094	0.094	0.094
IE	0.451	0.395	0.547	0.433	0.463
	0.451	0.512	0.351	0.470	0.438
	0.098	0.094	0.102	0.097	0.099
PP1	**0.485**	0.520	0.667	0.588	0.646
	0.429	0.392	0.289	0.320	0.266
	0.086	0.088	0.044	0.093	0.088

$$A^* = (\mu_{A^*_W}(x_j), v_{A^*_W}(x_j)) \text{ and } A^- = (\mu_{A^-_W}(x_j), v_{A^-_W}(x_j)) \quad (5)$$

where

$$\mu_{A^*_W}(x_j) = ((\max_i \mu_{A_iW}(x_j) \mid j \in B), (\min_i \mu_{A_iW}(x_j) \mid j \in C))$$

$$v_{A^*_W}(x_j) = ((\min_i v_{A_iW}(x_j) \mid j \in B), (\max_i v_{A_iW}(x_j) \mid j \in C))$$

$$\mu_{A^-_W}(x_j) = ((\min_i \mu_{A_iW}(x_j) \mid j \in B), (\max_i \mu_{A_iW}(x_j) \mid j \in C))$$

$$v_{A^-_W}(x_j) = ((\max_i v_{A_iW}(x_j) \mid j \in B), (\min_i v_{A_iW}(x_j) \mid j \in C))$$

In the case problem, B = {IGLM, IE, PP1, PP2, PP3, PC1, PC2, PC3, PT1, PT2, AC} and C = {PO, CC}. To obtain IFPIS and IFNIS, eq. (5) was applied on the IFNs of the AWIFDM decision matrix. The IFPIS and IFNIS were determined as follows:

A^* = ([0.599, 0.304, 0.097], [0.547, 0.351, 0.102], [0.667, 0.289, 0.044], [0.692, 0.215, 0.093], [0.667, 0.235, 0.099], [0.602, 0.307, 0.090], [0.564, 0.334, 0.102], [0.532, 0.378, 0.090], [0.657, 0.244, 0.099], [0.641, 0.288, 0.072], [0.562, 0.338, 0.100], [0.476, 0.437, 0.087], [0.446, 0.459, 0.095])

A^- = ([0.517, 0.390, 0.094], [0.395, 0.512, 0.094], [0.485, 0.429, 0.086], [0.536, 0.372, 0.092], [0.614, 0.284, 0.102], [0.539, 0.373, 0.088], [0.424, 0.481, 0.095], [0.445, 0.468, 0.087], [0.525, 0.377, 0.097], [0.513, 0.391, 0.096], [0.435, 0.468, 0.097], [0.613, 0.296, 0.091], [0.605, 0.293, 0.101])

Step 6: Calculate the separation between the alternative PPMIS. Next, the separation measures S_{i^*} and S_{i^-} were calculated for each candidate system A_i from IFPPIS and IFNIS, respectively. As a distance measure, the normalized Euclidean distance was adopted, since it has been proved to be a reliable distance measure that takes into account not only membership and non-membership but also the hesitation part of IFNs [29]. For each alternative these two separation values can be calculated as follows:

$$S^* = \sqrt{\frac{1}{2n} \sum_{j=1}^n [(\mu_{A_iW}(x_j) - \mu_{A^*_W}(x_j))^2 + (v_{A_iW}(x_j) - v_{A^*_W}(x_j))^2 + (\pi_{A_iW}(x_j) - \pi_{A^*_W}(x_j))^2]}$$

$$\quad (6)$$

$$S^- = \sqrt{\frac{1}{2n} \sum_{j=1}^n [(\mu_{A_iW}(x_j) - \mu_{A^-_W}(x_j))^2 + (v_{A_iW}(x_j) - v_{A^-_W}(x_j))^2 + (\pi_{A_iW}(x_j) - \pi_{A^-_W}(x_j))^2]}$$

By utilizing equations (6), the positive and negative separation for the five alternatives were calculated, shown in columns (1) and (2) of Table 8, respectively.

Table 8. Separation measures and relative closeness coefficient of each PPMIS

PPMIS	S^* (1)	S^- (2)	C^* (3)
A_1	0.076	0.074	0.495
A_2	0.091	0.074	0.448
A_3	0.041	0.116	0.737
A_4	0.069	0.074	0.520
A_5	0.088	0.085	0.490

Step 7: Determine the final ranking of PPMIS. The final score of each system was derived by calculating the corresponding relative closeness coefficient with respect to the intuitionistic fuzzy ideal solution. For each alternative A_i, the relative closeness coefficient C_{i^*} with respect to the IFPIS is defined as follows:

$$C_{i^*} = \frac{S_{i^-}}{S_{i^*} + S_{i^-}} \tag{7}$$

where $0 \le C_{i^*} \le 1$. Eq. (7) was used to calculate these coefficients (final scores) listed in column (3) of Table 8. The alternative PPMIS were ranked in a descending order of these scores as $A_3 > A_4 > A_1 > A_5 > A_2$, from where it can be deduced that alternative A_3 is the most dominant PPMIS for the present case study.

Table 9. Sensitivity analysis results

Exp.	Criteria Weights	Scores of PPMIS					Ranking
		A_1	A_2	A_3	A_4	A_5	
1	$w_{1\text{-}13} = [0.10,0.90]$	0.497	0.447	0.730	0.514	0.494	$A_3{>}A_4{>}A_1{>}A_5{>}A_2$
2	$w_{1\text{-}13} = [0.35,0.60]$	0.500	0.450	0.728	0.518	0.494	$A_3{>}A_4{>}A_1{>}A_5{>}A_2$
3	$w_{1\text{-}13} = [0.50,0.45]$	0.499	0.449	0.729	0.517	0.494	$A_3{>}A_4{>}A_1{>}A_5{>}A_2$
4	$w_{1\text{-}13} = [0.75,0.20]$	0.498	0.448	0.729	0.516	0.494	$A_3{>}A_4{>}A_1{>}A_5{>}A_2$
5	$w_{1\text{-}13} = [0.90,0.10]$	0.497	0.447	0.730	0.514	0.494	$A_3{>}A_4{>}A_1{>}A_5{>}A_2$
6	$w_1 = [0.90,0.10], w_{2\text{-}13} = [0.10,0.90]$	0.674	0.712	0.389	0.247	0.278	$A_2{>}A_1{>}A_3{>}A_5{>}A_4$
7	$w_2 = [0.90,0.10], w_{1,3\text{-}13} = [0.10,0.90]$	0.388	0.147	0.910	0.291	0.458	$A_3{>}A_5{>}A_1{>}A_4{>}A_2$
8	$w_3 = [0.90,0.10], w_{1\text{-}2,4\text{-}13} = [0.10,0.90]$	0.152	0.257	0.909	0.595	0.763	$A_3{>}A_5{>}A_4{>}A_2{>}A_1$
9	$w_4 = [0.90,0.10], w_{1\text{-}3,5\text{-}13} = [0.10,0.90]$	0.294	0.171	0.896	0.597	0.804	$A_3{>}A_5{>}A_4{>}A_2{>}A_1$
10	$w_5 = [0.90,0.10], w_{1\text{-}4,6\text{-}13} = [0.10,0.90]$	0.650	0.413	0.516	0.349	0.477	$A_1{>}A_3{>}A_5{>}A_2{>}A_4$
11	$w_6 = [0.90,0.10], w_{1\text{-}5,7\text{-}13} = [0.10,0.90]$	0.333	0.656	0.821	0.601	0.320	$A_3{>}A_2{>}A_4{>}A_1{>}A_5$
12	$w_7 = [0.90,0.10], w_{1\text{-}6,8\text{-}13} = [0.10,0.90]$	0.521	0.158	0.850	0.473	0.819	$A_3{>}A_5{>}A_1{>}A_4{>}A_2$
13	$w_8 = [0.90,0.10], w_{1\text{-}7,9\text{-}13} = [0.10,0.90]$	0.487	0.683	0.318	0.784	0.718	$A_4{>}A_5{>}A_2{>}A_1{>}A_3$
14	$w_9 = [0.90,0.10], w_{1\text{-}8,10\text{-}13} = [0.10,0.90]$	0.426	0.363	0.885	0.651	0.210	$A_3{>}A_4{>}A_1{>}A_2{>}A_5$
15	$w_{10} = [0.90,0.10], w_{1\text{-}9,11\text{-}13} = [0.10,0.90]$	0.400	0.250	0.881	0.197	0.703	$A_3{>}A_5{>}A_1{>}A_2{>}A_4$
16	$w_{11} = [0.90,0.10], w_{1\text{-}10,12\text{-}13} = [0.10,0.90]$	0.593	0.603	0.885	0.717	0.211	$A_3{>}A_4{>}A_2{>}A_1{>}A_5$
17	$w_{12} = [0.90,0.10], w_{1\text{-}11, 13} = [0.10,0.90]$	0.821	0.791	0.464	0.592	0.196	$A_1{>}A_2{>}A_4{>}A_3{>}A_5$
18	$w_{13} = [0.90,0.10], w_{1\text{-}12} = [0.10,0.90]$	0.818	0.806	0.908	0.606	0.170	$A_3{>}A_1{>}A_2{>}A_4{>}A_5$

Step 8: Sensitivity analysis. Sensitivity analysis is concerned with 'what-if' kind of scenarios to determine if the final answer is stable to changes (experiments) in the inputs, either judgments or weights of criteria. In the present case, sensitivity analysis

was performed by examining the impact of criteria weights (i.e., the weights of users' requirements from a PPMIS) on the final ranking. Of special interest was to see if criteria weights' changes alter the order of the alternatives. 18 experiments were conducted in a similar way with the approach presented in [30]. The details of all experiments are shown in Table 9, where w_1, w_2,.., w_{13} denote respectively the weights of criteria IGLM, IE, PP1, PP2, PP3, PC1, PC2, PC3, PT1, PT2, AC, PO, CC. In experiments 1-5, weights of all criteria were set equal to [0.10,0.90], [0.35,0.60], [0.50,0.45], [0.75,0.20] and [0.90,0.10], respectively. These IFNs correspond to the linguistic terms VU, U, M, I and VI, respectively (see Table 2). In experiments 6-18, the weight of each of the 13 criteria was set equal to the highest IFN [0.90,0.10], one by one, and the weights of the rest of criteria were set all equal to the lowest IFN [0.10,0.90]. The results show that PPMIS A_3 remains the dominant alternative in 14 out of the 18 experiments (this represents a clear "majority" equal to 77.77%). PPMIS A_1 was first in 2/18 experiments, namely in exp. 10 and in exp. 17, where the highest weights were assigned, respectively, to criterion PP3 (project planning) and criterion PO (total price for purchasing/ownership). System A_2 had the highest score in exp. 6, where the highest weight was assigned to criterion PP1 (portfolio planning), while system A_4 had the highest score in exp. 13, where the highest value was assigned to the weight of PC3 (portfolio controlling).

It should be noted that in the presented approach we have used the IFNs proposed in [18] to represent the linguistic terms of Table 2 and Table 3. Sensitivity analysis on the final ranking can be easily performed by changing these IFN values. In addition, further generalization of the approach requires the use of a parameterised form of the hesitation degree. This can be performed in two ways: i) by following a Positive-Confidence or a Negative-Confidence approach [17], and ii) by utilizing interval-valued intuitionistic fuzzy numbers [19]. We have plans to investigate these two solutions in a future research.

5   Conclusions

The paper presented, through a real case study, an approach that applied a group-based multi criteria decision making (MCDM) method for the evaluation and final selection of an appropriate Project and Portfolio Management Information System (PPMIS). The approach utilized a method that jointly synthesizes intuitionistic fuzzy sets and TOPSIS [18]. The benefit from this combination in a PPMIS selection is twofold: First, the selection approach actively involves decision makers and PPMIS users in the decision making process and aggregates their opinions to support agreement upon the final selection. Second, the approach considers that they both express their judgments under inherent uncertainty. More significantly, the approach handles adequately the degree of indeterminacy that characterizes decision makers and users in their evaluations. This is very important when an organization needs to decide upon the selection of any new, multi-functional information system, as in our case is a suitable PPMIS, since decision makers often cannot have full knowledge of the extend that each candidate system will support the user requirements. System users, on the other hand, can be unfamiliar with the processes supported by the required system, and thus, they cannot judge with certainty the importance of their needs.

The presented approach not only validated the method, as it was originally defined in [18], in a new application field that is the evaluation of PPMIS (where other application examples of MCDM methods are rather limited in the literature), but also considered a more extensive list of benefit and cost oriented criteria, suitable for PPMIS selection. In addition, final results were verified by applying sensitivity analysis. We should mention that the method underlying computations are not transparent to the problem stakeholders which utilise linguistic terms to state evaluations/preferences. Actually, we implemented the method in a spreadsheet program that helps to effectively and practically apply the method with a variety of inputs.

The study raises several issues that could spark further research. For example, an interesting idea could be to validate the approach applicability in addressing the selection of other types of software packages. We are now investigating the selection of e-learning management systems for the case organization (i.e., the Hellenic Open University). In addition, treating more with uncertainties would further strengthen the proposed approach in deriving more precise results. Therefore, we have plans to examine more powerful models in the same domain, such as the interval-valued intutionistic fuzzy sets [19], [20].

References

1. Liberatore, M.J., Pollack-Johnson, B.: Factors Influencing the Usage and Selection of Project Management Software. IEEE Transactions on Engineering Management 50(2), 164–174 (2003)
2. Project Management Institute: A Guide to the Project Management Body of Knowledge. 4th edn. Project Management Institute (PMI), Newtown Square (2008)
3. Meyer, M.M., Ahlemann, F.: Project Management Software Systems, 6th edn. Business Application Research Center, Wurzburg (2010)
4. Stang, D.B.: Magic Quadrant for IT Project and Portfolio Management. Gartner RAS Core Research Note, Gartner Research (2010)
5. Jadhav, A.S., Sonar, R.M.: Evaluating and Selecting Software Packages: a Review. Information and Software Technology 51(3), 555–563 (2009)
6. Vaidya, O.S., Kumar, S.: Analytic Hierarchy Process: an Overview of Applications. European Journal of Operational Research 169(1), 1–29 (2006)
7. Ho, W.: Integrated Analytic Hierarchy Process and its Applications – A Literature Review. European Journal of Operational Research 186(1), 211–218 (2008)
8. Ahmad, N., Laplante, P.: Software Project Management Systems: Making a Practical Decision using AHP. In: 30th Annual IEEE/NASA Software Engineering Workshop, pp. 76–84. IEEE Press, Los Alamitos (2006)
9. Gerogiannis, V.C., Fitsilis, P., Voulgaridou, D., Kirytopoulos, K.A., Sachini, E.: A Case Study for Project and Portfolio Management Information System Selection: a Group AHP-Scoring Model Approach. International Journal of Project Organisation and Management 2(4), 361–381 (2010)
10. Bozdag, C.E., Kahraman, C., Ruan, D.: Fuzzy Group Decision Making for Selection among Computer Integrated Manufacturing Systems. Computers in Industry 51(1), 3–29 (2003)

11. Chang, D.Y.: Application of Extent Analysis Method on Fuzzy AHP. European Journal of Operational Research 95(3), 649–655 (1996)
12. Chen, C.T.: Extensions of the TOPSIS for group decision-making under fuzzy environment. Fuzzy Sets and Systems 114(1), 1–9 (2000)
13. Cochran, J.K., Chen, H.N.: Fuzzy Multi-Criteria Selection of Object-Oriented Simulation Software for Production System Analysis. Computers & Operations Research 32(1), 153–168 (2005)
14. Lin, H.Y., Hsu, P.Y., Sheen, G.J.: A Fuzzy-based Decision-Making Procedure for Data Warehouse System Selection. Expert System with Applications 32(3), 939–953 (2007)
15. Atanassov, K.T.: Intuitionistic Fuzzy Sets. Fuzzy Sets and Systems 20(1), 87–96 (1986)
16. De, S.K., Biswas, R., Roy, A.R.: An Application of Intuitionistic Fuzzy Sets in Medical Diagnosis. Fuzzy Sets and Systems 117(2), 209–213 (2001)
17. Wang, P.: QoS-aware Web Services Selection with Intuitionistic Fuzzy Set under Consumer's Vague Perception. Expert Systems with Applications 36(3), 4460–4466 (2009)
18. Boran, F.E., Genc, S., Kurt, M., Akay, D.: A Multi-Criteria Intuitionistic Fuzzy Group Decision Making for Supplier Selection with TOPSIS Method. Expert Systems with Applications 36(8), 11363–11368 (2009)
19. Park, J.H., Park, I.Y., Kwun, Y.C., Tan, X.: Extension of the TOPSIS Method for Decision Making Problems under Interval-Valued Intuitionistic Fuzzy Environment. Applied Mathematical Modelling 35(5), 2544–2556 (2011)
20. Chen, T.Y., Wang, H.P., Lu, Y.Y.: A Multicriteria Group Decision-Making Approach based on Interval-Valued Intuitionistic Fuzzy Sets: a Comparative Perspective. Expert Systems with Applications 38(6), 7647–7658 (2011)
21. Raymond, L., Bergeron, F.: Project Management Information Systems: an Empirical Study of their Impact on Project Managers and Project Success. International Journal of Project Management 26(2), 213–220 (2008)
22. Project Management Institute: Project Management Software Survey. Project Management Institute (PMI), Newtown Square (1999)
23. Ali, A.S.B., Money, W.H.: A Study of Project Management System Acceptance. In: Proceedings of the 38th Hawaii International Conference on System Sciences, pp. 234–244. IEEE Press, Los Alamitos (2005)
24. Gomez, D.C., Alexander, A., Anderson, D., Cook, D., Poole, K., Findlay, O.: NASA Project Management Tool Analysis and Recommendations White Paper. Technical report, Project Management Tool Working Group, NASA Glenn Research Center (2004), http://km.nasa.gov/pdf/54927main_pm-tool-paper.pdf
25. Ahlemann, F.: Towards a Conceptual Reference Model for Project Management Information Systems. International Journal of Project Management 27(1), 19–30 (2009)
26. Business Application Research Center (BARC), Project Management Software Systems Report, http://www.pm-software-report.com
27. Xu, Z.S., Yager, R.R.: Some Geometric Aggregation Operators based on Intuitionistic Fuzzy Sets. International Journal of General Systems 35(4), 417–433 (2006)
28. Xu, Z.S.: Intuitionistic Fuzzy Aggregation Operators. IEEE Transactions on Fuzzy Systems 15(6), 1179–1187 (2007)
29. Szmidt, E., Kacprzy, K.J.: Distances between Intuitionistic Fuzzy Sets. Fuzzy Sets and Systems 114(3), 505–518 (2000)
30. Awasthi, A., Chauhan, S.S., Goyal, S.K.: A Fuzzy Multicriteria Approach for Evaluating Environmental Performance of Suppliers. International Journal of Production Economics 126(2), 370–378 (2010)

Fuzzy and Neuro-Symbolic Approaches to Assessment of Bank Loan Applicants

Ioannis Hatzilygeroudis[1] and Jim Prentzas[2]

[1] University of Patras, School of Engineering,
Department of Computer Engineering & Informatics,
26500 Patras, Greece
ihatz@ceid.upatras.gr
[2] Democritus University of Thrace, School of Education Sciences,
Department of Education Sciences in Pre-School Age, Laboratory of Informatics,
68100 Nea Chili, Alexandroupolis, Greece
dprentza@psed.duth.gr

Abstract. In this paper, we present the design, implementation and evaluation of intelligent methods that assess bank loan applications. Assessment concerns the ability/possibility of satisfactorily dealing with loan demands. Different loan programs from different banks may be proposed according to the applicant's characteristics. For each loan program, corresponding attributes (e.g. interest, amount of money that can be loaned) are also calculated. For these tasks, two separate intelligent systems have been developed and evaluated: a fuzzy expert system and a neuro-symbolic expert system. The former employs fuzzy rules based on knowledge elicited from experts. The latter is based on neurules, a type of neuro-symbolic rules that combine a symbolic (production rules) and a connectionist (adaline unit) representation. Neurules were produced from available patterns. Evaluation showed that performance of both systems is close although their knowledge bases were derived from different types of source knowledge.

1 Introduction

An important task of every bank involves the assessment of applications for bank loans. Such an assessment is important due to the involved risks for both clients (individuals or corporations) and banks. In recent financial crisis, banks suffered losses from a steady increase of customers' defaults on loans [12]. So, banks should avoid approving loans for applicants that eventually may not comply with the involved terms. It is significant to approve loans that satisfy all critical requirements and will be able to afford corresponding demands. To this end, various parameters related to applicant needs should be considered. Furthermore, different loan programs from different banks should also be taken into account, to be able to propose a loan program best tailored to the specific applicant's status. Computer-based systems for evaluating loan applicants and returning the most appropriate loan would be useful since valuable assessment time would be spared and potential risks could be reduced. Banking authorities encourage developing models to better quantify financial risks [9].

L. Iliadis et al. (Eds.): EANN/AIAI 2011, Part II, IFIP AICT 364, pp. 82–91, 2011.

Due to the complexity/significance of the assessment process, intelligent assessment systems have been used to reduce the cost of the process and the risks of bad loans, to save time and effort and generally to enhance credit decisions [1]. Such systems are mainly based on neural networks (e.g. [2]), but approaches such as genetic algorithms (e.g. [1]) and support vector machines (e.g. [12]) and other methods (e.g. [9]) have been applied too. There are some requirements in designing such a system. First, the system needs to include loan programs offered by different banks, to be able to return the most suitable one(s) for the applicant. Second, experience of banking staff specialized in loans is useful in order to outline loan attributes, applicant attributes, assessment criteria and the stages of the assessment process. Third, available cases from past loan applications are required to design and test the system.

In this paper, we present the design, implementation and evaluation of two intelligent systems that assess bank loan applications: a fuzzy expert system and a neuro-symbolic expert system. To construct those systems different types of source knowledge were exploited. The fuzzy expert system is based on rules that represent expert knowledge regarding the assessment process. The neuro-symbolic expert system is based on neurules, a type of hybrid rules integrating a symbolic (production rules) and a connectionist representation (adaline unit) [4]. Neurules exhibit advantages compared to pure symbolic rules such as, improved inference performance [4], ability to reach conclusions from unknown inputs and construct knowledge bases from alternative sources (i.e. symbolic rules or empirical data) [4], [5]. Neurules were produced from available cases (past loan applications). Evaluation of the fuzzy expert system encompassing rule-based expert knowledge and the neurule-based expert system encompassing empirical knowledge showed that their performance was close.

The rest of the paper is organized as follows. Section 2 introduces the domain knowledge involved and the stages of the assessment process. Section 3 discusses development issues of the fuzzy expert system. Section 4 briefly presents neurules and discusses development issues of the neurule-based expert system. Section 5 presents evaluation results for both systems. Finally, Section 6 concludes.

2   Domain Knowledge Modeling

In this section we discuss issues involving the primary loan and applicant attributes modeled in the systems as well as the basic stages of the inference process. The corresponding knowledge was derived from experts.

2.1   Modeled Loan Attributes

Each loan involves a number of attributes that need to be taken into account during inference. A basic attribute is the type of loan. Various types of loans exist. The two main types involve loans addressed to individuals and loans addressed to corporations.

Each main type of loan is further discerned to different categories. Loans addressed to individuals are discerned to personal, consumer and housing loans. Loans addressed to corporations are discerned to capital, fixed installations and other types of loans. The type of loan affects other attributes such as the amount of money that can be approved for loaning, the amount of installments and the interest.

Moreover, according to the type of loan, different requirements and applicant characteristics are taken into account. For instance, in the case of personal and consumption loans, a main applicant attribute taken into account is the net annual income. The maximum amount of money that can be approved for loaning is up to 70% of the annual income subtracting the amounts of existing loans in any bank. For example, an applicant with annual income €20,000 may borrow up to €13,000. In case there is a pending loan of €5,000, then he/she may borrow up to €8,000.

Applications for housing loans are thoroughly examined since the involved amount of money is usually large and the risks for banks are high. Various applicant attributes need to be considered such as property status and net annual income. For instance, the net annual income should be at least €15,000 and annual installments should not exceed 40% of the net annual income.

So, loan attributes such as the following are considered: *type of loan*, *the reason for applying*, *supporting documents*, *name of bank*, *type of interest* (i.e. fixed, floating), *commencement and termination of loan payment*, *amount of money loaned*, *loan expenses*, *way of payment* (e.g. monthly installments).

2.2 Modeled Applicant Attributes

To assess an applicant's ability to deal satisfactorily with loan demands, various applicant parameters are considered. The most significant attributes are the following:

- *Net annual income*. Expenses, obligations (e.g. installments) are excluded.
- *Financial and property status*. Possession of property is considered important (or even obligatory) for certain types of loans. This involves available bank accounts, bonds, stocks, real estate property, etc.
- *Personal attributes*. Personal attributes such as age, number of depending children, trade are considered important. Trade is considered a parameter of the applicant's social status. The number of depending children corresponds to obligations. Banks usually do not loan money to persons younger than twenty and older than seventy years old due to high risks (from the perspective of banks).
- *Warrantor*. A primary parameter for the overall assessment of an applicant is also the warrantor, who accepts and signs the bank's terms. In case the client cannot comply with obligations concerning the loan, the warrantor will undertake all corresponding responsibilities.

2.3 Inference Process Stages

The inference process involves four main stages outlined in the following. Table 1 summarizes the outputs produced in each stage.

In the first stage, basic inputs are given to the inference process concerning the requested loan. More specifically, the type of loan and the reason for loan application are given by responding to relevant system queries. Both inputs are stored as facts in the systems and are used to retrieve relevant loan programs from the Loan Programs

Base. The approach involves loan programs from four different banks. The retrieved loan programs are used in subsequent stages.

Table 1. Summary of outputs for the four stages of the inference process

Stage	Outputs	
Stage 1	*Retrieves relevant loan programs* from the Loan Programs Base according to basic inputs (e.g. type of loan and reason for loan application)	
Stage 2	*Applicant assessment* *Warrantor assessment*	*Overall assessment*
Stage 3	*Restrictions* involving the funding of a loan are taken into account	
Stage 4	*Relevant loan programs are returned.* For each loan program return: interest, approved amount of money, installment and loan payment period.	

Table 2. Summary of variables involved in applicant/warrantor assessment (second stage)

Variable	*Applicant assessment*	*Warrantor assessment*
Net annual income (bad, fair, good)	×	×
Overall financial status (bad, fair, good)	×	×
Number of depending children (few, fair, many)	×	
Age (young, normal, old)	×	×
Social status (bad, fair, good)		×

The second stage involves an overall assessment of the applicant. This stage consists of three tasks: (a) applicant assessment, (b) warrantor assessment and (c) overall applicant assessment. The third task takes as input the results of the other two tasks. Such a process is applied since an overall applicant assessment is based on assessment of both applicant and warrantor. Rule-based inference performs all tasks. The variables involved in applicant and warrantor assessment are summarized in Table 2 (where '×' means 'depends on').

Applicant assessment is considered an intermediate variable. Its evaluation is based on the values of input variables such as net annual income, overall financial status, number of depending children and age. Variable 'net annual income' can take three values: bad, fair and good. Variable 'overall financial status' can take three values: bad, fair and good. Variable 'number of depending children' can take three values: few children (corresponding to 0-2 children), fair number of children (corresponding to 3-5 children) and many children (corresponding to at least six children). Finally, variable 'age' can take three values: young, normal and old. Based on the values of these variables, the value of the intermediate variable 'applicant assessment' is evaluated. This variable takes three values: bad, fair and good. The design of systems has also taken into consideration that variables 'net annual income' and 'overall financial status' are more important in performing applicant assessment compared to the other two input variables (i.e. 'number of depending children' and 'age').

Warrantor assessment is also considered an intermediate variable. Such an assessment is significant in approving an application for a loan. The role of a warrantor is considered important in case of loans involving high risks. A warrantor is acceptable in case certain criteria (mainly financial) are satisfied. Evaluation of intermediate variable 'warrantor assessment' is based on warrantor attributes (input variables) such as net annual income, overall financial status, social status and age. Variables 'net annual income', 'overall financial status' and 'age' have similar representations as in the case of the applicant. Variable 'social status' depends on two parameters: monthly income and trade. This variable can take three values: bad, fair and good. Based on the values of the aforementioned variables, the value of the intermediate variable 'warrantor assessment' is evaluated. This variable takes three values: bad, fair and good.

The values of the intermediate variables ('applicant assessment' and 'warrantor assessment') are used to evaluate the value of the variable 'overall applicant assessment'. This variable takes five values: very bad, bad, fair, good and very good.

In the third stage, restrictions involving the funding of a loan are taken into account. As mentioned in a previous section, in case of personal and consumption loans, the maximum amount of money that can be approved for loaning is up to 70% of the annual income subtracting the amounts of existing loans in any bank. In case of home loans, annual installments should not exceed 40% of the annual income. Furthermore, applicant obligations (e.g. existing loans) are given as input.

In the fourth stage, results of all previous stages are taken into account. The maximum amount of money that can be approved for loaning is calculated. In case it is assessed that not all of the desired amount of money can be loaned to the applicant, the maximum possible amount of money that can be loaned is calculated. Relevant loan programs are produced as output. For each relevant loan program, all corresponding attributes are returned: interest, approved amount of money, installment and loan payment period.

3 The Fuzzy Expert System

Domain knowledge is characterized by inaccuracy since several terms do not have a clear-cut interpretation. Fuzzy logic makes it possible to define inexact domain entities via fuzzy sets. One of the reasons is that fuzzy logic provides capabilities for approximate reasoning, which is reasoning with inaccurate (or fuzzy) variables and values, expressed as linguistic terms [11]. All variables involved in the second stage of the inference process are represented as fuzzy variables.

The developed fuzzy expert system has the typical structure of such systems as shown in Fig. 1. The fact base contains facts given as inputs or produced during inference. The rule base of the expert system contains fuzzy rules. A fuzzy rule includes one or more fuzzy variables. Definition of each fuzzy variable consists of definitions of its values. Each fuzzy value is represented by a fuzzy set, a range of crisp (i.e. non-linguistic) values with different degrees of membership to the set. The degrees are specified via a membership function. Fuzzy values and corresponding membership functions have been determined with the aid of the expert. We used triangles and trapezoids to represent membership functions. The system also includes the loan programs base containing loan programs from different banks.

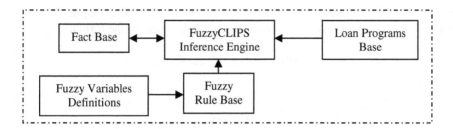

Fig. 1. Architecture of the fuzzy expert system

Reasoning in such a system includes three stages: fuzzification, inference, defuzzification. In fuzzification, the crisp input values (from the working memory) are converted to membership degrees (fuzzy values). In the inference stage, the MIN method is used for the combination of a rule's conditions, to produce the membership value of the conclusion, and the MAX method is used to combine the conclusions of the rules. In defuzzification, the centroid method is used to convert a fuzzy output to a crisp value, where applicable.

The system has been implemented in the FuzzyCLIPS expert system shell [13], an extension to CLIPS that represents fuzziness. Finally, about 55 fuzzy rules have been constructed. These rules were constructed with the aid of experts from four different banks specialized in loan programs. An example fuzzy rule is the following:

"**If** *net_annual_income* is bad **and** *number_of_depending_children* is few **and** *age* is young **and** *overall_financial_status* is bad **then** *applicant_assessment* is bad".

4   The Neurule-Based Expert System

Neurules are a type of hybrid rules integrating symbolic rules with neurocomputing giving pre-eminence to the symbolic component. Neurocomputing is used within the symbolic framework to improve the inference performance of symbolic rules [4]. In contrast to other hybrid approaches (e.g. [3]), the constructed knowledge base retains the modularity of production rules, since it consists of autonomous units (neurules), and also retains their naturalness in a great degree, since neurules look much like symbolic rules. The inference mechanism is a tightly integrated process resulting in more efficient inferences than those of symbolic rules [4] and other hybrid approaches [8]. Explanations in the form of if-then rules can be produced [6].

The form of a rule is depicted in Fig.2a. Each condition C_i is assigned a number sf_i, called its *significance factor*. Moreover, each rule itself is assigned a number sf_0, called its *bias factor*. Internally, each rule is considered as an adaline unit (Fig.2b). The *inputs* C_i ($i=1,...,n$) of the unit are the *conditions* of the rule. The weights of the unit are the significance factors of the rule and its bias is the bias factor of the neurule. Each input takes a value from the following set of discrete values: [1 (true), -1 (false), 0 (unknown)]. This gives the opportunity to easily distinguish between the falsity and the absence of a condition in contrast to symbolic rules. The *output D*, which represents the *conclusion* (decision) of the rule, is calculated via the formulas:

$$D = f(\mathbf{a}) , \quad \mathbf{a} = sf_0 + \sum_{i=1}^{n} sf_i \ C_i$$

$$f(\mathbf{a}) = \begin{cases} 1 & if \ \mathbf{a} \geq 0 \\ -1 & otherwise \end{cases}$$

where **a** is the *activation value* and *f(x)* the *activation function*, a threshold function. Hence, the output can take one of two values ('-1', '1') representing failure and success of the rule respectively. The general syntax of a condition C_i and the conclusion *D* is:

<condition>::= <variable> <l-predicate> <value>
<conclusion>::= <variable> <r-predicate> <value>

where <variable> denotes a *variable*, that is a symbol representing a concept in the domain, e.g. 'net annual income', 'age' etc, in a banking domain. <l-predicate> denotes a symbolic or a numeric predicate. The symbolic predicates are {is, isnot} whereas the numeric predicates are {<, >, =}. <r-predicate> can only be a symbolic predicate. <value> denotes a value. It can be a *symbol* or a *number*. The significance factor of a condition represents the significance (weight) of the condition in drawing the conclusion(s). Table 3 presents an example neurule, from a banking domain.

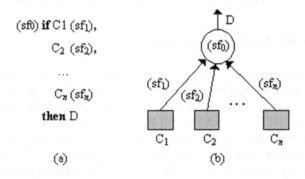

(sf_0) **if** $C1$ (sf_1),

C_2 (sf_2),

...

C_n (sf_n)

then D

(a)   (b)

Fig. 2. (a) Form of a neurule (b) a neurule as an adaline unit

Variables are discerned to input, intermediate or output ones. An input variable takes values from the user (input data), whereas intermediate or output variables take values through inference since they represent intermediate and final conclusions respectively. We distinguish between intermediate and output neurules. An intermediate neurule is a neurule having at least one intermediate variable in its conditions and intermediate variables in its conclusions. An output neurule is one having an output variable in its conclusions.

Table 3. An example neurule

N1

(-7.7) **if** social_status is good (6.6),
        net_annual_income is bad (3.4),
        age is normal (3.4),
        property_status is bad (3.3),
        property_status is fair (2.7)
    then warrantor_assessment is fair

Neurules can be constructed either from symbolic rules thus exploiting existing symbolic rule bases [4] or from empirical data (i.e. training patterns) [5]. In each process, an adaline unit is initially assigned to each intermediate and final conclusion and the corresponding training set is determined. Each unit is individually trained via the Least Mean Square (LMS) algorithm (e.g. [3]). When the training set is inseparable, more than one neurule having the same conclusion are produced.

In Fig. 3, the architecture of the neurule-based expert system is presented. The run-time system (in the dashed shape) consists of the following modules: the *working memory (WM)*, the *neurule-based inference engine (NRIE)*, the *explanation mechanism (EXM)*, the *neurule base (NRB)* and the *loan programs base (LPB)*. The neurule base contains neurules. These neurules are produced off-line from available empirical data concerning past loan cases (see Section 5). The construction process is performed by the *neurule construction mechanism (NCM)* [5].

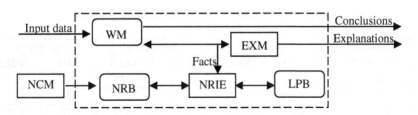

Fig. 3. Architecture of the neurule-based expert system

5 Evaluation of the Systems

To evaluate the fuzzy expert system (FES) and the neurule-based expert system (NBES), we used 100 past loan cases mainly deriving from the Bank of Greece and the Bank of Cyprus. 30% of those cases were randomly chosen and used to test both systems. Random choice of test cases was performed in a way that an equal number of test cases corresponded to each of the five values of overall applicant assessment (i.e. very bad, bad, fair, good, very good). The rest 70% of the available cases were used to construct the neurules contained in NBES. Thus, the knowledge base contents of the systems were derived from different source types (i.e. expert rules and cases).

Evaluation results for applicant assessment, warrantor assessment and overall applicant assessment are presented in Tables 4, 5 and 6 respectively. In each table, we present separate results for each one of the involved classes as well as average results for all classes. As mentioned in Section 2, applicant and warrantor assessment involve three classes whereas overall applicant assessment involves five classes. We use 'accuracy' accompanied by 'specificity' and 'sensitivity' as evaluation metrics:

$$accuracy = (a + d)/(a + b + c + d), \ sensitivity = a/(a + b), \ specificity = d/(c + d)$$

where a is the number of positive cases correctly classified, b is the number of positive cases that are misclassified, d is the number of negative cases correctly classified and c is the number of negative cases that are misclassified. By 'positive' we mean that a case belongs to the corresponding assessment class and by 'negative' that it doesn't. Results show that performance of both systems is comparable considering all metrics. In overall applicant assessment, NBES performed slightly better in terms of accuracy and much better as far as sensitivity is concerned.

Table 4. Evaluation results for applicant assessment

Applicant Assessment	Accuracy		Sensitivity		Specificity	
	FES	*NBES*	*FES*	*NBES*	*FES*	*NBES*
Bad	0.90	0.83	0.80	0.87	0.95	0.81
Fair	0.73	0.76	0.60	0.60	0.80	0.85
Good	0.86	0.86	0.80	0.83	0.85	0.88
Average	*0.83*	*0.82*	*0.73*	*0.77*	*0.87*	*0.85*

Table 5. Evaluation results for warrantor assessment

Warrantor Assessment	Accuracy		Sensitivity		Specificity	
	FES	*NBES*	*FES*	*NBES*	*FES*	*NBES*
Bad	0.93	0.93	0.80	0.80	1.00	1.00
Fair	0.76	0.76	0.70	0.80	0.80	0.75
Good	0.83	0.83	0.80	0.70	0.85	0.90
Average	*0.84*	*0.84*	*0.77*	*0.77*	*0.88*	*0.88*

Table 6. Evaluation results for overall applicant assessment

Overall Applicant Assessment	Accuracy		Sensitivity		Specificity	
	FES	*NBES*	*FES*	*NBES*	*FES*	*NBES*
Very bad	0.93	0.90	0.66	0.83	1.00	0.90
Bad	0.96	0.93	0.83	0.83	1.00	0.95
Fair	0.86	0.90	0.50	0.83	0.95	0.91
Good	0.80	0.90	0.83	0.50	0.79	1.00
Very good	0.96	0.96	0.50	1.00	0.95	0.95
Average	*0.90*	*0.92*	*0.66*	*0.80*	*0.94*	*0.94*

6 Conclusions

In this paper, we present the design, implementation and evaluation of two intelligent systems for assessing bank loan applicants. Loan programs from different banks are taken into consideration. The intelligent systems involve a fuzzy expert system and a neuro-symbolic expert system constructed from expert rules and available cases respectively. The knowledge base of the neuro-symbolic expert system contains neurules, a type of hybrid rules integrating symbolic rules with neurocomputing.

Evaluation results for both systems are comparable, despite the different types of knowledge sources. In certain aspects, one system performs better than the other. This means that both types of knowledge sources can be exploited in producing outputs. Based on the results, an integrated (or hybrid) approach could be developed. In this perspective, there are two directions for future research. One involves development of an hybrid system involving both systems as separate cooperating modules. The other direction involves development of an integrated system the exploits both types of knowledge sources. Research (e.g. [10]) has shown that synergies from using both (rule-based) domain theory and empirical data may result in effective systems. An approach as in [7] combining neurules and cases could be investigated.

References

1. Abdou, H.A.: Genetic Programming for Credit Scoring: The Case of Egyptian Public Sector Banks. Expert Systems with Applications 36, 11402–11417 (2009)
2. Eletter, S.F., Yaseen, S.G., Elrefae, G.A.: Neuro-Based Artificial Intelligence Model for Loan Decisions. American Journal of Economics and Business Administration 2, 27–34 (2010)
3. Gallant, S.I.: Neural Network Learning and Expert Systems. MIT Press, Cambridge (1993)
4. Hatzilygeroudis, I., Prentzas, J.: Neurules: Improving the Performance of Symbolic Rules. International Journal on AI Tools 9, 113–130 (2000)
5. Hatzilygeroudis, I., Prentzas, J.: Constructing Modular Hybrid Rule Bases for Expert Systems. International Journal on AI Tools 10, 87–105 (2001)
6. Hatzilygeroudis, I., Prentzas, J.: An Efficient Hybrid Rule-Based Inference Engine with Explanation Capability. In: Proceedings of the 14th International FLAIRS Conference, pp. 227–231. AAAI Press, Menlo Park (2001)
7. Hatzilygeroudis, I., Prentzas, J.: Integrating (Rules, Neural Networks) and Cases for Knowledge Representation and Reasoning in Expert Systems. Expert Systems with Applications 27, 63–75 (2004)
8. Hatzilygeroudis, I., Prentzas, J.: Neurules: Integrated Rule-Based Learning and Inference. IEEE Transactions on Knowledge and Data Engineering 22, 1549–1562 (2010)
9. Min, J.H., Lee, Y.-C.: A Practical Approach to Credit Scoring. Expert Systems with Applications 35, 1762–1770 (2008)
10. Prentzas, J., Hatzilygeroudis, I.: Categorizing Approaches Combining Rule-Based and Case-Based Reasoning. Expert Systems 24, 97–122 (2007)
11. Ross, T.J.: Fuzzy Logic with Engineering Applications. John Wiley & Sons, Chichester (2010)
12. Zhou, L., Lai, K.K., Yu, L.: Least Squares Support Vector Machines Ensemble Models for Credit Scoring. Expert Systems with Applications 37, 127–133 (2010)
13. http://awesom.eu/~cygal/archives/2010/04/22/ fuzzyclips_downloads/index.html

Comparison of Fuzzy Operators for IF-Inference Systems of Takagi-Sugeno Type in Ozone Prediction

Vladimír Olej and Petr Hájek

Institute of System Engineering and Informatics,
Faculty of Economics and Administration,
University of Pardubice, Studentská 84,
532 10 Pardubice, Czech Republic
{vladimir.olej,petr.hajek}@upce.cz

Abstract. The paper presents IF-inference systems of Takagi-Sugeno type. It is based on intuitionistic fuzzy sets (IF-sets), introduced by K.T. Atanassov, fuzzy t-norm and t-conorm, intuitionistic fuzzy t-norm and t-conorm. Thus, an IF-inference system is developed for ozone time series prediction. Finally, we compare the results of the IF-inference systems across various operators.

Keywords: Fuzzy inference systems, IF-sets, IF-inference system of Takagi-Sugeno type, t-norm, t-conorm, intuitionistic fuzzy t-norm and t-conorm, time series, ozone prediction.

1 Introduction

The IF-sets theory [1] represents one of the generalizations, the notion introduced by K.T. Atanassov [2]. The concept of IF-sets can be viewed as an alternative approach to define a fuzzy set in cases where available information is not sufficient for the definition of an imprecise concept by means of a conventional fuzzy set. In this paper IF-sets will be presented as a tool for reasoning in the presence of imperfect facts and imprecise knowledge. The IF-sets have been for example applied for ozone prediction [3] and air quality modelling as they provide a good description of object attributes by means of membership functions μ and non-membership functions v. They also present a strong potential to express uncertainty. In our previous paper [3] a novel IF-inference system of Takagi-Sugeno type was presented for time series prediction. In this paper, the outputs of the IF-inference system are compared across various operators in if-then rules.

This paper presents basic notions of Takagi-Sugeno type fuzzy inference systems (FIS) for time series prediction. Based on the FIS defined in this way and IF-sets, t-norms and t-conorms, intuitionistic fuzzy t-norms and t-conorms are defined. Next, IF-inference systems of Takagi-Sugeno type are introduced. Further, the paper includes a comparison of the prediction results obtained by the FIS characterized by membership functions μ, by the FIS characterized by non-membership functions v, and by the IF-inference system. The comparison is realized for the example of ozone prediction in the Pardubice micro region, the Czech Republic.

L. Iliadis et al. (Eds.): EANN/AIAI 2011, Part II, IFIP AICT 364, pp. 92–97, 2011.

2   IF-Inference System of Takagi-Sugeno Type

General structure of FIS is defined in [4]. It contains a fuzzification process of input variables by membership functions μ, the design of the base of if-then rules or automatic extraction of if-then rules from input data, application of operators (AND,OR,NOT) in if-then rules, implication and aggregation within these rules, and the process of defuzzification of gained values to crisp values.

In [4] there are mentioned optimization methods of the number of if-then rules. Operator AND between elements of two fuzzy sets can be generalized by t-norm and operator OR between elements of two fuzzy sets can be generalized by t-conorm [5]. The concept of IF-sets is the generalization of the concept of fuzzy sets, the notion introduced by L.A. Zadeh [5]. The theory of IF-sets is well suited to deal with vagueness. Recently, in this context, IF-sets have been used for intuitionistic classification and prediction models which can accommodate imprecise information. Based on [2], using membership function $\mu_A(x)$, non-membership function $v_A(x)$ and intuitionistic fuzzy index (IF-index) $\pi_A(x)$ it is possible to define an IF-system. The value $\pi_A(x)$ denotes a measure of non-determinancy, it may cater to either membership value or non-membership value, or both.

Let there exist a general IF-system defined in [6]. Then it is possible to define its output y^η as $y^\eta = (1 - \pi_A(x)) \times y^\mu + \pi_A(x) \times y^v$, where y^μ is the output of the FIS$^\mu$ using the membership function $\mu_A(x)$, y^v is the output of the FIS^v using the non-membership function $v_A(x)$. For the IF-inference system designed in this way, the following facts hold. If IF-index [3]:

- $\pi_A(x)=0$, then the output of IF-inference system $y^\eta=(1 - \pi_A(x)) \times y^\mu$ (Takagi-Sugeno type FIS is characterized by membership function μ).
- $\pi_A(x)=1$, then the output of IF-inference system $y^\eta=\pi_A(x) \times y^v$ (Takagi-Sugeno type FIS is characterized by non-membership function v).
- $0< \pi_A(x) <1$, then the output of IF-inference system $y^\eta = (1 - \pi_A(x)) \times y^\mu + \pi_A(x) \times y^v$ (Takagi-Sugeno type FIS is characterized by membership function μ and non-membership function v).

Let $x_1, x_2, \ldots, x_j, \ldots, x_m$ be input variables FIS$^\eta$ defined on reference sets $X_1, X_2, \ldots, X_j, \ldots, X_m$ and let y^η be an output variable defined on reference set Y. Then FIS$^\eta$ has m input variables $x_1, x_2, \ldots, x_j, \ldots, x_m$ and one output variable y^η, where $\eta=\mu$ are membership functions ($\eta=v$ are non-membership functions).

Further, each set X_j, $j=1,2, \ldots, m$, can be divided into $i=1,2, \ldots, n$ fuzzy sets which are represented by following way $\eta_{j,1}(x), \eta_{j,2}(x), \ldots, \eta_{j,i}(x), \ldots, \eta_{j,n}(x)$. Individual fuzzy sets, where $\eta=\mu$ are membership functions ($\eta=v$ are non-membership functions) represent a mapping of linguistic variables values which are related to sets X_j. Then the k-th if-then rule R_k in FIS$^\eta$ can be defined as follows

$$R_k: \text{if } x_1 \text{ is } A_{1,i(1,k)}{}^\eta \text{ AND } x_2 \text{ is } A_{2,i(2,k)}{}^\eta \text{ AND } \ldots \text{ AND } x_j \text{ is } A_{j,i(j,k)}{}^\eta \text{ AND } \ldots \text{ AND}$$
$$x_m \text{ is } A_{m,i(m,k)}{}^\eta \text{ then } y^\eta=h, \text{ or } y^\eta=f(x_1,x_2, \ldots, x_m), j=1,2, \ldots, m; i=1,2, \ldots, n, \tag{1}$$

where $A_{1,i(1,k)}{}^\eta, A_{2,i(2,k)}{}^\eta, \ldots, A_{j,i(j,k)}{}^\eta, \ldots, A_{m,i(m,k)}{}^\eta$ represent the values of linguistic variable for FIS$^\mu$ and FIS^v, h is constant, $f(x_1, x_2, \ldots, x_m)$ is a linear or polynomial function.

3 Intuitionistic Fuzzy t-Norms and t-Conorms

Let there be given two fuzzy sets which are defined in the same universe. Then it is possible to realize operations between them, i.e. such forms of projections (relations) where the result projects to the same universe. Within the fuzzy sets theory it is possible to define classes of operations where intersection is corresponded by t-norms and union by t-conorms. Therefore, definitions of the most important fuzzy sets operations are listed further [4,7]. As well as union and intersection are dual operations, also t-norms and t-conorms are dual. Their relation can be defined as

$$T(x,y)=1- S((1-x), (1-y)), S(x,y)=1- T((1-x), (1-y)). \tag{2}$$

Based on the introduced facts, it is possible to define stated t-norms and t-conorms which always create pairs (one t-norm and one t-conorm). Examples of t-norms and t-conorms are listed in Table 1. The analysis in [7] implies that t-norms and t-conorms are characterized by different rate of strictness, or tolerance, where T_D is the strictest t-norms and S_D the most tolerant t-conorm. They can be ordered in the following way

$$T_D \le T_L \le T_E \le T_A \le T_H \le T_M \text{ and } S_M \le S_H \le S_A \le S_E \le S_L \le S_D. \tag{3}$$

Fuzzy sets have associated a non-membership function $v_A(x)$ degree

$$A=\{\langle x, \mu_A(x), v_A(x)\rangle | \ x \in X\}=\{\langle x, \mu_A(x), 1- \mu_A(x)\rangle | \ x \in X\}. \tag{4}$$

Table 1. Examples of t-norms and t-conorms

Name		t-norm, t-conorm	Notes
Drastic	product	$T_D(x,y)= \min(x,y)$ if $\max(x,y)=1$ 0 otherwise	
	sum	$S_D(x,y)= \max(x,y)$ if $\min(x,y)=0$ 1 otherwise	
Bounded	difference	$T_L(x,y)=\max(0, x+y-1)$	Łukasiewicz t-norm
	sum	$S_L(x,y)=\min(1, x+y)$	Łukasiewicz t-conorm
Einstein	product	$T_E(x,y)=(x.y)/((2-(x+y-x.y))$	
	sum	$S_E(x,y)=(x+y)/(1+x.y)$	
Algebraic	product	$T_A(x,y)=x.y$	Algebraic product
Probabilistic	sum	$S_A(x,y)=x+y-x.y$	Probabilistic sum
Hamacher	product	$T_H(x,y)=(x.y)/(x+y-x.y)$	
	sum	$S_H(x,y)=(x+y-2.x.y)/(1-x.y)$	
MIN, MAX	min	$T_M(x,y)=\min(x,y)$	Gödel t-norm
	max	$S_M(x,y)=\max(x,y)$	Gödel t-conorm

Let be given an automorphism [8] of the unit interval, i.e. every function ψ: $[0,1] \to [0,1]$, that is continuous and strictly increasing such that $\psi(0)=0$ and $\psi(1)=1$. Further, let be given a function n: $[0,1] \to [0,1]$ in such a way that it holds $n(0) =1$ and $n(1) =0$. It is called a strong negation and it is always strictly decreasing, continuous and involutive. Then, as proved by [8], n: $[0,1] \to [0,1]$ is a strong negation if and only if there exists an automorphism ψ of the unit interval such that $n(x) = \psi^{-1} (1 - \psi(x))$. Let L^* be a set for which

$$L^* = \{(x,y) | (x,y) \in [0,1] \times [0,1] \text{ and } x+y \leq 1\} \tag{5}$$

and the elements $0_{L^*} = (0,1)$ and $1_{L^*} = (1,0)$. Then $\forall((x,y),(z,t)) \in L^*$ it holds:

- $(x,y) \leq_{L^*} (z,t)$ iff $x \leq z$ and $y \geq t$. This relation is transitive, reflexive and antisymmetric.

- $(x,y) = (z,t)$ iff $(x,y) \leq_{L^*} (z,t)$ and $(z,t) \leq_{L^*} (x,y)$, $(x,y) \ll (z,t)$ iff $x \leq z$ and $y \leq t$.

The designed IF-inference system of Takagi-Sugeno type works with the inference mechanism, based on intuitionistic fuzzy t-norm and t-conorm, by means of t-norm and t-conorm [9] on interval [0,1]. A function $T: (L^*)^2 \to L^*$ is called intuitionistic fuzzy t-norm if it is commutative, associative, and increasing in both arguments with respect to the order $\leq_{L^*}$ and with neutral element 1_{L^*}. Similarly, a function $S: (L^*)^2 \to L^*$ is called intuitionistic fuzzy t-conorm if it is commutative, associative, and increasing with neutral element 0_{L^*}. Intuitionistic fuzzy t-norm T is called t-representable if and only if there exists a t-norm T and t-conorm S on interval [0,1] such that $\forall(x,y),(z,t) \in L^*$ it holds $T((x,y),(z,t)) = (T(x,z),S(y,t)) \in L^*$.

Intuitionistic fuzzy t-norm and t-conorm can be constructed using t-norms and t-conorms on [0,1] in the following way.

Let T be a t-norm and S a t-conorm. Then the dual t-norm S^* of S is defined by $S^*(x,y) = 1 - S(1-x, 1-y)$, $\forall x,y \in [0,1]$. If $T \leq S^*$, i.e. if $\forall x,y \in [0,1]$, $T(x,y) \leq S^*(x,y)$, then the mapping T defined by $T(x,y) = (T(x_1,y_1), S(x_2,y_2))$, $\forall x,y \in L^*$, is an intuitionistic fuzzy t-norm and the mapping S defined by $S(x,y) = (S(x_1,y_1), T(x_2,y_2))$, $\forall x,y \in L^*$, is an intuitionistic fuzzy t-conorm. Note that condition $T \leq S^*$ is necessary and sufficient for $T(x,y)$ and $S(x,y)$ to be elements of L^* $\forall x,y \in L^*$. It can be written that $T = (T,S)$ and $S = (S,T)$.

4 Modelling and Analysis of the Results

Verification of behaviour of various t-norms and t-conorms in designed IF-inference system of Takagi-Sugeno type is shown in ozone modelling. The data for our investigations was obtained from the Czech Hydro-meteorological Institute. This data contains the average daily ozone measurements and the average daily meteorological variables (such as temperature, wind speed, wind direction, humidity, air pressure and solar radiation), vehicle emission variables (NO_2, CO, NO, NO_x, SO_2, PM_{10} and $PM_{2.5}$), and other dummy variables (working day, month). All the measured variables used in this study are actual same day values. Then the formulation of the basic model is following: $y = f(x_1^t, x_2^t, \ldots, x_m^t)$, $m=22$, where y is daily average ozone level at time $t+1$. For further modelling it is necessary to minimize the number of parameters x_m^t, $m=22$, and to maximize function f.

Based on the minimization of the number of parameters x_m^t, $m=22$ using Pearson's correlation coefficient, it was shown that NO_x, NO, NO_2, month of measurement, humidity, solar radiation, and the ozone level at day ahead were important to predict daily average ozone levels. The general formulation of the model is as follows $y = f(x_1^t, x_2^t, \ldots, x_m^t)$, $m=7$, where y is daily average ozone level at time $t+1$, x_1^t is oxide of nitrogen, x_2^t is nitric oxide, x_3^t is nitrogen dioxide, x_4^t is dummy variable (month), x_5^t is humidity, x_6^t is solar radiation, x_7^t is daily average ozone level at time t.

Based on the given facts, such FIS^μ is designed which is characterized by means of membership function μ, and FIS^ν characterized by non-membership function ν. Input variables $x_1^t, x_2^t, \ldots, x_m^t$, $m=7$ in time t are represented by two bell-shaped membership functions μ for FIS^μ and two non-membership functions ν for FIS^ν. Membership function μ and non-membership function ν, and if-then rules were designed using subtractive clustering algorithm. After projecting the clusters onto the input space, the antecedent parts of the fuzzy if-then rules can be found. The consequent parts of the if-then rules are represented by polynomial functions $y=f(x_1, x_2, \ldots, x_r)$. In this way, one cluster corresponds to one if-then rule. To be specific, two if-then rules are designed for FIS^μ and FIS^ν respectively. The output level y_k of each of the k-th if-then rule R_k is weighted. The final outputs y^μ and y^ν of the FIS^μ and FIS^ν are the weighted averages of all the if-then rule R_k outputs y_k, $k=1,2, \ldots, N$. The output of IF-inference system is represented by the predicted value y^η in time $t+1$. Table 2 shows the quality of ozone prediction represented by Root Mean Squared Error ($RMSE^\eta$) for different values of μ_{max} [3], different values of IF-index π and for various t-norms. The resulting errors are comparable to those presented in other studies, for example [10].

Table 2. $RMSE^\eta$ on testing data O_{test} for different values of μ_{max}, IF-index π, and T_E, T_A, T_H, T_M

$\pi=0.1$									
μ_{max}	0.1	0.2	0.3	0.4	0.5	0.6	0.7	0.8	0.9
T_E	23.23	24.00	25.30	26.96	28.64	30.78	34.60	42.24	21.43
T_A	23.08	23.55	24.17	25.01	26.19	27.85	30.46	35.07	19.81
T_H	16.99	18.02	18.85	19.33	19.35	18.83	17.74	15.95	13.31
T_M	-	11.01	11.06	11.06	11.06	11.06	11.05	11.04	10.94
$\pi=0.2$									
T_E	22.09	24.05	27.27	31.29	36.04	43.92	60.82	30.27	
T_A	21.65	22.70	24.30	26.69	30.25	35.78	46.63	25.14	
T_H	16.33	17.37	18.30	18.98	19.21	18.78	17.12	13.41	
T_M	-	10.90	10.93	10.93	10.92	10.92	10.91	11.63	
...									
$\pi=0.8$									
T_E	42.26	112.1							
T_A	37.99	108.1							
T_H	15.64	25.08							
T_M	12.22	26.78							

Extreme situations of the IF-inference system are to be found when IF-index $\pi_A(x) = 0$ (Table 3), then $\mu_A(x) + \nu_A(x) = 1$ (fuzzy sets are considered as a particular case IF-sets) and if IF-index $\pi_A(x) = 1$, then $\mu_A(x) = 0$ and $\nu_A(x) = 0$ (complete ignorance of the problem). For $\pi_A(x) = 0$, $RMSE^\mu = RMSE^\eta$. Thus, only $RMSE^\mu$ are reported. Drastic T_D and bounded T_L t-norms are not suitable for this IF-inference system design. The problem with drastic T_D and bounded T_L t-norms is that it can happen that no rule is activated. When no rule is activated, the output of the system is constant, which can be interpreted as some kind of neutral output. Strictness of these t-norms is crucial for this type of data. Based on designed IF-inference systems with t-norms T_D and T_L, it is implied that $RMSE^\eta$ for FIS^η is constant. For other designed IF-inference systems with t-norms T_E, T_A, T_H, T_M, $RMSE^\eta$ for FIS^η are given by Table 2.

Table 3. RMSE on testing data O_{test} for different values of $\mu_{\max}$, IF-index $\pi=0$, and T_E,T_A,T_H,T_M

$\mu_{\max}$		0.1	0.2	0.3	0.4	0.5	0.6	0.7	0.8	0.9
T_E	RMSE$^{\mu}$	24.72	24.69	24.65	24.57	24.47	24.34	24.19	24.03	23.92
	RMSE^{v}	15.94	15.07	30.33	56.82	81.50	106.00	139.60	196.30	292.60
T_A	RMSE$^{\mu}$	24.72	24.72	24.72	24.72	24.72	24.72	24.72	24.72	24.71
	RMSE^{v}	16.65	16.02	19.93	29.51	42.78	59.53	81.41	115.70	169.90
T_H	RMSE$^{\mu}$	17.70	18.59	19.20	19.44	19.22	18.54	17.42	16.05	14.76
	RMSE^{v}	15.76	16.75	20.53	24.35	27.18	28.84	29.08	27.06	19.59
T_M	RMSE$^{\mu}$	-	11.29	11.29	11.29	11.29	11.29	11.29	11.29	11.29
	RMSE^{v}	-	13.92	13.25	13.36	13.33	13.30	13.31	13.40	13.70

5 Conclusion

The development of effective prediction models of ozone concentrations in urban areas is important. However, it is difficult because the meteorological parameters and photochemical reactions involved in ozone formation are complex. The article compares IF-inference systems of Takagi-Sugeno type [3] for various fuzzy operators in if-then rules with the help of RMSE$^{\eta}$ for FIS$^{\eta}$ on testing data O_{test}. RMSE$^{\eta}$ represents the error of IF-inference system, which consists of RMSE$^{\mu}$ for FIS$^{\mu}$, which is represented by membership function μ, and error RMSE^{v} for FIS^{v}, which is represented by non-membership function v. Further, extreme situations in the designed IF-inference systems from the viewpoint of IF-index are shown.

Acknowledgments. This work was supported by the scientific research project of the Ministry of Environment, the Czech Republic under Grant No: SP/4i2/60/07.

References

[1] Dubois, D., Gottwald, S., Hajek, P., Kacprzyk, J., Prade, H.: Terminological Difficulties in Fuzzy Set Theory-The case of Intuitionistic Fuzzy Sets. Fuzzy Sets and Systems 156, 485–491 (2005)

[2] Atanassov, K.T.: Intuitionistic Fuzzy Sets. Fuzzy Sets and Systems 20, 87–96 (1986)

[3] Olej, V., Hájek, P.: IF-Inference Systems Design for Prediction of Ozone Time Series: The Case of Pardubice Micro-Region. In: Diamantaras, K., Duch, W., Iliadis, L.S. (eds.), pp. 1–11. Springer, Heidelberg (2010)

[4] Pedrycz, W.: Fuzzy Control and Fuzzy Systems. John Wiley and Sons, New York (1993)

[5] Zadeh, L.A.: Fuzzy Sets. Inform. and Control 8, 338–353 (1965)

[6] Montiel, O., et al.: Mediative Fuzzy Logic: A new Approach for Contradictory Knowledge Management. Soft Computing 20, 251–256 (2008)

[7] Klement, E.P., Mesiar, R., Pap, E.: Triangular Norms. Position Paper I: basic Analytical and Algebraic Properties. Fuzzy Sets and Systems 143, 5–26 (2004)

[8] Barrenechea, E.: Generalized Atanassov's Intuitionistic Fuzzy Index. Construction Method. In: IFSA-EUSFLAT, Lisbon, pp. 478–482 (2009)

[9] Deschrijver, G., Cornelis, C., Kerre, E.: On the Representation of Intuitionistic Fuzzy t-norm and t-conorm. IEEE Transactions on Fuzzy Systems 12, 45–61 (2004)

[10] Comrie, A.C.: Comparing Neural Networks and Regression Models for Ozone Forecasting. Journal of the Air and Waste Management Association 47, 653–663 (1997)

LQR-Mapped Fuzzy Controller Applied to Attitude Stabilization of a Power-Aided-Unicycle

Ping-Ho Chen, Wei-Hsiu Hsu, and Ding-Shinan Fong

Department of Electrical Engineering, Tungnan University, Taiwan
phc@mail.tnu.edu.tw

Abstract. Analysis of attitude stabilization of a power-aided unicycle points out that a unicycle behaves like an inverted pendulum subject to power constraint. An LQR-mapped fuzzy controller is introduced to solve this nonlinear issue by mapping LQR control reversely through least square and Sugeno-type fuzzy inference. The fuzzy rule surface after mapping remains optimal.

Keywords: LQR, Fuzzy, Inverted pendulum, Unicycle, Stabilization.

1 Introduction

This paper introduces the control of unicycle stabilization using an inversed optimal fuzzy controller [1][2][3] mapped from a LQR controller. Since control of unicycle stabilization is constrained by power and torque limitation due to motor specification, engineers will meet difficulty in implementing LQR [4][5][6] control surface owing to its nonlinearity. Dynamics of unicycle stabilization [7] and fuzzy transformation from LQR, will be dealt with in depth equivalently in this paper.

The goal of unicycle stabilization is to minimize attitude and rate of attitude of the rider, or equivalently the seat, by rider's mechanical power and fuzzy electrical power simultaneously. Proposed in this paper is an EM (electrical/mechanical) unicycle that has mass center right above the top of the wheel as if it were an inverted pendulum referred to Fig. 1(a)-(b). Since the seat is no longer as high as the traditional one, shortening of torque arm reduces manual torque and accordingly deteriorates maneuverability. With the help of auxiliary electrical power, it would be beneficial to compensate for torque reduction and furthermore to increase the stabilization effects.

Fig. 1(a)-(b) shows the seat plane (above wheel axle) and pedal platform (underneath wheel axle). Mass M (with arm L) and mass m (with arm l) represent masses distributed above and underneath the wheel axle respectively. The total mass center $M+m$ located at l_p above the axle behaves as if it were an inverted pendulum. The pendulum arm is

$$l_p = (ML - ml)/m_p, L = l_s + l_r, m_p = M + m \qquad (1)$$

where l_s : length between seat and wheel; l_r : length between mass center of rider's body and seat. Fig. 1(c) shows riding on the move.

L. Iliadis et al. (Eds.): EANN/AIAI 2011, Part II, IFIP AICT 364, pp. 98–103, 2011.

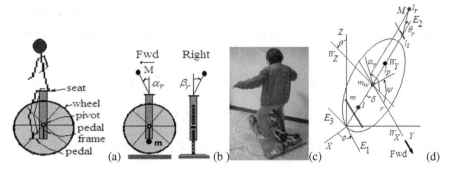

Fig. 1. (a) Schematic diagram of EM unicycle. (b) Pendulum mass M and m viewed from side and back. (c) Rider driving on the move. (d) Geometry of unicycle.

Control of the unicycle combines rider's manual controller and a complementary fuzzy controller. Manual torques τ_m and fuzzy torque τ_f are outputs of both controllers defined by $\tau_m : [\tau_{mP}\ \tau_{mR}]^t$, τ_{mR} : roll torque, τ_{mP} : pitch torque, $\tau_f : [\tau_{fw}\ \tau_{fP}\ \tau_{fR}]^t$ (fuzzy torque), τ_{fw} : fuzzy wheel torque, τ_{fP} : fuzzy pitch torque, τ_{fR} : fuzzy roll torque.

2 Dynamics of Unicycle

There are five coordinates involved in the dynamics, i.e. inertia coordinate $[X, Y, Z]^t$ → Euler coordinate $[E_1, E_2, E_3]^t$ → wheel coordinate $[W_X, W_Y, W_Z]^t$ → pendulum coordinate $[P_1, P_2, P_3]^t$ → rider coordinate $[R_1, R_2, R_3]^t$. Note that Euler coordinate $[E_1, E_2, E_3]^t$ is transformed from inertia coordinate $[X, Y, Z]^t$ by turning an angle of ϕ along axis Z and then tilting an angle of θ along axis E_1.

Application of Newton's second law [7] in matrix form gives

$$\dot{H} + \omega H = \tau \cdot \omega = \begin{bmatrix} 0 & -\omega_3 & \omega_2 \\ \omega_3 & 0 & -\omega_1 \\ -\omega_2 & \omega_1 & 0 \end{bmatrix}, H = [h_1\ h_2\ h_3]^t = [I_1\Omega_1, I_2\Omega_2, I_3\Omega_3]^t \quad (2)$$

$$I_1 = I_{wx} + m_w r^2 + I_{p1} + m_p(r+l_p)^2\ I_2 = I_{wy} + I_{p2}\ I_3 = I_{wz} + m_w r^2 + I_{p3} + m_p(r+l_p)^2 \quad (3)$$

For $i = 1, 2, 3$, H : angular momentum of unicycle; I_i : moment of inertia of unicycle along Euler axis E_i ; I_{wx} : moment of inertia of wheel along wheel axis W_x ; m_w : wheel mass; r : wheel radius; I_{pi} : moment of inertia of total pendulum along pendulum axis P_1 ; Ω_i : inertia angular rate along Euler axis E_i ; ω : precession matrix; τ : applied torque in Euler coordinate; ψ : wheel rotation; α_r : rider's pitch angle along rider axis R_3; θ_r :rider's roll tilt angle along rider axis R_1; δ : platform pitch angle; l : length between mass center of lower pendulum and wheel axle; g :gravity. Further derivation yields

$$I_1\dot{\Omega}_1 - I_2\omega_3\Omega_2 + I_3\omega_2\Omega_3 = \tau_1 = \tau_{v1} + \tau_{mR} + \tau_{fR}$$

$$\tau_{v1} = \tau_u \cos\theta, \ \tau_u = -[m_w r + m_p(r+l_p)]g \tag{4}$$

$$I_2\dot{\Omega}_2 - I_3\omega_1\Omega_3 + I_1\omega_3\Omega_1 = \tau_{v2}, \tau_{v2} = \tau_u \sin\theta \tag{5}$$

$$I_3\dot{\Omega}_3 - I_1\omega_2\Omega_1 + I_2\omega_1\Omega_2 = \tau_{fW} \tag{6}$$

Where τ_{v1} and τ_{v2} comes from gravity effect accompanied with wheel tilt. Eq. (4), (5) and (6) correspond respectively to the dynamics of roll, heading and wheel drive.

3   Stabilization of Pitch Based on LQR

Stabilization of pitch based on LQR depends on the analysis of pitch–wheel coupling. Governing equations, based on force and torque are:

$$(m_w + m_p)\ddot{x} + b\dot{x} - m_p l_p \ddot{\delta}\cos\delta + m_p l_p \dot{\delta}^2 \sin\delta = \frac{\tau_{fW}}{r} \tag{7}$$

$$(I_{p3} + m_p l_p{}^2)\ddot{\delta} - m_p l_p \ddot{x}\cos\delta = m_p g l_p \sin\delta + \tau_{mP} + \tau_{fP} \tag{8}$$

Where I_{p3}: moment of inertia of inverted pendulum along mass center located at l_p above the axle. While stabilized, $\delta \approx 0$ and $\dot{\delta} \approx 0$, Eq. (7) and (8) are represented in matrix form as

$$\begin{bmatrix} \ddot{x} \\ \dot{\delta} \\ \ddot{\delta} \end{bmatrix} = \begin{bmatrix} -b/\hat{m} & gm_p{}^2 l_p{}^2/\hat{m}\hat{I} & 0 \\ 0 & 0 & 1 \\ -m_p l_p b/(\hat{m}\hat{I}) & m_p g l_p/\hat{I} & 0 \end{bmatrix}\begin{bmatrix} \dot{x} \\ \delta \\ \dot{\delta} \end{bmatrix} + \begin{bmatrix} 1/\hat{m}r & m_p l_p/\hat{m}\hat{I} \\ 0 & 0 \\ m_p l_p/\overline{m}\hat{I}r & 1/\hat{I} \end{bmatrix}\begin{bmatrix} \tau_{fW} \\ \tau_{mP} + \tau_{fP} \end{bmatrix} \tag{9}$$

Where $\hat{I} = \overline{I} - m_p{}^2 l_p{}^2/\overline{m}$, $\hat{m} = \overline{m} - m_p{}^2 l_p{}^2/\overline{I}$, $\overline{m} = m_w + m_p$, $\overline{I} = I_{p3} + m_p l_p{}^2$
Obviously, three eigen values of transition matrix in Eq. (9) are $\lambda_1 < 0, \lambda_2 = 0$, $\lambda_3 > 0$. This results in an unstable pendulum system in the presence of drive-pitch coupling. Eq. (9) can be expressed by

$$\dot{X}_1 = A_1 X_1 + B_1 u_1, \quad Y_1 = C_1 X_1 \text{ (Subscript 1 for pitch and 2 for roll)} \tag{10}$$

In Eq. (10), X_1 , u_1 and Y_1 are state, control and measurement respectively. Assigning $M = 60kg, m = 15kg, m_w = 10kg, r = 0.52m, b = 0.5nt/m/s, l = 0.2m,$

$l_s = 0.18m, g = 9.8m/s^2$, it follows $m_p = 75kg, l_p = 0.52m, I_{p3} = 9.7kg\text{-}m^2$, $\bar{I} = 30kg\text{-}m^2$, $\overline{m} = 85kg, \hat{m} = 24.3kg, \hat{I} = 12.1kg\text{-}m^2$.

Eq. (9) becomes

$$X_1 = \begin{bmatrix} \dot{x} \\ \delta \\ \dot{\delta} \end{bmatrix}, \quad u_1 = \begin{bmatrix} \tau_{fW} \\ \tau_{mP} + \tau_{fP} \end{bmatrix}, A_1 = \begin{bmatrix} -0.02 & 50.69 & 0 \\ 0 & 0 & 1 \\ -0.02 & 31.59 & 0 \end{bmatrix}, B_1 = \begin{bmatrix} .079 & .133 \\ 0 & 0 \\ .073 & .083 \end{bmatrix}, C_1 = I_{3\times3} \quad (11)$$

Since Eigen values of A_1 are 0.0121, -5.6365, 5.6044, it implies the original system of unicycle is unstable. If control torque u_1 is substituted by state feedback - $K_1 X_1$ with $Q_1 = I_{3\times3}$, $R_1 = I_{2\times2}$, the optimal gain K_1 will be

$$K_1 = \begin{bmatrix} -0.4144 & 377.1436 & 67.5795 \\ 1.0834 & 431.9048 & 74.9096 \end{bmatrix} \quad (12)$$

Eigen values of the LQR are -0.04, -5.73 and -5.52 that guarantee the system to be stable.

4 Stabilization of Roll Based on LQR

Substituting $\Omega_1 = \omega_1 = \dot{\theta}$, $\Omega_2 = \omega_2 = \dot{\phi}\sin\theta, \Omega_3 = \omega_3 + \dot{\psi} = \dot{\phi}\cos\theta + \dot{\psi}$ and $\dot{\psi} = constant, \dot{\phi} = constant, \theta = \alpha + \pi/2, \alpha$: roll angle ($\alpha \approx 0$) into Eq. (4), it follows

$$I_1\ddot{\alpha} + k\alpha = \tau_h + \tau_{mR} + \tau_{fR}, k = \tau_u - (I_3 - I_2)\dot{\phi}^2, \tau_h = -I_3\dot{\psi}\dot{\phi} \quad (13)$$

The term τ_h is a reaction torque caused by precession of angular momentum. Assigning that wheel speed is set at 5 km/hr (i.e. 1.39 m/s), then making a quarter turn needs 11.31sec. It yields $\dot{\phi} = 0.139$ rad/s and $\dot{\psi} = 2.67$ rad/s. Else parameters are: $I_1 = 95.5kg\text{-}m^2, I_2 = 6.15kg\text{-}m^2, I_3 = 94.8\,kg\text{-}m^2, \tau_u = 815.4\,nt\text{-}m, k = 813.7nt\text{-}m$.

$$X_2 = \begin{bmatrix} \alpha \\ \dot{\alpha} \end{bmatrix}, \quad u_2 = \tau_{mR} + \tau_{fR}, A_2 = \begin{bmatrix} 0 & 1 \\ -119.7 & 0 \end{bmatrix}, B_2 = \begin{bmatrix} 0 \\ 0.01 \end{bmatrix}, C_2 = I_{2\times2} \quad (14)$$

Since eigen values of A_2 are 10.9407i and -10.9407i, it implies the original system of unicycle is critical stable. If control torque u_2 is substituted by state feedback - $K_2 X_2$ with $Q_2 = 500 I_{2\times2}, R_2 = 0.01$, the optimal gain K_2 will be

$$K_2 = [2.1 \quad 225] \quad (15)$$

Eigen values of the LQR are -1.1227+10.8840i and -1.1227-10.8840i that guarantees the system to be stable.

5 Inversed Fuzzy Controller Mapped from LQR [8]

Taking roll stabilization as an instance, the LQR control is $u_2 = 225\dot{\alpha} + 2.1\alpha$
constrained by

$$P_2 = \tau_2 \, \dot{\alpha} = u_2\dot{\alpha} = \tau_{fR}\dot{\alpha} = (K_2 X_2)\dot{\alpha} = (k_{21}\alpha + k_{22}\dot{\alpha})\dot{\alpha} \leq P_{2\max} \qquad (16)$$

For $P_{2\max} = 100$, $u_2\dot{\alpha} \leq 100$ (Inequality constraint) $\qquad (17)$

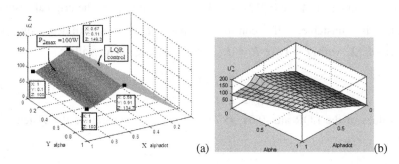

Fig. 2. (a) LQR control with power constraint (b) Inversed fuzzy controller after mapping

Preprocessing data assumption: 1) α and $\dot{\alpha}$, as fuzzy variables are normalized,
i.e. $\alpha = 1rad$ and $\dot{\alpha} = 1rad/s$ 2) Maximum power $P_{2\max}$ is $100w$, i.e. $\tau_{2\max} = 100nt\text{-}m$
at $\dot{\alpha} = 1rad/s$. 3) Apex of hyperboloid surface due to power constraint is indicated in
Fig. 2.(a). 4) Plane p_1 (LQR) $u_2 = 225\dot{\alpha} + 2.1\alpha$ is also shown in Fig. 2.(a). 5)
Plane p_2, simplified from hyperboloid surface by least square [8], is:

$$u_2 = \Lambda \, [\dot{\alpha} \ \ \alpha \ \ 1]^{\mathrm{T}}, \Lambda = [-108.3 \ \ -12.3 \ 216.1] \qquad (18)$$

The fuzzy rule surface, as a minimum of p_1 and p_2, is able to be implemented by
reversely finding fuzzy membership and inference using Sugeno model [9] as follows.

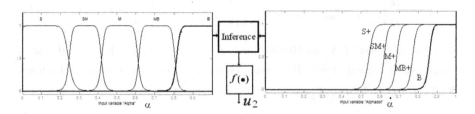

Fig. 3. Membership function of α and $\dot{\alpha}$

Membership functions of fuzzy variables α and $\dot{\alpha}$ are given in Fig. 3 that also shows Sugeno FIS (Fuzzy inference system) [8]. Sugeno linear models is employed in the inference as given below:

1. If (α is S) and ($\dot{\alpha}$ is B) then (u is $p2$);   2. If (α is SM) and ($\dot{\alpha}$ is MB+) then (u is $p2$)

3. If (α is M) and ($\dot{\alpha}$ is M+) then (u is $p2$); 4. If (α is MB) and ($\dot{\alpha}$ is SM+) then (u is $p2$)

5. If (α is B) and ($\dot{\alpha}$ is S+) then (u is $p2$); 6. If (α is S) and ($\dot{\alpha}$ is not B) then (u is $p1$)

7. If (α is SM) and ($\dot{\alpha}$ is not MB+) then (u is $p1$); 8. If (α is M) and ($\dot{\alpha}$ is not M+) then (u is $p1$); 9. If (α is MB) and ($\dot{\alpha}$ is not SM+) then (u is $p1$); 10. If (α is B) and ($\dot{\alpha}$ is not S+) then (u is $p1$).

Where linear surfaces $p_1 = 225\dot{\alpha} + 2.1\alpha$ and $p_2 = -108.3\dot{\alpha} - 12.3\alpha + 216.1$

The fuzzy rule surface, thus obtained as given in Fig. 2.(b), is very close to LQR control surface in Fig. 2.(a). Therefore the inversed fuzzy controller mapped from LQR remains optimal although it needs more massage. Fig. 2 and Fig. 3 are plotted by using Matlab fuzzy toolbox [10].

6 Conclusion

Through the analysis of unicycle dynamics, we find that stabilization control of a power-aided unicycle, in pitch and roll, meets difficulty in using LQR control under power and torque constraints. This paper applies the approach, called "inversed fuzzy controller mapped from LQR controller" to solve the issue of stabilization control of unicycle. A Sugeno fuzzy control surface, almost equivalent LQR control surface, is built by using fuzzy membership and inference. Through this kind of transformation, we have the equivalent control surface, still remaining optimal, but using fuzzy-logic-controller (FLC) to implement.

References

1. Zadeh, L.A.: The concept of a linguistic variable and its application to approximate reasoning. Parts 1, 2, and 3. Information Sciences 8, 199–249, 8, 301–357, 9, 43–80 (1975)
2. Mamdani, E.H., Assilian, S.: An Experiment in Linguistic Synthesis with Fuzzy Logic Controller. Int. J. Man Mach. Studies 7(1), 1–13 (1975)
3. Mamdani, E.H.: Advances in the linguistic synthesis of fuzzy controllers. International Journal of Man-Machine Studies 8, 669–678 (1976)
4. Zadeh, L.A.: A Theory of Approximating Reasoning, pp. 149–194. Elsevier, Reading (1979)
5. Lewis, F.L.: Applied Optimal Control & Estimation. Prentice Hall, Englewood Cliffs (1992)
6. Burl, J.B.: Linear Optimal Control. Addison-Wesley, Reading (1999)
7. Bert, C.W.: Dynamics and stability of unicycles and monocycles. Dynamical Systems 5(1), 1468–9375 (1990)
8. Chen, P.-H., Lian, K.-Y.: Optimal Fuzzy Controller Mapped from LQR under Power and Torque Constraints. In: 7th AIAI Conferences (2011)
9. Sugeno, M.: Fuzzy measures and fuzzy integrals: a survey. In: Gupta, M.M., Saridis, G.N., Gaines, B.R. (eds.) Fuzzy Automata and Decision Processes, pp. 89–102. North-Holland, NY (1977)
10. Matlab: toolbox: Fuzzy (2010)

Optimal Fuzzy Controller Mapped from LQR under Power and Torque Constraints

Ping-Ho Chen[1] and Kuang-Yow Lian[2]

[1] Department of EE, Tungnan University, Taipei, Taiwan
phc@mail.tnu.edu.tw
[2] Department of EE, National Taipei University of Technology, Taipei, Taiwan
kylian@ntut.edu.tw

Abstract. Dealing with a LQR controller surface subject to power and torque constraints, is an issue of nonlinear problem that is difficult to implement. This paper employs a fuzzy controller surface to replace the LQR surface subject to power and torque constraints by using class stacking, least square and Sugeno-type fuzzy inference mode. Through this type of transformation, called "Optimal fuzzy controller mapped from LQR", control of the system remains optimal.

Keywords: LQR, FLC, Fuzzy, Sugeno, Optimal, Class stacking.

1 Linear Quadratic Regulator (LQR)

The development of optimal control for a MIMO system [1][2] is mature nowadays. Approach of linear quadratic regulator (LQR) offers Kalman gain for state feedback to regulate a MIMO system in state-space form. But subject to power and torque limitation of motor drive, implementation of system becomes difficult using LQR approach. This section introduces some basis for LQR approach for further improvement.

1.1 LQR Approach

For a MIMO system in state-space form

$$\dot{X} = AX + Bu \tag{1}$$

where $X \in R^n$ is the state vector. The control $u \in R^w$ includes state-feedback torque in motor application, command for desired state output and disturbance that the system might encounter. Control u is thus expressed by

$$u = -KX + u_{com} + u_{dist} \tag{2}$$

where K : Kalman gain. u_{dist} : disturbance torque (colored noise). u_{com} : command for desired state output. Eq. (2) is a general form of input in LQG (Linear Quadratic Gaussian) problems. When $u_{dist} = o$ and $u_{com} \neq o$, the system becomes a tracking problem expressed by

L. Iliadis et al. (Eds.): EANN/AIAI 2011, Part II, IFIP AICT 364, pp. 104–109, 2011.
© IFIP International Federation for Information Processing 2011

$$\dot{X} = (A\text{-}BK)X + Bu_{com} \tag{3}$$

At steady state it satisfies

$$O = (A\text{-}BK)X + Bu_{com} \tag{4}$$

In case $u_{com} = o$ in Eq. (4), the system becomes a regulator problem further called"linear quadratic regulator (LQR)", if we select a gain K, called Kalman gain such that $u = -KX$ to minimize the cost J in quadratic form as

$$J = \frac{1}{2}X^{T}(T)S(T)X + \int_{0}^{T}[\frac{1}{2}(X^{T}QX + u^{T}Ru)$$
$$+ \lambda^{T}(AX + Bu - \dot{X})]dt \tag{5}$$

where Q: dynamic state weighting, R: control weighting S: terminal state weighting and λ: Lagrange multiplier. After several algebra operations to minimize the cost, we have the following Hamiltonian matrix presenting as the transition matrix in Eq. (6):

$$\begin{bmatrix} \dot{X} \\ \dot{\lambda} \end{bmatrix} = \begin{bmatrix} A & -BR^{-1}B^{T} \\ -Q & -A^{T} \end{bmatrix} \begin{bmatrix} X \\ \lambda \end{bmatrix} \tag{6}$$

$$\lambda(t) = S(t)X(t) \tag{7}$$

$$\lambda(T) = S(T)X(T) \tag{8}$$

$$U(t) = -R^{-1}B^{T}\lambda(t) = -R^{-1}B^{T}S(t)X(t) = -K(t)X(t) \tag{9}$$

The steady gain is

$$K = \lim_{t \to \infty}K(t) = R^{-1}B^{T}\lim_{t \to \infty}S(t) = R^{-1}B\psi_{21}(\psi_{11})^{-1} \tag{10}$$

where ψ_{21} and ψ_{11} are the first-column block matrix of Ψ which is the modal form of the Hamiltonian matrix in Eq. (6).

1.2 Constraints of LQR Control

While implementing the optimal control u aforementioned, we always meet difficulty when subject to constraints of torque limit u_l and power limit p_l. The torque constraint, caused by stall current in motor application, is expressed by

$$u \le u_l, \quad u_l = k_i i_s \tag{11}$$

where k_i : current constant of motor. i_s : stall current of motor. As usual the control u is associated with states of position and rate. Mechanical power p_m, constrained by maximum power p_l of motor, is represented by

$$p_m = u^T \dot{\theta} \leq p_l \qquad (12)$$

where $\dot{\theta}$ is rate-related state vector.

2 Class Stacking

From Eq. (2), the LQR control is $u = -KX = [u_1 \, u_2 \cdots u_w]^t$ where $K = [k_{ji}] \in R^{w \times n}$ and

$u_j = -\sum_{i=1}^{n} k_{ji} x_i$ for $j = 1, \cdots, w$. Since each component u_j of u will be solved one by

one by the same approach of class stacking, u_j will be replaced by $u = \sum_{i=1}^{n} k_i x_i$ for

illustration without loss of generality. Obviously for j^{th} component u_j of u, $k_i = -k_{ji}$.

The form of $u = \sum_{i=1}^{n} k_i x_i$ is a hyper plane of n-D. All the states x_i for $i = 1, \cdots, n$ are

sorted out into two classes. One class, having m states, is position-related and piled up
by

$$\upsilon_1 = \sum_{i=1}^{m} k_i x_i \qquad (13)$$

and the other class, having $n-m$ states, is rate-related and piled up by

$$\upsilon_2 = \sum_{i=m+1}^{n} k_i x_i \qquad (14)$$

Then $\qquad\qquad u = \upsilon_1 + \upsilon_2 \qquad\qquad\qquad\qquad (15)$
The torque constraint is

$$\upsilon_1 + \upsilon_2 = u \leq u_l \qquad (16)$$

The power constraint $u \upsilon_2 \leq p_l$ from Eq. (12) can be rewritten by

$$(\upsilon_1 + \upsilon_2)\upsilon_2 \leq p_l \qquad (17)$$

Eq. (13)~(15) thus reduce the system from n-D into 3-D with variables υ_1 and υ_2.
Eq. (16) indicates a plane $\upsilon_1 + \upsilon_2 = u$ with known LQR control u but upper bounded
by u_l. Eq. (17) indicates a nonlinear constraint. The process, called "class stacking",
thus make it feasible to plot a hyper plane of n-D in terms of 3-D. An example using
computer-aided plot is given as follows. A LQR control u, with number of total states
n=5 and number of position-related states m=3, is expressed by

$$u = \sum_{i=1}^{5} k_i x_i = \upsilon_1 + \upsilon_2, \ \upsilon_1 = \sum_{i=1}^{3} k_i x_i, \upsilon_2 = \sum_{i=4}^{5} k_i x_i \qquad (18)$$

where $|x_i| \le 1$ (normalized) and $k_i = i$ for $i = 1, \cdots, 5$. Obviously maximum of υ_1 is 6 and maximum of υ_2 is 9, subject to constraints: $\upsilon_1 + \upsilon_2 \le 12$ (Upper bound) and $(\upsilon_1 + \upsilon_2)\upsilon_2 \le 75$ (Inequality constraint).

By using computer-aided plot [3], Fig. 1(a)-(d) shows a sequence of transformations from LQR surface to cases of constraints are obtained. Fig. 1(a) shows the interception of a LQR surface, a torque-constrained surface and a hyperboloid due to power constraint. Fig. 1(b) shows LQR surface bounded by all constraints.

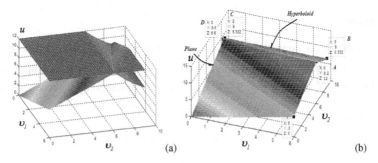

Fig. 1. Computer-aided plot (a) Interception of LQR plane, torque constraint plane and power constraint hyperboloid (b) LQR surface bounded by all constraints

3 Fuzzy Logic Controller (FLC)

Although the approach of class stacking reduces the hyper plane from the LQR controller of *n*-D into a bounded and constrained plane in 3-D, the 3-D controller surface shown in Fig. 1.(b) is still nonlinear. Therefore we may employ a fuzzy-logic-controller (FLC) to implement this 3-D controller surface shown in Fig. 1.(b).

3.1 Fuzzy Inference System (FIS)

A fuzzy logic controller (FLC) [4][5][6] is configured by processing a sequence of fuzzification, inference and defuzzification as follows. The fuzzy inference system (FIS) to be selected is Sugeno model [7] given in Fig. 2 with υ_1 and υ_2 as input fuzzy variables and the control u as known output variable which is bounded, constrained and expressed by plane $s1$: $u = \upsilon_1 + \upsilon_2$, plane $s2$: $u = 12$ (Upper bound) and hyperboloid: $u\upsilon_2 \le 75$ (Inequality constraint).

3.2 Fuzzification

Input fuzzy variables are υ_1 and υ_2. υ_1 has five triangle membership functions, equally partitioned and spaced as shown in Fig. 2. Membership functions are defined by linguistic terms, i.e. S, SM, M, MB and B. Output variable is the control u that has five membership functions, i.e. S+, SM+, M+, MB+ and B as shown in Fig. 2.

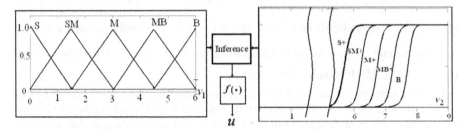

Fig. 2. Fuzzy inference system (FIS) using Sugeno model

3.3 Inference

Output variable is the control u that is composed of plane $s1$ and a hyperboloid. The plane $s1$ is $u = v_1 + v_2$. The hyperboloid $ABCD$ in Fig. 1(b) is approximately replaced by another plane in Fig. 3.(a), named $s2$ and obtained by least square with

$$s2: \; u = \Lambda \begin{bmatrix} v_2 & v_1 & 1 \end{bmatrix}^{\mathrm{T}} \tag{19}$$

$$\Lambda = [-0.8025 \quad 0.2566 \quad 14.7253] \tag{20}$$

Λ is obtained by

$$\Lambda^{\mathrm{T}} = (Y^{\mathrm{T}}Y)^{-1}(Y^{\mathrm{T}}Z) \tag{21}$$

with

$$Y = \begin{bmatrix} v_{2A} & v_{1A} & 1 \\ v_{2B} & v_{1B} & 1 \\ v_{2C} & v_{1C} & 1 \\ v_{2D} & v_{1D} & 1 \end{bmatrix}, Z = \begin{bmatrix} u_A \\ u_B \\ u_C \\ u_D \end{bmatrix} \tag{22}$$

The selected Λ is able to minimize error $\|Z - Y\Lambda\|$.

After finishing the definition of linear planes, we are able to develop the inference that includes ten rules. The former half governs the surface $s2$ and the latter half governs the surface $s1$. Rules are written as follows.

1. If (v_1 is S) and (v_2 is B) then (u is s2); 2. If (v_1 is SM) and (v_2 is MB+) then (u is s2)
3. If (v_1 is M) and (v_2 is M+) then (u is s2); 4. If (v_1 is MB) and (v_2 is SM+) then (u is s2)
5. If (v_1 is B) and (v_2 is S+) then (u is s2); 6. If (v_1 is S) and (v_2 is not B) then (u is s1)
7. If (v_1 is SM) and (v_2 is not MB+) then (u is s1); 8. If (v_1 is M) and (v_2 is not M+) then (u is s1); 9. If (v_1 is MB) and (v_2 is not SM+) then (u is s1); 10. If (v_1 is B) and (v_2 is not S+) then (u is s1).

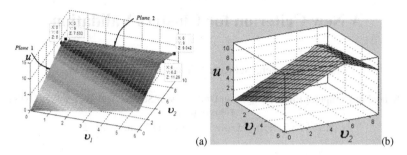

Fig. 3. (a) Hyperboloid is approximately replaced by plane s2 (b) Rule surface after fuzzy defuzzification

3.4 Defuzzification

A rule surface after defuzzification is generated as shown in Fig. 3(b) which is observed to be an approximation of Fig. 3.(a).

4 Conclusion

LQR control surface in hyperspace, subject to power and torque constraints, is difficult to implement. This paper employs class stacking to transform LQR surface in hyper space into a surface in 3-D and further applies a Sugeno-type fuzy logic controller to plot the 3-D nonlinear surface. This approach makes LQR feasible and visible.

References

1. Lewis, F.L.: Applied Optimal Control & Estimation. Prentice Hall, Englewood Cliffs (1992)
2. Burl, J.B.: Linear Optimal Control. Addison-Wesley, Reading (1999)
3. Matlab: toolbox: Fuzzy (2010)
4. Zadeh, L.A.: The concept of a linguistic variable and its application to approximate reasoning. Parts 1, 2, and 3. Information Sciences 8, 199–249, 8, 301–357, 9, 43–80 (1975)
5. Mamdani, E.H., Assilian, S.: An Experiment in Linguistic Synthesis with Fuzzy Logic Controller. Int. J. Man Mach. Studies 7(1), 1–13 (1975)
6. Mamdani, E.H.: Advances in the linguistic synthesis of fuzzy controllers. International Journal of Man-Machine Studies 8, 669–678 (1976)
7. Sugeno, M.: Fuzzy measures and fuzzy integrals: a survey. In: Gupta, M.M., Saridis, G.N., Gaines, B.R. (eds.) Fuzzy Automata and Decision Processes, pp. 89–102. North-Holland, NY (1977)

A New Criterion for Clusters Validation

Hosein Alizadeh, Behrouz Minaei, and Hamid Parvin

School of Computer Engineering, Iran University of Science and Technology (IUST),
Tehran, Iran
{halizadeh,b_minaei,parvin}@iust.ac.ir

Abstract. In this paper a new criterion for clusters validation is proposed. This new cluster validation criterion is used to approximate the goodness of a cluster. The clusters which satisfy a threshold of this measure are selected to participate in clustering ensemble. For combining the chosen clusters, a co-association based consensus function is applied. Since the Evidence Accumulation Clustering method cannot derive the co-association matrix from a subset of clusters, a new EAC based method which is called Extended EAC, EEAC, is applied for constructing the co-association matrix from the subset of clusters. Employing this new cluster validation criterion, the obtained ensemble is evaluated on some well-known and standard data sets. The empirical studies show promising results for the ensemble obtained using the proposed criterion comparing with the ensemble obtained using the standard clusters validation criterion.

Keywords: Clustering Ensemble, Stability Measure, Cluster Evaluation.

1 Introduction

Data clustering or unsupervised learning is an important and very difficult problem. The objective of clustering is to partition a set of unlabeled objects into homogeneous groups or clusters [3]. Clustering techniques require the definition of a similarity measure between patterns. Since there is no prior knowledge about cluster shapes, choosing a specific clustering method is not easy [12]. Studies in the last few years have tended to combinational methods. Cluster ensemble methods attempt to find better and more robust clustering solutions by fusing information from several primary data partitionings [8].

Fern and Lin [4] have suggested a clustering ensemble approach which selects a subset of solutions to form a smaller but better-performing cluster ensemble than using all primary solutions. The ensemble selection method is designed based on quality and diversity, the two factors that have been shown to influence cluster ensemble performance. This method attempts to select a subset of primary partitions which simultaneously has both the highest quality and diversity. The Sum of Normalized Mutual Information, SNMI [5]-[7] and [13] is used to measure the quality of an individual partition with respect to other partitions. Also, the Normalized Mutual Information, NMI, is employed for measuring the diversity among partitions. Although the ensemble size in this method is relatively small, this method achieves

L. Iliadis et al. (Eds.): EANN/AIAI 2011, Part II, IFIP AICT 364, pp. 110–115, 2011.

significant performance improvement over full ensembles. Law et al. proposed a multi objective data clustering method based on the selection of individual clusters produced by several clustering algorithms through an optimization procedure [10]. This technique chooses the best set of objective functions for different parts of the feature space from the results of base clustering algorithms. Fred and Jain [7] have offered a new clustering ensemble method which learns the pairwise similarity between points in order to facilitate a proper partitioning of the data without the a priori knowledge of the number of clusters and of the shape of these clusters. This method which is based on cluster stability evaluates the primary clustering results instead of final clustering.

Alizadeh et al. have discussed the drawbacks of the common approaches and then have proposed a new asymmetric criterion to assess the association between a cluster and a partition which is called Alizadeh-Parvin-Moshki-Minaei criterion, APMM. The APMM criterion compensates the drawbacks of the common method. Also, a clustering ensemble method is proposed which is based on aggregating a subset of primary clusters. This method uses the Average APMM as fitness measure to select a number of clusters. The clusters which satisfy a predefined threshold of the mentioned measure are selected to participate in the clustering ensemble. To combine the chosen clusters, a co-association based consensus function is employed [1].

2   Proposed Method

The main idea of our proposed clustering ensemble framework is utilizing a subset of best performing primary clusters in the ensemble, rather than using all of clusters. Only the clusters which satisfy a stability criterion can participate in the combination. The cluster stability is defined according to Normalized Mutual Information, NMI.

The manner of computing stability is described in the following sections in detail. After, a subset of the most stable clusters is selected for combination. This is simply done by applying a stability-threshold to each cluster. In the next step, the selected clusters are used to construct the co-association matrix. Several methods have been proposed for combination of the primary results [2] and [13]. In our work, some clusters in the primary partitions may be absent (having been eliminated by the stability criterion). Since the original EAC method [5] cannot truly identify the pairwise similarity while there is only a subset of clusters, we present a new method for constructing the co-association matrix. We call this method: Extended Evidence Accumulation Clustering method, EEAC. Finally, we use the hierarchical single-link clustering to extract the final clusters from this matrix.

Since goodness of a cluster is determined by all the data points, the goodness function $g_j(C_i,D)$ depends on both the cluster C_i and the entire dataset D, instead of C_i alone. The stability as measure of cluster goodness is used in [9]. Cluster stability reflects the variation in the clustering results under perturbation of the data by resampling. A stable cluster is one that has a high likelihood of recurrence across multiple applications of the clustering method. Stable clusters are usually preferable, since they are robust with respect to minor changes in the dataset [10].

Now assume that we want to compute the stability of cluster C_i. In this method first a set of partitionings over resampled datasets is provided which is called the reference set. In this notation D is resampled data and $P(D)$ is a partitioning over D. Now, the problem is: "How many times is the cluster C_i repeated in the reference partitions?" Denote by NMI($C_i,P(D)$), the Normalized Mutual Information between the cluster C_i and a reference partition $P(D)$. Most previous works only compare a *partition with another partition* [13]. However, the stability used in [10] evaluates the similarity between a *cluster and a partition* by transforming the cluster C_i to a partition and employing common partition to partition methods. To illustrate this method let $P_1 = P^a = \{C_i, D/C_i\}$ be a partition with two clusters, where D/C_i denotes the set of data points in D that are not in C_i. Then we may compute a second partition $P_2 = P^b = \{C^*, D/C^*\}$, where C^* denotes the union of all "positive" clusters in $P(D)$ and others are in D/C^*. A cluster C_j in $P(D)$ is positive if more than half of its data points are in C_i. Now, define NMI($C_i,P(D)$) by NMI(P^a,P^b) which is defined in [6]. This computation is done between the cluster C_i and all partitions available in the reference set. NMI_i stands for the stability of cluster C_i with respect to the i-th partition in reference set. The total stability of cluster C_i is defined as:

$$Stability(C_i) = \frac{1}{M} \sum_{i=1}^{M} NMI_i \qquad (1)$$

where M is the number of partitions available in reference set. This procedure is applied for each cluster of every primary partition.

Here a drawback of computing stability is introduced and an alternative approach is suggested which is named Max method. Fig. 1 shows two primary partitions for which the stability of each cluster is evaluated. In this example K-means is applied as the base clustering algorithm with K=3. For this example the number of all partitions in the reference set is 40. In 36 partitions the result is relatively similar to Fig 1a, but there are four partitions in which the top left cluster is divided into two clusters, as shown in Fig 1b. Fig 1a shows a true clustering. Since the well separated cluster in the top left corner is repeated several times (90% repetition) in partitionings of the reference set, it has to acquire a great stability value (but not equal to 1), however it acquires the stability value of 1. Because the two clusters in right hand of Fig 1a are

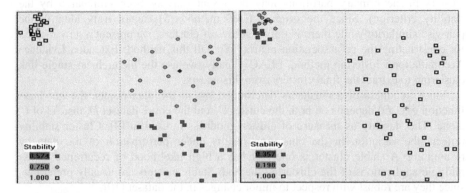

Fig. 1. Two primary partitions with k=3. (a) True clustering. (b) Spurious clustering.

relatively joined and sometimes they are not recognized in the reference set as well, they have less stability value. Fig. 1.b shows a spurious clustering which the two right clusters are incorrectly merged. Since a fixed number of clusters are forced in the base algorithm, the top left cluster is divided into two clusters. Here the drawback of the stability measure is apparent rarely. Although it is obvious that this partition and the corresponding large cluster on the right reference set (10% repetition), the stability of this cluster is evaluated equal to 1. Since the NMI is a symmetric equation, the stability of the top left cluster in Fig 1.a is exactly equal to the large right cluster in Fig 1.b; however they are repeated 90% and 10%, respectively. In other words, when two clusters are complements of each other, their stabilities are always equal. This drawback is seen when the number of positive clusters in the considered partition of reference set is greater than 1. It means when the cluster C* is obtained by merging two or more clusters, undesirable stability effects occur.

To solve this problem we allow only one cluster in reference set to be considered as the C* (i.e. only the most similar cluster) and all others are considered as D/C*. In this method the problem is solved by eliminating the merged clusters.

In the following step, the selected clusters are used to construct the co-association matrix. In the EAC method the m primary results from resampled data are accumulated in an $n{\times}n$ co-association matrix. Each entry in this matrix is computed from this equation:

$$C(i, j) = \frac{n_{i,j}}{m_{i,j}} \qquad (2)$$

where n_{ij} counts the number of clusters shared by objects with indices i and j in the partitions over the primary B clusterings. Also m_{ij} is the number of partitions where this pair of objects is simultaneously present. There are only a fraction of all primary clusters available, after thresholding. So, the common EAC method cannot truly recognize the pairwise similarity for computing the co-association matrix. In our novel method (Extended Evidence Accumulation Clustering, or EEAC) each entry of the co-association matrix is computed by:

$$C(i, j) = \frac{n_{i,j}}{\max(n_i, n_j)} \qquad (3)$$

where n_i and n_j are the number present in remaining (after stability thresholding) clusters for the i-th and j-th data points, respectively. Also, n_{ij} counts the number of remaining clusters which are shared by both data points indexed by i and j, respectively.

3 Experimental Results

This section reports and discusses the empirical studies. The proposed method is examined over 5 different standard datasets. Brief information about the used datasets is available in [11]. All experiments are done over the normalized features. It means each feature is normalized with mean of 0 and variance of 1, N(0, 1). All of them are reported over means of 10 independent runs of algorithm. The final performance of the clustering algorithms is evaluated by re-labeling between obtained clusters and the ground truth labels and then counting the percentage of the true classified samples. Table 1 shows the performance of the proposed method comparing with most common base and ensemble methods.

The first four columns of Table 1 are the results of some base clustering algorithms. The results show that although each of these algorithms can obtain a good result over a specific dataset, it does not perform well over other datasets. The four last columns show the performance of some ensemble methods in comparison with the proposed one. Taking a glance at the last four columns in comparison with the first four columns shows that the ensemble methods do better than the simple based algorithms in the case of performance and robustness along with different datasets. The first column of the ensemble methods is the results of an ensemble of 100 K-means which is fused by EAC method. The 90% sampling from dataset is used for creating diversity in primary results. The sub-sampling (without replacement) is used as the sampling method. Also the random initialization of the seed points of K-means algorithm helps them to be more diverse. The single linkage algorithm is applied as consensus function for deriving the final clusters from co-association matrix. The second column from ensemble methods is the full ensemble which uses several clustering algorithms for generating the primary results. Here, 70 K-means with the above mentioned parameters in addition to 30 linkage methods provide the primary results. The third column of *Ensemble Methods* is consensus partitioning using EEAC algorithm of top 33% stable clusters, employing NMI method as measure of stability. The fourth column of the ensemble methods is also consensus partitioning using EEAC algorithm of top 33% stable clusters, employing max method as measure of stability.

Table 1. Experimental results

Dataset	Simple Methods (%)				Ensemble Methods (%)			
	Single Linkage	Average Linkage	Complete Linkage	Kmeans	Kmeans Ensemble	Full Ensemble	Cluster Selection by NMI Method	Cluster Selection by max Method
Wine	37.64	38.76	83.71	96.63	96.63	97.08	**97.75**	97.44
Breast-C	65.15	70.13	94.73	95.37	95.46	95.10	95.75	**96.49**
Yeast	34.38	35.11	38.91	40.20	45.46	47.17	47.17	**51.27**
Glass	36.45	37.85	40.65	45.28	47.01	47.83	**48.13**	47.35
Bupa	57.68	57.10	55.94	54.64	54.49	55.83	58.09	**58.40**

4   Conclusion and Future Works

In this paper a new clustering ensemble framework is proposed which is based on a subset of total primary spurious clusters. Also a new alternative method for common

NMI is suggested. Since the quality of the primary clusters are not equal and presence of some of them can even yield to lower performance, here a method to select a subset of more effective clusters is proposed. A common cluster validity criterion which is needed to derive this subset is based on normalized mutual information. In this paper some drawbacks of this criterion is discussed and a method is suggested which is called max mehod. The experiments show that the proposed framework commonly outperforms in comparison with the full ensemble; however it uses just 33% of primary clusters. Also the proposed max criterion does slightly better than NMI criterion generally. Because of the symmetry which is concealed in NMI criterion and also in NMI based stability, it yields to lower performance whenever symmetry is also appeared in the dataset. Another innovation of this chapter is a method for constructing the co-association matrix where some of clusters and respectively some of samples do not exist in partitions. This new method is called Extended Evidence Accumulation Clustering, EEAC.

References

1. Alizadeh, H., Minaei-Bidgoli, B., Parvin, H., Mohsen, M.: An Asymmetric Criterion for Cluster Validation. In: International Conference on Industrial, Engineering and Other Applications of Applied Intelligent Systems (IEA/AIE 2011). LNCS. Springer, Heidelberg (2011)
2. Ayad, H., Kamel, M.S.: Cumulative Voting Consensus Method for Partitions with a Variable Number of Clusters. IEEE Trans. on Pattern Analysis and Machine Intelligence 30(1), 160–173 (2008)
3. Faceli, K., Marcilio, C.P., Souto, D.: Multi-objective Clustering Ensemble. In: Proceedings of the Sixth International Conference on Hybrid Intelligent Systems, HIS 2006 (2006)
4. Fern, X.Z., Lin, W.: Cluster Ensemble Selection. In: Jonker, W., Petković, M. (eds.) SDM 2008. LNCS, vol. 5159. Springer, Heidelberg (2008)
5. Fred, A., Jain, A.K.: Data Clustering Using Evidence Accumulation. In: Proc. of the 16th Intl. Conf. on Pattern Recognition, ICPR 2002, Quebec City, pp. 276–280 (2002)
6. Fred, A., Jain, A.K.: Combining Multiple Clusterings Using Evidence Accumulation. IEEE Trans. on Pattern Analysis and Machine Intelligence 27(6), 835–850 (2005)
7. Fred, A., Jain, A.K.: Learning Pairwise Similarity for Data Clustering. In: Proc. of the 18th Int. Conf. on Pattern Recognition, ICPR 2006 (2006)
8. Fred, A., Lourenco, A.: Cluster Ensemble Methods: from Single Clusterings to Combined Solutions. Studies in Computational Intelligence (SCI) 126, 3–30 (2008)
9. Lange, T., Braun, M.L., Roth, V., Buhmann, J.M.: Stability-based model selection. In: Advances in Neural Information Processing Systems, vol. 15. MIT Press, Cambridge (2003)
10. Law, M.H.C., Topchy, A.P., Jain, A.K.: Multiobjective data clustering. In: Proc. of IEEE Conference on Computer Vision and Pattern Recognition, vol. 2, pp. 424–430 (2004)
11. Newman, C.B.D.J., Hettich, S., Merz, C.: UCI repository of machine learning databases (1998), http://www.ics.uci.edu/~mlearn/MLSummary.html
12. Roth, V., Lange, T., Braun, M., Buhmann, J.: A Resampling Approach to Cluster Validation. In: Intl. Conf. on Computational Statistics, COMPSTAT (2002)
13. Strehl, A., Ghosh, J.: Cluster ensembles - a knowledge reuse framework for combining multiple partitions. Journal of Machine Learning Research 3, 583–617 (2002)

Modeling and Dynamic Analysis on Animals' Repeated Learning Process

Mu Lin[1,2,*], Jinqiao Yang[3], and Bin Xu[2]

[1] School of Applied Mathematics, Central University of Finance and Economics,
Beijing, 100081, China
[2] Department of Mathematical Sciences, Tsinghua University, Beijing, 10084, China
mlin@math.tsinghua.edu.cn
[3] School of Humanities, Dongbei University of Finance and Economics, Dalian,
116025, China

Abstract. Dynamical modeling is used to describe the process of animals' repeated learning. Theoretical analysis is done to explore the dynamic property of this process, such as the limit sets and their stability. The scope of variables is provided for different practical purpose, cooperated with necessary numerical simulation.

Keywords: Dynamical modeling, animals' learning, animal training, piecewise-smooth system.

1 Introduction

Learning is always a hot issue in pedagogy and psychology research. J. Dewey considered learning as the process of individual activities, from the perspective of Pragmatism [1]. He proposed "learning by doing", and emphasized the effect of specific situation on the individual learning behavior. However, from the perspective of Constructivism, J. Piaget considered learning as a kind of interaction process between individual and the environment surrounding him, which makes his cognitive structure develop constantly and gradually [2].

We can see different learning behaviors everywhere. One common adopted way is repeated learning, such like reciting foreign language words repeatedly to remember them, and doing exercises frequently to learn driving. [3]. This process can make some adaptable changes on the mental level. And one of the most important changes is, learners can abstract an "schema" from the operating process of the skill, and this is so called "transfer of learning" [4].

In modern psychology, learning is such a kind of behavior which is not only owned by human beings, but also by animals. Lots of researchers studied animals' learning behavior. In experiments, E. Thorndike observed the escaping behavior of a cat which was trapped in a box. He argued that the cat keep running aimlessly at first but gradually drop the ways that cannot open the box and finally acquire the right way to escape [5]. B. Skinner observed the bar-pressing

---

[*] Corresponding author.

L. Iliadis et al. (Eds.): EANN/AIAI 2011, Part II, IFIP AICT 364, pp. 116–121, 2011.

behavior of a mouse in the box, which can serve itself food, and argued that the animals reacted to the reinforcement in particular ways. And this "reinforcement - reaction" connection is learned by the animals themselves [6].

There are a lot of studies of learning behavior of animals, although the learning contents maybe different, but the way of learning is always repeated learning, for it is widely believed that animals' learning ability is poor. Furthermore, lots of studies seem to tell us that based on this level of learning ability, the animals can perform the transfer of learning like human being. For example, in the Pavlovian classical experiments, the dog acquired the conditioned reflex of "bell ringing -salivating" [7]. So, if we make an animal learn a kind of behavior repeatedly, it indeed can influence the generation of other related behaviors. But if this influence exists, how it works? And how strong it can be? Both of these questions are important because they can help us know the nature of learning deeper. The present study will answer these questions, and try to provide a mathematical model of relationship between these two behaviors.

Mathematical modeling is in psychology and society has rapid development these years [8]. Social mechanisms are analyzed by analytical and numerical approaches. For example, He et al. found a cellular automata model to describe the learning process [9]. Dynamic analysis is used for modeling the problems that have definite dynamic relation or feedback relationship. It has been widely used to solve ecology and demography problems. There are obviously two sub-processes in this repeated learning problem, i.e., the learning are forgetting states. It belongs to the piecewise-smooth problems [10].

The remainder of this paper is as follows. Section 2 is the mathematical modeling process, and three types of switching rules are enumerated in Section 3. More precise dynamic analysis including some numerical simulation is done in Section 4. Section 5 is the conclusion.

2 Modeling Framework

We establish a dynamic model in this section first. As mentioned above, for artificial animals' repeated learning process, there are obviously two discrete states, the training condition and forgetting condition, noted as $Q = 1$ and $Q = -1$ separately. Note x as the main memory ability, and y as the relevance memory ability, $0 \le x, y \le 1$.

In the training condition, following the logistic model, the dynamic function of the main memory can be described as

$$\dot{x} = \alpha_1 x(1 - x). \tag{1}$$

While in the forgetting condition, by Ebbinghaus forgetting curve, the dynamic function of main memory ability can be described as

$$\dot{x} = -\beta_1 x. \tag{2}$$

The variation of the relevance memory ability is related with the main memory ability. In the training condition, it can be described as

$$\dot{y} = \alpha_2(x - l_1)y(1 - y), \tag{3}$$

where $\alpha_2 > 0$. There is one extra term $(x - l_1)$ describing the influence of the main memory ability to the related memory ability. More concretely, there is a separation $x = l_1$, $0 < l_1 < 1$, such that the relevance memory ability increases when the main memory ability $x > l_1$ and reduces when $x < l_1$. Similarly, the equation is

$$\dot{y} = -\beta_2(l_2 - x)y(1 - y) \tag{4}$$

in the forgetting condition, where $0 < l_1 < l_2 < 1, \beta_2 > 0$.

The explicit solution of the simultaneous equations (1)-(4) is

$$\begin{cases} x(t) = 1 - \frac{1}{1+e^{\alpha_1 t + c_1}} \\ y(t) = 1 - \frac{1}{1+(1+e^{\alpha_1 t + c_1})^{\frac{\alpha_2}{\alpha_1}} e^{-\alpha_2 l_1 t + c_2}} \end{cases} \quad \text{when } Q = 1;$$

$$\begin{cases} x(t) = c_3 e^{-\beta_1 t} \\ y(t) = 1 - \frac{1}{1+e^{-\beta_2 l_2 t - c_3 \frac{\beta_2}{\beta_1} e^{-\beta_1 t} + c_4}} \end{cases} \quad \text{when } Q = -1, \tag{5}$$

where $c_i, i = 1, 2, 3, 4$ are arbitrary constants.

In order to get a complete deterministic system, a switching rule between the training and forgetting conditions is needed. Actually, there are four following types, as listed in the following section.

3   Dynamic Analysis under Different Switching Rules

3.1   Quantitative Control Switching Rule

The quantitative control switching rule contains two preestablished bounds w_1 and w_2. During the training condition($Q = 1$), the system switches to the forgetting condition($Q = -1$) when x reaches w_2. Vice versa. So it makes a periodic learning cycle consist of one training condition and one forgetting condition, See equation (6).

$$Q(t^+) = \begin{cases} -Q(t) & \text{if } Q(t) = -1 \text{ and } x(t) \leq w_1, \text{ or } Q(t) = 1 \text{ and } x(t) \geq w_2 \\ Q(t) & \text{else} \end{cases},$$

$$\tag{6}$$

Composed the above switching rule and equation (1)-(4), a deterministic piecewise-smooth system is obtained. Usually, the analysis of piecewise-smooth systems is quite complicated, while our model is a providential one for it has an explicit solution. Assume the point (w_1, y) changes to $(w_1, f(y))$ in a whole training-forgetting period. By computation ,we get $f(y) = \frac{Ay}{(A-1)y+1}$, where $A = [(\frac{w_1}{w_2})^{l_2} e^{w_2-w_1}]^{\beta} [(\frac{w_1}{w_2})^{l_1} (\frac{1-w_1}{1-w_2})^{1-l_1}]^{\alpha} > 0$, and $\alpha = \frac{\alpha_2}{\alpha_1}, \beta = \frac{\beta_2}{\beta_1}$. So $f'(y) = \frac{A}{[(A-1)y+1]^2}$.

Consider the equilibrium point of map f, i.e. the point y satisfies $f(y) = y$. The equilibrium point of f indicates the period solution of the whole system, meaning that the collateral memory ability changes periodically with the main memory ability. Solving the equation $f(y) = y$, we get $A = 1$ or $y = 0, 1$. For $y = 0$ and $y = 1$ are two trivial solutions, we focus on the case $A = 1$, then

$$\alpha(1-l_1)\ln(1-w_1)+(\alpha l_1+\beta l_2)\ln w_1-\beta w_1 = \alpha(1-l_1)ln(1-w_2)+(\alpha l_1+\beta l_2)\ln w_2-\beta w_2$$

Note $g(w) = \alpha(1-l_1)\ln(1-w)+(\alpha l_1+\beta l_2)\ln w-\beta w$, then $g(0) = g(1) = -\infty$. The solution of $g'(w) = 0$ is $w_0 = l_2 - \frac{\sqrt{(\alpha+\beta-\beta l_2)^2+4\alpha\beta(l_2-l_1)}-(\alpha+\beta-\beta l_2)}{2\beta}$. A group of parameters $\alpha_i, \beta_i, l_i, i = 1, 2$ guaranteeing $A = 1$ may exist for w_1, w_2 satisfied $0 < w_1 < w_0 < w_2 < 1$ can be easily chosen. That means $f(y) \equiv y, \forall 0 < y < 1$, i.e., the collateral memory ability can keep a stable circle at any ability.

Otherwise if $A > 1$, then $f'(0) = A > 1$, so $y = 0$ is an unstable equilibrium point, while $f'(1) = 1/A < 1$, so $y = 1$ is a stable equilibrium point. That is to say, under this quantitative control switching rule, orbits start from every initial point except $x = 0$ will converge to $y = 1$ after sufficient times of training-forgetting cycling, i.e. the collateral memory ability will converge to the upper bound. Oppositely, if $A < 1$, $y = 0$ is a stable equilibrium point while $y = 1$ is an unstable equilibrium point. So the collateral memory vanishes no matter how the initial ability was.

3.2 Timing Control Switching Rule

Another common switching rule is based on timing control. In this case, each time length of training and forgetting condition is fixed, noted as τ_1 and τ_2 separately. In most cases $\tau_2 > \tau_1 > 0$.

The Poincaré map of the system can also be computed explicitly in this case. The system starts from (x_0, y_0) and changes to

$$x(\tau_1 + \tau_2) = \frac{e^{-\beta_1\tau_2}}{1+(\frac{1}{x_0}-1)e^{-\alpha_1\tau_1}} = x_0 \cdot \frac{1}{e^{\beta_1\tau_2}[x_0(1-e^{-\alpha_1\tau_1})-(e^{-\beta_1\tau_2}-e^{-\alpha_1\tau_1})]+1}$$

and $y(\tau_1 + \tau_2) = \frac{M}{M+N(\frac{1}{y_0}-1)}$, where $M = e^{-\beta_2 l_2\tau_2 - \frac{\beta_2}{\beta_1}x_1 e^{-\beta_1\tau_2}}$,

$$N = \frac{(x_0)^{\frac{\alpha_2}{\alpha_1}} e^{\frac{\beta_2}{\beta_1}x_1}}{(1+(\frac{1}{x_0}-1)e^{-\alpha_1\tau_1})^{\frac{\alpha_2}{\alpha_1}} e^{\alpha_2(1-l_1)\tau_1}}, \text{ after a period.}$$

So the equilibrium point is $x = 0$ or $x = \frac{e^{-\beta_1\tau_2}-e^{-\alpha_1\tau_1}}{1-e^{-\alpha_1\tau_1}}$. For the model located in $x \geq 0$, there is only one equilibrium point $x = 0$ and it is stable, when $\alpha_1\tau_1 \leq \beta_1\tau_2$. There are one unstable equilibrium point $x = 0$ and one stable equilibrium point $x(0) = \frac{e^{-\beta_1\tau_2}-e^{-\alpha_1\tau_1}}{1-e^{-\alpha_1\tau_1}}$.

In this case, $y(0) = y(\tau_1 + \tau_2), \forall y(0)$ if and only if $M - N = 0$. Otherwise the system has two equilibrium points $y = 0$ and $y = 1$. If $M > N$, 0 is unstable and 1 is stable, while if $M < N$, 0 is stable and 1 is unstable.

3.3 Semi-timing Control Switching Rule

Composing the above two rules, the semi-timing control switching rule is also widely used in actual. This rule fixed simultaneously the upper bound (or lower bound) and the forgetting time length (or the training time length). The corresponding Poincare map is:

(1)The system starts from (w_1, y_0), goes through a τ_1 time training condition and reaches (x_1, y_1), then goes back to (w_1, y_2) through a forgetting condition. By computation, we get $y_2 = \frac{M}{M+N(\frac{1}{y_0}-1)}$, where $M = (w_1 + (1 - w_1)e^{-\alpha_1\tau_1})^{(\frac{\beta_2 l_2}{\beta_1}+\frac{\alpha_2}{\alpha_1})}e^{\alpha_2(1-l_1)\tau_1}$, $N = e^{-\frac{\beta_2}{\beta_1}\frac{1}{w_1+(1-w_1)e^{-\alpha_1\tau_1}}}$.

(2)The system starts from (w_2, y_0), goes through a τ_2 time forgetting condition and reaches (x_1, y_1), then goes back to (w_2, y_2) through a training condition. Here $y_2 = \frac{M}{M+N(\frac{1}{y_0}-1)}$, where

$$M = (1+(\frac{e^{\beta_1\tau_2}}{w_2}-1)e^{-\alpha_1\tau_1})^{\frac{\alpha_2}{\alpha_1}}(\frac{1-w_2}{e^{\beta_1\tau_2}-w_2})^{-\frac{\alpha_2}{\alpha_1}(1-l_1)}, N = x_0^{-\frac{\alpha_2}{\alpha_1}}e^{\beta_2 l_2\tau_2+\frac{\beta_1}{\beta_2}w_2(e^{-\beta_1\tau_2}-1)}.$$

In both two subcases, the dynamic property of y is similar with the one in the timing control switching rule.

4   The Ultimate Stable States and Bifurcation Analysis

Although infinite multiple stationary states are existence in some cases of the above analysis, most of them are unstable in fact, for the observation is discrete. More specifically, throughout the above analysis, we assume that the system switches instantaneously whenever the main memorial ability reaches the preestablished value in Section 3.1 and 3.3. Actually, this instant transformation cannot be realized for the estimation of memorial ability must be discrete, furthermore, it costs time.

To describe the influence of the discrete observation, the switching rule can be rewritten as: (take the quantitative control switching rule in Section 3.1 for instance)

$$Q(t^+) = \begin{cases} Q(t) & \text{if } t_n \leq t < t_{n+1} \\ -Q(t) & \text{else} \end{cases}, \tag{7}$$

where $t_i, i = 0, 1, \ldots$ is a group of switching time series. Usually they are determined mainly depending on a presupposed admissible error $\varepsilon > 0$. And the

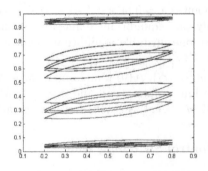

Fig. 1. The ultimate stable states under the perturbation of measurement error

system switches when the observed variable enters a presupposed neighborhood of the border line. That is to say, if Δt is the observed interval time length,

$$t_i = k_i \Delta t, s.t. |x_i(k\Delta t) - w_1| < \varepsilon, \text{ or } |x_i(k\Delta t) - w_2| < \varepsilon, k_i \in \mathbb{N}.$$

Under this switching rule, the ultimate stable stationary states can be drawn by numerical simulation, as in Figure 1. Unlike the results given in Section 3.1, there are only finite stable stationary states, means only these states can exist under the discrete observation.

5   Conclusion

Mathematical model is established to discuss the relationship between the main memory ability and the relevance memory ability. Different switching rules are introduced and under each rule, equilibria and their stability are discussed. Although with some particular parameters, every orbit can be a periodic invariable one, but there can only be finite periodic orbits, due to the discrete observation, as explained concretely in Section 4. The results of dynamic analysis can be used for animal training. Further observational study can be done to demonstrate the models.

Acknowledgements. This study was funded by the National Natural Science Foundation of China(11072274).

References

1. Dewey, J.: Democracy and Education: An Introduction to the Philosophy of Education. Macmilan Company, NYC (1916)
2. Piaget, J.: The Psychology of Intelligence. Taylor and Francis Press, Abington (2001)
3. Schmidt, R.A.: A schema theory of discrete motor skill learning. Psychological Review 82(4), 225–260 (1975)
4. Royer, J.M.: Theories of the transfer of learning. Educational Psychologist 14(1), 53–69 (1979)
5. Thorndike, E.: Animal Intelligence. Thoemmes Press (1911)
6. Charles, B., Fester, B.F., Skinner, C.D., Cheney, W.H., Morse, P.B.D.: Schedules of Reinforcement. B. F. Skinner Reprint Series. Copley Publishing Group (1997)
7. Chance, P.: Learning and Behavior: Active Learning Edition. Wadsworth Publishing Co. Inc., Belmont (2008)
8. Burghes, D.N., Borrie, M.S.: Modelling with Differential Equations. Horwood, Chichester (1981)
9. He, M., Deng, C., Feng, L., Tian, B.: A cellular automata model for a learning process. Advances in Complex Systems 07(03-04), 433–439 (2004)
10. Kunze, M.: Non-smooth Dynamical Systems. Springer, Heidelberg (2000)

Generalized Bayesian Pursuit: A Novel Scheme for Multi-Armed Bernoulli Bandit Problems

Xuan Zhang[1], B. John Oommen[2,1,*], and Ole-Christoffer Granmo[1]

[1] Dept. of ICT, University of Agder, Grimstad, Norway
[2] School of Computer Science, Carleton University, Ottawa, Canada

Abstract. In the last decades, a myriad of approaches to the multi-armed bandit problem have appeared in several different fields. The current top performing algorithms from the field of Learning Automata reside in the Pursuit family, while UCB-Tuned and the ε-greedy class of algorithms can be seen as state-of-the-art regret minimizing algorithms. Recently, however, the Bayesian Learning Automaton (BLA) outperformed all of these, and other schemes, in a wide range of experiments. Although seemingly incompatible, in this paper we integrate the foundational learning principles motivating the design of the BLA, with the principles of the so-called Generalized Pursuit algorithm (GPST), leading to the *Generalized Bayesian Pursuit* algorithm (GBPST). As in the BLA, the estimates are truly Bayesian in nature, however, instead of basing exploration upon direct sampling from the estimates, GBPST explores by means of the arm selection probability vector of GPST. Further, as in the GPST, in the interest of higher rates of learning, a *set of arms* that are currently perceived as being optimal is pursued to minimize the probability of pursuing a wrong arm. It turns out that GBPST is superior to GPST and that it even performs better than the BLA by controlling the learning speed of GBPST. We thus believe that GBPST constitutes a new avenue of research, in which the performance benefits of the GPST and the BLA are mutually augmented, opening up for improved performance in a number of applications, currently being tested.

Keywords: Bandit Problems, Estimator Algorithms, Generalized Bayesian Pursuit Algorithm, *Beta* Distribution, Conjugate Priors.

1 Introduction

The multi-armed Bernoulli bandit problem (MABB) is a classical optimization problem that captures the exploration-exploitation dilemma. The MABB setup consists of a gambling machine with multiple arms and an agent that sequentially pulls one of the arms, with each pull resulting in either a *reward* or a *penalty*[1]. The sequence of rewards/penalties obtained from each arm i forms a Bernoulli process with an *unknown* reward probability d_i, and a penalty probability $1 - d_i$. The dilemma is this: Should the arm that so far seems to provide the highest chance of reward be pulled once more, or

---

[*] *Chancellor's Professor*; *Fellow: IEEE* and *Fellow: IAPR*. The Author also holds an *Adjunct Professorship* with the Dept. of ICT, University of Agder, Norway.

[1] A *penalty* may also be perceived as the absence of a *reward*. However, we choose to use the term *penalty* as is customary in the LA and RL literature.

L. Iliadis et al. (Eds.): EANN/AIAI 2011, Part II, IFIP AICT 364, pp. 122–131, 2011.

should an inferior arm be pulled to learn more about *its* reward probability? Sticking prematurely with the arm that is presently considered to be the best one, may lead to not discovering which arm is truly optimal. On the other hand, lingering with an inferior arm unnecessarily, postpones the harvest that can be obtained from the optimal arm.

1.1 Existing Solutions to Multi-Armed Bernoulli Bandit Problem

Bandit like problems involve two highly related yet distinct fields: the field of Learning Automata and the field of Bandit Playing Algorithms. A myriad of approaches have been proposed within these two fields. Classical exact solutions for discounted rewards take advantage of Gittins Indices [1, 2, 3] — by always pulling the arm with the largest Gittins index (measuring the value associated with the state of a stochastic process), the expected amount of discounted rewards obtained is maximized. Calculating Gittins Indices in a computationally efficient manner is far from trivial [4,5], and this problem is currently being pursued [6,7]. Because of the computational difficulties associated with exact solutions based on Gittins Indices, a number of approximate solution techniques has also been proposed. The ε-*greedy* strategy, first described by Watkins [8], represents an early *approximate* solution to the bandit problem, in which the arm so far being perceived as the best is pulled with probability $1 - \varepsilon$, and a randomly chosen action is pulled with probability ε. Thus, the expected frequency of exploring a random action is determined by the parameter of ε. A variant of ε-*greedy* strategy is the ε-*decreasing* strategy [9, 10], which gradually shifts focus from exploration to exploitation by slowly decreasing ε. Recently, Tokic proposed an *adaptive* ε-greedy strategy based on reward value differences (VDBE) [11]. In this strategy, ε decreases on the basis of changes in the reward value estimates.

Another direction for solving the bandit problem is confidence interval based algorithms. They estimate confidence intervals for the reward probabilities, and identify an "optimistic" reward probability estimate for each arm. The arm with the most optimistic reward probability estimate is then greedily selected [12, 13]. Furthermore, Auer et al. has shown that variants of confidence interval based algorithms, the so-called UCB and UCB-Tuned schemes, provide a logarithmic regret bound [10].

There are also algorithms for solving MABB problems that are based on so-called Boltzmann exploration. These introduce the parameter τ as the "temperature" of the exploration. Related schemes include EXP3 [14]. The "Price of Knowledge and Estimated Reward" (POKER) algorithm proposed in [12] takes into consideration pricing of uncertainty, exploiting the arm reward distributions, and the horizon of the problem. The Linear Reward-Inaction (L_{RI}) learning automaton [15] and the Pursuit Scheme based on Linear Reward-Inaction philosophy (PST_{RI}) [16] [17] are known for their ε-optimality. Besides, PST_{RI}, which uses Maximum Likelihood (ML) reward probability estimates to pursue the currently optimal action, is known for its pioneering role in the estimator Learning Automata family. A thorough comparison of several of the above mentioned schemes can be found in [18, 12].

There also exists several algorithms based on Bayesian reasoning [19], with [20, 21] being a few examples. In general, Bayesian reasoning is in many cases computational intractable for bandit like problems [19]. However, based on the Thompson sampling principle [22], the *Bayesian Learning Automata (BLA)* reported in [23, 18] and extended in [24] to deal with non-stationary environments, are inherently Bayesian in nature, yet avoids computational intractability by relying simply on updating the hyper-parameters of sibling conjugate distributions, and on simultaneously sampling randomly from the respective posteriors. As seen in [18] and [24], BLA demonstrate significant performance advantage compared to a number of competing MABB solution schemes.

1.2   Contributions and Paper Organization

In this paper, we propose a new Bayesian algorithm for MABB problems, which we refer to as the Generalized Bayesian Pursuit algorithm (GBPST). GBPST augments the philosophy of BLA with the principles behind the GPST [25] and is able to outperform *both* the GPST as well as the BLA schemes in extensive experiments[2]. This augmentation is achieved as follows: Firstly, as in the BLA, the estimates are truly Bayesian (as opposed to ML in GPST) in nature. However, the arm selection probability vector of GPST is used for exploration purposes. Secondly, as opposed to the ML estimate, which is usually a single value - the one which maximizes the likelihood function - the use of a posterior distribution permits us to choose any one of a *spectrum* of values in the posterior, as the appropriate estimate. In the interest of being concrete, we have chosen a 95% percentile value of the posterior (instead of the mean) to pursue promising arms. Thirdly, as in the GPST, after each arm pull, all arms currently being associated with a higher reward estimate than the currently pulled arm, are pursued. Finally, the pursuit is done using the Linear Reward-Inaction philosophy, leading to the corresponding GBPST$_{RI}$ scheme[3]. To the best of our knowledge, all these contributions are novel to the field of MABB problem, and we thus believe that the GBPST constitutes a new avenue of research, in which the performance benefits of the GPST and the BLA are mutually augmented. We also believe that the theoretical contributions of this paper could lend itself to practical solutions improving performance in a number of applications, some of which are currently being tested.

The paper is organized as follows. In Section 2 we give an overview of GPST and BLA. Then, in Section 3, we present the new Bayesian estimator algorithm – the *Generalized Bayesian Pursuit* algorithm – by incorporating GPST and BLA. In Section 4, we provide extensive experimental results demonstrating that the GBPST is truly superior to GPST. The BLA scheme is also outperformed by appropriately choosing a learning speed parameter for the GBPST$_{RI}$. Finally, in Section 5, we report opportunities for further research, in addition to providing concluding remarks.

2   The Generalized Pursuit Algorithm and Bayesian Learning Automata

We here briefly review the selected schemes upon which the Generalized Bayesian Pursuit scheme builds.

---

[2] The theoretical results concerning the formal properties of the family of GBPST are currently being compiled.

[3] The pursuit can also be conducted using the Linear Reward-Penalty philosophy (GBPST$_{RP}$), as advocated in [17]. In the interest of brevity, we here report the best performing scheme, which is GBPST$_{RI}$.

Linear Updating Schemes: The more notable and well-used traditional LA approaches include the family of linear updating schemes, with the Linear Reward-Inaction (L_{RI}) automaton being designed for stationary environments [15]. In short, the L_{RI} maintains an arm selection probability vector $\bar{p} = [p_1, p_2, ..., p_r]$, with $\sum_{i=1}^{r} p_i = 1$ and r being the number of arms. The question of which arm is to be pulled is decided randomly by sampling from $\bar{p}$. Initially, $\bar{p}$ is uniform. The following linear updating rules summarize how rewards and penalties affect $\bar{p}$ with p'_i and p'_j being the resulting updated arm selection probabilities:

$$p'_j = (1 - \lambda) \times p_j, 1 \le j \le r, j \ne i$$
$$p'_i = 1 - \sum_{j \ne i} p'_j \text{ if pulling Arm } i \text{ results in a reward.}$$
$$p'_j = p_j, 1 \le j \le r \text{ if pulling Arm } j \text{ results in a penalty.}$$

In the above, the parameter λ ($0 < \lambda < 1$) governs the learning speed. As seen, after arm i has been pulled, the associated probability p_i is increased using the linear updating rule upon receiving a reward, with $p_j(j \ne i)$ being decreased correspondingly. Note that $\bar{p}$ is left unchanged upon a penalty.

Pursuit Schemes: A Pursuit scheme (PST) makes the updating of $\bar{p}$ more goal-directed in the sense that it maintains ML estimates $(\widehat{d_i})$ of the reward probabilities (d_i) associated with each arm. In brief, a Pursuit scheme increases the arm selection probability p_i associated with the currently largest ML estimate $\widehat{d_i}$, instead of the arm actually producing the reward. Thus, unlike L_{RI}, in which the reward from an inferior arm can cause unsuitable probability updates, in the Pursuit scheme, these rewards will not influence the learning progress in the short term, except by modifying the estimate of the reward vector. This, of course, assumes that the ranking of the ML estimates are correct, which is what it will be if each arm is chosen a "sufficiently large number of times". Accordingly, a Pursuit scheme consistently outperforms the L_{RI} in terms of its rate of convergence.

Generalized Pursuit Schemes: The Generalized Pursuit schemes (GPST) generalizes PST by allowing several arms to be pursued at the same time. Instead of only pursuing the arm with the highest reward estimate, the whole *set* of arms that possess higher reward estimates than the arm actually pulled is pursued. In PST, when the arm with the maximum reward estimate is not the one with the highest reward probability, the incorrect arm is pursued, thus potentially derailing the pursuit of the optimal action. The probability of this happening is reduced in GPST.

Bayesian Learning Automata: A unique feature of the *Bayesian Learning Automaton* (BLA) is its computational simplicity, achieved by relying *implicitly* on Bayesian reasoning principles. In essence, at the heart of the BLA we find the *Beta distribution*, which is the conjugate prior for the Bernoulli distribution. Its shape is determined by two positive parameters, denoted by a and b, producing the following probability density function:

$$f(x;a,b) = \frac{x^{a-1}(1-x)^{b-1}}{\int_0^1 u^{a-1}(1-u)^{b-1}\,du}, \quad x \in [0,1]. \tag{1}$$

Essentially, the BLA uses the *Beta* distribution for two purposes. First of all, it is used to provide a *Bayesian estimate* of the reward probabilities associated with each of the available arms - the latter being valid by virtue of the conjugate prior nature of the Binomial parameter. Secondly, a novel feature of the BLA is that it uses the *Beta* distribution as the basis for an *Order-of-Statistics*-based *randomized* selection mechanism.

3   The Generalized Bayesian Pursuit Algorithm (GBPST)

Bayesian reasoning is a probabilistic approach to inference which is of significant importance in machine learning because it allows for the *quantitative* weighting of evidence supporting alternative hypotheses, with the purpose of allowing optimal decisions to be made. Furthermore, it provides a framework for analyzing learning algorithms [26]. We present here a completely new estimator algorithm that builds upon the GPST framework. However, rather than utilizing ML reward probability estimates, optimistic Bayesian estimates are used to pursue the arms currently perceived to be *potentially* optimal. We thus coin the algorithm *Generalized Bayesian Pursuit* (GBPST).

As in the case of the BLA, the GBPST estimates the reward probability of each arm based on the *Beta* distribution. These Bayesian estimates allow us to accurately calculate an optimistic reward probability x_i that provides a 95% upper bound for the reward probability of arm i, by means of the respective cumulative distribution $F(x_i;a,b)$.

$$F(x_i;a,b) = \frac{\int_0^{x_i} v^{a-1}(1-v)^{b-1}\,dv}{\int_0^1 u^{a-1}(1-u)^{b-1}\,du}, \quad x_i \in [0,1]. \tag{2}$$

The following algorithm contains the essence of the GBPST approach.

Algorithm: GBPST$_{RI}$
Parameters:
α: The arm chosen by LA.
p_i: The ith element of the arm selection probability vector P.
λ: The learning speed, where $0 < \lambda < 1$.
a_i, b_i: The two positive parameters of the *Beta* distribution.
x_i: The ith element of the Bayesian estimate vector X, given by the 95% upper bound of the cumulative distribution function of the corresponding *Beta* distribution.
R: The response from the environment, where $R = 0$ (reward) or $R = 1$ (penalty).
Initialization:
1. $p_i(t) = 1/r$, where r is the number of arms.
2. Set $a_i = b_i = 1$. Then repeat Step 1 and Step 2 in "Method" below a small number of times (i.e., in this paper $10*r$ times) to get initial estimates for a_i and b_i.
Method: For t:=1 to N **Do**

1. Pick $\alpha(t)$ randomly according to arm selection probability vector $P(t)$. Suppose $\alpha(t) = \alpha_i$.

2. Based on the Bayesian nature of the conjugate distributions, update $a_i(t)$ and $b_i(t)$ according to the response from the environment:
   If $R(t) = 0$ **Then** $a_i(t) = a_i(t-1) + 1; b_i(t) = b_i(t-1);$
   Else $a_i(t) = a_i(t-1); b_i(t) = b_i(t-1) + 1;$
3. Identify the upper 95% reward probability bound of $x_i(t)$ for each arm i as:

$$\frac{\int_0^{x_i(t)} v^{(a_i-1)}(1-v)^{(b_i-1)} dv}{\int_0^1 u^{(a_i-1)}(1-u)^{(b_i-1)} du} = 0.95$$

4. If $M(t)$ is the number of arms with higher upper 95% reward probability bound than the pulled arm at time t, update the arm selection probability vector $P(t+1)$ according to the following rule:
   If $R(t) = 0$ **Then**
   $p_j(t+1) = (1-\lambda)p_j(t) + \frac{\lambda}{M(t)}$, for $\forall j \neq i$ such that $x_j(t) > x_i(t);$
   $p_j(t+1) = (1-\lambda)p_j(t)$, for $\forall j \neq i$ such that $x_j(t) \leq x_i(t);$
   $p_i(t+1) = 1 - \sum_{j \neq i} p_j(t+1).$
   Else
   $P(t+1) = P(t).$

End Algorithm: GBPST$_{RI}$

Observe that the GBPST is quite similar to the GPST in the sense that both of them pursue the currently perceived *potentially* optimal arms, and update the arm selection probability vector based on a linear updating rule. The difference is that instead of using ML estimates for the reward probabilities, in the GBPST the estimation is Bayesian, allowing the calculation of 95% upper bounds, $\{x_i\}$.

It is crucial that the salient features of the GBPST and the BLA are highlighted. The reader should observe that they both rely on the *Beta distribution* for reward probability estimation. However, the BLA does not perform any Bayesian computations explicitly. Instead, when it comes to arm selection and exploration, the BLA chooses an arm based on sampling directly from the *Beta* distributions, while the GBPST samples the arm selection space based on the arm selection probability vector. Also, by calculating the 95% upper bound, x_i, the GBPST is able to decide which arms are most promising to pursue.

4   Empirical Results

In this section, we evaluate the computational efficiencies of GBPST$_{RI}$ by comparing it with the GPST$_{RI}$ and the BLA. Although we have conducted numerous experiments using various reward distributions, we report here, for the sake of brevity, results based on the experimental configurations listed in Table 1.

In the experiments considered, Configurations 1 and 4 form the simplest environments, possessing a low reward variance and a large difference between the reward probabilities of the arms. This is because by reducing the difference between the arms, we increase the learning difficulty of the environment. Configurations 2 and 5 achieve this task. The challenge of Configurations 3 and 6 is their high variance combined with the small difference between the arms.

For these configurations, an ensemble of 1,000 independent replications with different random number streams was performed to minimize the variance of the reported results. In each replication, 100,000 arm pulls were conducted in order to examine both the short term and the limiting performance of the evaluated algorithms.

Since both the two schemes, the GBPST$_{RI}$ and the GPST$_{RI}$ depend on an external parameter λ, we measure performance using a wide range of learning speeds: $\lambda = 0.05$, $\lambda = 0.01$ and $\lambda = 0.005$. We report the best performing learning speeds.

Table 2 reports the average probability of pulling the optimal arm after 10, 100, 1,000, 10,000 and 100,000 rounds of arm pulls, for each configuration. As seen, in Configurations 1, 2 and 4, all three schemes converge to the optimal arm with high accuracy, with the BLA being the fastest scheme. The GBPST$_{RI}$ and the GPST$_{RI}$ perform comparably.

Table 1. Bernoulli distributed rewards used in 4-armed and 10-armed bandit problems

Config./Arm	1	2	3	4	5	6	7	8	9	10
1	0.90	0.60	0.60	0.60	-	-	-	-	-	-
2	0.90	0.80	0.80	0.80	-	-	-	-	-	-
3	0.55	0.45	0.45	0.45	-	-	-	-	-	-
4	0.90	0.60	0.60	0.60	0.60	0.60	0.60	0.60	0.60	0.60
5	0.90	0.80	0.80	0.80	0.80	0.80	0.80	0.80	0.80	0.80
6	0.55	0.45	0.45	0.45	0.45	0.45	0.45	0.45	0.45	0.45

Table 2. Probability of pulling the optimal arm after 10, 100, 1000, 10 000, and 100 000 rounds

Configuration	Algorithm	10	100	1000	10000	100000
*Conf.*1	**GBPST$_{RI}$ 0.05**	**0.2655**	**0.5949**	**0.9438**	**0.9941**	**0.9993**
	GPST$_{RI}$ 0.05	0.2649	0.5966	0.9437	0.9941	0.9994
	BLA	0.3512	0.7130	0.9540	0.9942	0.9993
*Conf.*2	**GBPST$_{RI}$ 0.005**	**0.2514**	**0.2869**	**0.6812**	**0.9645**	**0.9963**
	GPST$_{RI}$ 0.005	0.2507	0.2844	0.6852	0.9652	0.9956
	BLA	0.2852	0.4583	0.8190	0.9712	0.9960
*Conf.*3	**GBPST$_{RI}$ 0.01**	**0.2507**	**0.2800**	**0.6312**	**0.9552**	**0.9953**
	GPST$_{RI}$ 0.005	0.2505	0.2647	0.5387	0.9419	0.9933
	BLA	0.2761	0.3856	0.6942	0.9419	0.9915
*Conf.*4	**GBPST$_{RI}$ 0.05**	**0.1027**	**0.3136**	**0.8677**	**0.9854**	**0.9985**
	GPST$_{RI}$ 0.05	0.1035	0.3244	0.8735	0.9862	0.9986
	BLA	0.1407	0.4187	0.8707	0.9826	0.9978
*Conf.*5	**GBPST$_{RI}$ 0.01**	**0.0998**	**0.1204**	**0.5065**	**0.9368**	**0.9924**
	GPST$_{RI}$ 0.005	0.0996	0.1106	0.4194	0.9187	0.9819
	BLA	0.1103	0.1870	0.5492	0.9163	0.9878
*Conf.*6	**GBPST$_{RI}$ 0.005**	**0.0992**	**0.1035**	**0.2357**	**0.8494**	**0.9817**
	GPST$_{RI}$ 0.005	0.0995	0.1041	0.2596	0.8523	0.9581
	BLA	0.1075	0.1493	0.3572	0.8347	0.9752

In Configuration 3, the GBPST$_{RI}$ outperforms the GPST$_{RI}$. The BLA learns faster than GBPST$_{RI}$ scheme at the beginning but was caught up with and surpassed by the GBPST$_{RI}$ in the final 10,000 to 100,000 rounds.

Configurations 5 and 6 are the two most challenging experimental set-ups, in which the superiority of the GBPST$_{RI}$ over the GPST$_{RI}$ is more obvious than in previous configurations. As compared with the BLA, the GBPST$_{RI}$ are not as fast as the BLA at the beginning, but again outperforms the BLA from around the time index 10,000.

We now consider the so-called *Regret* of the algorithms. The *Regret* is *the difference between the sum of rewards expected after N successive arm pulls, and what would have been obtained by only pulling the optimal arm*. Assuming that a *reward* amounts to the value (utility) of unity (i.e., 1), and that a *penalty* possesses the value 0, the *Regret* can be defined as:

$$d_{opt} \cdot N - \sum_{i=1}^{N} d_i, \tag{3}$$

where d_{opt} is the reward probability of the optimal action and d_i is the reward probability of the selected action i.

The *Regret* offers the advantage that it does not overly emphasize the importance of pulling the best arm. In fact, pulling one of the inferior arms will not necessarily affect the overall amount of rewards obtained in a significant manner if, for instance, the reward probability of the inferior arm is relatively close to the optimal reward probability. For *Regret* it turns out that the performance characteristics of the algorithms are mainly decided by the reward distributions, and not by the number of arms. Thus, we now consider configuration 4, 5, and 6 only.

The plots in Fig. 1 illustrate the accumulation of the *Regret* of each algorithm with the number of rounds of pulling arms. As seen in Fig. 1(a), early in the learning phase, the BLA is clearly better than the other two schemes, but the GBPST$_{RI}$ and the GPST$_{RI}$ catch up later with the GBPST$_{RI}$ increasing slowly and the GPST$_{RI}$ converging to yield a constant *Regret*. In the more challenging configurations as shown in Fig. 1(b) and Fig. 1(c), none of the schemes converge to yield a constant *Regret* because of their low learning accuracy. However, the *Regret* of the GBPST$_{RI}$ is much lower and increases more slowly than the others, showing its superiority to the other schemes.

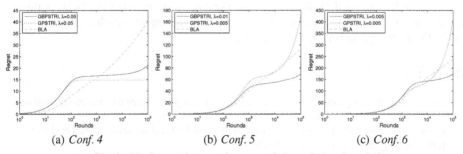

(a) *Conf. 4*   (b) *Conf. 5*   (c) *Conf. 6*

Fig. 1. The *Regret* for experiment conf. 4, conf. 5 and conf. 6

From the above results we draw the following conclusions:

1. The GBPST$_{RI}$ is superior to the GPST$_{RI}$, although the GBPST$_{RI}$ performs *slightly* worse than GPST$_{RI}$ in the simplest configuration. Furthermore, in the other configurations, the GBPST$_{RI}$ provides much better performance than the GPST$_{RI}$, suggesting the former's superiority.
2. By tuning the learning parameter λ, the GBPST$_{RI}$ provides better performance compared to the BLA. On the other hand, in several cases, the BLA initially improves performance faster. This difference in behavior can be explained by their respective distinct strategies for pulling arms. The BLA pulls arms based on the *magnitude* of a random sample drawn from the posterior distribution of the reward estimate, while the GBPST$_{RI}$ chooses arms based on the maintained arm selection probability vector. In fact, with some deeper insight one can see that the initial performance gap can be traced back to the initialization phase of the algorithms, where each arm is pulled to provide initial reward estimates.

5   Conclusion and Further Work

In this paper we have presented the Generalized Bayesian Pursuit Algorithm (GBPST) based on the Reward-Inaction philosophy (GBPST$_{RI}$). The GBPST$_{RI}$ maintains an arm selection probability vector that is used for the selection of the arms. However, it utilizes Bayesian estimates for the reward probabilities associated with each available arm, and adopts a reward-inaction linear updating rule for arm selection probability updating. Also, because we have used the posterior distributions, we are able to utilize a 95% upper bound of the estimates (instead of the mean) to pursue a set of potentially optimal arms. Thus, to the best of our knowledge, the GBPST is the first MABB solution scheme built according to the GPST strategy that also takes advantage of a Bayesian estimation scheme. Our reported extensive experimental results demonstrate the advantages of the GBPST over the GPST scheme. The GBPST also provides better performance than the BLA by choosing λ suitably.

Based on these results, we thus believe that the GBPST forms a new avenue of research, in which the performance benefits of the GPST and the BLA can be combined. In our further work, we intend to investigate how our Generalized Bayesian Pursuit strategy can be extended to the Discretized Pursuit family of schemes. Besides this, we are currently working on convergence proofs, including the results for games of GBPSTs, involving multiple interacting GBPSTs.

References

1. Gittins, J.C.: Bandit processes and dynamic allocation indices. Journal of the Royal Statistical Society. Series B (Methodological) 41(2), 148–177 (1979)
2. Gittins, J.C., Jones, D.M.: A dynamic allocation index for the discounted multiarmed bandit problem. Biometrika 66(3), 561–565 (1979)
3. Whittle, P.: Multi-armed bandits and the gittins index. Journal of the Royal Statistical Society. Series B (Methodological) 42(2), 143–149 (1980)
4. Varaiya, P., Walrand, J., Buyukkoc, C.: Extensions of the multiarmed bandit problem. IEEE Trans. Autom. Control 30, 426–439 (1985)

5. Katehakis, M., Veinott, A.: The multi-armed bandit problem: decomposition and computation. Math. Oper. Res. 12(2), 262–268 (1987)
6. Sonin, I.: A generalized gittins index for a markov chain and its recursive calculation. Statistics and Probability Letters 78, 1526–1533 (2008)
7. Nino-Mora, J.: A (2/3)n3 fast-pivoting algorithm for the gittins index and optimal stopping of a markov chain. INFORMS Journal of Computing 19(4), 596–606 (2007)
8. Watkins, C.J.C.H.: Learning from delayed rewards. Ph.D. thesis. Cambridge University (1989)
9. Cesa-Bianchi, N., Fischer, P.: Finite-time regret bounds for the multiarmed bandit problem. In: ICML1998, Madison, Wisconsin, USA, pp. 100–108 (1998)
10. Auer, P., Cesa-Bianchi, N., Fischer, P.: Finite time analysis of the multiarmed bandit problem. Machine Learning 47, 235–256 (2002)
11. Tokic, M.: Adaptive ε-greedy exploration in reinforcement learning based on value differences. In: Dillmann, R., Beyerer, J., Hanebeck, U.D., Schultz, T. (eds.) KI 2010. LNCS, vol. 6359, pp. 203–210. Springer, Heidelberg (2010)
12. Vermorel, J., Mohri, M.: Multi-armed bandit algorithms and empirical evaluation. In: Gama, J., Camacho, R., Brazdil, P.B., Jorge, A.M., Torgo, L. (eds.) ECML 2005. LNCS (LNAI), vol. 3720, pp. 437–448. Springer, Heidelberg (2005)
13. Kaelbling, L.P.: Learning in embedded systems. PhD thesis, Stanford University (1993)
14. Auer, P., Cesa-Bianchi, N., Freund, Y., Schapire, R.E.: Gambling in a rigged casino: the adversial multi-armed bandit problem. In: the 36th Annual Symposium on Foundations of Computer Science (FOCS 1995), Milwaukee, Wisconsin, pp. 322–331 (1995)
15. Narendra, K.S., Thathachar, M.A.L.: Learning Automat: An Introduction. Prentice Hall, Englewood Cliffs (1989)
16. Thathachar, M., Sastry, P.: Estimator algorithms for learning automata. In: The Platinum Jubilee Conference on Systems and Signal Processing, Bangalore, India, pp. 29–32 (1986)
17. Oommen, B., Agache, M.: Continuous and discretized pursuit learning schemes: various algorithms and their comparison. IEEE Trans. on Systems, Man, and Cybernetics, Part B: Cybernetics 31(3), 277–287 (2001)
18. Norheim, T., Brådland, T., Granmo, O.C., Oommen, B.J.: A generic solution to multi-armed bernoulli bandit problems based on random sampling from sibling conjugate priors. In: Filipe, J., Fred, A., Sharp, B. (eds.) ICAART 2010. CCIS, vol. 129, pp. 36–44. Springer, Heidelberg (2011)
19. Sutton, R.S., Barto, A.G.: Reinforcement learning: An introduction. MIT Press, Cambridge (1998)
20. Wyatt, J.: Exploration and inference in learning from reinforcement. PhD thesis, University of Edinburgh (1997)
21. Dearden, R., Friedman, N., Russell, S.: Bayesian q-learning. In: The 15th National Conf. on Artificial Intelligence, Madison, Wisconsin, pp. 761–768 (1998)
22. Thompson, W.R.: On the likelihood that one unknown probability exceeds another in view of the evidence of two samples. Biometrika 25, 285–294 (1933)
23. Granmo, O.: Solving two-armed bernoulli bandit problems using a bayesian learning automaton. International Journal of Intelligent Computing and Cybernetics 3(2), 207–234 (2010)
24. Granmo, O.C., Berg, S.: Solving Non-Stationary Bandit Problems by Random Sampling from Sibling Kalman Filters. In: García-Pedrajas, N., Herrera, F., Fyfe, C., Benítez, J.M., Ali, M. (eds.) IEA/AIE 2010. LNCS, vol. 6098, pp. 199–208. Springer, Heidelberg (2010)
25. Agache, M., Oommen, B.J.: Generalized pursuit learning schemes: new families of continuous and discretized learning automata. IEEE Trans. on Systems, Man, and Cybernetics, Part B: Cybernetics 32(6), 738–749 (2002)
26. Mitchell, T.M.: Machine Learning. McGraw-Hill, New York (1997)

A Multivalued Recurrent Neural Network for the Quadratic Assignment Problem

Gracián Triviño, José Muñoz, and Enrique Domínguez

E.T.S. Ingeniería Informática
Universidad de Málaga
{gracian,nmunozp,enriqued}@lcc.uma.es

Abstract. The Quadratic Assignment Problem (QAP) is an NP-complete problem. Different algorithms have been proposed using different methods. In this paper, the problem is formulated as a minimizing problem of a quadratic function with restrictions incorporated to the computational dynamics and variables $S_i \in \{1,2,..., n\}$. To solve this problem a recurrent neural network multivalued (RNNM) is proposed. We present four computational dynamics and we demonstrate that the energy of the neuron network decreases or remains constant according to the Computer Dynamic defined.

Keywords: Quadratic assignment problem, neural network, combinatorial optimization.

1 Introduction

The Quadratic Assignment Problem (QAP) is a problem formed by Koopmans and Beckmann in 1957 [1] which gives facilities to localizations. With the goal to minimize the cost function associated to the distance between localization and the flow between the facilities along with the importance of localizing the facility in certain localization. Three matrices of size n x n denominated $F=(f_{ij})$, $D=(d_{kl})$, $C=(c_{im})$, f_{ij} being the flow between the i facility and j facility; d_{kl} is the distance between the localization k and the localization l; c_{im} is the importance of localizing the facility i to the localization m. Moreover they define a variable matrix $n \times n$ size, denominated X, with the following interpretation for an element i ,m:

$$x_{i,m} = \begin{cases} 1 \text{ if the facility } i \text{ is asigned to the localization m} \\ 0 \text{ otherwise} \end{cases} \quad (1)$$

Taking into account the previous definitions, the QAP is formulated:

$$\min \sum_{i=1}^{n}\sum_{j=1}^{n}\sum_{m=1}^{n}\sum_{l=1}^{n} f_{ij} d_{ml} x_{im} x_{jl} + \sum_{i=1}^{n}\sum_{m=1}^{n} c_{im} x_{im} \quad (2)$$

Subject to:

L. Iliadis et al. (Eds.): EANN/AIAI 2011, Part II, IFIP AICT 364, pp. 132–140, 2011.

$$\sum_{m=1}^{n} x_{im} = 1 \quad i = 1, 2, ..., n \tag{3}$$

$$\sum_{i=1}^{n} x_{im} = 1 \quad m = 1, 2, ..., n \tag{4}$$

The restrictions (3) and (4) prevent assigning a facility more than one localization or that localization is given more than one facility. Therefore the matrix X can be represented as a vector φ of n size in which each i position is given a value k with $k = 1, 2, ..., n$ (that is $\varphi(i) = k$) which we can interpret as the variable $x_{ik} = 1$. The vector φ corresponds to a permutation of the possible solutions to the problem Γ_n being the group of aforesaid permutations. With the previous ideas the QAP is reformulated:

$$\min_{\varphi \in \Gamma_n} \sum_{i=1}^{n} \sum_{j=1}^{n} f_{ij} d_{\varphi(i)\varphi(j)} x_{i\varphi(i)} x_{j\varphi(j)} + \sum_{i=1}^{n} c_{i\varphi(i)} x_{i\varphi(i)} \tag{5}$$

The number of permutations of φ vector is n! Increasing n the number of permutations increases nonlinearly. Therefore we are before an NP-complete problem, as Sahni and Gonzalez have demonstrated [2].

The term $\sum_{i=1}^{n} c_{i\varphi(i)} V_{i\varphi(i)}$ of (5) is eliminated adding the costs to the matrix f_{ij} with $i = j$. The objective function can be written as:

$$\min_{\varphi \in \Gamma_n} \sum_{i=1}^{n} \sum_{j=1}^{n} f_{ij} d_{\varphi(i)\varphi(j)} x_{i\varphi(i)} x_{j\varphi(j)} \tag{6}$$

QAP has been solved by different techniques among which standout branch and bound, construction methods, improvement methods and , tabu search , simulated annealing, genetic algorithms, greedy randomized adaptive search procedure , ant system and Hopfield neural network .

There are a series of well known combinatorial optimization problems which can be formulated like the QAP: graph partitioning, maximum clique, the traveling salesman problem [4], the linear arrangement problem, the minimum weight feedback arc set problem, placement of electronic modules, balancing hydraulic turbine.

RNNM is free adjustment parametric and incorporate the restrictions of the problem to the computational dynamics. The energy of the network decrease in each iteration. The results between executions of a problem have a deviation little with time of executing minor that other techniques of the literature.

The paper is organized as follows. In section 2, a new formulation for the QAP is proposed. In section 3 we design four computational dynamic inspired in the Hopfield's synaptic potential (section 3.1) and the increment of Energy (section 3.2,

3.3 and 3.4). Experimental results comparing the performance of the simulated neural approach provided in section 4. Finally, conclusions are presented in the last section 5.

2 Multivaluated Recurrent Networks

In this work a multivalued recurrent network is proposed where the state of neuron i is given by the variable S_i. So, $S_i = m$ if the facility i is assigned to location $m \in \{1,2,..., n\}$. For the k iteration the state of the network is defined by the vector of state $S = \{S_1(t), S_2(t), ..., S_n(t)\}$ which represents the state of each of the neurons. Associated to each state of the net, an energy function is defined:

$$E = -1/2 \sum_{i=1}^{n} \sum_{j=1}^{n} w_{ij} g(S_i, S_j) + \sum_{i=1}^{n} \theta_i(S_i) \qquad (7)$$

where $W = (w_{ij})$ is an $n \times n$ matrix, denominated synoptic weight matrix; $\theta_i(S_i)$ is the threshold function and g is a similarity function. With the computational dynamics described in [5], the network evolves decreasing its energy function until stabilizing, reaching a configuration of the vector state S, so that whatever modification in one of its components, will produce an increase of the energy function or at least a state with equal energy.

The energy function defined in (7) corresponds to the QAP formulation described in (6) when $w_{ij}=-2f_{ij}$, $g(S_i, S_j) = d_{S_i S_j}$ and $\theta(S_i) = c_{iS_i}$.

Therefore, the decrease of the energy function of the network $E(k+1) \leq E(k)$ is guaranteed according to the computational dynamic defined in [5].

3 Computational Dynamics

The Computer Dynamics mentioned beforehand is generic and here we are going to set out 4 dynamics adapted to the QAP. We start from a vector of initial state S (0) which satisfies the restrictions (3) and (4) which we call pseudo feasible initial solution. The computer dynamic in each repetition must satisfy these restrictions, allowing them to be eliminated from the objective function. The first dynamic uses the concept of synoptic potential to establish if a pair of neurons should change state while the rest use the increase of energy.

3.1 Dynamic Based on Synaptic Potential u_i

The potential synoptic is defined as

$$u_i(t) = \sum_{j=1}^{n} f_{ij} d_{S_i(t) S_j(t)} \qquad (8)$$

The dynamic of the network $\forall i \in M$ and choosing two neurons with index a and b is

$$S_a(t+1) = \begin{cases} S_a(t) \text{ if } u_a(t+1) + u_b(t+1) \geq u_a(t) + u_b(t) \\ S_b(t) \text{ if } u_a(t+1) + u_b(t+1) < u_a(t) + u_b(t) \end{cases}$$

$$S_b(t+1) = \begin{cases} S_b(t) \text{ if } u_a(t+1) + u_b(t+1) \geq u_a(t) + u_b(t) \\ S_a(t) \text{ if } u_a(t+1) + u_b(t+1) < u_a(t) + u_b(t) \end{cases} \quad (9)$$

Theorem I. The energy of the neuron network decreases or remains constant according to the Computer Dynamic defined by the expression (9).

Proof. The energy of the neuron network in the k iteration is

$$E(t) = \sum_{i=1}^{n} \sum_{j=1}^{n} f_{ij} d_{S_i(t)S_j(t)} = \sum_{j=1}^{n} f_{aj} d_{S_a(t)S_j(t)} + \sum_{i=1 \atop i \neq a}^{n} f_{ia} d_{S_i(t)Sa(t)} +$$

$$+ \sum_{j=1}^{n} f_{bj} d_{S_b(t)S_j(t)} + \sum_{i=1 \atop i \neq b}^{n} f_{ib} d_{S_i(t)S_b(t)} + \sum_{i=1 \atop i \neq a \atop i \neq b}^{n} \sum_{j=1 \atop j \neq a \atop j \neq b}^{n} f_{ij} d_{S_i(t)S_j(t)} = \quad (10)$$

$$= u_a(t) + u_b(t) + \sum_{i=1 \atop i \neq a \atop i \neq b}^{n} (f_{ia} d_{S_i(t)Sa_b(t)} + f_{ib} d_{S_i(t)S_b(t)}) + \sum_{i=1 \atop i \neq a \atop i \neq b}^{n} \sum_{j=1 \atop j \neq a \atop j \neq b}^{n} f_{ij} d_{S_i(t)S_j(t)}$$

Given that

$$\sum_{i=1 \atop i \neq a \atop i \neq b}^{n} (f_{ia} d_{S_i(t)Sa_b(t)} + f_{ib} d_{S_i(t)S_b(t)}) = \sum_{i=1 \atop i \neq a \atop i \neq b}^{n} (f_{ai} d_{S_a(t)S_i(t)} + f_{bi} d_{S_b(t)S_i(t)}) = u_a + u_b \quad (11)$$

because f y d are symmetric functions. So, we can write the energy of the neuron network.

$$E(t) = 2(u_a(t) + u_b(t)) + \sum_{i=1 \atop i \neq a \atop i \neq b}^{n} \sum_{j=1 \atop j \neq a \atop j \neq b}^{n} f_{ij} d_{S_i(t)S_j(t)} \quad (12)$$

The energy of the neuron network in the (t+1) iteration is

$$E(t+1) = 2(u_a(t+1) + u_b(t+1)) + \sum_{i=1 \atop i \neq a \atop i \neq b}^{n} \sum_{j=1 \atop j \neq a \atop j \neq b}^{n} f_{ij} * d_{S_i(k+1)S_j(k+1)} \quad (13)$$

We observe that:

$$\sum_{i=1 \atop i \neq a \atop i \neq b}^{n} \sum_{j=1 \atop j \neq a \atop j \neq b}^{n} f_{ij} d_{S_i(t)S_j(t)} = \sum_{i=1 \atop i \neq a \atop i \neq b}^{n} \sum_{j=1 \atop j \neq a \atop j \neq b}^{n} f_{ij} * d_{S_i(t+1)S_j(t+1)} \quad (14)$$

The difference of the computation energy between the iteration k+1 and k is

$$E(t+1) - E(t) = 2(u_a(t+1) + u_b(t+1) - (u_a(t) + u_b(t))) \quad (15)$$

According to the dynamics of computing (9) $u_a(t+1) + u_b(t+1) < u_a(t) + u_b(t)$ and therefore $E(t+1) < E(t)$ ∎

3.2 Dynamic Based on Increment of the Energy $\Delta E_{(a,b)}$

The energy increase ΔE is defined between two configurations of the vector state $S(t+1)$ and $S(t)$ as $\Delta E = E(t+1)-E(t)$. If $\Delta E \leq 0$, the energy reduces or remains constant and $S(t+1)$ has an energy less than $S(t)$ associated. So that $E(t+1) \leq E(t)$.

Let be two neurons a and b so that $S_a(t) = l$ y $S_b(t) = m$ are the state of these neurons.

The following state of the net will be $S_a(t+1) = m$, $S_b(t+1) = l$ y $S_i(t+1) = S_i(t), \forall i \notin G$ if $E(t+1) \leq E(t)$. Otherwise $S_i(t+1) = S_i(t), \forall i \in M$.

The contribution to the energy of the network of two neurons a and b in the t-th step and vector of state $S(t)$ is defined by

$$H_{a,b}^{(t)} = \sum_{j=1}^{n} f_{aj} d_{S_a(t)S_j(t)} + \sum_{\substack{j=1\\i\neq a}}^{n} f_{ia} d_{S_i(t)Sa(t)} + \sum_{j=1}^{n} f_{bj} d_{S_b(t)S_j(t)} + \sum_{\substack{j=1\\i\neq b}}^{n} f_{ib} d_{S_i(t)S_b(t)} \tag{16}$$

The difference between these contributions $H_{a,b}^{(t+1)} - H_{a,b}^{(t)}$, can be interpreted as the energy difference between the state $S(t+1)$ with $S_a(t+1) = m$ and $S_b(t+1) = l$ and $S(t)$ with $S_a(t) = l$ and $S_b(t) = m$.

So we can write the expression $E(t+1) - E(t) = H_{a,b}^{(t+1)} - H_{a,b}^{(t)} = \Delta E_{(a,b)}$.

If, $E(t+1) \leq E(t)$ the vector state $S(t+1)$ is accepted as an entrance to the following repetition. Otherwise the vector state $S(t+1)$ is rejected and $S(t)$ is kept as the entrance to the following iteration.

The dynamic of the network $\forall i \in M$ and choosing two neurons with index a and b is:

$$S_a(t+1) = \begin{cases} S_a(t) \text{ if } \Delta E_{(a,b)} > 0 \\ S_b(t) \text{ if } \Delta E_{(a,b)} \leq 0 \end{cases}$$

$$S_b(t+1) = \begin{cases} S_b(t) \text{ if } \Delta E_{(a,b)} \leq 0 \\ S_a(t) \text{ if } \Delta E_{(a,b)} > 0 \end{cases} \tag{17}$$

Theorem II: The energy of the neuron network decreases or remains constant according to the Computer Dynamic defined by the expression (17).

Proof: Let be

$$E(t) = \sum_{i=1}^{n} \sum_{j=1}^{n} f_{ij} d_{S_i(t)\ S_j(t)} = \sum_{i=1}^{n} \sum_{\substack{j=1\\(i\neq j)}}^{n} H_{ij}^{(t)} \text{ and}$$

$$E(t+1) = \sum_{i=1}^{n} \sum_{j=1}^{n} f_{ij} d_{S_i(t+1)\ S_j(t+1)} = \sum_{i=1}^{n} \sum_{\substack{j=1\\(i\neq j)}}^{n} H_{ij}^{(t+1)} . \text{ The increase of energy can}$$

be written

$$E(t+1) - E(t) = \sum_{\substack{i=1 \\ (i \neq j)}}^{n} \sum_{j=1}^{n} H_{ij}^{(t+1)} - \sum_{\substack{i=1 \\ (i \neq j)}}^{n} \sum_{j=1}^{n} H_{ij}^{(t)} \tag{18}$$

If we update the neurons a–th and b-th so that $G = \{a,b\} \subset M$, the permutations $S(t+1)$ and $S(t)$ will coincide in the rest of the index, meaning $S_i(t+1) = S_i(t) \ \forall i \in M - G$. So we can write:

$$E(t+1) - E(t) = \sum_{\substack{i=1 \\ (i \neq j)}}^{n} \sum_{j=1}^{n} H_{ij}^{(t+1)} - \sum_{\substack{i=1 \\ (i \neq j)}}^{n} \sum_{j=1}^{n} H_{ij}^{(t)} = H_{ab}^{(t+1)} - H_{ab}^{(t)} = \Delta E_{(a,b)} \tag{19}$$

Because $H_{ij}^{(t)} = H_{ij}^{(t+1)} \ \forall i,j \in M \land ((i = a \land j \neq b) \lor (i \neq a \land j = b))$. If an update has been produced in the state vector it is because $\Delta E_G \leq 0$ therefore $E(t+1) \leq E(t)$.   ∎

3.3 Dynamic Based on Maximum Increment of a Neuron ΔE_{Min_e}

The dynamic obtains the maximum decrease between two consecutive updates setting a c neuron.

Let be $c \in M$ the c-th neuron of the randomly chosen recurrent neuronal network and let be $e \in M - \{c\}$ the e-th neuron of the network obtained from the rest of the indexes.

The minimum energy increase of the update group {c, e} is $\Delta E_{Min_{c,e}} = \min_{e \in M - \{c\}} \{H_{ce}^{(t+1)} - H_{ce}^{(t)}\}$ which corresponds to the maximum energy $\left| \Delta E_{Min_{c,e}} \right|$ with which the networks can decrease exchanging the c-th neuron with the e-th.

The dynamic of the network and choosing a neuron with $c \in M$ index will be:

$$S_i(t+1) = \begin{cases} S_e(t) & \text{if } \Delta E_{Min_{c,e}} \leq 0 \land (i = c) \\ S_c(t) & \text{if } \Delta E_{Min_{c,e}} \leq 0 \land (i = e) \\ S_i(t) & \text{otherwise} \end{cases} \tag{20}$$

Theorem III: i) The energy of the neuronal network decreases or remains permanent according to the dynamics of the computation defined by the expression (20).

ii) The slope of the energy E is maximum for the neuron which is taken as reference in each iteration.

Proof:

i) Similar to demonstration theorem I

ii) Let be t-th iteration and c the c-th neuron which we take as reference. So we can write:

$$\frac{E(t+1)-E(t)}{\Delta t} = \min_{e \in M-\{c\}} \{H_{ce}^{(t+1)} - H_{ce}^{(t)}\} = \Delta E_{Min_{c,e}} \geq \Delta E_{(a,b)} \qquad (21)$$

.

3.4 Dynamic Based on Maximum Increment of Two Configurations of the State Vector ΔE_{Min}

The dynamic obtains the maximum decrease between the two consecutive updates given whatever configuration of vector state **S**.

The minimum energy increase for a configuration of vector state S is $\Delta E_{MinMax} = \min_{c \in M \wedge e \in M-\{c\}} \{H_{ce}^{(t+1)} - H_{ce}^{(t)}\}$ which corresponds with the maximum energy $|\Delta E_{MinMax}|$ with which the network can decrease exchanging two neurons.

The dynamic of the $\forall i \in M$ network will be:

$$S_i(t+1) = \begin{cases} S_e(t) \text{ if } \Delta E_{Min} \leq 0 \wedge (i = c) \\ S_c(t) \text{ if } \Delta E_{Min} \leq 0 \wedge (i = e) \\ S_i(t) \text{ otherwise} \end{cases} \qquad (22)$$

Theorem IV:
 i) The energy of the neuronal network decreases or remains constant according to the dynamics of the computation defined by the expression (22).
ii) The slope of the E energy is the maximum in each iteration.
  Proof:
   i)   Similar to theorem I demonstration.
   ii) Similar to the demonstration of Theorem II section II

4 Experimental Results

The figures 1-2 show the evolution of energy for two classical problems in the literature [7] called *nug30* and *sko100a*. The running time is limited while the neural network converges to a local minimum before this deadline. For example, in Figure 1 ends at 0.36759 seconds after 1000 iterations. The minimum reached using the computational dynamic defined in the section 3.1 is 6382 which compared with the best literature solution is 6124 to get a gap = 4.2129.

In the table 1, summarize the description above for the rest of dynamics and the problems *nug30* and *sko100a*. The dynamic 3.1 and 3.2 obtain similar results but are conceptually different. The dynamic 3.3 and 3.4 are based on 3.2 and how we show in the Fig. 5 to 8 the slope of the energy is locally maximum for a neuron in iteration or the energy is locally maximum in each iteration, respectively.

The results (table 3) were compared with the results obtained by Ji et al. [6], Ahuja et al [3] and Drezner [8]. So, the gaps obtained by the RNNM are worst for the two problems. But, our algorithm reached solutions near the better solution of the literature in less time than the genetic algorithm defined in this work. By example, the problem denominate *nug30* finish in 21.6 s. with Ji´s algorithm versus 11.8443 s. with the RNMM.

This difference is major with *sko100a*: 10447.8 s. with Ji´s algorithm versus 203.4079 with the RNMM. These results are obtained in all the QAPLIB' problems.

Table 2, shows the gap and the time for problem with $n \geq 90$ and we once again deduct the same conclusion: the relations between the solution reached and the time that RNNM needed to reach is the best of the literature.

Table 1. Comparation result's betwen the differents dynamic with the problem *nug30* and *sko100a*

Dynamic	Problem	Energy Initial	Energy Final	Gap (%)	Iterations	Time (s)
Dynamic 3.1	nug30	8170	6382	4.2129	1000	0.36759
Dynamic 3.2	nug30	8002	6374	4.0823	2000	1.2107
Dynamic 3.3	nug30	8426	6272	2.4167	1000	11.8443
Dynamic 3.4	nug30	7940	6240	1.8942	35	5.9817
Dynamic 3.1	sko100a	177946	161082	5.9736	1000	5.0183
Dynamic 3.2	sko100a	176472	159684	5.0539	3001	24.0721
Dynamic 3.3	sko100a	177904	177904	3.0671	151	20.0509
Dynamic 3.4	sko100a	178534	160904	5.8565	32	203.4079

Table 2. Gap and the time for problem with $n \geq 90$

Problem	Gap (%)	Time (s)
will100	1.9902	22.7115
tai256c	4.9068	123.032
tho150	7.0377	50.9777
esc128	3.125	129.6653

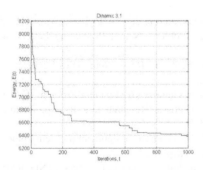

Fig. 1. Evolution of the energy for *nug30*

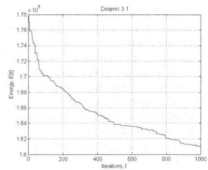

Fig. 2. Evolution of the energy for *sko100a*

Table 3. Comparative times RNMM versus other algorithm

Problem	JI	Ahuja	Drezner	RNMM
nug30	21.6	177.1	22,2	0.36759
Sko100a	10447.8	8304. 1	2013	5.0183

5 Conclusions

In this article, we discussed a multi-valued neural network to solve the quadratic assignment problem (QAP) using four dynamic starting from pseudo feasible solutions in relation to the constraints of the problem and updating two neurons reaches   finest solutions and execution times lower than those obtained with other methods. For small values of n the possibility that the network reaches a configuration which needs more than two changes in the values of the neurons increases and the network falls into a local minimum. However, for large n the possibility of reaching state configurations requiring more than two changes is less or may not occur. In the next work we will study the update of more than two neurons to avoid falling into local minimum.

In addition, the neural network is free of adjustment parameters. It is recommended for quadratic problems where finding an acceptable solution quickly is more appropriate than otherwise, or as a method to find an initial solution for further optimization with another method.

References

1. Koopmans, T.C., Beckmann, M.J.: Assignment problems and the location of economic activities. Econometrica 25, 53–76 (1957)
2. Bazaraa, M.S., Kirca, O.: Branch and bound based heuristic for solving the quadratic assignment problem. Naval Research Logistics Quarterly 30, 287–304 (1983)
3. Ahuja, R.K., Orlin, J.B., Tivari, A.: A greedy genetic algorithm for the quadratic assignment problem. In: Working paper 3826-95, Sloam School of Management. MIT, Cambridge (1995)
4. Merida Casermeiro, E., Galan Marin, G., Muñoz Perez, J.: An Efficient Multivalued Hopfield Network for the Travelling Salesman Problem. Neural Processing Letters 14, 203–216 (2001)
5. Domínguez, E., Benítez, R., Mérida, E., Muñoz Pérez, J.: Redes Neuronales Recurrentes para la resolución de Problemas de Optimización
6. Ji, P., Wu, Y., Liu, H.: A Solution Method for the QuadraticAssignment Problem (QAP). In: The Sixth International Symposium on Operations Research and Its Applications (ISORA 2006), Xinjiang, China, August 8-12 (2006)
7. Berkard, R.E., Karisch, S.E., Rendl, F.: QAPLIB – a quadratic assignment problem library, http://www.opt.math.tu-graz.ac.at/qaplib
8. Drezner, Z.: A new genetic algorithm for the quadratic assignment problem. INFORMS Journal on Computing 115, 320–330 (2003)

Employing a Radial-Basis Function Artificial Neural Network to Classify Western and Transition European Economies Based on the Emissions of Air Pollutants and on Their Income

Kyriaki Kitikidou and Lazaros Iliadis

Democritus University of Thrace, Department of Forestry and Management of the
Environment and Natural Resources, Pandazidou 193, 68200, Orestiada, Greece
kkitikid@fmenr.duth.gr

Abstract. This paper aims in comparing countries with different energy
strategies, and demonstrate the close connection between environment and
economic growth in the ex-Eastern countries, during their transition to market
economies. We have developed a radial-basis function neural network system,
which is trained to classify countries based on their emissions of carbon,
sulphur and nitrogen oxides, and on their Gross National Income. We used
three countries representative of ex-Eastern economies (Russia, Poland and
Hungary) and three countries representative of Western economies (United
States, France and United Kingdom). Results showed that the linkage between
environmental pollution and economic growth has been maintained in ex-
Eastern countries.

Keywords: Artificial Neural Networks, Atmospheric pollutants, Radial-Basis
Function, Transition economies.

1 Introduction

There is a general assumption that the former socialist economies failed, as regards
environmental protection. Nevertheless, failure makes sense only if one has a success
to oppose it with. It is interesting to see if the environmental damage associated with
the economic growth of socialist systems is comparable with the negative externalities
that, since the industrial evolution, have accompanied the economic development of
Western countries. Likewise, during the transition to market economies, we should
look for a similar development of the forces that have allowed the improvement of
environmental performances in Western countries [15].

Transition to a market economy, for all its long-term rewards, has not been an easy
process. In addition to the necessary but often difficult social, political and economic
adjustments, the countries of Central and Eastern Europe have had to cope with the
environmental legacy of inefficient industries, obsolete and polluting technologies,
and weak environmental policies ([4]; [5]; [10]).

In 1993, many serious environmental problems have been identified, several that
directly threatened human health and required immediate and urgent attention. These
included ([12]; [13]):

L. Iliadis et al. (Eds.): EANN/AIAI 2011, Part II, IFIP AICT 364, pp. 141–149, 2011.
© IFIP International Federation for Information Processing 2011

-   high levels of airborne particulates in urban and industrial "hot spots" from coal combustion by domestic users, small-scale enterprises, power and heating plants, and metallurgical plants;
-   high levels of sulphur dioxide and other air pollutants in "hot spot" areas from the combustion high-sulphur coal and fuel oil by these and other sources;
-   lead concentrations in air and soil linked to airborne emissions from industry and from the use of leaded gasoline; and
-   contamination of drinking water, in particular by nitrates in rural well water.

World Bank analyses of the role of economic and environmental reforms in reducing air pollution levels in ex-Eastern countries have used econometric models to adjust for changes in GDP levels, industrial output and energy use. Overall, these analyses suggest that economic reforms, environmental policies, and other developments over the transition have produced real environmental benefits in the advanced reform countries, beyond emissions reductions resulting from declines in output and energy use. In slower reform countries, such declines appear to be the main factors behind the emissions reductions. If these countries were to return to growth without further economic reforms and effective environmental policies, their air pollution emissions could increase accordingly ([21]; [22]).

The aim of this paper is to allow us to understand the nature of environmental paths in socialist countries. It would be interesting to see if there is a connection between transition of the economies of ex-Eastern countries and environmental pollution, in opposition to Western economies. This analysis is based on the comparison of atmospheric pollutant emissions from energy use of major emission intensity components, using artificial neural networks (ANN). Countries are categorized by energy patterns. The innovative part of this research effort lies in the use of a soft computing machine learning approach like the ANN to predict group memberships of countries with different energy strategies.

2 Materials and Methods

We have selected three ex-Eastern countries, Russia, Poland and Hungary, and three OECD (Organisation for Economic Co-operation and Development) countries, United States, France and the United Kingdom, representative of different energy strategies [20]. The emissions of atmospheric pollutants in ex-Eastern countries, for the year 1994, were estimated from an independent study, because of the inadequacy of international statistics as regards ex-Eastern countries [20]. All other data are taken from the OECD libraries [11]. The three atmospheric pollutants under investigation here are carbon dioxide (CO_2 measured in t/toe), nitrogen oxides (NOx measured in kg/toe) and sulphur dioxide (SO_2 measured in t/toe). Emissions were estimated by four intensity components (coal, natural gas, oil, and electricity-heat). The six countries, and the four intensity components, resulted in 24 cases in our data. The Gross National Income (GNI) per capita was measured in PPP international dollars, i.e. a common currency that eliminates the differences in price levels between countries allowing meaningful volume comparisons of GNI between countries.

Descriptive statistics for all variables are given in **Table 1**.

Table 1. Descriptive statistics for all variables used in the analysis

		CO_2 (t/toe)	NOx (kg/toe)	SO_2 (t/toe)	GNI (PPP int. $)
Ex-Eastern countries	Mean	3.5	12.7	33.0	6910
	Standard deviation	1.8	9.4	38.9	982
	Minimum	2.0	2.3	0.0	6020
	Maximum	8.4	38.0	124.3	8210
OECD countries	Mean	3.2	11.3	19.8	21510
	Standard deviation	1.8	9.7	22.9	3503
	Minimum	0.7	2.0	0.0	18750
	Maximum	6.7	34.6	65.5	26230

For the performance of the analysis, the radial-basis function (RBF) network model was used ([1]; [6]; [14]; [17]; [19]), from menu of the SPSS v.19 statistical package [8]. We specified that the relative number of cases assigned to the training:testing:holdout samples should be 6:2:1. This assigned 2/3 of the cases to training, 2/9 to testing, and 1/9 to holdout. The architecture of the developed ANN included three neurons in the hidden layer. The transfer functions (hidden layer activation functions and output function) determine the output by depicting the result of the distance function ([2]; [9]). The schematic representation of the neural network with transfer functions is given in Fig. 1.

3   Results-Discussion

From the analysis, 15 cases (62.5%) were assigned to the training sample, 6 (25.0%) to the testing sample, and 3 (12.5%) to the holdout sample. No cases were excluded from the analysis. The number of units in the input layer is the number of covariates plus the total number of factor levels; a separate unit is created for each category of emission intensity component and GNI, and none of the categories are considered redundant. Likewise, a separate output unit is created for each country category, for a total of 2 units in the output layer. Three units were chosen in the hidden layer.

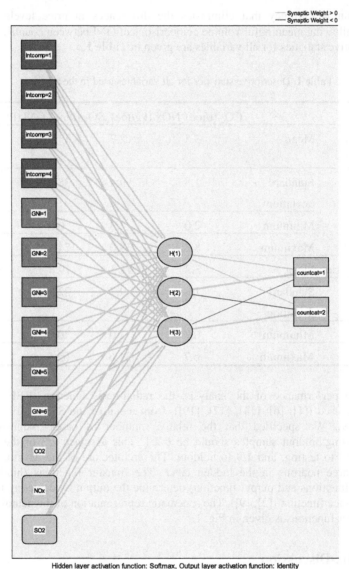

Hidden layer activation function: Softmax, Output layer activation function: Identity

intcomp = intensity component, countcat = country category

Fig. 1. Radial-basis function network structure

Table 2 shows the practical results of using the network. For each case, the predicted response is the category with the highest predicted pseudo-probability. Cells on the diagonal are correct predictions, while cells off the diagonal are incorrect predictions. The RBF network gets 80.0% of the countries. In particular, the RBF model excels at identifying ex-Eastern countries (100.0% for the training, testing and holdout samples). The holdout sample helps to validate the model; here, 66.7% of these cases were correctly classified by the model. This suggests that the RBF model is in fact correct about three out of five times. The lack of correct classification for the OECD countries to the testing sample (0.0%) must be due to the small data set available, which naturally limits the possible degree of complexity of the model [3].

Table 2. Confusion matrix

Sample	Observed	Predicted		Classified correctly
		Ex-Eastern country	OECD country	
Training	Ex-Eastern country	9	0	100.0%
	OECD country	3	3	50.0%
	Overall Percent	80.0%	20.0%	80.0%
Testing	Ex-Eastern country	2	0	100.0%
	OECD country	4	0	0.0%
	Overall Percent	100.0%	0.0%	33.3%
Holdout	Ex-Eastern country	1	0	100.0%
	OECD country	1	1	50.0%
	Overall Percent	66.7%	33.3%	66.7%

For categorical dependent variables, the predicted-by-observed chart displays clustered boxplots of predicted pseudo-probabilities for the combined training and testing samples. The x axis corresponds to the observed response categories, and the legend corresponds to predicted categories. The leftmost boxplot shows, for cases that have observed category ex-Eastern country, the predicted pseudo-probability of category ex-Eastern country. The next boxplot to the right shows, for cases that have observed category ex-Eastern country, the predicted pseudo-probability of category OECD country. The third boxplot shows, for cases that have observed category OECD country, the predicted pseudo-probability of category OECD country. Lastly, the fourth boxplot shows, for cases that have observed category OECD country, the predicted pseudo-probability of category OECD country.

From looking at a portion of cases in one boxplot, and the corresponding location of those cases in another boxplot, we determine that ex-Eastern countries are better classified than the OECD countries (the third and fourth boxplots are roughly equivalent). This was the conclusion extracted from the confusion matrix also.

146   K. Kitikidou and L. Iliadis

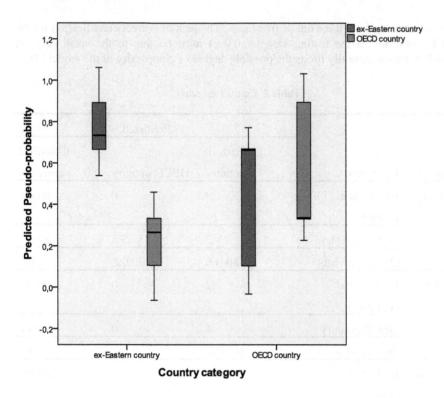

Fig. 2. Predicted-by-observed chart

Receiver operating characteristics (ROC) graphs are useful for organizing classifiers and visualizing their performance [16]. ROC graphs are commonly used in medical decision making, and in recent years have been used increasingly in machine learning and data mining research. The numbers along the major diagonal represent the correct decisions made, and the numbers of this diagonal represent the errors – the confusion – between the various classes. The true positive rate (sensitivity) of a classifier is estimated as:

$$\text{tp rate} \approx \frac{\text{Positives correctly classified}}{\text{Total positives}}$$

while the false positive rate of the classifier is:

$$\text{fp rate} \approx \frac{\text{Negatives incorrectly classified}}{\text{Total negatives}}.$$

Specificity of the ROC curve is defined as:

$$\text{specificity} = \frac{\text{True negatives}}{\text{False positives} + \text{True negatives}}.$$

The ROC curve in Fig. 3 gives us a visual display of the sensitivity by specificity for all possible classification cutoffs. The chart shown here displays two curves, one for each category of the target variable (country category). It seems that, for each category, the probability that the predicted pseudo-probability of a country being in that category is equal for a randomly chosen country in that category to a randomly chosen country not in that category.

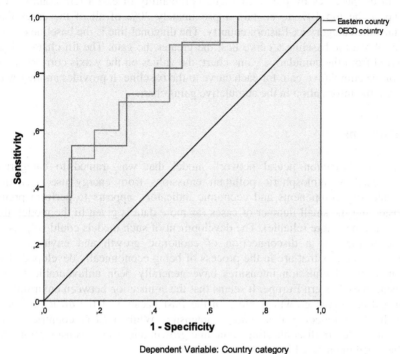

Dependent Variable: Country category

Fig. 3. ROC curves

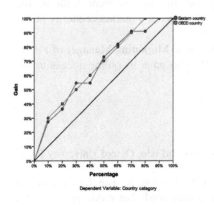

Fig. 4a. Gains chart

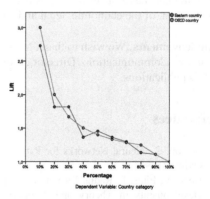

Fig. 4b. Lift chart

The cumulative gains chart in Fig. 4a shows the percentage of the overall number of cases in a given category gained by targeting a percentage of the total number of cases. For example, the first point on the curve for the ex-Eastern country category is approximately at (10%, 30%), meaning that if we score a dataset with the network and sort all of the cases by predicted pseudo-probability of ex-Eastern country, we would expect the top 10% to contain approximately 30% of all of the cases that actually take the category ex-Eastern country. The diagonal line is the baseline curve; the farther above the baseline a curve lies, the greater the gain. The lift chart in Fig. 4b is derived from the cumulative gains chart; the values on the y axis correspond to the ratio of the cumulative gain for each curve to the baseline. It provides another way of looking at the information in the cumulative gains chart.

4 Conclusions

The radial-basis function neural network model that was trained to categorize countries, based on atmospheric pollutant emissions from energy use of major emission intensity components and economic indicators, appears to perform pretty well, considering the small number of cases (as more data is given to the model, the prediction becomes more reliable). The development of such models could help us to investigate if there is a disconnection of economic growth and environmental pollution, for countries that are in the process of being economically developed. The four components of emission intensities have generally been unfavourable to the environment in ex-Eastern Europe. It seems that the connection between environment pollution and economic growth is close in the East because of the difficulty these countries have in reducing their energy intensity. While OECD countries have gradually succeeded in disconnecting economic growth and energy consumption, the linkage has been maintained in ex-Eastern countries [20].

Increasing energy prices, macroeconomic stabilization programmes, industrial restructuring and increasing industrial investments will force countries in transition to change their environmental habits [7]. However, environmental measures and foreign aid will not have a lasting, positive impact on the environment without the development of the economic, legal and social institutions of a market economy [18].

Aknoledgements. We wish to thank Mr James Kitchen, Marketing Manager of Public Affairs & Communications Directorate of OECD, who gave us online access to the OECD publications.

References

1. Bishop, C.: Neural Networks for Pattern Recognition, 3rd edn. Oxford University Press, Oxford (1995)
2. Bors, A., Pitas, I.: Radial Basis function networks. In: Howlett, R., Jain, L. (eds.) Recent Developments in Theory and Applications in Robust RBF Networks, pp. 125–153. Physica-Verlag, Heidelberg (2001)

3. Dendek, C., Mańdziuk, J.: Improving Performance of a Binary Classifier by Training Set Selection. Warsaw University of Technology, Faculty of Mathematics and Information Science, Warsaw, Poland (2008)
4. European Bank for Reconstruction and Development (EBRD): Transition Report Update, London (1998)
5. Gaddy, C., Ickes, B.: Russia's Virtual Economy. Foreign Affairs (1998)
6. Haykin, S.: Neural Networks: A Comprehensive Foundation, 2nd edn. Macmillan College Publishing, New York (1998)
7. Hughes, G.: Are the costs of cleaning up Eastern Europe exaggerated? Economic Reform and the Environment. Oxford Review of Economic Policy 7(4), 106–136 (1991)
8. IBM SPSS Neural Networks 19. SPSS Inc. (2010)
9. Iliadis, L.: Intelligent Information Systems and Applications in Risk Management, Stamoulis edn. Thessaloniki, Greece (2007)
10. Lovei, M., Levy, B.: Lead Exposure and Health in Central and Eastern Europe. In: Lovei, M. (ed.) Phasing out Lead from Gasoline in Central and Eastern Europe. Health Issues, Feasibility, and Policies. World Bank, Washington DC (1997)
11. OECD/IEA: Energy balances of OECD countries 1993-1994, Paris (1996)
12. OECD: Economic Survey: Russia, Paris (1998)
13. OECD: Environmental Data Compendium, Paris (1997)
14. Ripley, B.: Pattern Recognition and Neural Networks. Cambridge University Press, Cambridge (1996)
15. Shahgedanova, M., Burt, T.: New data on air pollution in the former Soviet Union. Global Environmental Change 4(3), 201–227 (1994)
16. Streiner, D., Cairney, J.: What's Under the ROC? An Introduction to Receiver Operating Characteristics Curves. The Canadian Journal of Psychiatry 52, 121–128 (2007)
17. Tao, K.: A closer look at the radial basis function (RBF) networks. In: Singh, A. (ed.) Conference Record of the Twenty-Seventh Asilomar Conference on Signals, Systems, and Computers. IEEE Computational Society Press, Los Alamitos (1993)
18. Toman, M., Simpson, R.: Environmental policies, economic restructuring and institutional development in the former Soviet Union. Resources 116, 20–23 (1994)
19. Uykan, Z., Guzelis, C., Celebi, M., Koivo, H.: Analysis of input-output clustering for determining centers of RBFN. IEEE Transactions on Neural Networks 11, 851–858 (2000)
20. Viguier, L.: Emissions of SO2, NOx and CO2 in transition economies: emission inventories and Divisia index analysis. The Energy Journal 20(2), 59–88 (1999)
21. World Bank: National Biodiversity Strategies and Action Plans. Preliminary Summary of Findings from Eastern Europe and Central Asia. World Bank, Washington DC (1998a)
22. World Bank: Transition Toward a Healthier Environment: Environmental Issues and Challenges in the Newly Independent States. World Bank, Washington DC (1998b)

Elicitation of User Preferences via Incremental Learning in a Declarative Modelling Environment

Georgios Bardis[1], Vassilios Golfinopoulos[1], Dimitrios Makris[1],
Georgios Miaoulis[1,2], and Dimitri Plemenos

[1] Department of Informatics, TEI of Athens, Ag. Spyridonos St., 122 10 Egaleo, Greece
[2] XLIM Laboratory, University of Limoges, 83 rue d'Isle, Limoges, 87000, France
{gbardis,golfinopoulos,demak,gmiaoul}@teiath.gr

Abstract. Declarative Modelling environments exhibit an idiosyncrasy that demands specialised machine learning methodologies. The particular characteristics of the datasets, their irregularity in terms of class representation, volume, availability as well as user induced inconsistency further impede the learning potential of any employed mechanism, thus leading to the need for adaptation and adoption of custom approaches, expected to address these issues. In the current work we present the problems encountered in the effort to acquire and apply user profiles in such an environment, the modified boosting learning algorithm adopted and the corresponding experimental results.

Keywords: Incremental Learning, Irregular Datasets, Declarative Modelling.

1 Introduction

A declarative scene modelling environment offers the designer/user the ability to produce a 3D scene without delving into its explicit geometric properties. User input is a set of declarative objects, associations and properties, practically representing constraints for the geometric properties of the final scene's objects, yet expressed in an intuitive manner ("a large bedroom", "kitchen adjacent to dining room"). The typical declarative scene modelling environment has to map these abstract notions to concrete constraints and, subsequently, resolve these constraints in order to achieve one or more geometric representations of the submitted input. Such a scenario implies numerous possible outcomes, not all of equal interest for the user. Hence, the question of user profiling and intelligent solution evaluation is a considerable concern. The issues that arise in this effort are identified and analysed in the following section. Next, a framework for capturing and applying user preferences in a declarative modelling environment is presented, including the modified version of an incremental learning algorithm. The performance of the proposed solution is exhibited through a number of experimental results which are discussed and analysed for future research directions.

2 Related Work

The ability to capture and apply user preferences in a declarative modelling environment is inherently inhibited by a number of particularities exhibited by the declarative design

L. Iliadis et al. (Eds.): EANN/AIAI 2011, Part II, IFIP AICT 364, pp. 150–158, 2011.

process, originating, in turn, from the nature of the declarative modelling methodology [3]. A series of approaches have been suggested addressing this issue. The effort presented in [8] offers user preference capture which is isolated per scene, gradually training a dedicated neural network upon user feedback. This approach, while efficient in pinpointing the user interpretation of the scene at hand, does not allow for a universal user profile applicable to newly submitted scenes, since it has to go through training phases corresponding to each new scene. The approach in [4] comprises a hybrid supervised/unsupervised concept acquisition, which constructs similarity classes in an unsupervised manner, presenting class representatives to the user in order to guide the mechanism towards the selection of the target concept of the original description. This approach operates under the assumption that the similarity of solutions grouped together during the unsupervised phase reflects the user's notion of similarity. This, however, may imply contradictions with actual user preferences in the later stages.

The widest scope is covered by [1] which allows for horizontal user profiling, creating for each user a profile that is applicable to all future scenes (s)he may submit. The advantage of this approach becomes evident during the Scene Understanding phase of the Declarative Modelling methodology [8] where the user may be presented with those geometric representatives of the generated solutions that are closer to previously acquired profile. This approach relies on the mapping of the generated solutions to a set of observed attributes, thus relying on the participation of the latter to the formation of actual user preferences. In the following we discuss the corresponding mechanism and present new experimental results based upon an extended attribute set.

3   Machine Learning in a Declarative Modelling Environment

The typical scenario of use for a declarative modelling environment starts with a rough idea of the scene to be designed on behalf of the user. It is often the case that at this stage of design, i.e. early-phase design, the user has a grasp of only an abstraction of the final expected result. Hence, the repertoire of building blocks for this initial input consists of declarative objects, properties and associations, leading to a declarative description similar to the one appearing in Fig. 1, represented by three corresponding sets.

$O = \{$
    [Living Room],
    [Guest Room],
    [Parents' Room],
    [Childrens' Room],
    [Corridor],
    [Bathroom]
$\}$

$A = \{$
    [Living Room adjacent west *Guest Room]*,
    [Living Room longer than *Guest Room]*,
    [Parents' Bedroom adjacent west *Bathroom]*,
    [Children's Bedroom adjacent east *Bathroom]*,
    [Kitchen adjacent south *Corridor]*
$\}$

$P = \{$
    [Corridor is very narrow]
$\}$

Fig. 1. Example declarative description of the spatial arrangement of a house

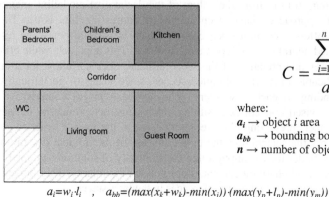

$$C = \frac{\displaystyle\sum_{i=1}^{n} a_i}{a_{bb}}$$

where:
$a_i \rightarrow$ object i area
$a_{bb} \rightarrow$ bounding box area
$n \rightarrow$ number of objects

$a_i = w_i \cdot l_i$, $a_{bb} = (max(x_k + w_k) - min(x_j)) \cdot (max(y_p + l_p) - min(y_m))$

Fig. 2. Example observed attribute *Compactness*

Once the user has submitted the declarative description, a mechanism is responsible for translating it into a set of constraints and resolving it, thus yielding – typically numerous – geometric interpretations of the original input, i.e. solutions for the constraint set.

The effort to acquire and apply user preferences to the generated solutions has to consider properties of interest to the user when assessing the generated results. In other words, a minimal set of observed attributes have to be assumed and calculated for each solution as a first step towards the effort to mimic the user's evaluation. Depending on the nature of the scene, these attributes can be, and usually are, suggested by the corresponding domain experts. The only requirement for these attributes, in regard to the automation of the learning process, is their computational interconnection and extraction feasibility with respect to available scene information, namely, its declarative description and its geometric equivalents. In the current work we assume the existence and availability of a well-defined and computationally feasible set of observed attributes, directly related to available scene information. An example attribute and its connection to the solutions' geometric characteristics appears in Fig. 2.

4 Declarative Modelling Machine Learning Requirements

Once the solutions have been mapped to a set of attributes and evaluated by the user we find ourselves in the familiar grounds of classified data vectors which may subsequently be used to train and/or evaluate a machine learning mechanism to be used for automatic classification of future generated solutions. Such a mechanism is expected to have acquired and therefore be able to apply user's preferences on previously unseen samples. A number of reasons, originating from the nature of the declarative modelling process and the corresponding environments prevent us, however, from the straightforward application of traditional classifier approaches. These reasons may be summarised as follows:

Original solutions may require large storage capacities. It may be the case that each solution is a complicated 3D model, a high resolution image or other volume intensive data form [1],[9]. In this case, storage capacity required for maintaining all previously evaluated solutions is prohibited, which, in turn, diminishes the value of preserving the solutions only in the encoded vector form.

Solutions are generated at distinct and usually distant times. The typical session of use of a declarative modelling environment comprises submission of a description on behalf of the user, solution generation on behalf of the environment and global or partial evaluation the outcome population, on behalf of the user. This functionality suggests that samples are available in bursts and in abnormal time intervals.

Solutions populations are of varied size and class representation. A declarative description may range from highly restrictive, thus leading to only a limited number of solutions, to highly abstract, thus yielding hundreds of thousands of alternative interpretations. Depending on the case, the user may or may not be willing to evaluate the entire population of outcomes since user's evaluation may be interrupted after a small number of approvals, thus leading to imbalanced datasets. In addition, depending on input declarative description, some attributes or some classes may not be included in the final outcome and, hence, not be present in the training (sub)set.

Even small data perturbations may imply classification divergence. When user evaluation is involved, small differences in the evaluated samples may cause radically different classification. For example, in terms of the example description of Fig. 1, the misalignment of the edges of two rooms, regardless how insignificant geometrically, may have an important impact on user's preference.

For the aforementioned reasons, many traditional classifiers can not be successfully applied in such an environment. The gradual and limited availability of evaluated samples suggests the need for a mechanism that will be able to maintain knowledge without having to maintain the supporting data for this knowledge. In other words, an incrementally learning mechanism is required to efficiently grow both in terms of size and complexity as well as in terms of knowledge absorption. Even mechanisms able to fulfil this requirement for adaptive learning [5], fail to successfully address the highly unstable nature of the data due to their inherent assumption of data stability, i.e. small perturbations of the input (data sample) imply small perturbation of the output (data classification). Last but not least, the varied and largely unknown distribution of the attribute values restricts the use of formal statistical methods. The requirements for the mechanism to be employed in such an environment may be summarised as follows:

- The mechanism must be able to acquire new knowledge through additional training without exhibiting "catastrophic forgetting".
- The mechanism must be able to capture divergent classifications of samples often exhibiting only subtle differences.
- The mechanism must be able to conform to varied sizes of additional training sets.
- The mechanism must be able to address potential training sets imbalance.

The mechanism employed to fulfil these requirements relies on a committee of weak classifiers, in which new members are gradually trained and added at the rate of newly produced samples. It uses the algorithm presented in [7], which in turn is inspired by [6], applying the boosting learning methodology as its basis, customised to the aforementioned requirements.

5 Algorithm Discussion

The idea of boosting is based on the fact that a large percentage of training time is dedicated in perfecting an already adequately performing mechanism. Hence, it is proposed that instead of one large highly trained mechanism, a committee of under-trained mechanisms, the *weak learners,* are used in order to achieve integral representation of the concept under investigation. This approach is implemented in [6], yet it relies on the existence of the entire training set for its successful application. This notion is carried further in [7], by allowing partial availability of the training set in the form of subsets at the cost of minimal previous knowledge degradation, yet requiring adequate class and attribute representation in each subset. The approach used in [1], moves one step further, accepting the training data subsets without any previous assumption regarding class representation since it is not realistic to expect so in such an environment. This is achieved by modifying the original algorithm to be able to capture, to an adequate degree, new knowledge based on data subsets of varying quality and size, at the cost of higher previous knowledge degradation which, however, is maintained at acceptable levels as demonstrated by experimental results.

The algorithm starts its operation considering all samples of equal weight, the latter representing their classification difficulty. For each new sample subset, a corresponding subcommittee comprising a predefined number of members is constructed. Back-propagation multi-layered neural networks are used as members of the committee, allowing, however, a high error margin during their training due to their expected role as *weak* rather than *strong* learners in the mechanism. During the subcommittee training, samples not successfully classified receive increased weight. Each new member is conclusively accepted to the subcommittee if it does not degrade overall performance below a given threshold. Not all members are created equal however: each bears a degree of importance, reflecting the weights of samples it successfully classified during its subcommittee creation. This allows later, during the prediction phase, for successful classification of hard-to-classify samples based on a minority vote of high-importance members.

One of the points modified to meet the current environment's requirements was subcommittee creation termination criteria. The original algorithm still requires half the remaining population to be used to train each new committee member. In that phase, the increased weight of hard-to-classify samples just increases their probability to be included in the training set due to the weighted random selection of samples. Creating a dataset including the possibly few remaining hard-to-classify samples and a lot of easy samples mislead the training towards the easy ones. In the customised version used herein we focus, instead, on the samples' weights, using *half of the sum of weights* of remaining samples as threshold instead of *half of the number* of remaining samples. This change, combined with the original algorithm's idea of weighted sample selection, has accelerated generation of adequate weak learners especially in advanced stages of the training process where the original threshold lead to weak learners of small contribution to overall performance. The second major difference is that the original algorithm generates new subcommittee members until the predefined number is reached. We have relaxed this restriction using a threshold for overall error instead, in order to accommodate for varied sizes of new training subsets. Hence, if the new version of the entire committee perfectly classifies all current samples, the subcommittee creation is terminated regardless of current number of members. Despite the undecidability of the general neural network loading problem [10], the computational cost of the applied algorithm is kept low due to weak learning requirement for the committee members and low overall error threshold for successful committee extension, further reduced by pre-processed, fine tuned weak learners' parameters. The algorithm used appears in Fig. 3., a modified version of [7] and a notation index is presented in Table 1.

```
for each new population of n samples s₁, s₂, …, sₙ    // i.e. for each new subcommittee k
    if k>1 then                          // in case the MLC already contains weak learners
        Evaluate new population according to MLCₖ₋₁ and update sample weights wᵢ
    else
        Initialise sample weights wᵢ to 1/n
    p=1
    repeat     // for all weak learners to be created in the subcommittee
        repeat  // check new weak learner until overall performance is acceptable
            repeat  //check new weak-learner until its performance is acceptable
                Create ts as a set of samples with at least half the total weight.
                Create the es with the remaining samples.
                Train a new weak learner using ts.
                Calculate eₚ for the current weak learner for ts and es.
            until eₚ≤0.5
            Add current weak learner to the Cₖ
            Evaluate current population according to MLCₖ
            Calculate E for the current form of the committee
        until E≤0.5
        Store bₚ=eₚ/(1-eₚ) for the current weak learner and increment p by 1
        Update sample weights according to recent evaluation and normalize
    until p>m or E=0
end of for                        // subcommittee k creation loop
```

Fig. 3. Algorithm used

Table 1. Algorithm parameters description

Param.	Description
m	pre-defined maximum number of members of a subcommittee
p	p∈{1..m}=the current number of weak learners in the current subcommittee
n	the number of solutions in the current population, i∈{1..n}
k	number of subcommittees created up to now
C_k	current subcommittee
MLC_k	{C₁,C₂,…,Cₖ}= overall committee, including current form of current subcommittee
g_i	grade vector of solution i in the current population
u_i	user evaluation for solution i
s_i	(gᵢ,uᵢ) = sample in the current population
w_i	sample weight
ts	current training set – created for every weak learner based on weight distribution.
es	current evaluation set – solutions of current population not belonging to the ts.
e_p	product of weights of all solutions classified incorrectly by the current weak learner
b_p	=eₚ/(1-eₚ) *importance* of weak learner's p vote during the overall voting
E	product of weights of all solutions classified incorrectly by the current committee

6 Experimental Results

In the following we briefly present some characteristic experimental results, based on actual user data from the MultiCAD environment. Generalisation error is presented with respect to gradual training of the mechanism with bursts of data originating from 5 descriptions. The mechanism was incrementally trained with evaluated solutions from one description at a time simulating actual system use. Generalisation error is measured against user evaluated solutions, previously unseen by the mechanism. An enhanced set of 8 prominent attributes was used for solution representation in vector form, thus demonstrating significant improvement against previous results.

It is interesting to notice in the results the degradation of the mechanism in the effort to acquire new knowledge by incremental training with a new subset at a time, maintaining, nevertheless, an error margin around 0.1. The generalisation error, concentrating on previously unseen solutions, is also maintained at acceptable levels.

The effort to acquire and automatically apply user preferences is bound to exhibit additional anomalies due to user discrepancies. Table 3 summarises the results for an alternative user. Here we notice that the consideration of training subsets from 2 additional scenes do not improve performance, in contrast with the data subset from scene 5. In the next experiment we alter the sequence of training after scene 2, submitting scene 5 immediately afterwards. The result is a steeper improvement in the generalisation error revealing the potentially more complex nature of scene 5.

Table 2. Error Evolution – User 1

	Initial Training Scene 1	Incr. Training Scene 2	Incr. Training Scene 3	Incr. Training Scene 4	Incr. Training Scene 5
Error for Scene 1	1.94%	14.93%	7.61%	9.40%	6.87%
Error for Scene 2	23.84%	1.95%	7.30%	16.30%	15.33%
Error for Scene 3	9.34%	14.01%	3.38%	13.04%	10.31%
Error for Scene 4	17.36%	3.96%	10.38%	1.89%	5.66%
Error for Scene 5	8.41%	41.68%	17.22%	13.31%	6.85%
Generalisation	14.74%	19.88%	13.80%	13.31%	-
Overall	12.18%	15.31%	9.18%	10.79%	9.00%

Table 3. Error Evolution – User 2

	Initial Training Scene 1	Incr. Training Scene 2	Incr. Training Scene 3	Incr. Training Scene 4	Incr. Training Scene 5
Error for Scene 1	1.49%	5.82%	5.82%	5.82%	2.24%
Error for Scene 2	5.35%	0.24%	0.24%	0.24%	0.97%
Error for Scene 3	24.48%	0.16%	0.16%	0.16%	0.00%
Error for Scene 4	12.64%	0.38%	0.38%	0.38%	0.19%
Error for Scene 5	0.98%	4.89%	4.89%	4.89%	0.20%
Generalisation	10.86%	1.81%	2.63%	4.89%	-
Overall	8.99%	2.30%	2.30%	2.30%	0.72%

Table 4. Error Evolution – User 2/Alternative Sequence

	Initial Training Scene 1	Incr. Training Scene 2	Incr. Training Scene 5	Incr. Training Scene 3	Incr. Training Scene 4
Error for Scene 1	1.49%	5.82%	2.54%	2.24%	5.82%
Error for Scene 2	5.35%	0.24%	0.97%	0.97%	0.24%
Error for Scene 5	0.98%	4.89%	0.20%	0.20%	4.89%
Error for Scene 3	24.48%	0.16%	0.16%	0.00%	0.16%
Error for Scene 4	12.64%	0.38%	2.45%	2.45%	0.38%
Generalisation	10.86%	1.81%	1.31%	2.45%	-
Overall	8.99%	2.30%	1.26%	1.17%	2.30%

7 Conclusion

In the current work we have examined and demonstrated the issues arising in the effort to apply machine learning for the acquisition and exploitation of user preferences in declarative modelling. The particular domain requirements, implied by the declarative modelling methodology and the environments implementing it, include datasets of varying sizes, class representation imbalance, user inconsistencies, inherent data instability and irregular availability. We have customised and applied an incremental learning algorithm based on the notion of boosting demonstrating the flexibility and applicability of this neural-network based approach. The results have revealed adequate generalisation performance, given the aforementioned domain characteristics, while largely maintaining previously acquired knowledge. In particular, the adopted mechanism has maintained acceptable levels of generalisation and overall errors, mainly varying with corresponding user consistency. It also appears to be sensitive to the training sets sequence and is, therefore, affected by user choices in system use. Nevertheless, results demonstrate the applicability of such a mechanism in the specific context.

Due to the idiosyncrasy of the domain, several approaches may be followed to improve the presented results. Identifying and utilizing attributes that capture a wider range of potential user evaluation criteria would offer improved mapping of data properties. In the same category of improvements, methods to ensure the absence of user inconsistencies, possibly in the form of user defined similarity metrics, would allow higher confidence in the available datasets. In the context of the machine learning methodology applied, reduction of the generalisation error could be achieved through guided user evaluation of selected scenes and solutions depicting wide variations in the observed attributes and their values. Lastly, the lack of adequate and consistent user feedback can be overcome through the availability of such a platform in a web environment, thus allowing a wider spectrum of evaluation contributions.

References

[1] Bardis, G.: Intelligent Personalization in a Scene Modeling Environment. In: Miaoulis, G., Plemenos, D. (eds.) Intelligent Scene Modelling Information Systems. SCI, vol. 181, pp. 973–978. Springer, Heidelberg (2009) ISBN 978-3-540-92901-7

[2] Bidarra, R., et al.: Integrating semantics and procedural generation: key enabling factors for declarative modeling of virtual worlds. In: Proceedings of the FOCUS K3D Conference on Semantic 3D Media and Content, France (February 2010)

[3] Bonnefoi, P.-F., Plemenos, D., Ruchaud, W.: Declarative Modelling in Computer Graphics: Current Results and Future Issues. In: Bubak, M., van Albada, G.D., Sloot, P.M.A., Dongarra, J. (eds.) ICCS 2004. LNCS, vol. 3039, pp. 80–89. Springer, Heidelberg (2004)

[4] Champciaux, L.: Classification: a basis for understanding tools in declarative modelling. Computer Networks and ISDN Systems 30, 1841–1852 (1998)

[5] Doulamis, A., et al.: On Line Retrainable Neural Networks: Improving the Performance of Neural Network in Image Analysis problems. IEEE Trans. on Neural Networks 11(1), 137–155 (2000)

[6] Freund, Y., et al.: A decision-theoretic generalization of on-line learning and an application to boosting. Journal of Computer and System Sciences 55(1), 119–139 (1997)

[7] Lewitt, M., et al.: An Ensemble Approach for Data Fusion with Learn++. In: Windeatt, T., Roli, F. (eds.) MCS 2003. LNCS, vol. 2709. Springer, Heidelberg (2003)

[8] Plemenos, D., et al.: Machine Learning for a General Purpose Declarative Scene Modeller. In: International Conference GraphiCon 2002, Nizhny Novgorod, Russia (2002)

[9] Smelik, R.M., et al.: Declarative Terrain Modeling for Military Training Games. International Journal of Computer Games Technology (2010)

[10] Wiklicky, H.: The Neural Network Loading Problem is Undecidable. In: Proceedings of Euro-COLT 1993, Conference on Computational Learning Theory, pp. 183–192. University Press (1994)

Predicting Postgraduate Students' Performance Using Machine Learning Techniques

Maria Koutina and Katia Lida Kermanidis

Department of Informatics, Ionian University,
7 Pl. Tsirigoti, 49100 Corfu, Greece
{c09kout,kerman}@ionio.gr

Abstract. The ability to timely predict the academic performance tendency of postgraduate students is very important in MSc programs and useful for tutors. The scope of this research is to investigate which is the most efficient machine learning technique in predicting the final grade of Ionian University Informatics postgraduate students. Consequently, five academic courses are chosen, each constituting an individual dataset, and six well-known classification algorithms are experimented with. Furthermore, the datasets are enriched with demographic, in-term performance and in-class behaviour features. The small size of the datasets and the imbalance in the distribution of class values are the main research challenges of the present work. Several techniques, like resampling and feature selection, are employed to address these issues, for the first time in a performance prediction application. Naïve Bayes and 1-NN achieved the best prediction results, which are very satisfactory compared to those of similar approaches.

Keywords: Machine Learning, Student Performance Prediction, Class imbalance.

1 Introduction

The application of machine learning techniques to predicting students' performance, based on their background and their in-term performance has proved to be a helpful tool for foreseeing poor and good performances in various levels of education. Thereby tutors are enabled to timely help the weakest ones, but also, to promote the strongest. Apart from this, detecting excellent students can be very useful information for institutions and so forth for allocating scholarships. Even from the very beginning of an academic year, by using students' demographic data, the groups that might be at risk can be detected [1]. The diagnosis process of students' performance improves as new data becomes available during the academic year, such as students' achievement in written assignments and their in-class presence and participation. It has been claimed that the most accurate machine learning algorithm for predicting weak performers is the Naïve Bayes Classifier [1]. Other studies tried to detect attributes, including personality factors, intelligence and aptitude tests, academic achievement and previous college achievements, in order to make an accurate prediction about the final grade of the student. Some of the most significant factors in dropping out, as

L. Iliadis et al. (Eds.): EANN/AIAI 2011, Part II, IFIP AICT 364, pp. 159–168, 2011.

shown in these studies, are: sex, age, type of pre-university education, type of financial support, father's level of education, whether the student is a resident of the university town or not [2], [3], [4], [5]. Although these attributes may change as students move from bachelor to master studies, even more features may differ among students which come from different departments of high-level educational institutes.

This research has several contributions. First, some of the most well-known learning algorithms were applied in order to predict the performance of an MSc student in Informatics (Department of Informatics of the Ionian University in Greece) will achieve in a course taking into account not only his demographic data but also his in-term performance and in-class behaviour. Secondly, the impact of these features, and how it varies for different courses is studied with interesting findings.

Thirdly, an intriguing research aspect of the generated data is the disproportion that arises among the instances of the various class values. This problem, known as class imbalance, is often reported as on obstacle for classifiers [6]. Based on this, a classifier will almost always produce poor accuracy results on an imbalanced dataset [7], as it will be biased in favor of the overrepresented class, against the rare class. Researchers have contemplated many techniques to overcome the class imbalance problem, including resampling, new algorithms and feature selection [7], [8], [9] Resampling, feature selection and combinations of these approaches were applied to address the imbalance. To date, and to the authors' knowledge, no previous research work in marks prediction has combined feature selection with learning algorithms in order to achieve the best performance of the classifiers and to address the class imbalance problem. The results show that the use of learning algorithms combined with feature selection and resampling techniques provide us with more accurate results than simply applying some of the most well-known algorithms.

Finally, providing this information to tutors enables them to adjust their lesson depending on students' demographic and curriculum data and take timely precautions to support them.

The rest of this paper is organized as follows: Section 2 describes the data of our study. Section 3 describes the used learning algorithms and techniques of our study. Section 4 presents the experimental results. Section 5 follows with the results analysis. Finally, Section 6 ties everything together with our concluding remarks and some future research recommendations.

2 Data Description

The Department of Informatics of the Ionian University in Corfu launched a Postgraduate program in the year 2009, under the title "Postgraduate Diploma of Specialization in Informatics". The innovation of this program is that it poses no restrictions to the candidates' previous studies, and accepts graduates from any department (e.g. psychology, physics, economics/management departments e.t.c.). The main goal of this program is to educate graduate students of universities (AEI) and Technological Educational Institutes (TEI) in specialized areas of knowledge and research, in order to enable them to conduct primary scientific research and development tasks in the field of Information Technology.

A total of 117 instances have been collected. The demographic data were gathered from MSc students using questionnaires. Moreover, the in-term performance data of every student were given by the tutor of every course. The data of this study came from three courses of the first semester during the year 2009-2010, namely "Advanced Language Technology" (ALT) (11 instances), "Computer Networks" (CN) (35 instances) and "Information Systems Management" (ISM) (35 instances). Furthermore, the data were enriched with two more courses from the first semester of the year 2010-2011, namely "Advanced Language Technology" (ALT2) (8 instances) and "Computer Networks" (CN2) (28 instances),. Every course is an independent dataset, considering that in-term performance estimation differed among the courses. In some, a student had to submit at least one written midterm assignment, while, in others, midterm assignments were more than one. At this point it should be stressed that some written assignments were team-based while others were not.

The attributes (features) of the datasets are presented in Table 1 and 2 along with the values of every attribute. The demographic attributes represent information collected through the questionnaires from the MSc students themselves, concerning sex, age, marital status, number of children and occupation. Moreover, prior education in the field of Informatics, and the association between the students' current job and computer knowledge were additionally taken into consideration. For example, if a student had an ECDL (European Computer Driving License) that clarifies that (s)he is computer literate, then (s)he would qualify as a 'yes' in computer literacy. Furthermore, students who use software packages in their work (such as a word processor) and students who were graduates of Informatics departments are signified with a 'yes' in their job association with computers, whether they work part-time or full-time.

In addition to the above attributes, others, denoting possession of a second MSc degree, Informatics department graduates, and students who had a four- or five-year University degree or a four-year TEI degree, were also taken into account.

Table 1. Demographic attributes used and their values

Demographic attributes	Value of every attribute
Sex	male, female
Age group	A) [21-25]
	B) [26-30]
	C) [31-35]
	D) [36- ..]
Marital Status	single, married
Number of children	none, one, two or more
Occupation	no, part-time, fulltime
Job associated with computers	no, yes
Bachelor	University, Technological Educational Institute
Another master	no, yes
Computer literacy	no, yes
Bachelor in informatics	no, yes

Table 2. In-term performance attributes and their values

In-term performance attributes	Value of every attribute
1st written assignment	0-10
2nd written assignment	0-10
3rd written assignment	0-10
Presence in class	none, mediocre, good
Final grade	bad, good, very good

In-term performance attributes were collected from tutors' records concerning students' marks on written assignments and their presence in class. Finally, the results on the final examination were grouped into three categories: grades from zero to four were considered to be bad (failing grades), grades from five to seven were noted as good and grades from 8 to 10 were marked as very good.

3 Learning

Six classification algorithms have been used in our study, which are widely-used among the machine learning community [8]. All learners were built using WEKA (Waikato Environment for Knowledge Analysis[1]), a popular suite of machine learning software.

C4.5 decision tree learner was constructed [10] using J48 in WEKA, with the pruning parameter set both on and off. Three different k-nearest neighbors classifiers (denoted IBk in WEKA) were constructed, using k=1, 3 and 5 denoted 1NN 3NN, and 5NN. For these classifiers, all their parameters were left at default values. Experiments were also run using the Naïve-Bayes (NB) classifier, RIPPER (Repeated Incremental Pruning to Produce Error Reduction), which is a rule-based learner (JRIP in WEKA), Random Forest and, finally, a Support Vector Machines (SVMs) learner that uses the Sequential Minimal Optimization (SMO) algorithm for training and a polynomial kernel function.

3.1 The Class Imbalance Problem

During the experimental procedure the problem of class imbalance arose, which is a challenge to machine learning and it has attracted significant research in the last 10 years [7]. As mentioned above, every course is an individual dataset, thus, the smallest dataset has eight instances whereas the biggest has thirty five. Additionally, in some courses it was noticed that final grades either between zero to four or eight to ten had less instances than grades between five to seven. Consequently, classifiers produced poor accuracy results on the minority classes.

Researchers have crafted many techniques in order to overcome the class imbalance problem. One simple technique is to obtain more samples from the minority class, which is not the optimal solution in real-world applications because the imbalance is a congenital part of the data [11]. Other sampling techniques include randomly undersampling the majority class [12], [13], oversampling the minority

[1] http://www.cs.waikato.ac.nz/ml/weka/

class [14], or combining over- and undersampling techniques in a systematic manner. Furthermore, a wide variety of learning methods have been created especially for this problem. These learners achieve this goal by learning using only positive data points and no other background information. One of these learners are SVMs. Last but not least, feature selection is a very promising solution for class imbalance problems; the goal of feature selection is to select a subset of j features that allows a classifier to reach optimal performance, where j is a user-specified parameter [7].

4   Experimental Setup

The training phase was divided into five consecutive steps. The first step included the demographic data along with the in-term performance data and the resulting class (bad, good, very good) for all datasets. Table 3 shows the accuracy of the predictions when training the initial data. In the second step the resample function was applied to the initial data. The resample function in WEKA oversamples the minority class and undersamples the majority class in order to create a more balanced distribution for training algorithms. Table 4 shows the accuracy of the predictions when training with re-sampled datasets.

Step three is performed with attribute evaluation, using the OneR algorithm, in order to identify which attributes have the greatest impact on the class in every dataset. OneR attribute evaluation looks at the odds of a feature occurring in the positive class normalized by the odds of the feature occurring in the negative class [6]. Studies have shown that OneR attribute evaluation proved to be helpful in improving the performance of Naïve Bayes, Nearest Neighbor and SMO [7]. During this phase it was found that attributes such as Bachelor in Informatics, Presence in class, sex and age have a great impact on the class, whereas others, like Marital Status and Another Master degree, are considered redundant.

During the fourth step, the best features (according to the results of the previous step) were selected and the learning algorithms that had the best accuracy results in step 1 were run on them. Table 5 shows the accuracy of J48 (unpruned), 1-NN, NB, Random Forest and SMO in all the datasets.

In the final step the resample filter combined with 7 features that had great influence in all our datasets was applied (Table 6). The features with the highest influence in our datasets are: Bachelor in Informatics, presence in class, sex, age group, first written assignment, job association with Informatics, and number of children.

Table 3. Total accuracy (%) of the initial data

Modules	J48 (pruned)	J48 (unpruned)	1-NN	3-NN	5-NN	NB	RF	SMO	J-Rip
ALT	54.54	54.54	72.72	63.63	54.54	72.72	72.72	72.72	54.54
CN	57.41	40.00	54.28	62.85	62.85	71.42	57.14	71.42	51.42
ISM	51.42	54.28	57.14	60.00	57.14	60.00	51.42	51.42	51.42
ALT2	25.00	25.00	25.00	37.50	25.00	37.50	37.50	25.00	25.00
CN2	57.14	60.71	60.71	50.00	53.57	60.71	57.14	57.14	67.82

Table 4. Total accuracy (%) of re-sampled data

Modules	J48 (pruned)	J48 (unpruned)	1-NN	3-NN	5-NN	NB	RF	SMO	J-Rip
ALT	63.63	81.81	90.90	54.54	45.45	72.72	90.90	72.72	54.54
CN	65.71	71.42	85.71	80.00	74.28	80.00	88.51	82.85	71.42
ISM	80.00	82.85	85.71	60.00	42.85	85.71	88.57	82.85	80.00
ALT2	62.50	62.50	87.50	62.50	67.50	87.50	87.50	87.50	37.50
CN2	82.14	82.14	100.00	64.28	67.85	85.71	96.42	89.28	71.42

Table 5. Total accuracy (%) of best features along with the initial data

Modules	J48 (unpruned)	1-NN	NB	RF	SMO
ALT	54.54	72.72	81.81	63.63	72.70
CN	48.57	57.14	74.28	57.14	68.57
ISM	48.57	51.42	62.85	54.28	62.85
ALT2	25.00	12.50	50.00	12.50	25.00
CN2	42.85	53.57	57.14	42.85	67.85

Table 6. Total accuracy (%) of re-sample data and feature selection

Modules	J48 (unpruned)	1-NN	NB	RF	SMO
ALT	63.63	90.90	90.90	90.90	90.90
CN	65.71	71.42	82.85	74.28	60.00
ISM	68.57	80.00	80.00	80.00	74.28
ALT2	62.50	100.00	100.00	87.50	87.50
CN2	67.85	71.42	71.42	71.42	64.28

5 Results and Discussion

Unlike previous attempts related to grade prediction that focus on a single algorithm and do not perform any form of feature selection, the overall goal of this paper is to find the best combination of learning algorithms and selected features in order to achieve more accurate prediction in datasets with an imbalanced class distribution and a small number of instances.

One important conclusion that can be drawn from the results is that J48 prediction accuracy is much lower than that of NB and 1-NN, which shows that in small datasets NB and 1-NN can perform better than decision trees. Another interesting issue is that the accuracy of prediction, using either the re-sample datasets alone or feature selection combined with resampling, improves when the original data is much smaller in size (ALT, ALT2).

Comparing Table 3 with Table 4, the predictions using the re-sample dataset are significantly more accurate. Furthermore, there is a significant improvement in the accuracy of NB, 1-NN, Random Forest and SMO learning algorithms. For example, the accuracy of module ALT2 using 1-NN with the initial data is 25%, whereas, by using the re-sample technique accuracy rises up to 85.7%. The results (Table 5) show that using only the best features proves to be helpful for NB, but does not help the overall accuracy of the remaining classifiers. Additionally, comparing the results of Table 4 with Table 6, it is evident that NB is the only classifier that is being helped by feature selection. For example in module ALT2 accuracy is noticeably improved when using resampling and feature selection. The above results follow the findings of [2] on how different classifiers respond to different distributions of data.

5.1 Detailed Analysis of the Results

Trying to take a deeper look at the obtained results, it is presented in figure 1 to 5 in detail the f-measure for all class values in all our datasets, throughout the first, second and fifth steps, including only the learning algorithms with predominately best results.

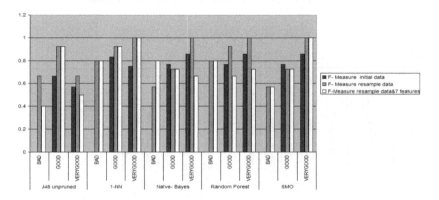

Fig. 1. F-Measure for module Advanced Language Technology (ALT)

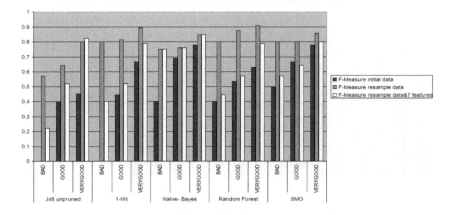

Fig. 2. F-Measure for module Computer Networks (CN)

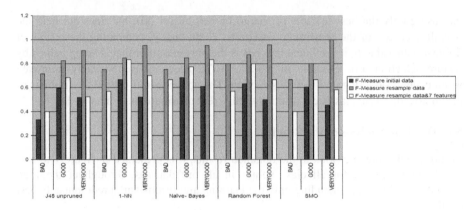

Fig. 3. F-Measure for module Information Systems Management (ISM)

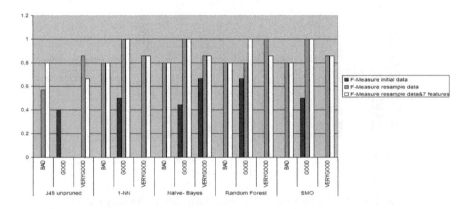

Fig. 4. F-Measure for module Advanced Language Technology 2 (ALT2)

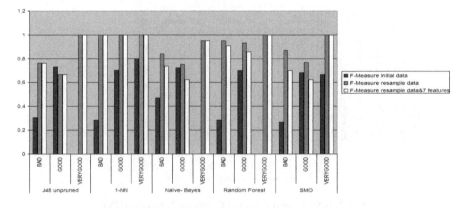

Fig. 5. F-Measure for module Computer Networks 2 (CN2)

The above figures show that in all cases both the resample technique and the combination of feature selection with resampling significantly improve the prediction of minority classes. For example, in module ALT both with NB and with 1-NN, the f-measure for class value *Bad*, using either resampling alone or in combination with feature selection, rises significantly from 0 to 0.8. Moreover, careful inspection of the f-measure in all datasets reveals that the highest scores are always achieved for the majority class label, which is *Good*.

5.2 Feature Selection Results

Apart from the most suitable method in order to predict postgraduate students' performance, it was also attempted to find which attributes influence the most the selected classifiers. The most important attributes in all the datasets were *Presence in class* and *Bachelor in Informatics,* which shows, first, that in-term performance of a student highly affects his final grade, and, secondly, that students who don't have a degree in Informatics are at risk. Another important issue is that in modules, ALT-ALT2 and ISM there is no significant influence from features like Job associated with computers and Computer literacy whereas those attributes were important in modules CN and CN2. Hence, it is believed that if actual predictions are required, it would be better splitting datasets into technical and non-technical lessons and apply on them the same algorithms but with difference selected features.

6 Conclusion

Machine learning techniques can be very useful in the field of grade prediction, considering that they enable tutors to identify from the beginning of the academic year the risk groups in their classes. Hence, this will help them adjust their lesson in order to help the weakest but also to improve the performance of the stronger ones.

An interesting finding from this research work is that NB and 1-NN, combined with resampling alone, or in combination with feature selection, accurately predict the students' final performance, given our datasets, especially when these include a small number of instances.

The overall prediction accuracy in our analysis varies from 85.71% to 100%, learning a discrete class that takes three values. Results are more than promising and enable the future implementation of a student performance prediction tool for the support of the tutors in the Informatics Department of the Ionian University. Furthermore, extending the above tool using regression methods which will predict the exact grade of the student may provide even more fine-grained support to tutors.

Another interesting issue in our study is that the average accuracy of the learning algorithms can be improved by feature selection. It was found that students' occupation, type of bachelor degree (AEI or TEI), and their possession of another master degree do not improve accuracy. Thus, this information is not necessary. However, students' presence in class and their possession of a Bachelor degree in Informatics proved to be very important for the classifiers.

Acknowledgments. The authors would like to thank the tutors and students who participated in this study for their significant contribution.

References

1. Kotsiantis, S., Pierrakeas, C., Pintelas, P.: Predicting Students Performance in Distance Learning Using Machine Learning Techniques. Applied Artificial Intelligence (AAI) 18(5), 411–426 (2004)
2. Parmentier, P.: La reussite des etudes universitaires: facteurs structurels et processuels de la performance academique en premiere annee en medecine. PhD thesis, Catholic University of Louvain (1994)
3. Touron, J.: The determination of factors related to academic achievement in the university: implications for the selection and counseling of students. Higher Education 12, 399–410 (1983)
4. Lassibille, G., Gomez, L.N.: Why do higher education students drop out? Evidence from Spain. Education Economics 16(1), 89–105 (2007)
5. Herzog, S.: Measuring determinants of student return vs. dropout/stopout vs. transfer: A first-to-second year analysis of new freshmen. In: Proc. of 44th Annual Forum of the Association for Institutional Research, AIR (2004)
6. Forman, G., Cohen, I.: Learning from Little: Comparison of Classifiers Given Little Training. In: Proc. Eighth European Conf. Principles and Practice of Knowledge Discovery in Databases, pp. 161–172 (2004)
7. Wasikowski, M., Chen, X.: Combining the Small Sample Class Imbalance Problem Using Feature Selection. IEE Computer Society 22, 1388–1400 (2010)
8. Hulse, J.V., Khoshgoftaar, T.M., Napolitano, A.: Experimental Perspectives on Learning from Imbalanced Data. Appearing in Proceedings of the 24th International Conference on Machine Learning, Corvallis, OR (2007)
9. Barandela, R., Valdovinos, R.M., Sanchez, J.S., Ferri, F.J.: The imbalanced training sample problem: Under or over sampling? In: Fred, A., Caelli, T.M., Duin, R.P.W., Campilho, A.C., de Ridder, D. (eds.) SSPR&SPR 2004. LNCS, vol. 3138, pp. 806–814. Springer, Heidelberg (2004)
10. Quinlan, J.R.: C4.5: Programs for machine learning. Morgan Kaufmann, San Mateo (1993)
11. Chawla, N., Japkowicz, N., Kotcz, A.: Editorial: Special Issue on Learning from Imbalanced Data Sets. ACM SIGKDD Explorations Newsletter 6(1), 1–6 (2004)
12. Kubat, M., Matwin, S.: Addressing the Curse of Imbalanced Data Sets: One Sided Sampling. In: Proc. 14th Int'l Conf. Machine Learning, pp. 179–186 (1997)
13. Chen, X., Gerlach, B., Casasent, D.: Pruning Support Vectors for Imbalanced Data Classification. In: Proc. Int'l Joint Conf. Neural Networks, pp. 1883–1888 (2005)
14. Kubat, M., Matwin, S.: Learning When Negative Examples Abound. In: van Someren, M., Widmer, G. (eds.) ECML 1997. LNCS, vol. 1224, pp. 146–153. Springer, Heidelberg (1997)

Intelligent Software Project Scheduling and Team Staffing with Genetic Algorithms

Constantinos Stylianou[1] and Andreas S. Andreou[2]

[1] Department of Computer Science,
University of Cyprus,
75 Kallipoleos Avenue, P.O. Box 20537,
Nicosia, 1678, Cyprus
cstylianou@cs.ucy.ac.cy
[2] Department of Electrical Engineering and Information Technology,
Cyprus University of Technology,
31 Archbishop Kyprianos Avenue,
P.O. Box 50329, Limassol, 3603, Cyprus
andreas.andreou@cut.ac.cy

Abstract. Software development organisations are under heavy pressure to complete projects on time, within budget and with the appropriate level of quality, and many questions are asked when a project fails to meet any or all of these requirements. Over the years, much research effort has been spent to find ways to mitigate these failures, the reasons of which come from both within and outside the organisation's control. One possible risk of failure lies in human resource management and, since humans are the main asset of software organisations, getting the right team to do the job is critical. This paper proposes a procedure for software project managers to support their project scheduling and team staffing activities – two areas where human resources directly impact software development projects and management decisions – by adopting a genetic algorithm approach as an optimisation technique to help solve software project scheduling and team staffing problems.

Keywords: Software project management, project scheduling, team staffing, genetic algorithms.

1 Introduction

A major problem that still exists in the area of software engineering is the high rate of software project failures. According to the 2009 Standish Group CHAOS Report [1], only 32% of software projects are delivered on time, within budget and with the required functionality, whereas 44% are delivered late, over budget and/or with less than the required functionality. The remaining 24% of software projects are cancelled prior to completion or delivered and never used. These figures reveal that project success rates have fallen from the group's previous study and, more alarmingly, that project failures are at the highest they've been in the last decade. Consequently, as more projects continue to fail, questions need to be asked of project management.

L. Iliadis et al. (Eds.): EANN/AIAI 2011, Part II, IFIP AICT 364, pp. 169–178, 2011.

Questions such as "Was the team technically skilled to undertake the project?" or "Was it the project manager's fault for under-costing and bad planning?" provide a strong incentive to address the shortcomings of software project management by focusing on two activities that project managers undertake. In particular, project scheduling and team staffing are examined because these activities are closely tied with human resources and since human resources are the only major resource of software development organisations, they are areas that warrant investigation. The paper approaches project scheduling and team staffing as an optimisation problem and employs a genetic algorithm to perform optimisation.

The remainder of the paper is organised as follows: section 2 provides an overview of software project scheduling and team staffing activities. Subsequently, section 3 describes in brief how genetic algorithms work and presents how this optimisation technique was designed and adapted to solve the project scheduling and team staffing problem. Section 4 illustrates the experiments carried out on several test projects to evaluate the proposed approach followed by a discussion on the results obtained. Finally, section 5 concludes with a brief summary and notes on future work.

2   Literature Overview

2.1   Software Project Scheduling

One of the main responsibilities of a software project manager is to determine what work will be carried out, how and when it will be done. This responsibility consists of identifying the various products to be delivered, estimating the effort for each task to be undertaken, as well as constructing the project's schedule. Due to the importance of this activity, it should have priority over all others, and furthermore, a project's schedule needs to be updated regularly to coincide with the project's current status.

One of the major issues of project scheduling is that of representation [2]. Specifically, the representations that most project scheduling tools provide "*cannot model the evolutionary and concurrent nature of software development*". One of the most practical challenges project managers face in constructing project schedules is the fact that project scheduling problems are NP-complete. There are no algorithms that can provide an optimal solution in polynomial time, which means that brute force methods are basically inadequate [3, 4]. This problem is heightened due to the fact that software projects are intangible in nature and labour-intensive, and thus involve an even higher level of complexity and uncertainty. As a consequence of the uniqueness of software, project managers cannot completely depend on using experiences from previous projects nor can they use past project information unless used as indicative guidelines. Furthermore, goals may be different with respect to what type of optimisation is required (e.g., minimal cost, maximum resource usage, etc.). Finally, another issue is attributed to the level of information available to construct a project schedule. Similarly to software cost estimation, information available at the start of a project is usually missing or incomplete [2]. Therefore, any mechanism adopted must be able to provide a means to continuously update the schedule [5].

2.2 Software Team Staffing

People and their role in teams are highly important for project success, and they are even taken into account in many cost estimation models, as for instance in the COCOMO model [6]. Therefore, employees in software development organisations should be characterised as *"human capital"*, which is considered the most important asset of the company [7]. The more effectively this capital is managed, the higher the competitive benefit achieved over other organisations. Therefore, when a company undertakes the development of a new project, another of the main responsibilities of a project manager is to decide who will be working on the project. Team formation therefore is a task that, although seemingly easy at first, requires experience and careful execution. Not getting the right team to do the job could possibly lead to overrunning its schedule, exceeding its budget or compromising the necessary quality.

Recent changes in software development processes have also brought on changes into the way project managers view teamwork and this has lead to new approaches and methods being proposed for enhancing teamwork processes. An example of such attempt can be found in [8] who provide a Team Software Process (TSP) as an extension of the Personal Software Process (PSP). Another approach is proposed by [9] who employ a grey decision-making (fuzzy) approach that selects members based on their characteristics. The Analytical Hierarchy Process (AHP) is another method that can be used for team formation as a decision-making tool as shown in [10], who use multifunctional knowledge ratings and teamwork capability ratings. Recently, [11] have presented the Web Ontology Language (OWL) – a model that represents semantic knowledge – as a useful technique for team composition.

As stated in [12], the most common staffing methods available to software project managers rely heavily on the project manager's personal experiences and knowledge. However, these are highly biased techniques and subjectivity does not always yield the correct or best results. Another issue is the fact that because every project is unique, the application of a specific recruiting and staffing method on a project may not yield the expected results as it was applied on another project because of the differences in project characteristics [13]. And this links to the fact that skill-based and experience-based methods are not suitable enough for project managers to deal with interpersonal relationships and social aspects which strongly exist is software development organisations [14].

3   Methodology

3.1 Genetic Algorithm Overview

Project scheduling and team staffing may be considered optimisation problems, and as such will require specialised techniques to be solved. Genetic algorithms are one such optimisation technique, with which it is possible to adequately model the mathematical nature of project scheduling and team staffing.

Genetic algorithms, introduced by John Holland in 1975 [15], work iteratively with populations of candidate solutions competing as a generation, in order to achieve the individual (or the set of individuals) considered as an optimal solution to a problem. Based on the process of natural evolution, their aim is for fitter individual solutions to prevail over those that are less strong at each generation. To achieve this, the fitness of every individual solution is evaluated using some criteria relative to the problem, and subsequently those evaluated highly are more probable to form the population of the next generation. Promoting healthy, better-off individuals and discarding less suitable, weaker individuals in a given generation is aided by the use of variations of the selection, crossover, and mutation operators, which are responsible for choosing the individuals of the next population and altering them to increase fitness as generations progress – making thus the whole process resemble the concept of 'survival of the fittest'.

3.2 Representation and Encoding

For the problem of project scheduling and team staffing, the candidate solutions for optimisation need to represent two pieces of information. On the one hand, schedule constraint information, regarding when and in which order tasks are executed and, on the other hand, skill constraint information, concerning the assignment of employees to tasks based on skill sets and experience required for a task. Fig. 1 below gives an example of the representation of a software project schedule containing 4 tasks and 5 possible employees. As shown, the genetic algorithm uses a mixed-type encoding: schedule information is represented by a positive, non-zero integer symbolising the start day of the task, whereas employee assignment information is represented by a binary code, wherein each bit signifies whether an employee is (a value of 1) or is not (a value of 0) assigned to execute the task.

1	10100	11	00010	16	01001	31	00110

Fig. 1. Example of project schedule representation

In Fig. 1, the first task starts at day 1 and employee 1 and 3 will execute it, task 2 starts at day 11 with only employee 4 assigned to it, and so on.

3.3 Fitness Evaluation Process

The evaluation of the fitness of each individual solution consists of an assessment across the two constraint dimensions that are the focus of this research (i.e., schedule constraints and skill constraints), and for each constraint dimension a corresponding objective function was constructed.

Schedule constraint objective. This objective concerns assessing the degree to which the dependencies of the tasks in the software project are upheld as well as the unnecessary delays existing between dependent tasks. The objective requires a maximisation function and requires information only from the start days of the tasks since it is not affected by which employees are assigned to carry out the task. For each $task_t$ in the project, the fitness is calculated based on Eq. 1. If a task's start day does not satisfy all its predecessor dependencies then it is given a value of zero. Otherwise, the number of idle days between the task and its predecessor task is calculated. The lower the number, the higher the value allocated. In the case where $task_t$ has more than one predecessor tasks, the predecessor task that ends the latest is used for the value $predecessor\_end\_day_t$.

$$f_{dependencies}(task_t) = \begin{cases} 0 & start\_day_t \leq predecessor\_end\_day_t \\ \frac{1}{1+idle\_days} & start\_day_t > predecessor\_end\_day_t \end{cases} \quad (1)$$

where

$$idle\_days = start\_day_t - predecessor\_end\_day_t - 1 . \quad (2)$$

The values obtained for each task in the project are then averaged over the total number of tasks to give the final evaluation for this objective for the individual.

Skill constraint objective. Employees assigned to work on tasks are evaluated based on the degree of experience they possess in the skills required by the task. The objective function in Eq. 3 shows the formula used to make this evaluation. Experience in a skill is not considered as cumulative in this maximisation function and therefore is not presented as simply the summation of the skill experience value of all employees assigned. Instead, for each $skill_s$ required by a task, it makes use of the highest experience value (i.e., the level of the most experienced employee assigned with $skill_s$) and adds to that the mean level of experience of all the employees assigned to work on the task requiring $skill_s$.

$$f_{experience}(skill_s) = \max(experience\_levels_s) + avg(experience\_levels_s) \quad (3)$$

In this way, the objective function helps assign highly experienced employees to a task and simultaneously prevents the assignment of employees without the skills required (i.e., non-contributors) as the average experience of the team will be lowered. As with the previous objective function, the values obtained for each skill are then averaged over the total number of skills to produce an individual's final skill constraint evaluation.

Conflict objective. The two aforementioned objective functions individually target to realise the shortest project duration or to assign the most experienced employees respectively. However, when used together, there will be cases where conflicts will arise due to assigning one or more employees to work on tasks that have been scheduled to execute simultaneously. For this reason a third objective function was created to handle assignment conflicts by taking into account the number of days each $employee_e$ has been assigned to work and how many of these days they have been assigned to more than one permitted task.

$$f_{conflict}(employee_e) = 1 - \frac{conflicting\_days_e}{total\_working\_days_e} \quad (4)$$

Subsequently, the overall fitness for conflicts is computed as the average of all employees. Finally, adding all three evaluations gives an individual's total fitness.

3.4 Parameters and Execution

The parameters selected for the genetic algorithm are summarised in Table 1 below.

Table 1. Genetic algorithm parameters

Population size: 100 individuals	Selection method: Roulette wheel
Maximum number of iterations: 10000	Crossover rate: 0.25
	Mutation rate: $2/chromosome\_length$

The genetic algorithm is initialised with random individuals and is set to execute for a maximum number of iterations. The population is evaluated using the objective functions described in subsection 3.3. However, it should be noted that each objective value is given a weight so as to allow guidance of the genetic algorithm based on the preference of project managers. For instance, a project manager may want to focus primarily on the construction of a project schedule with the shortest possible duration and secondarily on the experience of employees. In such a case, a higher weight will be assigned to the objective function evaluating the schedule constraint and a lower weight will be assigned to the objective function evaluating the skill constraint. If the most experienced employee is assigned to work on two parallel tasks (i.e., leading to a conflict), then preference will be given to keeping the duration of the project the same but assigning the next most experienced to either one of the tasks. Conversely, if a project manager prefers to have a team that is the most experienced and gives lower priority to project duration then, when a conflict occurs, the project schedule will grow in duration so that the most experienced employee remains assigned to the two tasks. Subsequent to the population evaluation, those individual solutions ranked the highest in terms of fitness will be passed on to the next generation after being transformed using crossover and mutation operators. This is repeated until a suitable solution (or number of suitable solutions) has been found.

4 Application and Experimental Results

4.1 Design of Experiments

Experiments were carried out to validate the approach on several aspects. Foremost, it is essential to examine whether the objective functions used in the optimisation algorithm were sufficient enough to produce the correct results individually but also in their competitive environment. For this purpose, a number of test projects of different sizes were designed and constructed that also allowed investigation of the behaviour of the optimisation algorithm. The two test projects are depicted in the task precedence graph (TPG) in Fig. 2 below, which was used in [3]. The smaller test project comprises a subset of 10 tasks (T1-T10) of the TPG, whereas the larger test project contains all the tasks (T1-T15) in the TPG. All dependencies between tasks are finish-to-start, and the duration and the set of skills required for each task is given inside the respective tasks nodes. In addition, Table 2 provides the degree of experience that employees possess in the skills required by the project's tasks.

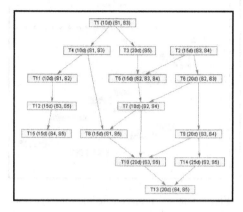

Fig. 2. Task precedence graph for test projects

Table 2. Employees' level of experience for the test projects

	S1	S2	S3	S4	S5
E1	0	0	0.4	0.8	0
E2	0.2	0	0.4	0	0
E3	0	0.8	0	0	0
E4	0	0	0.4	0.8	0.6
E5	0	0.6	0	0	0
E6	0	0.6	0.4	0.8	0
E7	0	0.4	0.4	0	0
E8	0	0.6	0.6	0.8	0
E9	0	0.2	0.4	0	0
E10	0.6	0.4	0	0	0.6

The genetic algorithm was executed thirty times for each test project in each experiment. The results reported in the following subsections contain examples of some of the best executions that converged or that came very close to converging to the optimal solution in each experiment. For Experiment 1, roughly 60% of the executions resulted in finding the optimal solution (i.e., the project schedule with the shortest makespan), whereas for Experiment 2 all executions obtained the optimal solution (i.e., the project schedule with the most experienced employees assigned). Finally, for Experiment 3, only around 5% of executions converged to the correct solution. It should also be noted here that genetic algorithms are random in nature with respect to initialisation and application of genetic operations. Because this randomness affects convergence, that is,. the number of iterations required to find the optimal solution, the time taken to complete an execution varied. Hence, any figures regarding execution time can only give a very rough indication of the overall behaviour of the genetic algorithm in terms of performance. Execution time ranged between 17 sec and 5 min, depending also on the size of the project.

4.2 Results and Discussion

Experiment 1. The first experiment involved executing the genetic algorithm on the test projects to evaluate the objective function for the schedule constraint only. This was done to assess that dependencies between tasks were in fact satisfied and no unnecessary delays existed. An example of the behaviour of the genetic algorithm, one for each test project, can be seen in Fig. 3. For the smaller test project the optimal solution was found around 1000 iterations whereas, as expected, for the larger test project the optimal solution required a higher number of iterations (roughly 3000). Construction of the corresponding optimal project schedules correctly shows that the shortest possible duration for the first test project is 90 days and for the second test project the shortest makespan is 110 days (Fig. 4).

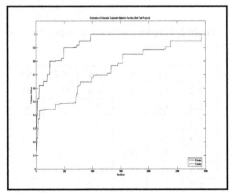 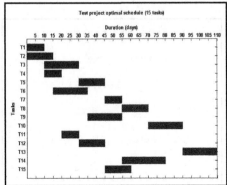

Fig. 3. Evolution of the best individual for the two test projects

Fig. 4. Optimal project schedule for large test project (15 tasks)

Experiment 2. The second experiment examined whether the genetic algorithm was indeed able to find the optimal project team with regards to experience by only evaluating the objective function for the skill constraint. For both test projects, the genetic algorithm successfully managed to assign to each task the most experienced employee or group of employees so that all skills were satisfied and no "idle" or "surplus" employees were used. Table 3 shows an example of a resulting assignment matrix that displays which employees will staff the small software test project.

Table 3. Employee-task assignment matrix (small test project)

	T1	T2	T3	T4	T5	T6	T7	T8	T9	T10
E4	0	0	1	0	0	0	0	0	0	1
E6	0	0	0	0	0	0	1	0	0	0
E8	1	1	0	1	1	1	0	0	1	0
E10	1	0	0	1	0	0	0	1	0	0

Experiment 3. The third and final experiment was carried out to investigate the behaviour of the genetic algorithm when all three objectives were included in the evaluation. Since the objective functions are considered to be competing with each other, each objective function was multiplied by its preference weight as explained in subsection 3.4.

Firstly in this series of experiments, greater preference was given to the schedule constraint than to the skill constraint. The results obtained showed that with all three objective functions active, the genetic algorithm was successfully able to find the optimal schedule for both test projects and, in order to avoid conflicts, managed to assign the next best employees in terms of experience to parallel tasks. An example of the assigned employees of the smaller test project is given in Table 4.

Table 4. Employee-task assignment matrix keeping shortest possible project schedule

	T1	T2	T3	T4	T5	T6	T7	T8	T9	T10
E2	1	0	0	1	0	0	0	0	0	0
E4	0	0	1	0	0	0	0	0	0	1
E6	0	0	0	0	1	0	1	0	0	0
E8	0	1	0	0	0	1	0	0	1	0
E10	1	0	0	1	0	0	0	1	0	0

Secondly, a higher preference was given to the skill constraint and a lower preference to the schedule constraint. This was done to examine whether the genetic algorithm was able to keep the most experienced employees assigned by lengthening the project duration. Results obtained here showed that the genetic algorithm found it difficult to reach an optimal solution. Specifically, runs carried out on the test projects show that, as expected, the genetic algorithm draws its attention to the skill constraint objective (as seen in the example in Fig. 5), and thus the project duration is prolonged (Fig. 6). However, a minor dependence violation in one task of the order of 1 day is caused due to the non-multiobjective nature of the algorithm.

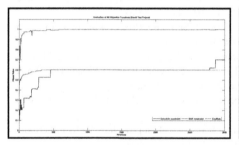 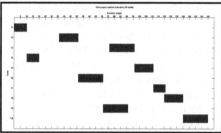

Fig. 5. Evolution of best individual for small test project (10 tasks)

Fig. 6. Project schedule for small test project (10 tasks)

5   Concluding Remarks

This paper presented an approach to solving the problem of software project scheduling and team staffing by adopting a genetic algorithm as an optimisation technique in order to construct a project's optimal schedule and to assign the most experienced employees to tasks. The genetic algorithm uses corresponding objective functions to handle constraints and the results obtained when using either one of the objective functions show that the genetic algorithm is capable of finding optimal solutions for projects of varying sizes. However, when the objective functions were combined, the genetic algorithm presents difficulties in reaching optimal solutions especially when having preference to assign the most experienced employees over the project's duration. Through observation of a number of executions, it was noticed that in this case the genetic algorithm couldn't reduce idle "gaps" or was not able to produce a conflict-free schedule. One possible reason for this observation is due to the competitive nature of the objective functions, and a definite improvement to the approach will be to use multi-objective optimisation rather than using the aggregation of individual objective functions, which is how the genetic algorithm presently works. This could possibly be a means to handle the competitiveness of the objective functions and also allow for a set of optimal solutions to be produced throughout both constraint dimensions simultaneously, thus removing the need for using weights to give preference to either one of the constraints. Also, refinements may be needed to help the algorithm escape from local optima and, thus, improve its convergence rate.

References

1. Standish Group: Standish Group CHAOS Report. Standish Group International, Inc., Boston (2009)
2. Chang, C.K., Jiang, H., Di, Y., Zhu, D., Ge, Y.: Time-Line Based Model for Software Project Scheduling with Genetic Algorithms. Inform. Software Tech. 50(11), 1142–1154 (2008)
3. Chang, C.K., Christensen, M.J., Zhang, T.: Genetic Algorithms for Project Management. Ann. Softw. Eng. 11(1), 107–139 (2001)
4. Pan, N., Hsaio, P., Chen, K.: A Study of Project Scheduling Optimization using Tabu Search Algorithm. Eng. Appl. Artif. Intel. 21(7), 1101–1112 (2008)
5. Joslin, D., Poole, W.: Agent-based Simulation for Software Project Planning. In: 37th Winter Simulation Conference, pp. 1059–1066. IEEE Press, New York (2005)
6. Boehm, B.W.: Software Engineering Economics. Prentice Hall Inc., New Jersey (1981)
7. Acuña, S.T., Juristo, N., Moreno, A.M., Mon, A.: A Software Process Model Handbook for Incorporating People's Capabilities. Springer, New York (2005)
8. Humphrey, W.S.: The Team Software Process^{SM} (TSP^{SM}). Technical Report, Carnegie-Mellon University (2000)
9. Tseng, T.-L., Huang, C.-C., Chu, H.-W., Gung, R.R.: Novel Approach to Multi-Functional Project Team Formation. Int. J. Proj. Manage. 22(2), 147–159 (2004)
10. Chen, S.-J., Lin, L.: Modeling Team Member Characteristics for the Formation of a Multifunctional Team in Concurrent Engineering. IEEE T. Eng. Manage. 51(2), 111–124 (2004)
11. Chi, Y., Chen, C.: Project Teaming: Knowledge-Intensive Design for Composing Team Members. Expert Sys. Appl. 36(5), 9479–9487 (2009)
12. Acuña, S.T., Juristo, N., Moreno, A.M.: Emphasizing Human Capabilities in Software Development. IEEE Softw. 23(2), 94–101 (2006)
13. Wi, H., Oh, S., Mun, J., Jung, M.: A Team Formation Model Based on Knowledge and Collaboration. Expert Sys. Appl. 36(5), 9121–9134 (2009)
14. Amrit, C.: Coordination in Software Development: The Problem of Task Allocation. In: 27th International Conference on Software Engineering, pp. 1–7. ACM, New York (2005)
15. Holland, J.H.: Adaptation in Natural and Artificial Systems. University of Michigan Press, Michigan (1975)

Comparative Analysis of Content-Based and Context-Based Similarity on Musical Data

C. Boletsis, A. Gratsani, D. Chasanidou, I. Karydis, and K. Kermanidis

Dept. of Informatics, Ionian University, Kerkyra 49100, Greece
{c10bole,c10grat,c10chas,karydis,kerman}@ionio.gr

Abstract. Similarity measurement between two musical pieces is a hard problem. Humans perceive such similarity by employing a large amount of contextually semantic information. Commonly used content-based methodologies rely on information that includes little or no semantic information, and thus are reaching a performance "upper bound". Recent research pertaining to contextual information assigned as free-form text (tags) in social networking services has indicated tags to be highly effective in improving the accuracy of music similarity. In this paper, we perform a large scale (20k real music data) similarity measurement using mainstream content and context methodologies. In addition, we test the accuracy of the examined methodologies against not only objective metadata but real-life user listening data as well. Experimental results illustrate the conditionally substantial gains of the context-based methodologies and a not so close match these methods with the real user listening data similarity.

1 Introduction

For a classic rock lover, Led Zeppelin's "Kashmir" and Deep Purple's "Perfect Strangers", may be two similar songs while for a hip-hop lover the very same songs may be completely different and an association of Led Zeppelin's "Kashmir" with Puff Daddy's "Come with me" is quite possible. The aforementioned example portrays just one scenario of the purely subjective nature of music similarity assessment and the problem that its measurement poses [27,9].

Despite the inherent difficulties in assessing musical similarity, its function is of high value to numerous areas of Music Information Retrieval (MIR) [9]. Based on music-similarity measures [9]: (a) listeners can query using performed or hummed parts, (b) music researchers can identify recurring parts in different works, (c) the music industry offers music discovery tools in order to support potential buyers, and (d) music professionals and amateurs can organise their music effectively.

Musical similarity depends on the characterising attributes of the musical data to be compared and thus has been focused on three key directions: the objective metadata accompanying the musical works, the actual musical content and the contextual information humans assign on everything music.

L. Iliadis et al. (Eds.): EANN/AIAI 2011, Part II, IFIP AICT 364, pp. 179–189, 2011.
© IFIP International Federation for Information Processing 2011

Objective metadata, such as the song title, the singer name, the composer name or the genre of a musical piece can be used to assess music similarity. However, methods using metadata are in some cases not effective since metadata may be unavailable, their use requires knowledge that is, in general, not conveyed by listening, and in addition have limited scope, as these rely on predefined descriptors [9].

Content-based similarity focuses on features extracted from the audio content. This task appears as a common process for humans due to the powerful ability of the brain to utilise an enormous amount of contextually semantic information for the process of identifying similarities and differences between sounds as well as classifying these sounds [6,21]. On the contrary, in automated computer systems, the equivalent process based on content extracted features is much more difficult as the attributes expressed by the extracted features are of very little or lacking any semantic meaning [21].

Contextual knowledge, on the other hand, is derived from the information that humans apply to music through the practice of appointment free-form text (a.k.a. tags) on musical data on the web. Based on the previously mentioned ability of the human brain to utilise contextual information for music similarity and the rich contextually semantic nature of the human-generated information that is assigned to the musical works, the important role of tagging in MIR comes as no surprise. Consequently, measurements of musical similarity based on tags are in cases [19,9] reported more accurate than content-based measurements. However, contextual information is no panacea, as far as music similarity is concerned and a number of issues are reported [12] to burden its use.

1.1   Contribution and Paper Organisation

In this paper, we compare and evaluate content-based versus context-based approaches for measuring music similarity. The contribution of this work is summarised as follows:

- Perform large scale (20k tracks) similarity measurement using mainstream content and context methodologies.
- Measure the accuracy of the examined methodologies against not only metadata but real-life user listening data.

The rest of the paper is organised as follows. Section 2 describes background and related work, Section 3 provides a complete account of the similarity measurement methods examined. Next, Sectiom 4 describes the context-based similarity approach examined herein. Subsequently, Section 5 presents and discusses the experimentation and results obtained, while the paper is concluded in Section 6.

2   Related Work

Music information retrieval has been under extensive research in the last decade and *similarity measurement* has been at the very core of the research [16,22,23,27,3,2,5] due to its importance to numerous areas of MIR.

Content-based similarity has been the corner-stone of automated similarity measurement method in MIR and most research [24,16,22,3,2,5,11] is focused in this direction. Content-based approaches assume that documents are described by features extracted directly from the content of musical documents. Accordingly, the selection of appropriate features is very important as meaningful features offer effective representation of the objects and thus accurate similarity measurements. The work of Pampalk [22,25] on Single Gaussian Combined, as submitted to the MIREX 2006 [20] is of high importance as it achieved the highest score and in addition, in current literature, spectral measures are receiving an ever growing interest as these describe aspects related to timbre and model the "global sound". In the direction of content-based feature usage and in order to alleviate the burden of programming for the extraction of features, McEnnis et al. [17,18] developed a feature extraction library.

In contrast to content-based attributes of the musical data, context-based information refers to semantic metadata appointed by humans. Initial research in this direction focused in mining information from the web [4,10] for the purposes of artist classification and recommendation. Nevertheless, the widespread penetration of "Web 2.0" enabled web users to change their previous role of music consumers to contributors [8] by simply assigning tags information on musical data. The increased appeal of the tagging process led to the assignment of large amounts of such information on everything musical. Accordingly, research [12,14,15] expanded in this direction in order to measure the similarity of musical content. Lamere [12] explores the use of tags in MIR as well as issues and possible future research directions for tags. Finally, Levy and Sandler [14] present a number of information retrieval models for music collections based on social tags.

3   Content-Based Similarity

Content-based approaches assume that documents are described by features extracted directly from the content of musical documents [11]. In our analysis, we experiment with two widely known cases: (a) content feature extraction based on the jAudio application [17] that produces a set of, generic for the purposes of MIR, features and (b) the more MIR specific Single Gaussian Combined method, as implemented in the MA Toolbox Matlab library [22], that was shown to perform more than adequately in the MIREX contests.

3.1   Generic Features

MIR processes depend heavily on the quality of the extracted audio features [18]. The performance of a classifier or other interpretive tool is defined by the quality of the extracted features. Thus, poor-quality features will result in the poor performance of the classifier. The extracted features can be portrayed as a "key" to the latent information of the original data source [18]. Since, in our study, we focus on the interpretive layer, we created and maintained a large array of features. For the extraction of these features the jAudio application was used.

jAudio is an application designed to extract features for use in a variety of MIR tasks [18]. It eliminates the need for reimplementing existing feature extraction algorithms and provides a framework that facilitates the development and deployment of new features [18].

jAudio is able to extract numerous basic features [17]. These features may be one-dimensional (e.g., RMS), or may consist of multi-dimensional vectors (e.g., MFCC's) [18]. Metafeatures are feature templates that automatically produce new features from existing features [18]. These new features function just like normal features-producing output on a per-window basis [18]. Metafeatures can also be chained together. jAudio provides three basic metafeature classes (Mean, Standard Deviation, and Derivative).

For the purposes of our experimentation we retained the following features: spectral centroid, spectral roll-off point, spectral flux, compactness, spectral variability, root mean square, fraction of low energy windows, zero crossings, strongest beat, beat sum, strength of strongest beat, first thirteen MFCC coefficients, first ten LPC coefficients and first five method of moments coefficients.

3.2 Targeted Features

In order to proceed to the extraction of targeted features, we utilised the feature extraction process based on the Single Gaussian Combined (G1C) [23]. Initially, for each piece of music the Mel Frequency Cepstrum Coefficients (MFCCs) are computed, the distribution of which is summarised using a single Gaussian (G1) with full covariance matrix [22]. The distance between two Gaussians is computed using a symmetric version of the Kullback-Leibler divergence. Then, the fluctuation patterns (FPs) of each song are calculated [22]. The FPs describe the modulation of the loudness amplitudes per frequency bands, while to some extent it can describe periodic beats. All FPs computed for each window are combined by computing the median of all patterns. Accordingly, two features are extracted from the FP of each song, the gravity (FP.G) which is the centre of gravity of the FP along the modulation frequency dimension and the bass (FP.B) which is computed as the fluctuation strength of the lower frequency bands at higher modulation frequencies [22]. For the four distance values (G1, FP, FP.B and FP.G) the overall similarity of two pieces is computed as a weighted linear combination (normalised in [0,1]) as described in detail in [23].

4   Context-Based Similarity

As far as contextual information is concerned, as tags are free-form text assigned by users, it requires preprocessing. Accordingly we employed Latent Semantic Analysis (LSA) [7], in order to alleviate the problem of finding relevant musical data from search tags [12]. The fundamental difficulty arises when tags are compared to find relevant songs, as the task eventually requires the comparisons of the meanings or concepts behind the tags. LSA attempts to solve this problem by mapping both tags and songs into a "concept" space and doing the comparison in this space. For this purpose, we used Singular Value Decomposition (SVD) in order to produce a reduced dimensional representation of the term-document matrix that emphasises the strongest relationships and reduces noise.

5   Performance Evaluation

In this section we experimentally compare the accuracy of the content and context based methods using as groundtruth both the metadata of the tracks and the similarity provided Last.fm [13] web service based on real-life user listening data. We initially describe the experimental set-up, then present the results and finally provide a short discussion.

5.1   Experimental Setup

For the purposes of performance evaluation of the alternative methods to compute similarity we accumulated two datasets from web services. The first dataset, henceforth titled *dataset A*, comprises of data selected for their high volume of contextual information, tags, as assigned in the Last.fm. The aforementioned web service does in addition provide, for most of the tracks, other tracks that are similar to them, based on user listening data. Thus, the second dataset, henceforth titled *dataset B*, comprises of tracks that are similar to the tracks of dataset A, following the information provided by Last.fm.

- **Audio:** Content data were harvested from iTunes [1] using the iTunes API. Track selection for dataset A was based on the cumulative highest popularity tags offered for a track in Last.fm by selecting the fifty top rank tracks for each top rank tag. Track selection for dataset B was based on their similarity to the tracks of dataset A following the information provided by Last.fm. The data gathered contain $5,460$ discrete tracks for dataset A and $14,667$ discrete tracks for dataset B, retaining only the first 10 most similar tracks for each track of dataset A. Each track is a 30 second clip of the original audio, an audio length commonly considered in related research [28,20].
- **Social tags:** For each track accumulated, the most popular tags assigned to it at Last.fm were gathered using the Last.fm API. The data gathered contain more than $165,000$ discrete tags. Although Last.fm had a very large number of tags per track, our selection was based on the number of times a specific tag has been assigned to a track by different users.
- **External metadata:** For each track gathered from iTunes, its respective metadata concerning the track's title, artist, album and genre were also stored. In contrast to the former two types of data, audio and social tags, the external metadata where merely used as a means for evaluating the accuracy of computed similarity. In following experimentation we focus on genre information, which is commonly used for evaluating similarity measures [20,9].

As far as the audio content data is concerned, the representation of tracks in our experimentation is based on the following two schemes: (a) Content features: spectral centroid, spectral roll-off point, spectral flux, compactness, spectral variability, root mean square, fraction of low energy windows, zero crossings, strongest beat, beat sum, strength of strongest beat, first thirteen MFCC coefficients, first ten LPC coefficients and first five method of moments coefficients, as described in Section 3.1. Extraction was achieved using the *jAudio* [18] application for each entire musical datum producing thus a single content feature point of 39 dimensions per track. (b) Content features: Single Gaussian Combined (G1C) as described in Section 3.2. Extraction was achieved through *MA Toolbox*, a collection of Matlab functions that implement G1C, as described in [23]. Throughout the remainder of this paper, the latter scheme is used except when explicitly stated otherwise.

For the social tags, each tag has been pre-processed, in order to remove stop words that offer diminished specificity, and additionally stemmed, in order to reduce inflected or derived words to their stem using the algorithm described by Porter [26]. Moreover, tags were further processed using the LSA method as already described in Section 4 in order to minimise the problem of finding relevant musical data from search tags. To this end, the SVD method has been used in order to produce a reduced dimensional representation of the term-document matrix that emphasises the strongest relationships and discards noise. Unless otherwise stated, the default value of dimensions for the SVD method was set to 50 dimensions.

Initially we tested the methodologies examined herein solely in dataset A. Accordingly, Figures 1 and 2 report results on similarity measurement accuracy just for dataset A. On the other hand, Figures 3, 4 and 5 present results concerning the incorporation of dataset B into the similarity measurement process, following the similarity results of Last.fm, in order to use it as a groundtruth. Thus, the intuitive result of using real user listening data as a groundtruth similarity is to observe the capability of the examined methodologies to measure similarity similarly to the manner real-life users would.

For the evaluation of the similarity between tracks, we used the precision resulting from the k nearest neighbors (k-NN) of a query song, i.e., for each query song we measured the fraction of its k-NN that share the same genre with the query song. In the cases that employ both datasets A & B, queries are selected from dataset A while similar matches are retrieved from both datasets.

5.2 Experimental Results

In the first experiment, Figure 1(left), we tested the accuracy of similarity measurement using solely the content of tracks from subset A. For this experiment we utilised the features extracted using the jAudio application representing thus each track by a 39 dimension vector. This experiment verifies that for a generic set of features, extracted from the content of a track, the mean precision is very low, serving thus as a key motivation factor for the development of methodologies that perform better. In the next experiment, we examined the attained accuracy

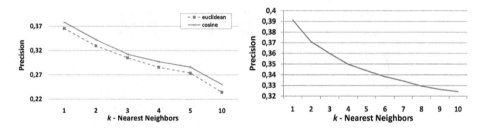

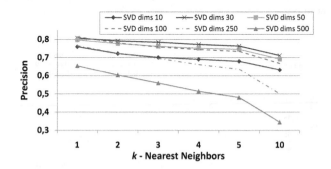

Fig. 1. Dataset A - content, mean precision vs. kNNs, using features extracted from jAudio (left) and from MA-Toolbox (right)

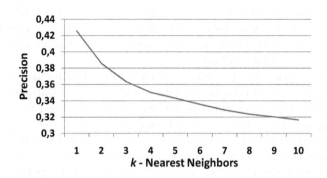

Fig. 2. Dataset A - context, mean precision vs. kNNs vs. SVD dims

in computed similarity utilising the features included in the MA-Toolbox. Figure 1 (right) presents the resulting precision for varying k number of nearest neighbors using the G1C features. As in the previous result, the initial setting accuracy provided by the MA-Toolbox is comparable to the accuracy provided by the generic set of features.

Fig. 3. Dataset A&B - content, mean precision vs. kNNs

Continuing further, the next experiment presents the accuracy of the similarity measurement using the contextual information of the dataset A tracks.

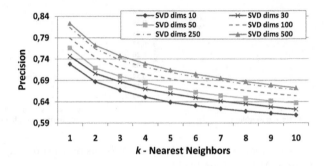

Fig. 4. Dataset A&B - context, mean precision vs. kNNs vs. SVD dimensions using the tracks' metadata.

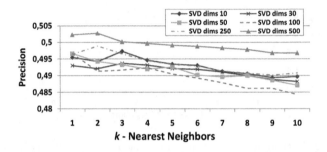

Fig. 5. Dataset A&B - context, mean precision vs. kNNs vs. SVD dimensions using the similarity by Last.fm.

Figure 2 clearly shows that the accuracy of similarity measurement in the tag feature space outperforms similarity in the audio feature space. In addition, the effect of the SVD dimensionality reduction can also be seen: an increase in the dimensions utilised in SVD has a clear augmenting impact on the precision of the resulting similarity. Still, for larger increase, the ability of SVD to emphasise the strongest relationships and discard noise in data, diminishes and so does the precision of the resulting similarity.

The following experiment, Figure 3 aims in providing further insight as to the attained accuracy in computed similarity utilising the features included in the MA-Toolbox using both datasets. Once again, the resulting precision is very low, following the previously mentioned result in Figure 1 (right).

In the next experiment, as shown in Figure 4, we tested the similarity measurement using the contextual information of both dataset A & B. Again, it is clearly shown that the accuracy of similarity measurement in the tag feature space outperforms similarity in the audio feature space, following the result of Figure 2.

Finally, we examined the accuracy in similarity measurement using both datasets relying on the contextual in formation of the tracks. The groundtruth in this case is the similarity based on real user listening data from Last.fm. As it can be seen in Figure 5 the contextual information provided by tags offers increased discriminating capability in comparison to the features extracted from the content of the track. Nevertheless, the examined methodology for the calculation of the similarity does not match closely the real user listening data similarity of and thus offering not as high accuracy.

5.3   Discussion

The presented performance evaluation results can be summarised as follows:

- The generic tag-based approach utilised herein outperforms the audio-based method for all k-NN values given the ample amount of tags per track. This result is in accordance with relevant research stating that the contextual information provided by tags is known to offer increased discriminating capability for the purposes of MIR.
- The similarity measurement methodologies examined herein fail to closely match the real user listening data similarity, providing motivation for techniques that will offer higher accuracy.
- The effect of the SVD dimensionality reduction is of importance to the accuracy of the examined methodology and thus requires tuning.

6   Conclusion

Measuring music similarity is a research area that is of great importance for the purposes of music information retrieval. Different directions exist as to which attributes of a musical datum to retain in order to estimate the similarity between songs. The most common approaches focus on datum metadata, content-based extracted features and "web 2.0" contextual information. Each alternative presents a number of advantages and disadvantages.

In this work, we examine the accuracy of commonly utilised methodologies to musical similarity calculation based on content-based extracted features and "web 2.0" contextual information of the musical data. In addition to common practice groundtruth based on objective metadata we also employ real-life user preference based similarity as provided by Last.fm web service. Experimental results indicate the superiority of the methods based on contextual information and in addition a not close match of these methods to the similarity as perceived by the real-life user preferences.

Future research directions include the examination of more methods that utilise contextual information for musical similarity, experimentation on the number of tags required per musical track in order to establish high accuracy results and the identification of methods that result to a closer match with user perceived similarity.

References

1. Apple: iTunes - Everything you need to be entertained, http://www.apple.com/itunes/
2. Aucouturier, J.J., Pachet, F.: Music similarity measures: What's the use? In: Proc. International Symposium on Music Information Retrieval, pp. 157–1638 (2003)
3. Aucouturier, J.J., Pachet, F.: Improving timbre similarity: How high is the sky? Journal of Negative Results in Speech and Audio Sciences 1 (2004)
4. Baumann, S., Hummel, O.: Using cultural metadata for artist recommendations. In: International Conference on Proc. Web Delivering of Music (2003)
5. Berenzweig, A., Logan, B., Ellis, D.P.W., Whitman, B.P.W.: A large-scale evaluation of acoustic and subjective music-similarity measures. Computer Music Journal 28, 63–76 (2004)
6. Byrd, D.: Organization and searching of musical information, course syllabus (2008), http://www.informatics.indiana.edu/donbyrd/Teach/I545Site-Spring08/SyllabusI545.html
7. Dumais, S.T., Furnas, G.W., Landauer, T.K., Deerwester, S.: Using latent semantic analysis to improve information retrieval. In: Proc. Conference on Human Factors in Computing, pp. 281–285 (1988)
8. Karydis, I., Laopodis, V.: Web 2.0 cultural networking. In: Proc. Pan-Hellenic Conference in Informatics (2009)
9. Karydis, I., Nanopoulos, A.: Audio-to-tag mapping: A novel approach for music similarity computation. In: Proc. IEEE International Conference on Multimedia & Expo (2011)
10. Knees, P., Pampalk, E., Widmer, G.: Artist classification with web-based data. In: Proc. International Symposium on Music Information Retrieval, pp. 517–524 (2004)
11. Kontaki, M., Karydis, I., Manolopoulos, Y.: Content-based information retrieval in streaming music. In: Proc. Pan-Hellenic Conference in Informatics, pp. 249–259 (2007)
12. Lamere, P.: Social tagging and music information retrieval. Journal of New Music Research 37, 101–114 (2008)
13. Last.fm: Listen to internet radio and the largest music catalogue online, http://www.last.fm
14. Levy, M., Sandler, M.: Learning latent semantic models for music from social tags. Journal of New Music Research 37(2), 137–150 (2008)
15. Levy, M., Sandler, M.: Music information retrieval using social tags and audio. IEEE Transactions on Multimedia 11, 383–395 (2009)
16. Logan, B., Ellis, D.P.W., Berenzweig, A.: Toward evaluation techniques for music similarity. In: Proc. International Conference on Multimedia & Expo 2003 (2003)
17. McEnnis, D., McKay, C., Fujinaga, I.: jAudio: A feature extraction library. In: Proc. International Conference on Music Information Retrieval (2005)
18. McEnnis, D., McKay, C., Fujinaga, I.: jAudio: Additions and improvements. In: Proc. International Conference on Music Information Retrieval, p. 385 (2006)
19. McFee, B., Barrington, L., Lanckriet, G.: Learning similarity from collaborative filters. In: International Society of Music Information Retrieval Conference, pp. 345–350 (2010)
20. MIREX: Music Information Retrieval Evaluation eXchange
21. Mitrovic, D., Zeppelzauer, M., Breiteneder, C.: Features for content-based audio retrieval. In: Advances in Computers: Improving the Web, vol. 78, pp. 71–150. Elsevier, Amsterdam (2010)

22. Pampalk, E.: Audio-based music similarity and retrieval: Combining a spectral similarity model with information extracted from fluctuation patterns. In: Proc. International Symposium on Music Information Retrieval (2006)
23. Pampalk, E.: Computational Models of Music Similarity and their Application in Music Information Retrieval. Ph.D. thesis, Vienna University of Technology, Vienna, Austria (2006)
24. Pampalk, E., Dixon, S., Widmer, G.: On the evaluation of perceptual similarity measures for music. In: Proc. International Conference on Digital Audio Effects, pp. 7–12 (2003)
25. Pampalk, E.: MA Toolbox, `http://www.pampalk.at/ma/`
26. Porter, M.F.: The porter stemming algorithm, `http://tartarus.org/~martin/PorterStemmer/`
27. Slaney, M., Weinberger, K., White, W.: Learning a metric for music similarity. In: Proc. International Conference on Music Information Retrieval, pp. 313–318 (2008)
28. Wang, D., Li, T., Ogihara, M.: Are tags better than audio? The effect of joint use of tags and audio content features for artistic style clustering. In: Proc. International Society for Music Information Retrieval, pp. 57–62 (2010)

Learning Shallow Syntactic Dependencies from Imbalanced Datasets: A Case Study in Modern Greek and English

Argiro Karozou and Katia Lida Kermanidis

Department of Informatics, Ionian University
7 Pl. Tsirigoti, 49100 Corfu, Greece
argykaroz@gmail.com, kerman@ionio.gr

Abstract. The present work aims to create a shallow parser for Modern Greek subject/object detection, using machine learning techniques. The parser relies on limited resources. Experiments with equivalent input and the same learning techniques were conducted for English, as well, proving that the methodology can be adjusted to deal with other languages with only minor modifications. For the first time, the class imbalance problem concerning Modern Greek syntactically annotated data is successfully addressed.

Keywords: shallow parsing, Modern Greek, machine learning, class imbalance.

1 Introduction

Syntactic analysis is categorized into full/deep parsing -where a grammar and a search strategy assign a complete syntactic structure to sentences- and shallow parsing - finding basic syntactic relationships between sentence elements [15]. Information extraction, machine translation, question-answering and natural language generation are widely known applications that require shallow parsing as a pre-processing phase.

Shallow parsing may be rule-based [1][12][2] or stochastic [8][5]. Rule-based approaches are expensive and labor-intensive. Shallow parsers usually employ techniques originating within the machine learning (or statistical) community [19]. Memory Based Sequence Learning (MBSL) [3][18], and Memory-based learning (MBL) [9][21] have been proposed for the assignment of subject-verb and object-verb relations. The authors in [9] (the approach closest to the one described herein) provide an empirical evaluation of the MBL approach to syntactic analysis on a number of shallow parsing tasks, using the WSJ Treebank corpus [17]. Their reported f-measure is 77.1% for subject detection and 79.0% for object detection. The same techniques have been implemented for Dutch [4]. Regarding Modern Greek (MG), there is meager work in parsing and most of it refers to full parsing. Chunking and tagging have been attempted using Transformation-based error-driven learning [20].

This led to the idea of creating a shallow parser for MG. The present work deals with finding subject-verb and object-verb syntactic relations in MG text. A unique label (tag) is assigned to each NP-VP pair in a sentence. In an attempt to research the language-independence of the methodology, it is applied to English text as well, and

L. Iliadis et al. (Eds.): EANN/AIAI 2011, Part II, IFIP AICT 364, pp. 190–195, 2011.

its performance in the two languages is compared. Furthermore, a basic problem is addressed, namely the imbalance of the learning examples of each class in the data, the so-called *class imbalance problem*. State-of-the-art techniques, like resampling, are employed for the first time to the authors' knowledge, to deal with this data disproportion in the task at hand, and the results are more than encouraging. Finally, the approach relies on limited resources, i.e. elementary morphological annotation and a chunker that uses two small keyword and suffix lexica to detect non-overlapping phrase chunks. Thereby the methodology is easily adaptable to other languages that are not adequately equipped with sophisticated resources.

MG has a rich morphology and does not follow the subject-verb-object (SVO) ordering schema. For instance, /ipia gala xthes/ (I drank milk yesterday), /gala ipia xthes/ and /xthes ipia gala/ are all syntactically correct and semantically identical. MG is a pro-drop (pronoun drop) language since the subject may be omitted. In MG verbs agree with their subject in gender and number. Subjects and predicates (nominals denoting a property of the subject, also called "copula") are in the nominative case, objects in the accusative and genitive case.

2 Data Collection

The MG text corpus used for the experimental process comes from the Greek daily newspaper "Eleftherotypia" (http://www.elda.fr/catalogue/en/text/W0022.html) and includes 3M words. A subset of the corpus (250K words) is morphologically annotated and automatically chunked by the chunker described in [20].The present methodology focuses on the identification of syntactic relations concerning the subject and object relations that NPs have with VPs in the same sentence. This restriction possibly excludes useful relations, the central meaning of a sentence, however, can be retrieved quite well considering just these relations.

Each NP-VP pair in a corpus sentence constitutes a learning instance. During feature extraction, i.e. the morphosyntactic features (21 in number) that represent the pair and its context and affect subject and object dependencies between an NP and a VP were selected: the case, the number and the person of the headword of the NP, the person, the number and the voice of the head word of the VP, the distance (number of intervening phrases) between the NP and VP, the verb type, i.e. whether it is connective (είμαι-to be, γίνομαι–to become, φαίνομαι–to seem), or impersonal (πρέπει-must, μπορεί-may, βρέχει-to rain, πρόκειται–about to be, etc.), the part of speech of the headword of the NP and VP, the number of commas, verbs, coordinating conjunctions and other conjunctions between the NP and VP, and the types of the phrases up to two positions before and after the NP and the VP. The dataset consisted of 20,494 learning instances, which were manually annotated with the correct class label ("Subject" for a subject-verb dependency, "Object" for a verb-object dependency, and "NULL" otherwise). The annotation process required roughly about 50 man-hours.

The same process was applied to an English corpus so as to compare and contrast it to the Greek shallow parser, and find out how well a parser, that utilizes equivalent resources and the same methodology can work for English. The corpus used was Susanne (www.cs.cmu.edu/afs/cs/project/ai-repository/ai/areas/nlp/corpora/susanne), about 130K words of written American English text. Syntactic tree structures were flattened into IOB-format sentence structures of consecutive non-overlapping chunks, using Buchholz's software (http://ilk.uvt.nl/team/sabine). Thereby, the input data is equivalent for both languages. 6,119 NP-VP pairs were transformed into learning instances, and were manually annotated with one of the three class labels. Features were selected to describe morphosyntactic information about the NP-VP pair, taking into account the relevant properties of the language: the voice of the VP, whether the VP is infinitive or has a gerund form, its number and person, the part of speech of the headword of the NP and VP, the number of the headword of the NP, the distance (number of intervening phrases) and the number of conjunctions and verbs between the NP and VP and the types of the phrases up to two positions before and after the NP and the VP.

A disproportion of the class distribution in both datasets was clearly noticeable. In the MG data, from the 20,494 total learning instances, 16,335 were classified as null and only 4,150 were classified as subject or object while in the English data, from the 6,119, 4,562 were classified as null and just 1,557 as subject or object. This problem, where one or more classes are under-represented in the data compared to other classes, is widely known as Class Imbalance.

Class imbalance is a challenge to machine learning and data mining, and is prevalent in many applications like risk management, fraud/intrusion detection, text classification, medical diagnosis/monitoring, etc. Numerous approaches have been proposed both at the data and algorithmic levels. Concerning the data level, solutions include different kinds of re-sampling [16] [22]. In under-sampling, a set of majority instances is removed from the initial dataset while all the minority instances are preserved. In over-sampling, the number of minority instances is increased, so that they reach the number of majority instances. Random over-sampling in general is among the most popular sampling techniques and provides competitive results.

In this work, we employed random over-sampling by duplicating minority examples, random under-sampling, and feature selection [23], using filtering techniques available in the Weka machine learning workbench [24]: Resample and Synthetic Minority Over-sampling Technique (SMOTE) [13]. For high-dimensional data sets, filters are used that score each feature independently based on a rule. Resample produces a random subsample of a dataset, namely sampling with replacement. SMOTE is an over-sampling method [6] that generates synthetic examples in a less application-specific manner, by operating in "feature space rather than data space". Our final results show that feature selection is a very competitive method, as proven by different approaches [7].

3   Experimental Setup, Results and Discussion

The experiments were conducted using WEKA. After experimentation with different algorithms, those leading to the best results were selected: C4.5 (decision tree learning) and k-*nearest-neighbor* (instance-based learning, with $1 \le k \le 17$). Grafting [11] is also applied as a post process to an inferred decision tree to reduce the prediction error. Post-pruning was applied to the inducted decision trees. Classifier performance was evaluated using precision, recall, and the f-measure. Validation was performed using *10 – fold cross validation*. The best results were obtained by the decision tree classifiers, and in particular with grafting.

As regards k-NN, the value of k affected significantly the results, while k increased more training instances that were relevant to the test instance were involved. Only after a relatively large value of k (k=17), the f-measure started dropping and performance was affected by noise. Results are listed below.

Table 1. First experiment for the MG and English corpora (classifiers evaluated by f-measure)

	C4.5	C4.5graft	1-NN	3-NN	9-NN	15-NN	17-NN
Subject (MG)	68.4	68.7	59.2	60.8	62	62.5	60.7
Object (MG)	77.1	77.4	60.7	64	65.5	69	65.8
Subject (Eng)	73.3	73.5	62.1	63.8	64.7	66.6	66.1
Object (Eng)	68.5	68.4	51.2	56.71	61.8	61	61.1

After the implementation of the methods that face the class imbalance problem, results had a significantly upward course. Concerning both datasets, the best results were achieved with C4.5 and are presented in Table 2.

Regarding the pre-processing phase, the outcome results are influenced strongly by the automatic nature of the chunking process. A small but concrete number of errors are attributed to erroneous phrase splitting, excessive phrase cut-up and erroneous phrase type identification.

Table 2. Overall comparative results for the MG and English corpora evaluated by f-measure

	First	Resample	Undersampling	Oversampling	SMOTE
Object (MG)	77.1	97.2	95.9	89.9	86
Subject	68.4	94.2	86.4	87.9	68
Object (Eng)	68.5	93.6	79.4	85.8	81.4
Subject	73.3	92.1	83	88.1	73.4

A significant difference between subject and object detection performance in the Greek corpus was also noticed. Object detection exceeded subject detection by almost 10%. This is due to the existence of noise. In many cases training instances had exactly the same feature description, but different classification. It is a general problem that pertains to the difficulty of the MG language and especially the problem of distinguishing the subject of a copular (linking) verb from its copula. The instances of the copula and the subject had the same features (nominative case, short distance from the verb, etc.) but different class values. In the MG sentence *NP [To θέμα της συζήτησης] VP[είναι] NP[τα σκουπίδια] (NP[the point of the conversation] VP[is] NP[the rubbish])*, the first NP is the subject and the second the copula. The same syntactic relation would hold even if the NPs were in different order.

Classifiers behaved almost in the same way on the English corpus. This shows that the shallow parser can be adjusted (using a very similar set of features and equivalent pre-processing) to cope with the English language. Better results were obtained for the subject recognition in English. This indicates that subject detection is easier in English, as it has a stricter and simpler structure. Compared to previous work [9], the results presented herein are very satisfying and outperform those of similar approaches (the authors in [9] report an accuracy of 77.1% for subject and 79% for object recognition).

In both languages, over-sampling led to the second-best results after the Resample method, except for the case of object detection in the MG corpus, meaning that during random under-sampling more noisy instances were removed. Generally, as over-sampling gave better results than under-sampling, it turns out that under-sampling is not always as effective as has been claimed [6].

During the experimental process for the MG data, a lexicalized version of the data was also used, according to the approach of Daelemans et al. [9]. The lemma of the head word of the VP was included in the data. However, the final results dropped significantly (the f-measure for C4.5 was 61.1% for subject and 68.3% for object detection). The verb type used in our initial feature set is sufficient for satisfactory classification, while lexicalization includes redundant information that is misleading.

Important future improvements could be: the application of other techniques to deal with the class imbalance problem (e.g. cost-sensitive classification, focused one-sided sampling etc.), an improvement of the second level of the shallow parser, experimentation with a larger number of training instances. The creation of a fourth class for the classification of NPs that are copulas could improve the final result if the feature set was altered accordingly, and force classifiers to learn to distinguish between these two types relations (subjects and copulas) and finally the recognition of other types of syntactic relations of the verb and simultaneously the inclusion of other phrase types, apart from NPs.

4 Conclusion

In this paper, a shallow parser for MG subject/object detection is created, using machine learning algorithms. The parser utilizes minimal resources, and does not require grammars or lexica of any kind. The parser can be adjusted to cope with English language text as well, with minor modifications, with impressive results. Additionally, for the first time to the authors' knowledge, the class imbalance problem in the data is successfully addressed and the final results climb up to 97.2% for object and 94.2% for subject detection in the MG corpus and 92.1% and 93.6% respectively in the English corpus.

References

1. Abney, S.: In Principle-Based Parsing: Computation and Psycholinguistics, pp. 257–278. Kluwer Academic Publishers, Dordrecht (1991)
2. Aït-Mokhtar, S., Chanod, P.: Subject and Object Dependency Extraction Using Finite-State Transducers. Rank Xerox Research Centre, France
3. Argamon, D.I., Krymolowski, Y.: A Memory-based Approach to Learning Shallow Natural Language Patterns. In: 36th Annual Meeting of the ACL, Montreal, pp. 67–73 (1998)
4. Canisius, S.: Memory-Based Shallow Parsing of Spoken Dutch. MSc Thesis. Maastricht University, The Netherlands (2004)

5. Charniak, E.: Statistical Parsing with a Context-free Grammar and Word Statistics. In: Proc. National Conference on Artificial Intelligence (1997)
6. Chawla, N., Japkowicz, N., Kolcz, A.: Special Issue on Learning from Imbalanced Data Sets. In: Sigkdd Explorations, Canada (2005)
7. Chen, Y.: Learning Classifiers from Imbalanced, Only Positive and Unlabeled Data Sets. Department of Computer Science Iowa State University (2009)
8. Collins, M.: Three Gneretive, Lexicalised Models for Statistical Parsing. Univerity of Pennsylvania, U.S.A (1996)
9. Daelemans, W., Buchholz, S., Veenstra, J.: Memory-based Shallow Parsing. ILK, Tilburg University (2000)
10. Evaluations and Language resources Distribution Agency, http://www.elda.fr/catalogue/en/text/W0022.html
11. Webb, G.I.: Decision Tree Grafting From the All-Tests-But-One Partition. Deakin University, Australia
12. Grefenstette, G.: Light parsing as finite-state filtering. In: Wahlster, W. (ed.) Workshop on Extended Finite State Models of Language, ECAI 1996, Budapest, Hungary. John Wiley & Sons, Ltd., Chichester (1996)
13. Hulse, J., Khoshgoftaar, T., Napolitano, A.: Experimental Perspectives on Learning from Imbalanced Data. Florida Atlantic University, Boca Raton (2007)
14. Journal of Machine Learning Research, http://jmlr.csail.mit.edu/papers/special/shallow_parsing02.html
15. Jurafsky, D., Martin, J.: Speech and Language Processing: An Introduction to Natural Processing. Computational Linguistics, and Speech Recognition (2000)
16. Ling, C.X., Li, C.: Data Mining for Direct Marketing: Problems and Solutions. American Association for Artificial Intelligence. Western Ontario University (1998)
17. Marcus, et al.: Building a large annotated corpus of English: The penn Treebank. Coputational Linguistics 19(2), 313–330 (1993)
18. Munoz, M., et al.: A Learning Approach to Shallow Parsing, Department of Computer Science University of Illinois at Urbana (1999)
19. Roth, D., Yih, W.: Probabilistic reasoning for entity & relation recognition. In: Proc. of COLING 2002, pp. 835–841 (2002)
20. Stamatatos, E., Fakotakis, N., Kokkinakis, G.: A Practical Chunker for Unrestricted Text. In: Proceedings of the Conference on Natural Language Processing, Patras, Greece, pp. 139–150 (2000)
21. Sang, T.K., Eibe, F., Veenstra, J.: Representing text chunks. In: Proceedings of EACL 1999, pp. 173–179 (1999)
22. Solberg, A., Solberg, R.: A Large-Scale Evaluation of Features for Automatic Detection of Oil Spills in ERS SAR Images. In: International Geoscience and Remote Sensing Symposium, Lincoln, NE, pp. 1484–1486 (1996)
23. Wasikowski, M.: Combating the Small Sample Class Imbalance Problem Using Feature Selection. In: 10th IEEE Transactions on Knowledge and Data Engineering (2010)
24. Witten, I., Eibe, F.: Data Mining: Practical Machine Learning Tools and Techniques. Department of Computer Science University of Waikato (2005)

A Random Forests Text Transliteration System for Greek Digraphia

Alexandros Panteli and Manolis Maragoudakis

University of the Aegean,
Samos, Greece
{icsd06136,mmarag}@aegean.gr

Abstract. Greeklish to Greek transcription does undeniably seem to be a challenging task since it cannot be accomplished by directly mapping each Greek character to a corresponding symbol of the Latin alphabet. The ambiguity in the human way of Greeklish writing, since Greeklish users do not follow a standardized way of transliteration makes the process of transcribing Greeklish back to Greek alphabet challenging. Even though a plethora of deterministic approaches for the task at hand exists, this paper presents a non-deterministic, vocabulary-free approach, which produces comparable and even better results, supports argot and other linguistic peculiarities, based on an ensemble classification methodology of Data Mining, namely Random Forests. Using data from real users from a conglomeration of resources such as Blogs, forums, email lists, etc., as well as artificial data from a robust stochastic Greek to Greeklish transcriber, the proposed approach depicts satisfactory outcomes in the range of 91.5%-98.5%, which is comparable to an alternative commercial approach.

Keywords: Greek Language, Transliteration, Data Mining, Random Forests, Non-Deterministic.

1 Introduction

Greeklish is a term which originates from the words *Greek* and *English*, signifying a writing style in which Greek words are written using the Latin alphabet. Other synonyms for Greeklish are *Latinoellinika* or *ASCII Greek*. This phenomenon is not only appearing within the Greek domain, it is linguistically identified as *Digraphia* [1]. Digraphia is either synchronic, in the sense that two writing systems coexist for the same language, or diachronic, meaning that the writing system has changed over time and has finally been replaced by a new one. Examples of digraphia are common in a variety of languages that do not adopt the Latin alphabet or Latin script (e.g. Greek, Serbian, Colloquial Arabic, Chinese, Japanese etc.). Serbian is probably the most noticeable modern instance of synchronic digraphia, in which Serbian texts is found to be written concurrently in the Cyrillic script and in an adapted Latin-based one. *Singlish*, a word similar to Greeklish refers to an English-based creole used in Singapore which employs transliteration practices, as well as vocabulary modifications and additions from the English language. As a final point, we should

L. Iliadis et al. (Eds.): EANN/AIAI 2011, Part II, IFIP AICT 364, pp. 196–201, 2011.

mention the case of the Romanian Language, where there has been a full adoption of Latin-based writing style instead of the original Cyrillic one. The same principle is also appearing in the Turkish and other Central Asian countries of the former Soviet Union.

There is a significant amount of research papers that deal with the application and the acceptability of Greeklish as a writing style. The most representative amongst them was introduced by [2], in which a sequence of issues is studies such as the degree of penetration of Greeklish in textual resources, the acceptance rate of them, etc. More specifically, some descriptive statistical results mention that 60% of users have been reported to use Greeklish in over 75% of the contexts they submit. In addition, 82% of the users accept Greeklish as an electronic communication tool while 53% consider this style as non-appealing, 24% concern it as a violation or even vandalism of the Greek Language and 46% have reported to face difficulties in the reading of such texts. As regards to the latter, other research works study the reading time for the comprehension of words and sentences written in the Greek and Greeklish texts. The results indicate that the response time is lower when the text is written in Greek (657ms mean value) than when it is written using characters of the Latin alphabet (886ms mean value) [3].

Although an official prototype has been proposed (ELOT 743:1982 [4]) and already approved and used by the British council, the majority of users follow empirical styles of Greeklish styles, mainly categorized into four distinct groups:

- *phonetic transcription*. Each letter or combination of letters is mapped into an expression with similar acoustic form.
- *optical transliteration*. Each letter or combination of letter is mapped into an expression that optically resembles the former. For example the Greek letter θ is usually mapped into 8, due to its optical similarity.
- *keyboard-mapping conversion*. in this group, many letters are mapped to Latin ones according to the QWERTY layout of the Greek keyboard. For example, θ is mapped to its corresponding key in the Greek/English keyboard which is u.

Additionally, Greeklish writing suffers from the presence of "iotacism", a phenomenon which is characterized by the use of the Latin character "i" for the transliteration of the Greek symbol sets "ι", "η", "υ", "ει", and "οι" since they are all pronounced as "I" according to the SAMPA phonetic alphabet.

Based on the aforementioned issues, the present work is a Data Mining approach towards an efficient Greeklish-to-Greek transliteration tool, based on a state-of-the-art ensemble classification algorithm of Random Forests. This is, according to our knowledge, the first attempt to this domain in a non-deterministic manner. The use of Greeklish is now considered of an issue of high controversy and it is banned from numerous web sites and forums (e.g. Greek Translation Forum, Athens Wireless Metropolitan Network Forum, etc.). Therefore, a robust and affective transcriber is considered of high importance in order for users not to be excluded from web discussions and other social networking activities.

2   Previous Works in Greek Digraphia

A lot of work has been done in the field of Greeklish-to-Greek conversion. The most representative approaches including E-Chaos [5], Greeklish Out [6], Greek to Greeklish by Innoetics [7], All Greek to me![8] and deGreeklish [9]. The first two approaches are not using a vocabulary and they are mainly based on manual rules, refined by the user and adjustable to include more in the future. The second implementation does not mention its scientific parameters as it is a commercial application, however, the company mention 98% using language models as the core mechanism. The third and fourth approaches are based on a more sophisticated methodology, namely Finite State Automata (FSA), which make the mapping of each letter more straightforward. The latter system is implemented as a web service in PHP and C++ and addresses a novel search strategy in the directed acyclic graph. Note that all of the above approaches use deterministic implementation, either using hand-coded rules or FSA, or other user-defined methods.

3   Random Forests

Nowadays, numerous attempts in constructing ensemble of classifiers towards increasing the performance of the task at hand have been introduced [10]. A plethora of them has portrayed promising results as regards to classification approaches. Examples of such techniques are Adaboost, Bagging and Random Forests. Random Forests are a combination of tree classifiers such that each tree depends on the values of a random vector sampled independently and with the same distribution for all trees in the forest. A Random Forest multi-way classifier $\Theta(x)$ consists of a number of trees, with each tree grown using some form of randomization, where x is an input instance [11]. The leaf nodes of each tree are labeled by estimates of the posterior distribution over the data class labels. Each internal node contains a test that best splits the space of data to be classified. A new, unseen instance is classified by sending it down every tree and aggregating the reached leaf distributions.

Each tree is grown as follows:

- If the number of cases in the training set is N, sample N cases at random but with replacement, from the original data. This sample will be the training set for growing the tree.
- If there are M input variables, a number m<<M is specified such that at each node, m variables are selected at random out of the M and the best split on these m is used to split the node. The value of m is held constant during the forest growing.
- Each tree is grown to the largest extent possible. Therefore, no pruning is applied.

4   Experimental Design

The training data (a set of Greeklish characters with the corresponding Greek characters) were created using Stochastic Greek2Greeklish Transcriber. For each character a separate instance is created. For the small scale experiments a data set of ~12000 instances was used, these instances were created from some random articles from in.gr The large scale experiments use "OpenThesaurus – Green synonyms thesaurus OpenOffice.org edition" (under GNU general public license), which consists of ~84000 Greek words (with duplicates), this thesaurus combined with the 12000 instance dataset yields a dataset of over 0.7million instances. Using multiples of words produces a model with higher accuracy. This happens because each instance of a word would have a slightly different Greeklish conversion thus creating a better prediction model. The training data is converted to vectors suitable for data mining using the n-grams method. For each pair of Greek, Greeklish characters (e.g. (g,G)) an k-dimensional vector is created using n preceding characters of the Greek character (g), the Greek character g and m proceeding characters (n+m = k-1). The corresponding supervisory signal is of course the character G.

Using a large k impacts not only the accuracy of the classifier (as explained later) but the training and classification time. For example if a tree based classification algorithm is used the dimensionality of the training data affects the size of the produced tree and the time it takes to build it. Fortunately there are some clues that guided us to choose an optimal vector length (discussed later).

As regards to the creation of instances, before the input data is converted to instances all characters are turned to lowercase, this does not affect in any way the process since the capitalization rules are known. Apart from the change in case, all diaeresis are removed since they are rare and could impact the accuracy more (of all other classes) than not using them. All punctuation marks and whitespace are ignored and will be preserved unaltered. All words are independent from each other in the sense that no instance has characters from more than one word, the value for characters beyond the current word are filled in with the character '*' which represents whitespace. The reasoning behind this is the same with the removal of diaeresis, since words with double word stresses are rare and context dependent and generally word stressing is independent for each word.

A training instance consists of a number of features which represent the next n characters in the word, the previous k characters and the current character. The corresponding class is the Greek character in the same position as the current Greeklish character. Special care has been taken for Greek character pairs (e.g. diphthongs) that are equivalent phonetically to one Latin character.

5   Experimental Evaluations

Using the dataset as mentioned above, a series of experiments was conducted using WEKA and RapidMiner benchmarks. For reasons of thorough evaluation, we have compared Random Forests (in practice, both implementations, either Random Input or Random Combination Forests proven to behave similarly, with little variation amongst them) against K-Nearest neighbor (IB1 and IB3 respectively), Decision Trees (J48), Naive Bayes (NB) and Bayesian Networks (BN) classification algorithms. The obtained the results are tabulated in the following table:

Table 1. The cell values represent correct prediction percentage using 10 fold cross validation

	Algorithm					
Window size	J48	RF	IB1	IB3	BN	NB
[+2-2]	87.59	90.25	85.94	84.28	85.03	84.80
[+2-4]	89.99	92.70	87.12	84.28	86.66	86.13
[+3-3]	89.90	93.34	88.02	85.11	87.02	86.27
[+4-2]	87.90	91.20	84.99	82.72	85.51	84.80
[+2-6]	90.13	92.63	82.76	80.31	86.31	85.85
[+3-5]	89.96	93.21	85.25	82.49	86.60	85.97
[+4-4]	90.08	97.4	86.73	83.25	86.75	85.92
[+5-3]	90.09	98.43	85.79	82.97	87.27	86.22
[+6-2]	88.18	91.31	81.62	79.71	85.29	84.55
[+3-6]	90.10	93.14	93.25	81.06	86.42	85.83
[+5-4]	90.15	93.36	85.25	82.31	86.63	85.95
[+4-5]	90.01	93.34	85.06	81.90	86.61	85.85
[+6-3]	90.08	93.31	83.72	81.46	86.84	86.12
[+5-5]	90.08	93.24	83.98	81.49	86.54	85.72

The results shown in the above table are visualized in the following figure (Fig.1).

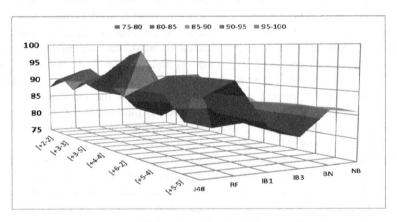

Fig. 1. Results on the set of benchmark algorithms, in terms of prediction accuracy

The dataset used for the creation of the classifier, as mentioned, consists of about 84000 Greek words (including duplicates). This dataset was obtained by using a window size of [+4-4] to extract the n-grams. Character classes with $1/10^{th}$ the number of instances are only 1% more accurate. As shown by a study done by Hatzigeorgiu et al. [12] the average length of a Greek word is between 6-7 characters, so using a much greater than this number of grams should yield worst results since the data would be sparse (remember that if the length of the word is smaller than the number of features whitespace is added). Our results confirm this by having an accuracy peak at 9 features (considering 3 previous characters and 5 next, plus the current character ([+5-3]). A second observation is that accuracy decreases if the number of characters considered is highly asymmetrical. This is to be expected since instances at the beginning or the end of the word (depending on if more next or previous characters are considered) will have a lot of whitespace, thus resulting in sparse data. The accuracy measurements obtained from the dataset persist when analyzing using the large dataset. Random Forest showed an overall accuracy of over 98%. This percentage of course refers to a single character being classified correctly.

6 Conclusions

This work dealt with the importance issue of implementing a Greeklish to Greek transliteration tool, which differs from existing approaches in two ways. The former lies to the fact that no vocabulary is used, therefore the proposed approach is robust to slang and other linguistic idioms, while the latter lies to the fact that it is a non-deterministic, Data Mining approach, which could encompass a variety of user's writing styles and be independent of manually defined, empirical rules. Evaluations against numerous other Data Mining classification approaches have supported our claim that Random Forests (using both of their existing utilizations) are well suited for the task at hand and behave competitive or better that existing deterministic, commercial implementations

References

1. Dale, I.R.H.: "Digraphia". International Journal of the Sociology of Language 26, 5–13 (1980)
2. Androutsopoulos, J.: Latin-Greek spelling in e-mail messages: Usage and attitudes. In: Studies in Greek Linguistics, pp. 75–86 (2000) (in Greek)
3. Tseliga, T., Marinis, T.: On-line processing of Roman-alphabeted Greek: the influence of morphology in the spelling preferences of Greeklish. In: 6th International Conference in Greek Linguistics, Rethymno, Crete, September 18-21 (2003)
4. ELOT, Greek Organisation of Standardization (1982)
5. e-Chaos: freeware Greeklish converter, http://www.paraschis.gr/files.php
6. Greek to Greeklish by Innoetics, http://services.innoetics.com/greeklish/
7. Chalamandaris, A., Protopapas, A., Tsiakoulis, P., Raptis, S.: All Greek to me! An automatic Greeklish to Greek transliteration system. In: Proceedings of the 5th Intl. Conference in Language Resources and Evaluation, pp. 1226–1229 (2006)
8. DeGreeklish, http://tools.wcl.ece.upatras.gr/degreeklish
9. Greeklish Out!, http://greeklishout.gr/main/
10. Breiman, L.: Random forests. Machine Learning Journal 45, 532 (2001)
11. Kononenko, I.: Estimating attributes: analysis and extensions of Relief. In: De Raedt, L., Bergadano, F. (eds.) ECML 1994. LNCS, vol. 784, pp. 171–182. Springer, Heidelberg (1994)
12. Hatzigeorgiu, N., Mikros, G., Carayannis, G.: Word length, word frequencies and Zipf's law in the Greek language. Journal of Quantitative Linguistics 8, 175–185 (2001)

Acceptability in Timed Frameworks with Intermittent Arguments

Maria Laura Cobo, Diego C. Martinez, and Guillermo R. Simari

Artificial Intelligence Research and Development Laboratory (LIDIA),
Department of Computer Science and Engineering, Universidad Nacional del Sur,
Av. Alem 1253 - (8000) Bahía Blanca - Bs. As. - Argentina
{mlc,dcm,grs}@cs.uns.edu.ar
http://www.cs.uns.edu.ar/lidia

Abstract. In this work we formalize a natural expansion of timed argumentation frameworks by considering arguments that are available with (possibly) some repeated interruptions in time, called *intermittent arguments*. This framework is used as a modelization of argumentation dynamics. The notion of acceptability of arguments is analyzed as the framework evolves through time, and an algorithm for computing intervals of argument defense is introduced.

1 Introduction

One of the main concerns in Argumentation Theory is the search for rationally based positions of acceptance in a given scenario of arguments and their relationships. This task requires some level of abstraction in order to study pure semantic notions. Abstract argumentation systems [10,17,2] are formalisms for argumentation where some components remain unspecified, being the structure of an argument the main abstraction. In this kind of system, the emphasis is put on the semantic notion of finding the set of accepted arguments. Most of these systems are based on the concept of *attack* represented as an abstract relation, and extensions are defined as sets of possibly accepted arguments. For two arguments $\mathcal{A}$ and $\mathcal{B}$, if $(\mathcal{A}, \mathcal{B})$ is in the attack relation, it is said that argument $\mathcal{A}$ *attacks* $\mathcal{B}$, implying that the acceptance of $\mathcal{B}$ is conditioned by the acceptance of $\mathcal{A}$, but not the other way around.

The simplest abstract framework is defined by Dung in [10]. It only includes a set of abstract arguments and a binary relation of attack between arguments, allowing the definition of several semantic notions. Dung's framework became the foundation of further research, either by extending the formalism, as in [2,5,13] or by elaborating new semantics [11,4].

In recent works [6,15,16], the dynamics of an argumentation framework is considered. The main subject of study is how the outcome changes when the set of arguments or the set of attacks are changed. In this paper we are interested in the overall evolution of an argumentation framework through time. The combination of time and argumentation is a novel research line. In [12] a calculus

L. Iliadis et al. (Eds.): EANN/AIAI 2011, Part II, IFIP AICT 364, pp. 202–211, 2011.

for representing temporal knowledge is proposed, and defined in terms of propositional logic. This calculus is then considered with respect to argumentation, where an argument is defined in the standard way: an argument is a pair constituted by a minimally consistent subset of a database entailing its conclusion. This work is thus related to [3].

In [7,8] a novel framework is proposed, called *Timed Abstract Framework* (TAF), combining arguments and temporal notions. In this formalism, arguments are relevant only in a period of time, called its *availability interval*. This framework maintains a high abstract level in an effort to capture intuitions related with the dynamic interplay of arguments as they become available and cease to be so. The notion of *availability interval* refers to an interval of time in which the argument can be legally used for the particular purpose of an argumentation process. In particular, the actions of adding and removing an argument in a framework, as considered in argumentation dynamics, can be naturally interpreted as a temporal availability of that argument. As several arguments are added or removed through time, the classical notion of acceptability of arguments requires a deeper analysis.

Timed abstract frameworks are suitable to model the evolution of argumentation. In this work we formalize a natural expansion of timed argumentation frameworks by considering arguments with more than one availability interval. These are called *intermittent arguments*, and it is a more accurate modelization of argument dynamics. These arguments are available with (possibly) some repeated interruptions in time. Using this extended timed argumentation framework, we analyze the notion of *acceptability* and we introduce a timed structure for arguments, called t-profiles, that allows a refined characterization of classical acceptability when new arguments are available or cease to be so.

This paper is organized as follows. In the next section we recall the notion of time-intervals [7,8] and the terminology used in this work. In Section 3, our Timed Abstract Argumentation Framework with intermittent arguments is introduced. Acceptability is formalized in Section 4 and a procedural refinement is introduced in Section 5. Finally, conclusions and future work are discussed.

2   Time Representation

In order to capture a time-based model of argumentation, we enrich the classical abstract frameworks with temporal information regarding arguments. In this work we use *temporal intervals of discrete time* as primitives for time representation [1,9,14], and thus only metric relations for intervals are applied.

Definition 1. *An interval is a pair build from $a, b \in \mathbb{Z} \cup \{-\infty, \infty\}$, in one of the following ways:*

- *$[a, a]$ denotes a set of time moments formed only by moment a.*
- *$[a, \infty)$ denotes a set of moments formed by all the numbers in $\mathbb{Z}$ since a (including a).*
- *$(-\infty, b]$ denotes a set of moments formed by all the numbers in $\mathbb{Z}$ until moment i (including b).*

- $[a, b]$ *denotes a set of moments formed by all the numbers in* $\mathbb{Z}$ *moment* i *until moment* j *(including both* a *and* b*).*
- $(-\infty, \infty)$ *a set of moments formed by all the numbers in* $\mathbb{Z}$.

The moments a, b *are called endpoints. The set of all the intervals defined over* $\mathbb{Z} \cup \{-\infty, \infty\}$ *is denoted* ι.

For example, $[5, 12]$ and $[1, 200]$ are intervals. If X is an interval then X^-, X^+ are the corresponding endpoints (*i.e.*, $X = [X^-, X^+]$). An endpoint may be a point of discrete time, identified by a natural number, or infinite.

We will usually work with sets of intervals (as they will be somehow related to arguments). Thus, we introduce several definitions and properties needed for semantic elaborations.

Definition 2. *Let* S *be a set of intervals and let* i *be a moment of time. The exclusion of* i *from* S, *denoted* $S \ominus i$, *is defined as follows:*

$$S \ominus i = \{I : I \in S \wedge i \notin I\} \qquad \cup$$
$$\{[I^-, i - 1] : I \in S \wedge i \in I, i \neq I^-\} \cup$$
$$\{[i + 1, I^+] : I \in S \wedge i \in I, i \neq I^+\}$$

The exclusion of interval I *from set* S, *denoted as* $S \textcircled{I} I$, *is recursively defined as follows:*

$$i. \ S \textcircled{I} I = S \ominus I^- \qquad\qquad\qquad\quad \text{if } I^- = I^+$$
$$ii. \ S \textcircled{I} I = (S \ominus I^-) \textcircled{I} [I^- + 1, I^+] \quad \text{if } I^- \neq I^+$$

The difference between sets of intervals, denoted $S_1 \stackrel{I}{-} S_2$ *is defined as* $\text{\m}(S_1 \textcircled{I} I)$, $\forall I \in S_2$.

Intersection is another relevant operation on intervals. The intersection of two intervals is the interval formed by all the common points in both of them. Its endpoints are the minimal and maximal time points in common.

Definition 3. *Let* I_1 *and* I_2 *be two intervals. The intersection is defined as:* $I_1 \cap I_2 = [x, y]$ *with* $x, y \in I_1$ *and* $x, y \in I_2$ *such that there are no* $w, z : w, z \in I_1$ *and* $w, z \in I_2$ *with* $w < x$ *or* $y < z$.

Definition 4. *Let* S_1 *and* S_2 *be two sets of intervals. The intersection of these sets, noted as* $S_1 \text{\m} S_2$, *is:* $S_1 \text{\m} S_2 = \{I : I = I_1 \cap I_2, I \neq [\ ], \forall I_1 \in S_1, I_2 \in S_2\}$.

The intersection of two sets of intervals S_1 and S_2 is formed by all the intersections of an interval of S_1 with an interval of S_2. The following definition formalizes a special relation of inclusion for sets of intervals. In order to do that, we define the set of time-points for a set of intervals. If S is a set of arguments, then $tp(S)$ is the set of time-points in S. For instance, if $S = \{[1, 3], [8, 10]\}$ then $tp(S) = \{1, 2, 3, 8, 9, 10\}$.

Definition 5. *Let* S_1 *and* S_2 *be two sets of intervals. The timed-inclusion, denoted* $\subseteq'$, *is defined as* $S_1 \subseteq' S_2 = tp(S_1) \subseteq tp(S_2)$.

The *partition* of a set of arguments *breaks* overlapped intervals into smaller intervals, and then there is no overlapping intervals in a partition. This is formalized as follows.

Definition 6. *Let S be a set of intervals. The partition of S, denoted* Part(S) *is defined as:*

- Part(S) $= S$ *if* $\forall I_1, I_2 \in S, I_1 \cap I_2 = \emptyset$.
- Part(S) $=$ Part$(S - \{I_1, I_2\} \cup \{I_1 - (I_1 \cap I_2), I_2 - (I_1 \cap I_2), I_1 \cap I_2\})$, *with* $I_1, I_2 \in$ S *and* $I_1 \cap I_2 \neq \emptyset$

The notion of partition simplifies semantic elaborations, since it discretizes the evolution of the framework according to moments where arguments start or cease to be available.

In the following section we present Timed Abstract Argumentation Frameworks with intermittent arguments.

3   Timed Argumentation Framework

As remarked before, in Timed Argumentation Frameworks the period of time in which an argument is available for consideration is modeled. The formal definition of our timed abstract argumentation framework follows.

Definition 7. *A timed abstract argumentation framework (TAF) is a 3-tuple* $\langle Args, Atts, Av \rangle$ *where Args is a set of arguments, Atts is a binary relation defined over Args and Av is the availability function for timed arguments, defined as* $Av : Args \rightarrow \wp(\iota)$.

Example 1. *The triplet* $\langle Args, Atts, Av \rangle$*, where* $Args = \{\mathcal{A}, \mathcal{B}, \mathcal{C}, \mathcal{D}, \mathcal{E}\}$*, Atts =* $\{(\mathcal{B}, \mathcal{A}), (\mathcal{C}, \mathcal{B}), (\mathcal{D}, \mathcal{A}), (\mathcal{E}, \mathcal{D})\}$ *and the availability function is defined as*

Args	Av	Args	Av
$\mathcal{A}$	$\{[10, 40], [60, 75]\}$	$\mathcal{B}$	$\{[30, 50]\}$
$\mathcal{C}$	$\{[20, 40], [45, 55], [60, 70]\}$	$\mathcal{D}$	$\{[47, 65]\}$
$\mathcal{E}$	$\{(-\infty, 44]\}$		

is a timed abstract argumentation framework.

The framework of Example 1 can be depicted as in Figure 1, using a digraph where nodes are arguments and arcs are attack relations. An arc from argument $\mathcal{X}$ to argument $\mathcal{Y}$ exists if $(\mathcal{X}, \mathcal{Y}) \in Atts$. Figure 1 also shows the time availability of every argument, as a graphical reference of the Av function. It is basically the framework's evolution in time. Endpoints are marked with a vertical line, except for $-\infty$ and ∞. For space reasons, only some relevant time points are numbered in the figure. The shorthand notation $\mathcal{A}^+$ and $\mathcal{A}^-$ are used to represent, respectively, the set of arguments attacked by $\mathcal{A}$ and the set of arguments attacking $\mathcal{A}$.

An attack to an argument may actually occur only if both the attacker and the attacked argument are available in the framework. An attack in such a condition is said to be *attainable*.

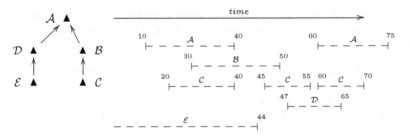

Fig. 1. Framework of Example 1

Definition 8. *Let $\Phi = \langle Args, Atts, Av \rangle$ be a TAF, and let $\{\mathcal{A}, \mathcal{B}\} \subseteq Args$ such that $(\mathcal{B}, \mathcal{A}) \in Atts$. The attack $(\mathcal{B}, \mathcal{A})$ is said to be attainable if $Av(\mathcal{A}) \cap Av(\mathcal{B})$ is not empty. The attack is said to be attainable in $Av(\mathcal{A}) \cap Av(\mathcal{B})$. The set of intervals where an attack $(\mathcal{B}, \mathcal{A})$ is attainable will be noted as $Av(\mathcal{B}, \mathcal{A})$.*

Example 2. *Consider the timed argumentation framework of Example 1. The attacks $(\mathcal{D}, \mathcal{A})$ and $(\mathcal{B}, \mathcal{A})$ are both attainable in the framework. Attack $(\mathcal{D}, \mathcal{A})$ is attainable since $[47, 65]$ overlaps $[60, 75]$ with $[47, 65] \in Av(\mathcal{D})$ and $[60, 75] \in Av(\mathcal{A})$. Attack $(\mathcal{B}, \mathcal{A})$ is attainable since $[30, 50]$ overlaps $[10, 40]$, in $[30, 40]$. Recall that $[30, 50] \in Av(\mathcal{B})$, $[10, 40] \in Av(\mathcal{A})$. The attack $(\mathcal{C}, \mathcal{B})$ is also attainable. Since $Av(\mathcal{C}) = \{[20, 40], [30, 50]\}$ and $Av(\mathcal{B}) = \{[30, 50]\}$ then, we can assure the attainability of the attack by one of the following relations: $[20, 40]$ overlaps $[30, 50]$, $[45, 55]$ overlaps $[30, 50]$. The attack is then attainable at $\{[30, 40], [45, 50]\}$, i.e. $Av(\mathcal{C}) \cap Av(\mathcal{B})$. The attack $(\mathcal{E}, \mathcal{D})$ is not attainable, since $[-\infty, 45]$ does not overlaps $[47, 65]$. The arguments involved in this attack are never available at the same time.*

Since an argument is defended by attacking its attackers, it is important to capture the set of intervals in which an argument is actually defended, as next.

Definition 9. *Let $\Phi = \langle Args, Atts, Av \rangle$ be a TAF, and let $\{\mathcal{A}, \mathcal{B}, \mathcal{C}\} \subseteq Args$ such that $(\mathcal{B}, \mathcal{A})$, $(\mathcal{C}, \mathcal{B}) \in Atts$. The overlapped availability of the attacks, denoted $Av(\mathcal{C}, \mathcal{B}, \mathcal{A})$, is defined as $Av(\mathcal{C}, \mathcal{B}, \mathcal{A}) = Av(\mathcal{A}) \cap Av(\mathcal{B}) \cap Av(\mathcal{C})$.*

The set $Av(\mathcal{C}, \mathcal{B}, \mathcal{A})$ captures all the intervals in which $\mathcal{C}$ actually defends $\mathcal{A}$ against $\mathcal{B}$. In the following section the notion of defense in timed frameworks is analyzed.

4   Defense through Time

An argument may be attacked in several intervals of time. It is defended in these *threat* intervals, only when another argument has an attainable attack to its attacker. Thus, unlike classic frameworks where an argument may be defended or not, here an argument may be defended or not in *some* moments of time. It is not enough to establish if a defense condition is present by looking attacks. It is mandatory to find out *when* these defenses may occur.

Definition 10. *Let* $\Phi = \langle Args, Atts, Av \rangle$ *be a TAF. The set of defense intervals for* $\mathcal{A}$ *against* $\mathcal{B}$, *denoted* $\mathcal{D}f(\mathcal{A}, \mathcal{B})$, *is defined as:* $\bigcup_{\mathcal{X} \in \mathcal{B}^-} Av(\mathcal{A}, \mathcal{B}, \mathcal{X})$.

For a set of arguments S, *the intervals restricted to defenses in* S, *denoted* $\mathcal{D}f(\mathcal{A}, \mathcal{B})_S$, *is defined as:* $\bigcup_{\mathcal{X} \in \mathcal{B}^- \cap S} Av(\mathcal{A}, \mathcal{B}, \mathcal{X})$.

The set $\mathcal{D}f(\mathcal{A}, \mathcal{B})$ captures all the intervals in which a defense is available, *i.e.* there is an overlapping between $\mathcal{A}$, its attacker $\mathcal{B}$ and every attacker of $\mathcal{B}$.

Now that the set of defense intervals is characterized, the classical notion of acceptability [10] can be addressed. An argument $\mathcal{A}$ is acceptable with respect to a set of arguments S if this set can provide attackers for every attacker of $\mathcal{A}$. In a timed, dynamic context it is natural to consider defenses for an argument $\mathcal{A}$ whenever $\mathcal{A}$ is available, that is, during $Av(\mathcal{A})$.

Definition 11. *Let* $\Phi = \langle Args, Atts, Av \rangle$ *be a TAF. An argument* $\mathcal{A}$ *is acceptable with respect to a set of arguments* S *if* $Av(\mathcal{A}, \mathcal{B}) \subseteq' \mathcal{D}f(\mathcal{A}, \mathcal{B})_S$ *for every attacker* $\mathcal{B}$ *of* $\mathcal{A}$.

This definition is close to the classical definition of acceptability. However, in a timed scenario is a very restrictive definition since it is possible for an argument to temporarily be out of defenses.

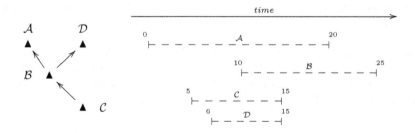

Fig. 2. Argument $\mathcal{D}$ acceptable with respect to $\{\mathcal{C}\}$

Consider the framework of Figure 2. Argument $\{\mathcal{C}\}$ is acceptable with respect to the $\emptyset$. Argument $\mathcal{D}$ is acceptable with respect to $\{\mathcal{C}\}$, since it is defended whenever $\mathcal{B}$ attacks $\mathcal{D}$. However, argument $\mathcal{A}$ is not acceptable with respect to the set $\{\mathcal{C}\}$ since it has no defense against $\mathcal{B}$ in the period $[15, 20]$.

Definition 11 does not take into account the fact that it may be some periods of time in where the argument is actually defended. It would be interesting to have an alternative, more refined definition of acceptability in order to capture proper intervals of defense. Even more, by characterizing specific intervals of provided defense, it is possible to analyze the acceptance of other arguments. Consider the framework depicted on Figure 3. The only argument that is acceptable with respect to $S = \emptyset$ is $\mathcal{C}$. Notice that in the period of time where $\mathcal{E}$ needs defense against $\mathcal{D}$, $\mathcal{A}$ is not attacked so the defense actually takes place. It would be interesting to capture the precise intervals in which an argument is provided

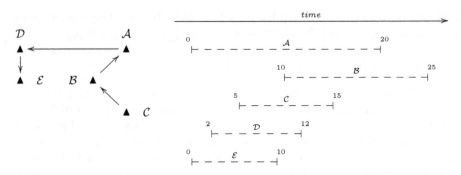

Fig. 3. Framework Simple

with a defense. For instance, if $S = \emptyset$ the only acceptable argument according to Definition 11 is $\mathcal{C}$. However, the argument $\mathcal{A}$ does not need defenders in $[0, 15]$.

A more refined version of acceptability can be provided, but we need a structure to associate arguments and intervals of time. This is called a *t-profile*, and it may have different particular meanings: the availability of the argument, the periods of time in which this argument is attacked, defended, or completely justified under some particular semantics. The definition follows.

Definition 12. *Let* $\Phi\langle Args, Atts, Av\rangle$ *be a TAF. A timed argument profile in* Φ, *or simply* t-profile, *is a pair* $[\mathcal{A}, T]$ *where* $\mathcal{A} \in Args$ *and* $T \subseteq Av(\mathcal{A})$.

The set of all the profiles definable from Φ will be noted as Θ. In the following section we present an algorithm that can be used to compute the set of intervals in which an argument is defended. This is a procedural refinement of acceptability as defined in Definition 11.

5   Computing Intervals of Defense

Following the general idea of acceptability of arguments, an algorithm is defined for the characterization of the those intervals of time in which a set of t-profiles defend a particular argument.

The algorithm receives an argument X and a set of t-profiles P. There are three different main procedures. The first one captures those moments in time where the argument was previously acceptable or unattacked. The second one determines all the particular defenses that P provides against every attacker in a particular interval. Finally, the intersection of them is calculated to establish *when* the argument is actually defended against all of its attackers in that interval. The last two parts are evaluated for all the intervals that are obtained from the partition of all the availability periods of the attacks against the argument in consideration.

Assuming that unions and intersections are computed in constant time, this algorithm has a cubic order of execution in the worst case. Particularly the order

---

Algorithm 1. DEFENDED PERIODS

Require: $\mathcal{X} \in Args$, P a set of t-profiles, $\Phi = \langle Args, Atts, \mathcal{A}v \rangle$ a TAF
Ensure: D the set of intervals where T defends $\mathcal{X}$.

1: $S = \bigcup_{Y \in X^-} \mathcal{A}v(\mathcal{Y}, \mathcal{X})$
2: **if** $[\mathcal{X}, U] \in P$ **then**
3:     $S = S \stackrel{I}{-} U$
4:     $D = U$
5: **else**
6:     $D = \mathcal{A}v(\mathcal{X}) \stackrel{I}{-} S$
7: **end if**
8: $S = \mathsf{Part}(S)$
9: **for all** $I \in S$ **do**
10:     $\Gamma = \emptyset$
11:     **for all** $\mathcal{Y} \in Args$ such that $\{I\} \subseteq \mathcal{A}v(\mathcal{Y}, \mathcal{X})$ **do**
12:         **for all** $[\mathcal{Z}, U] \in P$ such that $(\mathcal{Z}, \mathcal{Y}) \in Atts$ and $U \cap I \neq \emptyset$ **do**
13:             $\Gamma = \Gamma \cup [U \cap (\mathcal{A}v(\mathcal{Z}, \mathcal{Y}, \mathcal{X})])$
14:         **end for**
15:     **end for**
16:     $\Lambda = \{I\}$
17:     **for all** $X \in \Gamma$ **do**
18:         $\Lambda = \Lambda \cap \{X\}$
19:     **end for**
20:     $D = D \cup \Lambda$
21: **end for**

---

of execution is mnp being m, n and p the cardinality of the sets S, $Args$ and P respectively. The choice of relevant arguments and t-profiles in the inner loops helps to keep a low execution time.

Example 3. *Consider the framework depicted on Figure 4.*

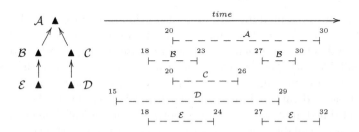

Fig. 4. Framework to ilustrate the use of Algorithm 1

Let us analyze the behavior of Algorithm DEFENDED PERIODS *when it is called with the argument* $\mathcal{A}$ *and the empty set as* P. *The first line of the algorithm builds a set* S *formed by the intervals in which* $\mathcal{A}$ *is attacked. In this case the set is* $S = \mathcal{A}v(\mathcal{B}, \mathcal{A}) \cup \mathcal{A}v(\mathcal{C}, \mathcal{A})$, *i.e.* $\{[20, 23], [27, 30], [20, 26]\}$. *The* set S is then partitionated in order to properly analyze defenses. The S is then $\{[20, 23], [24, 26], [27, 30]\}$. For each one of these intervals, a set of defenses is calculated. This partition is made since defense requirements are different in each one of them. In $[20, 23]$ a defense is needed against B and C, and in $[24, 26]$ against C, while in $[27, 30]$ it is needed against B. The set D is initialized with the periods on $Av(A) \perp S$, i.e. those periods of time in which A is not attacked. Following the example, the set D is $\{[31, 32]\}$, $\{[20, 30]\} \perp \{[20, 23], [24, 26], [27, 30]\}$. The inner loops are not relevant in this case since the are no defenders for A, since P is the emptyset. So D ends with $\{[31, 32]\}$.

If the algorithm is called again with A and $P = \{ [D, Av(D)], [E, Av(E)]\}$, the inner loops become relevant now. Set D starts with $\{[31, 32]\}$, i.e. those intervals in which A is not attacked. Lets see in detail how the algorithm evaluates Γ for $[20, 23]$. The for-cycle on line 11 chooses the arguments that attack A at $[20, 23]$. In this case there are two arguments: B and C. The inner for-cycle (line 12) chooses the t-profiles in P that can provide a defense, i.e. where the time in the profile overlaps $[20, 23]$ and they attack the attacker. The set Γ begins as an empty set. Then,

Argument	t-profile	Γ
B	$[E, Av(E)]$	$Av(E) \cap Av(E, B, A)$
		$\{[18, 24], [27, 32]\} \cap \{[20, 23], [27, 30]\}$
		$\{[20, 23], [27, 30]\}$
C	$[D, Av(D)]$	$Av(D) \cap Av(D, C, A)$
		$\{[15, 29]\} \cap \{[20, 26]\}$
		$\{[20, 26]\}$

Finally it makes the intersection of these sets with I, given that the defense provided by P in I is I.

6   Conclusions and Future Work

In this work we presented an extension of previously defined Timed Argumentation Frameworks in which arguments with more than one availability interval are considered. These arguments are called *intermittent arguments*, and are temporaly available with some repeated interruptions in time. Using this extended timed argumentation framework, we studied the notion of acceptability, which requires the consideration of time as a new dimension, leading to the definition of *t-profiles* of timed arguments. This is important in argumentation change, when arguments are available and cease to be so as the framework evolves through time.

Future work has several directions. A declarative formalization of acceptability semantics using t-profiles is being studied. Also, algorithms for other classical semantics such as stable extensions are being considered. We are also interested in the general study of the evolution of the framework through time. For instance, using timed frameworks it is possible to analyze the impact of a new argument, by examining how much the status of acceptance is affected for other arguments, and for how long they are affected. This may lead to future semantic elaborations that contribute to the research area of argumentation change.

References

1. Allen, J.: Maintaining knowledge about temporal intervals. Communications of the ACM 1(26), 832–843 (1983)
2. Amgoud, L., Cayrol, C.: On the acceptability of arguments in preference-based argumentation. In: 14th Conference on Uncertainty in Artificial Intelligence (UAI 1998), pp. 1–7. Morgan Kaufmann, San Francisco (1998)
3. Augusto, J.C., Simari, G.R.: Temporal defeasible reasoning. Knowl. Inf. Syst. 3(3), 287–318 (2001)
4. Baroni, P., Giacomin, M.: Resolution-based argumentation semantics. In: Proc. of 2nd International Conf. on Computational Models of Argument (COMMA 2008), pp. 25–36 (2008)
5. Bench-Capon, T.: Value-based argumentation frameworks. In: Proc. of Nonmonotonic Reasoning, pp. 444–453 (2002)
6. Cayrol, C., de Saint-Cyr, F.D., Lagasquie-Schiex, M.C.: Change in abstract argumentation frameworks: adding an argument. Journal of Artificial Intelligence Research 38, 49–84 (2010)
7. Cobo, M., Martinez, D., Simari, G.: An approach to timed abstract argumentation. In: Proc. of Int. Workshop of Non-monotonic Reasoning 2010 (2010)
8. Cobo, M., Martinez, D., Simari, G.: On admissibility in timed abstract argumentation frameworks. In: Coelho, H., Studer, R., Wooldridge, M. (eds.) ECAI. Frontiers in Artificial Intelligence and Applications, vol. 215, pp. 1007–1008 (2010)
9. Dechter, R., Meiri, I., Pearl, J.: Temporal constaints networks. In: Proceedings KR 1989, pp. 83–93 (1989)
10. Dung, P.M.: On the acceptability of arguments and its fundamental role in nonmonotonic reasoning, logic programming and n-person games. Artificial Intelligence 77(2), 321–358
11. Jakobovits, H.: Robust semantics for argumentation frameworks. Journal of Logic and Computation 9(2), 215–261 (1999)
12. Mann, N., Hunter, A.: Argumentation using temporal knowledge. In: Proc. of 2nd International Conf. on Computational Models of Argument (COMMA 2008), pp. 204–215 (2008)
13. Martínez, D.C., García, A.J., Simari, G.R.: Modelling well-structured argumentation lines. In: Proc. of XX IJCAI 2007, pp. 465–470 (2007)
14. Meiri, I.: Combining qualitative and quantitative contraints in temporal reasoning. In: Proceedings of AAAI 1992, pp. 260–267 (1992)
15. Rotstein, N.D., Moguillansky, M.O., Falappa, M.A., Garcia, A.J., Simari, G.R.: Argument theory change: Revision upon warrant. In: Proceeding of the 2008 Conference on Computational Models of Argument- COMMA 2008, pp. 336–347 (2008)
16. Rotstein, N.D., Moguillansky, M.O., Garcia, A.J., Simari, G.R.: An abstract argumentation framework for handling dynamics. In: Proc. of Int. Workshop of Non-monotonic Reasoning 2010, pp. 131–139 (2010)
17. Vreeswijk, G.A.W.: Abstract argumentation systems. Artificial Intelligence 90(1-2), 225–279 (1997)

Object Oriented Modelling in Information Systems Based on Related Text Data

Kolyo Onkov

Agricultural University,
Department of Computer Science and Statistics,
Mendeleev 12, 4000 Plovdiv, Bulgaria
kolyoonkov@yahoo.com

Abstract. Specialized applied fields in natural sciences – medicine, biology, chemistry etc. require building and exploring of information systems based on related text forms (words, phrases). These forms represent expert information and knowledge. The paper discusses the integration of two basic approaches – relational for structuring complex related texts and object oriented for data analysis. This conception is implemented for building of information system "Crop protection" in Bulgaria based on the complex relationships between biological (crops, pests) and chemical (pesticides) terms in textual form. Analogy exists between class objects in biology, chemistry and class objects and instances of object oriented programming. That fact is essential for building flexible models and software for data analysis in the information system. The presented example shows the potential of object oriented modelling to define and resolve complex tasks concerning effective pesticides use.

Keywords: expert data, natural sciences, key words, text data retrieval, relational database, object oriented modelling, crop protection.

1 Introduction

Text is the most common form for presenting expert information and knowledge. Paper guides and reference books using knowledge from natural sciences – medicine, biology, physics, chemistry etc., aim to impose the state rules and regulations in order to keep proofs of decisions made and to share and disseminate the useful information. As being heavily used by specialists it is difficult when the volume of text data is large and there are complex relationships between basic terms in textual form, for example the names of drugs and diseases. The following basic problems arise in the process of transformation text data containing related key words and phrases into intelligent computer based system: a) to retrieve related key words and phrases and to define relationships between them; b) to store text data and relationships in easy accessible database; c) to develop software for database analysis to meet information and knowledge requirements determined by specialists in the applied fields.

This paper reveals one conception for building of information systems based on related text data through integration of two approaches – relational for data structuring and object oriented for data modelling and analysis. This conception is implemented for the development of Bulgarian information system "Crop protection".

L. Iliadis et al. (Eds.): EANN/AIAI 2011, Part II, IFIP AICT 364, pp. 212–218, 2011.

2 Conception of Information System Development

The conceptual framework of building and exploring of information system based on related text data is presented in figure 1.

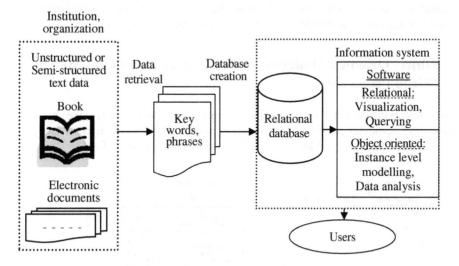

Fig. 1. Conception of building and exploring of information system based on related text objects

In many cases paper or electronic guides and reference books in pharmacology, veterinary, agronomy etc present knowledge from natural sciences in the form of semi-structured or unstructured text data [1], [14]. The core of these expert texts is a set of contextually related words and phrases (terms in the fields). The retrieval of "key words" and relationships among them is usually done through text mining methods and algorithms [2], [3], [4], [5], [6]. The process of verification and correction [7], [8] is needed to guarantee storing correct expert data in the database.

Figure 2 presents information model for treatment of subjects which is common for several natural sciences: medicine, biology, chemistry and pharmacology. The relationships between terms (key words, elements of the sets) exist in both sequences: $(M_1 \_ M_2 \_M_3)$ and $(M_3 \_ M_2 \_M_1)$. Coding of key words is applied for building

Fig. 2. Information model for treatment of subjects

entity-relationships database model [9]. Ontology-based extraction and structuring [10], conceptual modelling of XML data [11] and relational system between two databases [12] are developed for structuring this type of data.

Relational software provides easy access to data while object oriented approach is the proper paradigm for building flexible models. The objects classifications in natural sciences are usual form of knowledge presentation. That is additional and important argument for building object oriented software for data analysis.

3 Object Oriented Modelling in Bulgarian Information System "Crop Protection"

The Bulgarian ministry of agriculture prepares annually a reference book [1] presenting pesticides permitted for use in the country (figure 3). The book content is based on the state laws and knowledge of national experts from the field.

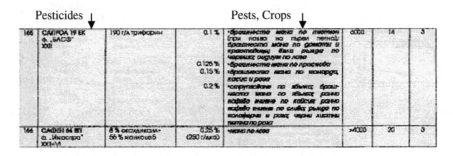

Fig. 3. Scan copy of a fragment from one sheet of the reference book

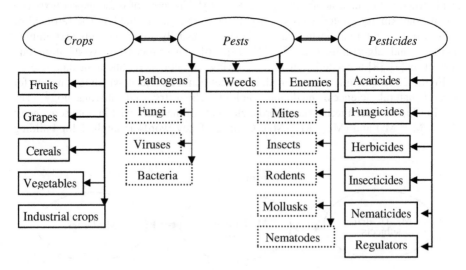

Fig. 4. Information model for crop protection

The conception from the previous chapter is applied to transform the reference book into information system "Crop protection". The relationships between basic biological (crops, pests) and chemical (pesticides) terms are implemented.

The information model for crop protection (figure 4) consists of three hierarchical structures. The nodes of the model are the sets "Crops", "Pests" and "Pesticides". Each node is built by class objects corresponding to the biological and chemical classification. Each class contains text data related with the text data from other class.

Key words and phrases belonging to different classes and subclasses are extracted from the reference book and coded by natural numbers. Then the codes are used as primary keys of table schemes in the relational database. Figure 5 illustrates data access and visualization through relationships in the sequence "crops–pests (pathogens) –pesticides". The quantitative and qualitative characteristics of each pesticide – active ingredient, dose and minimum lethal dose (LD) are also presented.

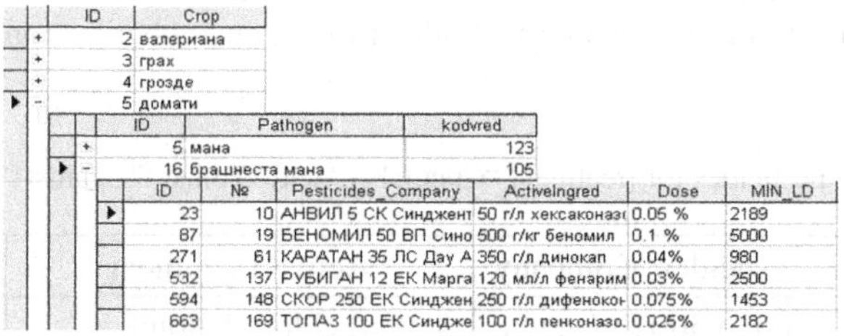

ID	Crop
2	валериана
3	грах
4	грозде
5	домати

ID	Pathogen	kodvred
5	мана	123
16	брашнеста мана	105

ID	№	Pesticides_Company	Activeingred	Dose	MIN_LD
23	10	АНВИЛ 5 СК Синджент	50 г/л хексаконазс	0.05 %	2189
87	19	БЕНОМИЛ 50 ВП Сино	500 г/кг беномил	0.1 %	5000
271	61	КАРАТАН 35 ЛС Дау А	350 г/л динокап	0.04%	980
532	137	РУБИГАН 12 ЕК Марга	120 мл/л фенарим	0.03%	2500
594	148	СКОР 250 ЕК Синджен	250 г/л дифенокон	0.075%	1453
663	169	ТОПАЗ 100 ЕК Синдже	100 г/л пенконазо	0.025%	2182

Fig. 5. Relational database – data access and visualization

Probably the easy access to data is enough for farmers and agronomists, but scientists and experts in biology, chemistry and phytopharmacy usually have to solve more complex tasks which require deep analysis of related text data. The complexity of data analysis descends from the close properties and at the same time variations of subjects from definite class, as well as variations among classes. That is the main reason to create flexible data models by using of object oriented approach. The coding of key words which is corresponding to the subjects from definite class gives opportunity for transition from relational to object oriented data structures. Each class object contains key words field and procedure for implementation of relationships based on their codes. The data models can include two or more related class objects. The process of information extraction, analysis and finally decision making through object oriented modelling has sequential character in spite of the hierarchical subjects' classification of the model (figure 4).

The presented model aims at analyzing similar properties of the objects from definite class objects as well as differences toward treatment of crops against one or more pests. The analysis of these variations supports decision making. The logical sequence "crops–pests–pesticides" is implemented.

Let's define: A: subclass of class "Crops"; B: subclass of class "Pests"; C: class objects "Pesticides". The data modelling can be presented as follows:

a) Input data: Choose N object instances of A; Choose M object instances of B.

b) Apply operations: $(A \longrightarrow B) \longrightarrow C$. These two operations indicate a code application for implementing relationships between instances of the defined subclasses *A, B* and class *C*. For each instance of *A* and each instance of *B* will be found corresponding subset of *C*. The result is a matrix with NxM sets:

$$D = \begin{vmatrix} D_{11} \ D_{12} \ ... \ D_{1M} \\ .... \\ D_{N1} \ D_{N2} \ ... \ D_{NM} \end{vmatrix}$$

The set D_{ij} (i=1, 2, ... N; j=1, 2, ... M) contains pesticides which are proper for use in treatment i^{-th} crop against jth pest. All sets of the matrix *D* are subsets of the set *C*.

c) Section (set *E*) and disjunction (set *F*) of the chosen sets of the matrix *D*. Set *E* shows the common pesticides for treatment of the chosen crops and pests. Set F refers to all pesticides.

This model can be applied for each subclass of objects and instances belonging to classes "Crops" and "Pests". Let's present an example. Two crops {"cucumber", "tomato"} are object instances of class objects "Vegetables" while pest {"mildew"} is instance of class objects "Fungi". Table 1 presents the final result of data processing – common pesticides for treatment the both crops {"cucumber", "tomato"} against {"mildew"}.

Table 1. Common pesticides for treatment "Cucumber" and "Tomato" against "mildew" (set *E*)

№	Code	Pesticides_Company	Active Ingredient	Dose
1	23	Bravo 500 Syngenta	500 g/l chlorothalonil	0.3%
2	51	Equation DuPont	225 g/kg famoxadone + 300 g/kg simoxanile	0.04% (40 g/da)
...	...	...	...	...
8	196	Champion Nufarm	77% cupric hydroxide	0.15%

The modelling based on reverse relationships "pesticides–pests–crops" is useful because of extraction information on available pesticides and variances for their practical use. The object oriented approach provides extending the information model (figure 4), e.g. to add new class objects in hierarchical structures. The model extension can be done in the sense of cognitive and application aims of the IS including biological, geographical and other factors. The flexibility of the model allows the development of application data models focused on different users groups:

• Agronomists and farmers responsible for crop protection measures. Instance level modelling allows working with real data on cultivated crops. The agronomists need to know the variations for the use of different pesticides against identified pests;

• Scientists and experts in the fields of biology, chemistry and phytopharmacy. They would be provoked by the instance level modelling and real data analysis in two directions: research and development of more effective chemicals for crop protection and improvement of the governmental rules, regulations and control mechanisms;

• Specialists who manage pesticides business and related technical and financial resources. The expert analytical information for needed pesticides is important for applying economical models: managing inventories, cost benefit and risk assessment.

The analogy between classes and subclasses in biology and chemistry and class objects and instances of object oriented approach is a base for flexible data modelling. The modern trends and problems coming from the field of crop protection require extending of the models through adding data and new class objects referring to the variations of pest resistance to a pesticide [13], changes in the nature (soil and climatic) etc. The object oriented models can be developed not only on the base of new classes of chemicals, but also by classes of natural products for crop protection.

4 Conclusion

This paper presents a conception for building and exploring of IS based on related text data through integration of two basic approaches – relational for data structuring and object oriented for data analysis. This conception is successfully applied for the creation of Bulgarian information system in the field of crop protection. The potential of the object oriented modelling consists of creating agile models based on related class objects. The developed solution provides working with easy retrieval database, flexible data models and object instances for data analysis. The extracted information will enhance expertise of the specialists and will facilitate decision-making process. Observing specific regulations referring to the pesticides use strategy, it can be concluded that data in the field of crop protection are different in different countries, but the relational database structuring and the software for object oriented modeling will be very similar. In this sense the presented ideas and experience can be useful.

References

1. Bulgarian Ministry of Agriculture: List of the permitted products for plant protection and fertilizers in Bulgaria, Sofia, Videnov & son (2009)
2. Feldman, R., Sanger, J.: The text mining handbook. In: Advanced Approaches in Analyzing Unstructured Data, Cambridge University Press, Cambridge (2007)
3. Bramer, M.: Principles of data mining. Springer-Verlag London Limited, Heidelberg (2007)
4. Crowsey, M., Ramstad, A., Gutierrez, D., Paladino, G., White, K.P.: An Evaluation of Unstructured Text Mining Software. In: IEEE Systems and Information Engineering Design Symposium, Charlottesville (2007)
5. Mahgoub, H., Rösner, D., Ismail, N., Torkey, F.: A Text Mining Technique Using Association Rules Extraction. International J. of Computational Intelligence 4, 21–27 (2007)
6. Mooney, R., Bunescu, R.: Mining Knowledge from Text Using Information Extraction, Natural language processing and text mining. ACM SIGKDD Explorations Newsletter 7, 3–10 (2005)
7. Lopresti, D.: Optical character recognition errors and their effects on natural language processing. In: Second workshop on Analytics for noisy unstructured text data, ACM Digital Library, pp. 9–16 (2008)
8. Onkov, K.: Effect of OCR-errors on the transformation of semi-structured text data into relational database. In: Third Workshop on Analytics for Noisy Unstructured Text Data, ACM Digital Library, pp. 123–124 (2009)

9. Dimova, D., Onkov, K.: An algorithm for automated creation of a PC database storing related text objects. Journal of Information Technologies and Control 5(2), 48–52 (2007)
10. Embley, D.W., Campbell, D.M., Smith, R.D., Liddle, S.W.: Ontology-Based Extraction and Structuring of Information from Data-Rich Unstructured Documents. In: Seventh International Conference on Information and Knowledge Management, pp. 52–59 (1998)
11. Necasky, M.: Conceptual Modeling for XML. IOS Press, Amsterdam (2008)
12. Kouno, T., Ayabe, M., Hitomi, H., Machida, T., Moriizumi, S.: Development of Relational System between Plant Pathology Database and Pesticide Database. In: Second Asian Conference for Information Technology in Agriculture, AFITA (2000),
   http://www.afita.org/files/web_structure/20110302115147_70153
   5/20110302115147_701535_11.pdf
13. United States Environmental Protection Agency, http://www.epa.gov/
14. Bulgarian Ministry of agriculture, Authorized institution for plant protection,
   http://www.stenli.net/nsrz/main.php?module=info&object=info&a
   ction=view&inf_id=21

Ranking Functions in Large State Spaces

Klaus Häming and Gabriele Peters

University of Hagen,
Chair of Human-Computer-Interaction,
Universitätsstr. 1, 58097 Hagen, Germany
{klaus.haeming,gabriele.peters}@fernuni-hagen.de

Abstract. Large state spaces pose a serious problem in many learning applications. This paper discusses a number of issues that arise when ranking functions are applied to such a domain. Since these functions, in their original introduction, need to store every possible world model, it seems obvious that they are applicable to small toy problems only. To disprove this we address a number of these issues and furthermore describe an application that indeed has a large state space. It is shown that an agent is *enabled* to learn in this environment by representing its belief state with a ranking function. This is achieved by introducing a new entailment operator that accounts for similarities in the state description.

Keywords: ranking functions, reinforcement learning, belief revision, computer vision.

1 Introduction

This paper discusses the applicability of ranking functions on problem domains which have a large state space. We discuss this issue with an autonomous agent in mind whose belief state is represented as such a function. This agent has to incorporate its experience into its belief state in order to learn and solve its task successfully. The considerations in this paper are assessed by implementing such an agent and expose him to a suitable task. This is done in an reinforcement learning [10] application, which has been augmented by a two-level representation of the agents belief. This two-level representation has been inspired by the work of Sun [9,8] and is based on psychological findings, e.g., of Reber [5] and Gombert [3]. For the largest part however, this work discusses ranking functions independently from a concrete application.

Ranking functions were introduced by Spohn [6] under the term of ordinal conditional functions. They were introduced to account for the dynamics of belief revision [1,2]. Traditionally, belief revision deals with belief sets, which capture the current belief of an agent. A belief set changes whenever an agent perceives new information it wants to include. A well accepted set of postulates which constrain the possible change of a belief set is given by the AGM theory of Alchourrón, Gärdenfors and Makinson [1]. Ranking functions are compliant

L. Iliadis et al. (Eds.): EANN/AIAI 2011, Part II, IFIP AICT 364, pp. 219–228, 2011.

with these postulates [7]. Ranking functions extend the belief set in a way that, additional to the actual belief, the disbelieved facts are also included. This property allows ranking functions to account effectively for changes in belief and also allows an elegant incorporation of conditional information.

Spohn's ranking functions are not to be confused with ranking functions that aid a search algorithms in a heuristic manner. An example for the latter kind of ranking function is the well-known tf–idf weight [11] which is often used in text mining. These ranking functions are a tool to *cope* with large state spaces, as shown in, e.g., [12]. The ranking function we discuss in this work rank world models and hence will suffer from large state spaces *themselves*, if we do not take steps against it.

The next section provides a brief review of Spohn's ranking functions, followed by a discussion of their problems with large state spaces. A proposal of how to handle this is given thereafter. Since uncertainty is also a problem for belief representations, we discuss this in Section 5 with our example application in mind, which is introduced thereafter. After that, a brief section with concluding remarks is given.

2   Ranking Functions

To define ranking functions, we need the notion of a *world model*. Given the set of all possible worlds $\mathfrak{W}$, a world model $M \in \mathfrak{W}$ describes exactly one instance. Therefore, to apply ranking functions, the world an agent lives in must be completely describable. We assume such a world.

A particular ranking function $\kappa : \mathfrak{W} \to \mathbb{N}$ assigns a non-negative integer value to each of the world models. These *ranks* reflect the amount of *disbelief* an agent shows for each model. The agent believes in a model M, iff

$$\kappa(M) = 0 \wedge \kappa(\overline{M}) > 0 . \tag{1}$$

Besides querying the rank of a model, we may ask for the rank of a formula F. Since a model M captures all aspects of the world the agent lives in, F partitions $\mathfrak{W}$ into two sets:

1. $\{M|M \models F\}$
2. $\{M|M \models \overline{F}\}$

This allows us to define the rank $\kappa(F)$ of a formula F as $\min\{\kappa(M)|M \models F\}$. So, if the agent believes in a particular world model which happens to entail F, it will also belief F.

Whenever the agent experiences new information about its world, new ranks are assigned to the models to reflect the agent's new belief state. This incorporation of new information is called *revision*. A peculiar property of ranking functions is their ability to not only belief in a certain proposition, but to belief in it with a given strength $\alpha \in \mathbb{N}$, which is enforced during revision.

We will discuss two types of revision. First, the revision with propositional information, and second the revision with conditional information. After each definition, a short explanation of its meaning is given.

Definition 1. *Given a proposition P, the revision $\kappa * (P, \alpha)$ is given by*

$$\kappa * (P, \alpha)(M) = \begin{cases} \kappa(M) & : \kappa(\overline{P}) \geq \alpha \\ \kappa(M) - \kappa(P) & : \kappa(\overline{P}) < \alpha \wedge M \models P \quad (2) \\ \kappa(M) + \alpha - \kappa(\overline{P}) & : \kappa(\overline{P}) < \alpha \wedge M \models \overline{P} \end{cases}$$

Essentially this definition states that

1. there is nothing to do, if P is already believed with strength α
2. otherwise two partitions of the set of models are relevant:
   (a) those, that agree on P, and whose ranks are modified so that we assign rank 0 to the least disbelieved model.
   (b) those, that agree on $\overline{P}$, and whose ranks are modified so that $\kappa(\overline{P}) \geq \alpha$ holds afterwards.

The rationale behind this partitioning is that the models in $\{M|M \models P\}$ are independent *conditional* on P and should therefore keep their *relative* ranks. The same argument is applied to $\{M|M \models \overline{P}\}$ and $\overline{P}$.

After this gentle introduction to revising with a proposition, we now focus on revising with a conditional $(B|A)$ with A the antecedent and B the consequent. As shown in [4], it is advisable to use the additional operator $\kappa[F] := \max\{\kappa(M)|M \models F\}$ to define this revision.

Definition 2. *Given a conditional $(B|A)$, the revision $\kappa * (B|A, \alpha)$ is given by*

$$\kappa * (B|A, \alpha)(M) :=$$
$$\begin{cases} \kappa(M) & : D \geq \alpha \\ \kappa(M) - \kappa(A \Rightarrow B) & : D < \alpha \wedge M \models (A \Rightarrow B) \quad (3) \\ \kappa[AB] - \kappa(A \Rightarrow B) + \alpha + \kappa(M) - \kappa(A\overline{B}) & : D < \alpha \wedge M \models (A\overline{B}) \end{cases}$$

with $D := \kappa(A\overline{B}) - \kappa[AB]$.

The general idea behind Definition 2 is the same as the one behind Definition 1: Partition the models into two groups and shift the ranks up and down until the required condition is reached! The seemingly more complex rules of Definition 2 emerge from the fact, that $\{M|M \models (A \Rightarrow B)\} = \{M|M \models \overline{A}\} \cup \{M|M \models AB\}$, while the postcondition $\kappa(A\overline{B}) - \kappa[AB] \geq \alpha$ only refers to $\kappa(A\overline{B})$ and $\kappa(AB)$.

The following example clarifies the belief change in the presence of conditional information. Suppose our world is described by three variables $a, b, c \in \{1, 2\}$. Suppose further that our current belief state represents a belief in $M_0 = (a = 2) \wedge (b = 1) \wedge (c = 1)$ which is consequently mapped to rank 0. The left column of the following table shows this scenario. The first row contains world models with rank 0, i.e., the believed ones, which is in this example just M_0. There are a few other models whose respective ranks are 1, 2, and 3. The models are represented by the values of the variables in the order a, b, c. The models with higher ranks are omitted for clarity.

$$
\begin{array}{c}
\kappa \\
\begin{Vmatrix}
2\,1\,1 \\ \hline
2\,1\,2 \\
2\,2\,1 \\
2\,2\,2 \\
\\
\vdots
\end{Vmatrix}
\end{array}
\xrightarrow{\;\kappa*((a=2)\Rightarrow(c=2),1)\;}
\begin{array}{c}
\kappa' \\
\begin{Vmatrix}
2\,1\,2 \\ \hline
2\,2\,2 \\
2\,1\,1 \\
\\
2\,2\,1 \\
\\
\vdots
\end{Vmatrix}
\end{array}
\tag{4}
$$

The right column of this example captures the belief after κ has been revised with $(a = 2) \Rightarrow (c = 2)$ using Definition 2. As you can see, the models have been partitioned into two groups ($\{2\,1\,2, 2\,2\,2\}$ and $\{2\,1\,1, 2\,2\,1\}$), which have been shifted such that κ obeys the postcondition $\kappa(A\overline{B}) - \kappa[AB] = 1$.

3   Large State Spaces

After the general mechanics of ranking functions have been described, we want to focus on the applicability of them on problems with a considerably large state space. Assume, for example, that a world is described by n boolean variables which describe binary properties of that world. A concatenation of these variables yields an n-digit binary number. The number of possible world models is therefore 2^n, which grows dramatically as n grows.

Therefore, for a sufficiently complex world, we cannot expect to be able to store all its possible models to represent κ:

$$
\begin{array}{c}
\kappa \\
\begin{Vmatrix}
M_1\; M_2\; M_3 \cdots \lightning \\
M_4\; M_5 \\
\vdots \\
\lightning
\end{Vmatrix}
\end{array}
$$

We may try to circumvent this problem by modeling a skeptic agent, one that initially disbelieves anything. We may think that this means we commit to a scenario, in which initially all models have rank ∞ and simply omit those. That would mean our initial ranking function is empty.

But there is a difficulty. As Spohn has shown, for each model M it must hold that

$$
\kappa(M) = 0 \vee \kappa(\overline{M}) = 0 \; .
\tag{5}
$$

If we are to put every single model initially to rank ∞, this rule will be violated. Fortunately, the actual rank is less important that the fact whether a model is believed or not. Re-consider equation 1. If neither a model M nor $\overline{M}$ has been experienced, we are also free to interpret this as

$$\kappa(M) = 0 \wedge \kappa(\overline{M}) = 0 , \tag{6}$$

which means that neither M nor $\overline{M}$ is believed. And once κ has been revised with M, the missing model $\overline{M}$ can be interpreted as $\kappa(\overline{M}) = \infty$. Hence, a skeptic agent indeed allows us to omit not yet experienced world models.

Concerning revision in large state spaces, we need to investigate whether the dynamics induced by Definition 1 and Definition 2 create additional problems. First, revising with propositions is not problematic, as long as there are not many models which entail it. Revising with propositional information is also not the common case for a learning agent, since the information is usually in conditional form as, e.g., in our example application below.

A Revision with conditional information, however, creates more serious problems. Let us assume that an agent needs to believe $A \Rightarrow B$, therefore the models $\{M | M \models \overline{A} \vee B\}$ have to be shifted towards rank 0. If we re-consider the example of the world described by binary variables, revising with a conditional which is currently not believed and whose antecedent consists of just one variable means, that at least half of the models have to be shifted towards rank 0. Hence, a lot of models not yet present in the ranking function have to be created to maintain their relative ranks. Furthermore, the operator $\kappa[AB]$ creates additional complications. If $\kappa[AB] = \infty$, we will also have to create all $\{M | M \models AB \ \wedge \kappa(M) = \infty\}$.

4   The Proposed Solution

If we want to use ranking functions on larger state spaces, we will obviously need a way around these issues—and we do have an option. The problem is caused by strictly requiring that within each of the partitions of $\mathfrak{W}$ (which are induced by a revision) the relative ranks stay the same. If we are allowed to bend this rule in order to instantiate the least amount of models, then we can proceed using ranking functions without hitting any serious computational barrier.

For a revision with propositional information, the postcondition $\kappa(\overline{P}) \geq \alpha$ must hold afterwards. Just taking P and those models into account that have been experienced so far obviously complies with this requirement. This is true, because either there are models from $\{M | M \models \overline{P}\}$ in the ranking function or not. If there are such models, the shifting of their ranks creates a ranking function as desired. If there are not such models, there is nothing to do because the postcondition already holds.

Similarly, for a revision with conditional information, we want the postcondition $\kappa(A\overline{B}) - \kappa[AB] \geq \alpha$ to hold afterwards. So, if κ is to be revised with $A \Rightarrow B$ and $\kappa(\overline{A} \vee B) = \infty$, then we may shift $\{M | M \models A \wedge B\}$ towards rank 0, regardless of other models in $\{M | M \models \overline{A} \vee B\}$. We also need to adjust the ranks of $\{M|M \models A\overline{B}\}$, of course, but these retain their ranks or are shifted to higher ranks, therefore posing no additional difficulties. We also modify $\kappa[AB]$ to take the form

$$\kappa[AB] = \max\left\{\kappa(M)|M \models AB \wedge \kappa(M) < \infty\right\}, \tag{7}$$

i.e., it returns the highest ranked *known* model of AB. Fortunately, this still leads to revisions after which the conditional $(B|A)$ is believed, because this simply requires that some model of $\overline{A} \vee B$ has rank 0. Of course, there still is the case $\kappa(AB) = \infty \wedge \kappa(\overline{A} \vee B) \leq \infty$. In this case $\{M|M \models AB\}$ is shifted towards the highest known rank incremented by one.

5   Partially Known or Noisy State Spaces

Describing a world in terms of logical propositions leads to problems once the world starts to exhibit uncertainty. This section discusses a particular scenario in which this is the case.

Assume that an agent's state description consists of visual information. In particular, assume the state description enumerates a number of visual features. This scenario may happen in a computer vision application which uses a feature detector on images.

The problem is, that feature detection is not reliable. The image is a quantized and noisy representation of the continuous world's electromagnetic signals. Viewing a scene from the same position may therefore yield slightly different descriptions of the visual input.

Assume the state description takes the form

$$S = v_1 \wedge v_2 \wedge v_3 \wedge v_4 \wedge \ldots v_n, \tag{8}$$

where v_i is either **TRUE** or **FALSE** depending on the visibility of the feature f_i. If the value of one of these v_i switches, the resulting state description will be completely different in terms of ranking functions. There is no definition of similarity.

Another problem is, that it may be impossible to enumerate all models. This is the case, if the possible features are not known beforehand but added as they appear:

$$\left\| \begin{matrix} \kappa \\ M_1 = v_1 \wedge \overline{v_2} \wedge v_3 \wedge v_4 \wedge \overline{v_5} \wedge \cdots \maltese \\ \vdots \end{matrix} \right\|$$

As a consequence, we cannot create the partitions of $\mathfrak{W}$ necessary for revision.

In the particular example of a features-perceiving agent, it is possible to use ranking functions in spite of these issues. Both difficulties, the absence of a similarity definition and the unknown set of features is addressed by a modified entailment operator.

Before we introduce this operator, we have to make sure that for a conditional $A \Rightarrow B$ we are able to partition the models of $\mathfrak{W}$ into the three subsets $\{M|M \models$

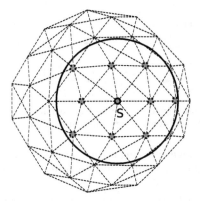

Fig. 1. Simulated Features. Assume, that the current state is S. Then, a visual perception can be simulated by using the grid nodes in the vicinity of S.

$\overline{A}\}$, $\{M|M \models AB\}$, and $\{M|M \models A\overline{B}\}$. This is required to apply Definition 2. To enforce this, we require the conditionals to be elements of the following set $\mathfrak{C}$:

Definition 3. $\mathfrak{C} := \{A \Rightarrow B | A \in \mathfrak{P} \neq \emptyset \land B \in \mathfrak{Q} \neq \emptyset \land \mathfrak{P} \cap \mathfrak{Q} = \emptyset\}$ *with* $\mathfrak{P}$ *and* $\mathfrak{Q}$ *chosen such that* $\forall M \in \mathfrak{W} : \exists P \in \mathfrak{P}, Q \in \mathfrak{Q} : M = P \land Q.$

So, the antecedents and consequents of the allowed conditionals are taken from different sets of variables which together contain all the world's variables. This allows us to handle the entailment differently for the antecedent and the consequent.

Back to our feature-based world, let us assume our models take the form $M = P \land Q$. P shall contain the unreliable, feature-based information, while Q covers everything else. Let us further assume that a function $F(P)$ exists which maps the feature-part P to all the features visible in M. This can be used to relate the visible features $F(P)$ and $F(P')$ of two models M and $M' = P' \land Q'$ to define an entailment operator $M \models_t M'$ such as in

Definition 4. $M \models_t M' :\Leftrightarrow \left(\frac{|F(P) \cap F(P')|}{|F(P) \cup F(P')|} > t \right) \land (Q \land Q').$

The left term thus defines a similarity on sets of visible features. But how should t be chosen? This is assessed in the next section.

6   Example Application

To assess the applicability of the aforementioned ideas, we use an experimental set-up which presents an environment to the agent that is perceived as a set of (unreliable) features as introduced in Section 5.

In this environment, the agent needs to find a spot on a spherical grid. The agent's state consists of a subset of all nodes present in a spherical neighborhood around its current location S on the grid, includeing S itself. Each of these nodes

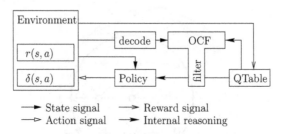

Fig. 2. Two-level reinforcement learning set-up. The ranking function ("OCF") acts as a filter on the possible actions a reinforcement learning agent may take.

is classified as "visible" with a preset probability. The purpose of this experiment is to mimic the situation in which an object is perceived by a camera and represented by visual features. The visibility probability models feature misses and occlusions.

The agent uses the information in its state signal about visible features to determine its position above the object. Fig. 1 shows a picture of this set-up. We used a grid with 128 nodes placed on the surface of a unit sphere. The radius of the sphere that defines the neighborhood has been set to 1.2, which results in approximately 50 enclosed "features" at each position. An action is represented by a number which is associated with a grid node.

We used the two-level-learning approach of [4], in which a ranking function acts as a filter on the possible actions a reinforcement learning [10] agent may take. Fig. 2 clarifies the architecture. There, δ is a transition function and r a reward function. The agent explores an environment through trial and error

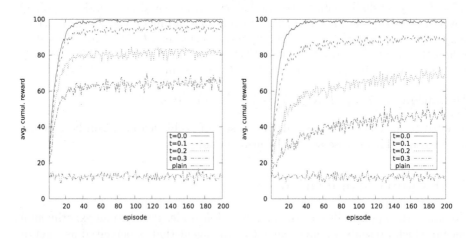

Fig. 3. Learning progress over 200 episodes, averaged over 500 runs. The graphs of the left figure have been calculated with a feature detection probability of 0.9 (cf. Fig. 1 and text), while the right figure shows the same experiments for a detection probability of 0.5.

until a goal state is reached. The arrival at the goal state triggers a reward of 100. All other actions are neither rewarded nor punished. The policy is an ϵ-greedy policy with $\epsilon = 0.1$. Fig. 2 also shows the presence a "decode" module. This module creates the symbolical state description from the state signal the environment provides. The Q-table is realized by applying a hash-function on the state description to create a suitable integer number.

The conditionals used for revision are created with the aid of this Q-table. In each state S a revision with the conditional $S \Rightarrow A$ takes place, if a best action A has been identified by the Q-table.

As shown in Fig. 3, the answer to the question asked in Section 5 about a good value of t in equation 4 is $t = 0$. This can be interpreted in such a way that the agent is able to identify its current state by recognizing one feature alone. Requiring it to recognize more features hinders its learning and therefore slows its progress. For comparison, we also included the graph of a simple Q-learner into Fig. 3 ("plain"). This Q-learner uses a Q-table without a supplementing ranking function. It is not able to show any progress at all.

7   Conclusion

This work discusses the applicability of ranking functions in situations where large state spaces seem to strongly discourage their usage. A number of potential problems such as the instantiation of a large number of models during revision and uncertainty in the state description have been discussed. ranking functions seem to be used rather sparingly which we attribute to these difficulties. To encourage their application we presented an example application from the domain of reinforcement learning. In this example, ranking functions allow a learner to succeed where a traditional Q-learner fails. This has been achieved by introducing a new entailment operator that accounts for similarities in the state description.

Acknowledgments. This research was funded by the German Research Association (DFG) under Grant PE 887/3-3.

References

1. Alchourron, C.E., Gardenfors, P., Makinson, D.: On the Logic of Theory Change: Partial Meet Contraction and Revision Functions. J. Symbolic Logic 50(2), 510–530 (1985)
2. Darwiche, A., Pearl, J.: On the Logic of Iterated Belief Revision. Artificial Intelligence 89, 1–29 (1996)
3. Gombert, J.E.: Implicit and Explicit Learning to Read: Implication as for Subtypes of Dyslexia. Current Psychology Letters 1(10) (2003)
4. Häming, K., Peters, G.: An Alternative Approach to the Revision of Ordinal Conditional Functions in the Context of Multi-Valued Logic. In: 20th International Conference on Artificial Neural Networks, Thessaloniki, Greece (2010)

228   K. Häming and G. Peters

5. Reber, A.S.: Implicit Learning and Tacit Knowledge. Journal of Experimental Psycology: General 3(118), 219–235 (1989)
6. Spohn, W.: Ordinal Conditional Functions: A Dynamic Theory of Epistemic States. Causation in Decision, Belief Change and Statistics, 105–134 (August 1988)
7. Spohn, W.: Ranking Functions, AGM Style. Internet Festschrift for Peter Grdenfors (1999)
8. Sun, R., Merrill, E., Peterson, T.: From Implicit Skills to Explicit Knowledge: A Bottom-Up Model of Skill Learning. Cognitive Science 25, 203–244 (2001)
9. Sun, R.: Robust Reasoning: Integrating Rule-Based and Similarity-Based Reasoning. Artif. Intell. 75, 241–295 (1995)
10. Sutton, R.S., Barto, A.G.: Reinforcement Learning: An Introduction. MIT Press, Cambridge (1998)
11. Wu, H.C., Luk, R.W.P., Wong, K.F., Kwok, K.L.: Interpreting TF-IDF Term Weights as Making Relevance Decisions. ACM Trans. Inf. Syst. 26, 13:1–13:37 (2008)
12. Xu, Y., Fern, A., Yoon, S.: Learning Linear Ranking Functions for Beam Search with Application to Planning. J. Mach. Learn. Res. 10, 1571–1610 (2009)

Modelling of Web Domain Visits by Radial Basis Function Neural Networks and Support Vector Machine Regression

Vladimír Olej and Jana Filipová

Institute of System Engineering and Informatics,
Faculty of Economics and Administration, University of Pardubice,
Studentská 84, 532 10 Pardubice, Czech Republic
{vladimir.olej,jana.filipova}@upce.cz

Abstract. The paper presents basic notions of web mining, radial basis function (RBF) neural networks and ε-insensitive support vector machine regression (ε-SVR) for the prediction of a time series for the website of the University of Pardubice. The model includes pre-processing time series, design RBF neural networks and ε-SVR structures, comparison of the results and time series prediction. The predictions concerning short, intermediate and long time series for various ratios of training and testing data. Prediction of web data can be benefit for a web server traffic as a complicated complex system.

Keywords: Web mining, radial basis function neural networks, ε-insensitive support vector machine regression, time series, prediction.

1 Introduction

Data modelling (prediction, classification, optimization) obtained by web mining [1] from the log files and data on a virtual server, as a complicated complex system that affects these data, there are a few problems worked out. The web server represents a complicated complex system; virtual system usually works with several virtual machines that operate over multiple databases. The quality of the activities of the system also affects the data obtained using web mining. Given virtual system is characterized by its operational parameters, which are changing over time, so it is a dynamic system. The data show a nonlinear characteristics, are heterogeneous, inconsistent, missing and uncertain. Currently there are a number of methods for modelling of the data obtained by web mining. These methods can generally be divided into methods of modelling with unsupervised learning [2,3] and supervised learning [4]. The present work builds on the modelling of web domains visit with uncertainty [5,6].

In the paper is presented a problem formulation with the aim of describing the time series web upce.cz (web presentation visits), including possibilities of pre-processing which is realized by means of simple mathematic-statistic methods. Next, there are introduced basic notions of RBFs [7] neural networks and ε-SVRs [4,8] for time series prediction. Further, the paper includes a comparison of the prediction results of

L. Iliadis et al. (Eds.): EANN/AIAI 2011, Part II, IFIP AICT 364, pp. 229–239, 2011.
© IFIP International Federation for Information Processing 2011

the model designed. The model represents a time series prediction for web upce.cz with the pre-processing done by simple mathematical statistical methods. The prediction is carried out through RBF neural networks and ε-SVRs for different durations of time series and various ratios $O_{train}:O_{test}$ training O_{train} and testing O_{test} dates ($O = O_{train} \cup O_{test}$). There is expected, that for shorter time series prediction it would be better to use RBF neural networks and for longer ε-SVRs.

2 Problem Formulation

The data for prediction of the time series web upce.cz over a given time period was obtained from Google Analytics. This web mining tool, which makes use of Java-Script code implemented for web presentation, offers a wide spectrum of operation characteristics (web metrics). Metrics provided by Google Analytics can be divided into following sections: visits, sources of access, contents and conversion. In the section 'visits' it can be monitored for example the number of visitors, the number of visits and number of pages viewed as well as the ratio of new and returning visitors. Indicator geolocation, i.e. from which country are visitors most often, is needed to be known because of language mutations, for example. In order to predict the visit rate to the University of Pardubice, Czech Republic website (web upce.cz) it is important monitoring the indicator of the number of visits within a given time period. One 'visit' here is defined as an unrepeated combination of IP address and cookies. A sub-metrics is absolutely unique visit defined by an unrepeatable IP address and cookies within given time period. The basic information obtained from Google Analytics about web upce.cz during May 2009 considered of the following: The total visit rate during given monthly cycles. A clear trend is obvious there, with Monday having the highest visit rate, which in turn decreases as the week progresses; Saturday has the lowest visit rate; The average number of pages visited is more than three; A visitor stays on certain page five and half a minutes on average; The bounce rate is approximately 60%; Visitors generally come directly to the website, which is positive; The favourite pages is the main page, followed by the pages of the Faculty of Economic and Administration and the Faculty of Philosophy.

The measurement of the visit rate of the University of Pardubice web page, (web upce.cz) took place time periods of regular, controlled intervals. The result represents a time series. The pre-processing of data was realized by means of simple mathematic-statistic methods: a simple moving average (SMA), a central moving average (CMA), a moving median (MM) along with simple exponential smoothing (SES) and double exponential smoothing (DES) at time t. Pre-processing was used with aim of smoothing the outliers, while maintaining the physical interpretation of data.

The general formulation of the model of prediction of the visit rate for upce.cz can be stated in thusly: $y' = f(x'_1, x'_2, \ldots, x'_m)$, $m=5$, where y' is the number of daily web visits in time $t+1$, y is the number of daily web visits in time t, x'_1 is SMA, x'_2 is CMA, x'_3 is MM, x'_4 is SES, and x'_5 is DES at time t. An example of the pre-processing of the time series of web upce.cz is represented in Fig. 1.

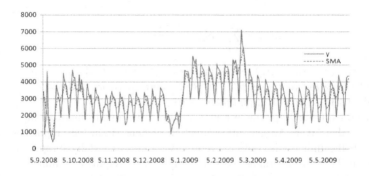

Fig. 1. The pre-processing of web upce.cz visits by SMA

On the basis of the different durational time series of web upce.cz, concrete prediction models for the visit rate to upce.cz web can be defined thusly

$$y' = f(TS^S_{SMA}, TS^S_{CMA}, TS^S_{MM}, TS^S_{SES}, TS^S_{DES}),$$
$$y' = f(TS^I_{SMA}, TS^I_{CMA}, TS^I_{MM}, TS^I_{SES}, TS^I_{DES}), \quad (1)$$
$$y' = f(TS^L_{SMA}, TS^L_{CMA}, TS^L_{MM}, TS^L_{SES}, TS^L_{DES}),$$

where TS - time series, S - short TS (264 days), I - intermediate TS (480 days), L - long TS (752 days) at time t.

3 Basic Notions of RBF Neural Networks and ε-SVR

The term RBF neural network [7] refers to any kind of feed-forward neural networks that uses RBF as an activation function. RBF neural networks are based on supervised learning. The output $f(x,H,w)$ RBF of a neural network can be defined this way

$$f(x, H, w) = \sum_{i=1}^{q} w_i \times h_i(x), \quad (2)$$

where $H = \{h_1(x), h_2(x), \ldots, h_i(x), \ldots, h_q(x)\}$ is a set of activation functions RBF of neurons (of RBF functions) in the hidden layer and w_i are synapse weights. Each of the m components of vector $x = (x_1, x_2, \ldots, x_k, \ldots, x_m)$ is an input value for the q activation functions $h_i(x)$ of RBF neurons. The output $f(x,H,w)$ of RBF neural network represents a linear combination of outputs from q RBF neurons and corresponding synapse weights w.

The activation function $h_i(x)$ of an RBF neural network in the hidden layer belongs to a special class of mathematical functions whose main characteristic is a monotonous rising or falling at an increasing distance from center c_i of the activation function $h_i(x)$ of an RBF. Neurons in the hidden layer can use one of several activation functions $h_i(x)$ of an RBF neural network, for example a Gaussian activation function (a one-dimensional activation function of RBF), a rotary Gaussian activation function (a two-dimensional RBF activation function), multisquare and inverse multisquare activation functions or Cauchy's functions. Results may be presented in this manner.

$$h(x,C,R)=\sum_{i=1}^{q}\exp\left(-\frac{\|\,x-c_i\,\|^2}{r_i}\right),$$ (3)

where $x=(x_1,x_2, \ldots ,x_k, \ldots ,x_m)$ represents the input vector, $C=\{c_1,c_2, \ldots ,c_i, \ldots ,c_q\}$ are centres of activation functions $h_i(x)$ an RBF neural network and $R=\{r_1,r_2, \ldots ,r_i, \ldots ,r_q\}$ are the radiuses of activation functions $h_i(x)$.

The neurons in the output layer represent only a weighted sum of all inputs coming from the hidden layer. An activation function of neurons in the output layer can be linear, with the unit of the output eventually being convert by jump instruction to binary form. The RBF neural network learning process requires a number of centres c_i of activation function $h_i(x)$ of the RBF neural networks to be set as well as for the most suitable positions for RBF centres c_i to be found. Other parameters are radiuses of centres c_i, rate of activation functions $h_i(x)$ of RBFs and synapse weights $W(q,n)$. These are setup between the hidden and output layers. The design of an appropriate number of RBF neurons in the hidden layer is presented in [4]. Possibilities of centres recognition c_i are mentioned in [7] as a random choice. The position of the neurons is chosen randomly from a set of training data. This approach presumes that randomly picked centres c_i will sufficiently represent data entering the RBF neural network. This method is suitable only for small sets of input data. Use on larger sets, often results in a quick and needless increase in the number of RBF neurons in the hidden layer, and therefore an unjustified complexity of the neural network. The second approach to locating centres c_i of activation functions $h_i(x)$ of RBF neurons can be realized by a K-means algorithm.

In nonlinear regression ε-SVR [4,8,9,10,11] minimizes the loss function $L(d,y)$ with insensitive ε [4,11]. Loss function $L(d,y)=|d-y|$, where d is the desired response and y is the output estimate. The construction of the ε-SVR for approximating the desired response d can be used for the extension of loss function $L(d,y)$ as follows

$$L_\varepsilon(d,y) = \begin{cases} |d-y|-\varepsilon & \text{for} |d-y| \ge \varepsilon \\ 0 & \text{else} \end{cases},$$ (4)

where ε is a parameter. Loss function $L_\varepsilon(d,y)$ is called a loss function with insensitive ε.

Let the nonlinear regression model in which the dependence of the scalar d vector x expressed by $d=f(x) + n$. Additive noise n is statistically independent of the input vector x. The function $f(.)$, and noise statistics are unknown. Next, let the sample training data (x_i,d_i), $i=1,2, \ldots ,N$, where x_i and d_i is the corresponding value of the output model d. The problem is to obtain an estimate of d, depending on x. For further progress is expected to estimate d, called y, which is widespread in the set of nonlinear basis functions $\varphi_j(x)$, $j=0,1, \ldots ,m_l$ this way

$$y = \sum_{j=0}^{m_l} w_j\varphi_j(x) = w^T\varphi(x),$$ (5)

where $\varphi(x)=[\varphi_0(x), \varphi_1(x), \ldots , \varphi_{m_l}(x)]^T$ and $w=[w_0,w_1, \ldots ,w_{m_l}]$. It is assumed that $\varphi_0(x)=1$ in order to the weight w_0 represents bias b. The solution to the problem is to minimize the empirical risk.

$$R_{emp} = \frac{1}{N} \sum_{i=1}^{N} L_\varepsilon(d_i, y_i), \qquad (6)$$

under conditions of inequality $\|w\|^2 \leq c_0$, where c_0 is a constant. The restricted optimization problem can be rephrased using two complementary sets of non-negative variables. Additional variables ξ and ξ' describe loss function $L_\varepsilon(d,y)$ with insensitivity ε. The restricted optimization problem can be written as an equivalent to minimizing the cost functional

$$\phi(w, \xi, \xi') = C\left(\sum_{i=1}^{N}(\xi_i + \xi'_i)\right) + \frac{1}{2} w^T w, \qquad (7)$$

under the constraints of two complementary sets of non-negative variables ξ and ξ'. The constant C is a user-specified parameter. Optimization problem (7) can be easily solved in the dual form. The basic idea behind the formulation of the dual-shaped structure is the Lagrangian function [9], the objective function and restrictions. Can then be defined Lagrange multipliers with their functions and parameters which ensure optimality of these multipliers. Optimization the Lagrangian function only describes the original regression problem. To formulate the corresponding dual problem a convex function can be obtained (for shorthand)

$$Q(\alpha_i, \alpha'_i) = \sum_{i=1}^{N} d_i(\alpha_i - \alpha'_i) - \varepsilon \sum_{i=1}^{N}(\alpha_i + \alpha'_i) -$$

$$\frac{1}{2}\sum_{i=1}^{N}\sum_{j=1}^{N}(\alpha_i - \alpha'_i)(\alpha_j - \alpha'_j)K(x_i, x_j), \qquad (8)$$

where $K(x_i, x_j)$ is kernel function defined in accordance with Mercer's theorem [4,8]. Solving optimization problem is obtained by maximizing $Q(\alpha, \alpha')$ with respect to Lagrange multipliers α and α' and provided a new set of constraints, which hereby incorporated constant C contained in the function definition $\phi(w, \xi, \xi')$. Data points covered by the $\alpha \neq \alpha'$, define support vectors [4,8,9].

4   Modelling and Analysis of the Results

The designed model in Fig. 2 demonstrates prediction modelling of the time series web upce.cz. Data pre-processing is carried out by means of data standardization. Thereby, the dependency on units is eliminated. Then the data are pre-processed through simple mathematical statistical methods (SMA,CMA,MM,SES, and DES). Data pre-processing makes the suitable physical interpretation of results possible. The pre-processing is run for time series of different duration namely (TS^S_{SMA}, $TS^S_{CMA}, TS^S_{MM}, TS^S_{SES}, TS^S_{DES}$),($TS^I_{SMA}, TS^I_{CMA}, TS^I_{MM}, TS^I_{SES}, TS^I_{DES}$),($TS^L_{SMA}, TS^L_{CMA}$, $TS^L_{MM}, TS^L_{SES}, TS^L_{DES}$). The prediction is made for the aforementioned pre-processed time series S, I, and L with the help of RBF neural networks, ε-SVR with a polynomial kernel function and ε-SVR with a RBF kernel function for various sets of training O_{train} and testing O_{test} data.

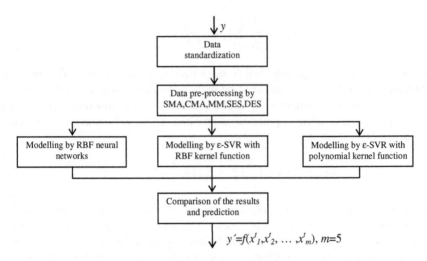

Fig. 2. Prediction modelling of time series for web.upce.cz

In Fig. 3 are shown the dependencies of Root Mean Squared Error (RMSE) on the number of neurons q in the hidden layer. The dependencies RMSE on the parameter μ are represented in Fig. 4. In Fig. 5 the dependencies RMSE are on the parameter v for training O_{train} and as well as the testing O_{test} set, where: • - O_{train_S}, ♦ - O_{train_I}, ■ - O_{train_L}, × - O_{test_S}, ▲- O_{test_I}, ∗ - O_{test_L}. The parameter μ allows for an overrun of the local extreme in the learning process and the following progress of learning. The parameter v represent the selection of centers RBFs as well as guarantees the correct allocation of neurons in the hidden layer for the given data entering the RBF neural network.

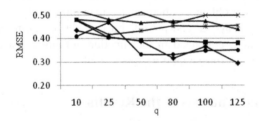

Fig. 3. RMSE dependencies on the parameter q

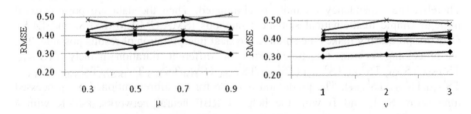

Fig. 4. RMSE dependencies on the parameter μ **Fig. 5.** RMSE dependencies on the parameter v

Conclusions presented in [4,7,10] are verified by the analysis of results (with 10-fold cross validation). $RMSE_{train}$ is lowered until it reaches to the value 0.3, when q neurons in the hidden layer increases in size 125 and with the greater ratio $O_{train}:O_{test}$. $RMSE_{test}$ decreases only when the number q RBF neurons increases at a significantly slower ratio than the ratio $O_{train}:O_{test}$. Minimum $RMSE_{test}$ moves right with the increasing ratio $O_{train}:O_{test}$. Next, determination of the optimal number of q neurons in the hidden layer is necessary. Respectively, a lower number q neurons represents the increasing $RMSE_{test}$. In Table 1, Table 2, and Table 3 are shown optimized results of the analysis of the experiments for different parameters of the RBF neural networks (with RBF activation function) with different durations of time series, various ratios of $O_{train}:O_{test}$ training O_{train} and testing O_{test} of data sets and same amount of learning at $p=600$ cycles. Table 2 builds on the best set of q in Table1 and Table 3 builds on the best set of q (Table 1) and μ (Table 2).

Table 1. Optimized results of RMSE analysis using the amount of neurons q in the hidden layer (μ and v are constant for given TS)

TS	$O_{train}:O_{test}$	Number q of neurons	μ	v	$RMSE_{train}$	$RMSE_{test}$
S	50:50	80	0.9	1	0.383	0.408
S	66:34	80	0.9	1	0.331	0.463
S	80:20	80	0.9	1	0.343	0.365
I	50:50	125	0.9	1	0.372	0.316
I	66:34	125	0.9	1	0.295	0.440
I	80:20	125	0.9	1	0.352	0.326
L	50:50	25	0.9	1	0.409	0.402
L	66:34	25	0.9	1	0.404	0.415
L	80:20	25	0.9	1	0.409	0.390

Table 2. Optimized results of RMSE analysis using parameter μ (q and v are constant for given TS)

TS	$O_{train}:O_{test}$	Number q of neurons	μ	v	$RMSE_{train}$	$RMSE_{test}$
S	50:50	80	0.5	1	0.473	0.385
S	66:34	80	0.5	1	0.341	0.445
S	80:20	80	0.5	1	0.311	0.408
I	50:50	125	0.3	1	0.313	0.344
I	66:34	125	0.3	1	0.302	0.428
I	80:20	125	0.3	1	0.343	0.365
L	50:50	25	0.3	1	0.400	0.387
L	66:34	25	0.3	1	0.398	0.409
L	80:20	25	0.3	1	0.406	0.391

Table 3. Optimized results of RMSE analysis using parameter v (q and μ are constant for given TS)

TS	O_{train}:O_{test}	Number q of neurons	μ	v	$RMSE_{train}$	$RMSE_{test}$
S	50:50	80	0.5	1	0.385	0.473
S	66:34	80	0.5	1	0.341	0.445
S	80:20	80	0.5	1	0.311	0.408
I	50:50	125	0.3	3	0.357	0.375
I	66:34	125	0.3	3	0.327	0.405
I	80:20	125	0.3	3	0.376	0.441
L	50:50	25	0.3	1	0.400	0.387
L	66:34	25	0.3	1	0.398	0.409
L	80:20	25	0.3	1	0.406	0.391

The dependencies RMSE on the parameter C are represented in Fig. 6, the dependencies RMSE on the parameter β in Fig. 7, the dependencies RMSE on the parameter γ in Fig. 8 for training O_{train} as well as testing O_{test} set, where: • - O_{train_S}, ♦ - O_{train_I}, ▪ - O_{train_L}, × - O_{test_S}, ▲ - O_{test_I}, * - O_{test_L}. The parameters C, ε are functions of kernel functions $k(x,x_i)$ [4,11] variations. In the learning process ε-SVR are set using 10-fold cross validation. Parameter C controls the trade off between errors of the ε-SVR of training data and margin maximization; ε [4] selects support vectors in the regression structures, and β represents the rate of polynomial kernel function $k(x,x_i)$. The coefficient γ characterizes polynomial and RBF kernel function.

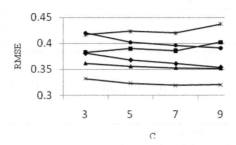

Fig. 6. RMSE dependencies on the parameter C

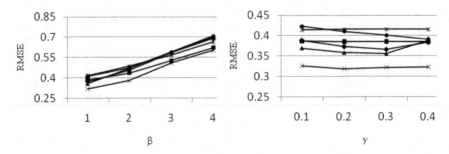

Fig. 7. RMSE dependencies on the parameter β **Fig. 8.** RMSE dependencies on the parameter γ

The confirmation of conclusions presented in [4,11] is verified by the analysis of results. $RMSE_{test}$ for the ε-SVR with RBF kernel function lowers toward zero value with decreasing C (in the case the user experimentation) and higher ratio of $O_{train}:O_{test}$. In the use of 10-fold cross validation, $RMSE_{test}$ moves towards zero with an increase of parameter γ for O_{train_S}. In the ε-SVR with polynomial kernel function $RMSE_{test}$ significantly decreases when the parameter β decreases (Fig. 7). Minimum $RMSE_{test}$ moves rights with an increase of the parameter γ (the minimum is between 0.2 to 0.3), whereas an indirect correlation between the ratio $O_{train}:O_{test}$ and $RMSE_{test}$ obtains.

In Table 4 and Table 5 are shown the optimized results of the analysis of the experiments for different parameters of the ε-SVR (with RBF and a polynomial kernel function) with different durations of time series, various ratios $O_{train}:O_{tes}$ training O_{train} and testing O_{test} of data sets and same amount of learning at $p=600$ cycles. The tables are not represented by the partial results for the various parameters, but only the resulting set of parameters for the TS and the ratio $O_{train}:O_{test}$.

Table 4. Optimized results of RMSE analysis using parameter C

TS	$O_{train}:O_{test}$	C	ε	γ	$RMSE_{train}$	$RMSE_{test}$
S	50:50	8	0.2	0.4	0.343	0.414
S	66:34	10	0.1	0.4	0.365	0.314
S	80:20	10	0.1	0.4	0.355	0.306
I	50:50	10	0.1	0.4	0.331	0.369
I	66:34	10	0.1	0.4	0.342	0.361
I	80:20	9	0.1	0.4	0.350	0.336
L	50:50	6	0.1	0.2	0.382	0.409
L	66:34	4	0.1	0.2	0.383	0.416
L	80:20	4	0.1	0.2	0.384	0.445

Table 5. Optimized results of RMSE analysis using parameters γ and β

TS	$O_{train}:O_{test}$	C	ε	γ	β	$RMSE_{train}$	$RMSE_{test}$
S	50:50	10	0.1	0.2	1	0.411	0.319
S	66:34	10	0.2	0.2	2	0.474	0.379
S	80:20	10	0.1	0.2	3	0.587	0.509
I	50:50	6	0.1	0.3	1	0.385	0.416
I	66:34	10	0.1	0.3	2	0.436	0.486
I	80:20	6	0.1	0.3	3	0.526	0.567
L	50:50	10	0.1	0.2	2	0.455	0.468
L	66:34	6	0.1	0.2	1	0.385	0.416
L	80:20	8	0.1	0.2	1	0.385	0.443

The original value of TS^S (O_{test}) y compared with predicted values of y' (in which time is $t+\Delta t$, $\Delta t=1$ day) using RBF neural network is displayed in Fig. 9. In Fig. 10 TS^S (O_{test}) y is then compared to predicted values y' using the ε-SVR (with RBF kernel function).

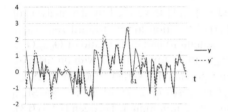   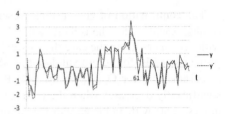

Fig. 9. TS^S with values predicated by RBF   **Fig. 10.** TS^S with values predicated by SVR

In Table 6, there is a comparison of the $RMSE_{train}$ and $RMSE_{test}$ on the training and testing set to other designed and analyzed structures. For example, it was used a fuzzy inference system (FIS) Takagi-Sugeno [5,6], intuitionistic fuzzy inference system (IFIS) Takagi-Sugeno [5,6], feed-forward neural networks (FFNNs), RBF neural networks and ε-SVR^1 (ε-SVR^2) with RBF (polynomial kernel function) with pre-processing input data by simple mathematical statistical methods.

Table 6. Comparison of the $RMSE_{train}$ and $RMSE_{test}$ on the training and testing data to other designed and analyzed structures of fuzzy inference systems and neural networks

	FIS	IFIS	FFNN	RBF	ε-SVR^1	ε-SVR^2
$RMSE_{train}$	0.221	0.224	0.593	0.311	0.331	0.385
$RMSE_{test}$	0.237	0.239	0.687	0.408	0.369	0.416

5 Conclusion

The proposed model consists of data pre-processing and actual prediction using RBF neural networks as well as ε-SVR with polynomial and RBF kernel functions. Furthermore, the modelling was done for time series of different lengths and different parameters of neural networks. The analysis results for various ratios of $O_{train}:O_{test}$ show trends for $RMSE_{train}$ and $RMSE_{test}$. From the analysis of all obtained results of modelling time series by RBF neural network (ε-SVR) shows that $RMSE_{test}$ takes minimum values [4,10,11] for TS^I (TS^S). Further direction of research in the area of modelling web domain visits (excluding the use of uncertainty [5,6] and modelling using RBF neural networks and machine learning using ε-SVR) is focused on different structures of neural networks. The crux of modelling are different lengths of time series, various ratios of $O_{train}:O_{test}$ and different techniques of their partitioning.

Prediction using web mining gives better characterization of webspace. On the basis of that system engineers can better characterize the load of complex virtual system and its dynamics. The RBF neural networks (ε-SVR) design was carried out in SPSS Clementine (STATISTICA).

Acknowledgments. This work was supported by the scientific research project of Ministry of Environment, the Czech Republic under Grant No: SP/4i2/60/07.

References

[1] Cooley, R., Mobasher, B., Srivistava, J.: Web Mining: Information and Pattern Discovery on the World Wide Web. In: 9th IEEE International Conference on Tools with Artificial Intelligence, ICTAI 1997, Newport Beach, CA (1997)

[2] Krishnapuram, R., Joshi, A., Yi, L.: A Fuzzy Relative of the K-medoids Algorithm with Application to Document and Snippet Clustering. In: IEEE International Conference on Fuzzy Systems, Korea, pp. 1281–1286 (1999)

[3] Pedrycz, W.: Conditional Fuzzy C-means. Pattern Recognition Letters 17, 625–632 (1996)

[4] Haykin, S.: Neural Networks: A Comprehensive Foundation. Prentice-Hall Inc., New Jersey (1999)

[5] Olej, V., Hájek, P., Filipová, J.: Modelling of Web Domain Visits by IF-Inference System. WSEAS Transactions on Computers 9(10), 1170–1180 (2010)

[6] Olej, V., Filipová, J., Hájek, P.: Time Series Prediction of Web Domain Visits by IF-Inference System. In: Mastorakis, N., et al. (eds.) Proc. of the 14th WSEAS International Conference on Systems, Latest Trends on Computers, Greece, vol. 1, pp. 156–161 (2010)

[7] Broomhead, D.S., Lowe, D.: Multivariate Functional Interpolation and Adaptive Networks. Complex Systems 2, 321–355 (1988)

[8] Cristianini, N., Shawe-Taylor, J.: An Introduction to Support Vector Machines and other Kernel-based Learning Methods. Cambridge University Press, Cambridge (2000)

[9] Smola, A., Scholkopf, J.: A Tutorial on Support Vector Regression. Statistics and Computing 14, 199–222 (2004)

[10] Niyogi, P., Girosi, F.: On the Relationship between Generalization Error, Hypothesis Complexity, and Sample Complexity for Radial Basis Functions. Massachusetts Institute of Technology Artificial Intelligence Laboratory, Massachusetts (1994)

[11] Vapnik, V.N.: The Nature of Statistical Learning Theory. Springer, New York (1995)

A Framework for Web Page Rank Prediction

Elli Voudigari[1], John Pavlopoulos[1], and Michalis Vazirgiannis[1,2,*]

[1] Department of Informatics, Athens University of Economics and Business, Greece
elliv@aueb.gr, annis.pavlo@gmail.com, mvazirg@aueb.gr
[2] Institut Télécom, Ecole de Télécom ParisTech,
Département Informatique et Réseaux, Paris, France

Abstract. We propose a framework for predicting the ranking position
of a Web page based on previous rankings. Assuming a set of successive
top-k rankings, we learn predictors based on different methodologies.

The prediction quality is quantified as the similarity between the predicted and the actual rankings. Extensive experiments were performed
on real world large scale datasets for global and query-based top-k rankings, using a variety of existing similarity measures for comparing top-k
ranked lists, including a novel and more strict measure introduced in this
paper. The predictions are highly accurate and robust for all experimental setups and similarity measures.

Keywords: Rank Prediction, Data Mining, Web Mining, Artificial
Intelligence.

1 Introduction

The World Wide Web is a highly dynamic structure continuously changing,
as Web pages and hyperlinks are created, deleted or modified. Ranking of the
results is a cornerstone process enabling users to effectively retrieve relevant
and important information. Given the huge size of the Web graph, computing
rankings of Web pages requires awesome resources-computations on matrices
whose size is of the order of Web size (10^9 nodes).

On the other hand the owner of the individual web page can see its ranking
only in the case of the web graph by submitting queries to the owner of the
graph (i.e. a search engine). Given a series of time-ordered rankings of the nodes
of a graph where each bears its ranking for each time stamp, we develop learning mechanisms that enable predictions of the nodes ranking in future times.
The predictions require only local feature knowledge while no global data are
necessary. Specifically, an individual node can predict its ranking only knowing
the values of its own ranking. In such a case the node could plan actions for
optimizing its ranking in future.

In this paper we present an integrated effort for a framework towards Web
page rank prediction considering different learning algorithms. We consider

* Partially supported by the DIGITEO Chair grant LEVETONE in France and the
Research Centre of the Athens University of Economics and Business, Greece.

L. Iliadis et al. (Eds.): EANN/AIAI 2011, Part II, IFIP AICT 364, pp. 240–249, 2011.

i) variable order Markov Models (MMs), ii) regression models and iii) an EM based approach with Bayesian learning. The final purpose is to represent the trends and predict future rankings of Web pages. All the models are learned from timeseries datasets where each training set corresponds to pre-processed rank values of Web pages observed over time.

For all methods, prediction quality is evaluated based on the similarity between the predicted and actual ranked lists, while we focus on the top-k elements of the Web page ranked lists, as top pages are usually more important in Web search.

Preliminary work on this topic was presented in [13] and [15]. The current work significantly differs and advances previous works of the authors in the following ways: a) Refined and careful re-engineering of the MMs' parameter learning procedure by using cross validation, b) Integration and elaboration of the results of [15] in order to validate the performance comparison between regression (boosted with clustering) and MM predictors, in large scale real world datasets, c) Namely we adopt: Linear Regression, random $1^{st}/2^{nd}/3^{rd}$ order Markov models proving the robustness of the model, d) A new top-k list similarity measure $(R - Sim)$ is introduced and used for the evaluation of predictors and more importantly, e) Additional, extensive and robust experiments took place using query based on top-k lists from Yahoo! and Google Search engine.

2 Related Work

The ranking of query results in a Web search-engine is an important problem and has attracted significant attention in the research community.

The problem of predicting PageRank is partly addressed in [9]. It focuses on Web page classification based on URL features. Based on this, the authors perform experiments trying to make PageRank predictions using the extracted features. For this purpose, they use linear regression; however, the complexity of this approach grows linearly in proportion to the number of features. The experimental results show that PageRank prediction based on URL features does not perform very well, probably because even though they correlate very well with the subject of pages, they do not influence page' s authority in the same way.

A recent approach towards page ranking prediction is presented in [13] generating Markov Models from historical ranked lists and using them for predictions.

An approach that aims at approximating PageRank values without the need of performing the computations over the entire graph is [6]. The authors propose an algorithm to incrementally compute approximations to PageRank, based on evolution of the link structure of Web graph (a set of link changes). Their experiments demonstrate that the algorithm performs well both in speed and quality and is robust to various types of link modifications. However, this requires continuous monitoring of the Web graph in order to track any link modifications. There has also been work in adaptive computation of PageRank ([8], [11]) or even estimation of PageRank scores [7].

In [10] a method called predictive ranking is proposed, aiming at estimating the Web structure based on the intuition that the crawling and consequently the ranking results are inaccurate (due to inadequate data and dangling pages). In this work, the authors do not make future rank predictions. Instead, they estimate the missing data in order to achieve more accurate rankings.

In [14] the authors suggest a new measure for ranking scientific articles, based on future citations. Based on publication time and author' s name, they predict future citations and suggest a better model.

3   Prediction Methods

In this section, we present a framework that aims to predict the future rank position of Web pages based on their trends shown the past. Our goal is to find patterns in ranking evolution of Web pages. Given a set of successive Web graph snapshots, for each page we generate a sequence of rank change rates that indicates the trends of this page among the previous snapshots. We use these sequences of previous snapshots of the Web graph as a training set and try to predict the trends of a Web page based on previous. The remaining of this section is organized as follows: In Sect. 3.1 we train MMs of various orders and try to predict the trends of a Web page. Section 3.2 discusses an approach that uses a separate linear regression model for each web page, while Sect. 3.3 combines linear regression with clustering based on an EM probabilistic framework.

Rank Change Rate. In order to predict future rankings of Web pages, we need to define a measure introduced in [12] suitable for measuring page rank dynamics. We briefly present its design.

Let G_{t_i} be the snapshot of the Web graph created by a crawl and $n_{t_i} = |G_{t_i}|$ the number of Web pages at time t_i. Then, $rank(p, t_i)$ is a function providing the ranking of a Web page $p \in G_{t_i}$, according to some criterion (i.e. PageRank values). Intuitively, an appropriate measure for Web pages trends is the rank change rate between two snapshots, but as the size of the Web graph constantly increases the trend measure should be comparable across different graph sizes. Thus, we utilize the normalized rank ($nrank$) of a Web page, as it was defined in [12].

For a page p ranked at position $rank(p, t_i)$: $nrank\,(p,\ t_i) = \frac{2 \cdot rank(p,\ t_i)}{n_{t_i}^2}$, which ranges between $2n_{t_i}^{-2}$ and $2n_{t_i}^{-1}$. Then, using the normalized ranks, the Rank Change Rate ($Racer$) is given by $racer\,(p,\ t_i) = 1 - \frac{nrank(p,\ t_{i+1})}{nrank(p,\ t_i)}$.

3.1   Markov Model Learning

Markov Models (MMs) [1] have been widely used for studying and understanding stochastic processes and behave very well on modeling and predicting values in various applications. Their fundamental assumption is that the future value depends on a number of m previous values, where m is the order of the MM.

They are defined based on a set of states $S = \{s_1, s_2, \ldots, s_n\}$ and a matrix T of transition probabilities t_i each of which represents the probability that a state s_i occurs after a sequence of states.

Our goal is to represent the Web pages ranking trends across different web graph snapshots. We use the *racer* values to describe the rank change of a Web page between two snapshots and we utilize *racer* sequences to learn MMs. Obviously, stable ranking across time is represented by a zero *racer* value, while all other trends by real numbers generating a huge space of discrete values. As expected (intuitively most pages are expected to remain stable for some time irrespective to their rank at the time), the zero value has an unreasonably high frequency compared to all other values which means that all states besides the zero one should be formed by inherent ranges of values instead of a single discrete. In order to ensure equal probability for transition between any pair of states, we guaranteed equiprobable states by forming ranges with equal cumulative frequencies (showing *racer* value within the range) with each other.

In order to calculate the state number for our MMs, we computed the relative cumulative frequency of the zero *racer* state $RF_{Racer=0}$ and used this to find the optimum number of states $n_s = \frac{l}{RF_{Racer=0}}$. Next, we formed n_s equiprobable partitions and used the ranges' mean average values as states to train our model. We should note that within the significantly high frequency of the zero *racer* values, are also considered pages initially obtained within the top-k list and then fell (and remained) out. We remove any bias from $RF_{Racer=0}$, excluding any values not corresponding to stable rank and obtaining $RF_{Racer=0} \approx 0.1$ which in turn suggested 10 equiprobable states.

Predictions with Racer. Based on the set of states mentioned above and formed to represent Web page trends, we are able to train MMs and predict the trend of a Web page in the future according to past trends. By assuming m+1 temporally successive crawls, resulting in respective snapshots, a sequence of m states (representative of racer values) are constructed for each Web page. These are used to construct an m-order MM. Note that the memory m is an inherent feature of the model. After computing transition probabilities for every path, using the generated states, the future states can be predicted by using the chain rule [1]. Thus, for an m-order Markov Model, the path probability of a state sequence is $P(s_1 \rightarrow \ldots \rightarrow s_m) = P(s_1) \cdot \prod_{i=2}^{m} P(s_i | s_{i-m}, \ldots, s_{i-1})$, where each s_i $(i \in \{1, 2, \ldots, n\})$ for any time interval may vary over all the possible states (ranges of racer values). Then, predicting the future trend of a page is performed by computing the most likely next state given the so far state path.

In specific, assuming m time intervals, the next most probable state X is computed as: $X = \arg\max_X P(s_1 \rightarrow \ldots \rightarrow s_{m-1} \rightarrow X)$.

Using that, we predict future states for each page. As each state is the mean of a Racer range, we compute back the future nrank. Therefore, we are able to predict future top-k ranking by sorting the racer of Web pages in ascending order.

3.2 Regression Models

Assume a set of N Web pages and observations of *nrank* values at m time steps. Let $x_i = (x_{i1}, \ldots, x_{im})$ be the *nrank* values for Webpage i at the time points $t = (t_1, \ldots, t_m)$, where the $(N \times m)$ design matrix X stores all the observed *nrank* values so that each row corresponds to a Webpage and each column to a time point. Given these values we wish to predict the *nrank* value x_{i*} for each Webpage at some time t_* which typically corresponds to a future time point $(t_* > t_i, i = 1, \ldots, m)$. Next, we discuss a simple prediction method based on linear regression where the input variable corresponds to time and the output to the *nrank* value.

For a certain Webpage i we assume a linear regression model having the form $x_{ik} = a_i t_k + b_i + \epsilon_k, k = 1, \ldots, m$ (ϵ_k denotes a zero-mean Gaussian noise). Note that the parameters (a_i, b_i) are Webpage-specific and their values are calculated using least squares. In other words, the above formulation defines a separate linear regression model for each Web page thus they treat independently. This can be restrictive since possible existing similarities and dependencies between different Web pages are not taken into account.

3.3 Clustering Using EM

We assume that the *nrank* values of each Web page fall into one of J different clusters. Clustering can be viewed as training a mixture probability model. To generate the *nrank* values x_i for Web page i, we first select the cluster type j with probability π_j (where $\pi_j \geq 0$ and $\sum_{j=1}^{J} \pi_j = 1$) and then produce the values x_i according to a linear regression model $x_{ik} = a_i t_k + b_i + \epsilon_k, k = 1, \ldots, m$, where ϵ_k is independent Gaussian noise with zero mean and variance σ_j^2. This implies that given the cluster type j the *nrank* values are drawn from the product of Gaussians $p(\mathbf{x_i} \mid j) = \prod_{k=1}^{m} N(x_{ik} \mid a_j t_k + b_j, \sigma_j^2)$.

The cluster type that generated the nrank values of a certain Web page is an unobserved variable and thus after marginalization we obtain a mixture unconditional density $p(x_i) = \sum_{j=1}^{J} \pi_j p(x_i \mid j)$ for the observation vector x_i. To train the mixture model and estimate the parameters $\theta = (\pi_j, \sigma_j^2, a_j, b_j)_{j=1,\ldots,J}$, we can maximize the log likelihood of the data $L(\theta) = log \prod_{i=1}^{N} p(x_i)$ by using the EM algorithm [2]. Given an initial state for the parameters, EM optimizes over θ by iterating between E and M steps:

The E step computes the posterior probabilities $R_j^i = \frac{\pi_j p(x_i \mid j)}{\sum_{\rho=1}^{J} \pi_\rho p(x_i \mid \rho)}$, for $j = 1, \ldots, J$ and $i = 1, \ldots, N$, (N is the total number of web pages).

The M step updates the parameters according to: $\pi_j = \frac{1}{N} \sum_{i=1}^{N} R_j^i$,

$$\sigma_j^2 = \frac{\sum_{i=1}^{N} R_j^i \sum_{k=1}^{m} (x_{ik} - a_j t_k - b_j)^2}{\pi_j} \text{ and } \begin{bmatrix} a_j \\ b_j \end{bmatrix} = \frac{1}{N_j} \begin{bmatrix} t^T t & t^T 1 \\ t^T 1 & m \end{bmatrix}^{-1} \begin{bmatrix} \sum_{i=1}^{N} R_j^i x_i^T t \\ \sum_{i=1}^{N} R_j^i x_i^T 1 \end{bmatrix},$$

$j = 1, \ldots, J$, $\mathbf{t}$ is the vector of all time points and $\mathbf{1}$ is the m-dimensional vector of ones.

Once we have obtained suitable values for the parameters, we can use the mixture model for prediction. Particularly, to predict the *nrank* value x_{i*} of Web page i at t_* given the observed values $x_i = (x_{i1}, \ldots, x_{im})$ at previous times, we express the posterior distribution $p(x_{i*}|x_i)$ using the Bayes rule $p(x_{i*} \mid x_i) = \sum_{j=1}^{J} R_j^i N\left(x_{i*} \mid a_j t_* + b_j, \ s_j^2\right)$, where R_j^i is computed according to E-step. To obtain a specific predictive value for x_{i*}, we can use the mean value of the above posterior distribution $x_{i*} = \sum_{j=1}^{J} R_j^i \left(a_j t_* + b_j\right)$ or the median estimate $x_{i*} = a_j\, t_* + b_j$, where $j = argmax_\rho R_\rho^i$ that considers a hard assignment of the Web page into one of the J clusters.

4 Top-k List Similarity Measures

In order to evaluate the quality of predictions, we need to measure the similarity of the predicted to the actual top-k ranking. For this purpose, we examine measures commonly used for comparing rankings, point out the shortcomings of existing and define a new similarity measure for top-k rankings, denoted as *RSim*.

4.1 Existing Similarity Measures

The first one, denoted as $OSim(A,B)$ [4] indicates the degree of overlap between the top-k elements of two sets A and B(each one of size k): $OSim\,(A, B) = \frac{|A \cap B|}{k}$.

The second, $KSim(A,B)$ [4], is based on Kendall's distance measure [3] and indicates the degree that the relative orderings of two top-k lists are in agreement: $KSim(A, B) = \frac{|(u,v):A',B',\ agree\ in\ order|}{|A \cup B|(|A \cup B|-1)}$, where A' is an extension of A resulting from appending at its tail the elements $x \in A \cup (B - A)$ and B' is defined analogously.

Another interesting measure introduced in Information Retrieval for evaluating the accumulated relevance of a top-k document list to a query is the *(Normalized) Discounted Cumulative Gain (N(DCG))* [5]. This measure assumes a top-k list, where each document is featured with a relevance score accumulated by scanning the list from top to bottom. Although DCG could be used for the evaluation of our predictions, since it takes into account the relevance of a top-k list to another, it exhibits some basic features that prevented us from using it in our experiments. It penalizes errors by maintaining an increasing value of cumulative relevance. While this is based on the rank of each document, the size k of the list is not taken into account – thus the length of the list is irrelevant in DCG. Errors in top ranks of a top-k list should be considered more important than errors in low-ranked positions. This important feature lacks from both DCG and NDCG measures. Moreover, DCG value for each rank in the top-k list is computed taking into account the previous values in the list.

Next, we introduce *Spearman's Rank Correlation Coefficient*, which was used during the experimental evaluation, consists a non-parametric (distribution-free) rank statistic proposed by Spearman (1904) measuring the strength of associations between two variables and is often symbolized by ρ. It estimates how well the relationship between two variables can be described using a monotonic function. If there are no repeated data values of these variables (like in ranking problem), a perfect Spearman correlation of +1 or -1 exists if each variable is a perfect monotone function of the other.

It is often confused with the Pearson correlation coefficient between ranked variables. However, the procedure used to calculate ρ is much simpler. If X and Y are two variables with corresponding ranks x_i and y_i, $d_i = x_i - y_i$, $i = 1, \ldots, n$, between the ranks of each observation on the two variables, then it is given by:

$$\rho = 1 - \frac{6 \cdot \sum_{i=1}^{n} d_i^2}{n(n^2 - 1)}.$$

4.2 RSim Quality Measure

The observed similarity measures do not cover sufficiently the fine grained requirements arising, comparing top-k rankings in the Web search context. So we need a new similarity metric taking into consideration: a)The absolute difference between the predicted and actual position for each Webpage as large difference indicates a less accurate prediction and b)The actual ranking position of a Web page, because failing to predict a highly ranked Webpage is more important than a low-ranked. Based on these observations, we introduce a new measure, named *RSim*. Every inaccurate prediction made incurs a certain penalty depending on the two noted factors. If prediction is 100% accurate (same predicted and actual rank), the penalty is equal to zero. Let B_i be the predicted rank position for page i and A_i the actual. The Cumulative Penalty Score (CPS) is computed as $CPS(A, B) = \sum_{i=1}^{k} |A_i - B_i| \cdot (k + 1 - A_i)$.

The proposed penalty score CPS represents the overall error (difference) between the involved top-k lists A and B and is proportional to $|A_i - B_i|$. The term $(k + 1 - A_i)$ increases when A_i becomes smaller so errors in highly ranked Web pages are penalized more. In the best case, rank predictions for all Web pages are completely accurate ($CPS = 0$), since $A_i = B_i$ for any value of i. In the worst case, the rank predictions for all Web pages not only are inaccurate, but also bear the greatest CPS penalty possible. In such a scenario, all the Web pages predicted to be in the top-k list, actually hold the position $k+1$ (or worse).

Assuming that we want to compare two rankings of length k, then the maximum CPS for even and odd values of k is equal to $\frac{2k^3 + 3k^2 + k}{6}$. The proof for CPS_{max} final form is omitted due to space limitations.

Based on the above we define a new similarity measure, *RSim*, to compare the similarity between top-k rank lists as follows:

$$RSim(A_i, B_i) = 1 - \frac{CPS(A_i, B_i)}{CPS_{max}(A_i, B_i)}. \tag{1}$$

In the best-case prediction scenario, *RSim* is equal to one, while in the worst-case *RSim* is equal to zero. So the closer the value of *RSim* is to one, the better and more accurate the rank predictions are.

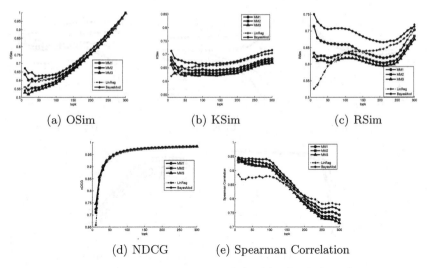

Fig. 1. Prediction accuracy vs Top-k list length - Yahoo dataset

5 Experimental Evaluation

In order to evaluate the effectiveness of our methods we performed experiments on two different real world datasets. These consist collections of top-k ranked lists for 22 queries over a period of 11 days as resulted from the Yahoo![1] and the Google search engines, produced in the same way. In our experiments, we evaluate the prediction quality in terms of similarities between the predicted and the actual top-k ranked lists using *OSim, KSim, NDCG, Spearman correlation* and the novel similarity measure *RSim*.

5.1 Datasets and Query Selection

For each dataset (Yahoo and Google) a wealth of snapshots were available, ensuring we have enough evolution to test our approach. A concise description of each dataset and query-based approach follow. The Yahoo and Google datasets consist of 11 consecutive daily top-1000 ranked lists computed using the Yahoo Search Web Services[2] and the Google Search engine respectively. These sets were picked from popular: a) queries appeared in Google Trends[3] and b) current queries (i.e. euro 2008 or Olympic games 2008).

5.2 Experimental Methodology

We compared all predictions among the various approaches and we next describe the steps assumed for both datasets. At first, we computed PageRank scores for

---

[1] http://search.yahoo.com
[2] http://developer.yahoo.com/search/
[3] http://www.google.com/trends

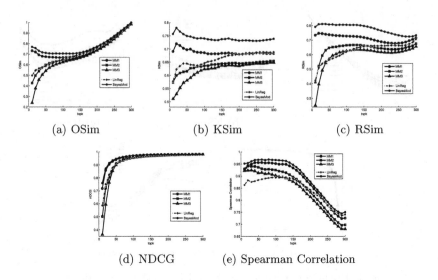

(a) OSim   (b) KSim   (c) RSim

(d) NDCG   (e) Spearman Correlation

Fig. 2. Prediction accuracy vs Top-k list length - Google dataset

each snapshot of our datasets and obtained the top-k rankings using the scoring function mentioned. Having computed the scores, we calculated the nrank (racer values for MMs) for each pair of consecutive graph snapshots and stored them in a matrix $nrank(racer) \times time$. Then, assuming an m-path of consecutive snapshots, we predict the $m + 1$ state. For each page p, we predict a ranking comparing it to actual by a 10-fold cross validation process (training 90% of dataset and testing on the remaining 10%).

In the case of the EM approach, we tested the quality of clustering results for clusters cardinality between 2 and 10 for each query and chose the one that maximized the overall quality of clustering. This was defined as a monotone combination of within-cluster w_c (sum of squared distances from each point to the center of cluster it belongs to) and between-cluster variation b_c (distance between cluster centers). As score function of clustering, we considered the ratio b_c/w_c.

5.3   Experimental Results

Regarding the Google and Yahoo! dataset results coming out of the experimental evaluation, one can see that the MMs prevail with very accurate results. Regression based techniques (LinReg) reach and outweigh MMs performance as the length of top-k list increases proving their robustness.

In both datasets experiments prove the superiority of EM approach (BayesMod) whose performance is very satisfying for all similarity measures. The MMs come next in the evaluation ranking, where as smaller the order is the better is the prediction accuracy, though one would think of the contrary.

Obviously (figures) the proposed framework offers incredibly high accuracy predictions and is very encouraging, as it ranges systematically between 70% and 100% providing a tool for effective predictions.

6   Conclusions

We have described predictor learning algorithms for Web page rank prediction based on a framework of learning techniques (MMs, LinReg, BayesMod) and experimental study showed that they can achieve overall very good prediction performance. Further work will focus in the following issues: a) Multi-feature prediction: we intend to deal with the internal mechanism that produces the ranking of pages (not only rank values) based on multiple features, b) Combination of such methods with dimensionality reduction techniques.

References

1. Kemeny, J.G., Snell, J.L.: Finite Markov Chains. Prinston (1963)
2. Dempster, A.P., Laird, N.M., Rubin, D.B.: Maximum Likelihood from Incomplete Data via the EM Algorithm. Royal Statistical Society 39, 1–38 (1977)
3. Kendall, M.G., Gibbons, J.D.: Rank Correlation Methods. Charles Griffin, UK (1990)
4. Haveliwala, T.H.: Topic-Sensitive PageRank. In: Proc. WWW (2002)
5. Jarvelin, K., Kekalainen, J.: Cumulated gain-based evaluation of IR techniques. TOIS 20, 422–446 (2002)
6. Chien, S., Dwork, C., Kumar, R., Simon, D.R., Sivakumar, D.: Link Evolution:Analysis and Algorithms. Internet Mathematics 1, 277–304 (2003)
7. Chen, Y.-Y., Gan, Q., Suel, T.: Local Methods for Estimating PageRank Values. In: Proc. of CIKM (2004)
8. Langville, A.N., Meyer, C.D.: Updating PageRank with iterative aggregation. In: Proc. of the 13th International World Wide Web Conference on Alternate Track Papers and Posters, pp. 392–393 (2004)
9. Kan, M.-Y., Thi, H.O.: Fast webpage classification using URL features. In: Conference on Information and Knowledge Management, pp. 325–326. ACM, New York (2005)
10. Yang, H., King, I., Lu, M.R.: Predictive Ranking:A Novel Page Ranking Approach by Estimating the Web Structure. In: Proc. of the 14th International WWW Conference, pp. 1825–1832 (2005)
11. Broder, A.Z., Lempel, R., Maghoul, F., Pedersen, J.: Efficient PageRank approximation via graph aggregation. Inf. Retrieval 9, 123–138 (2006)
12. Vlachou, A., Berberich, K., Vazirgiannis, M.: Representing and quantifying rank-change for the Web graph. In: Aiello, W., Broder, A., Janssen, J., Milios, E.E. (eds.) WAW 2006. LNCS, vol. 4936, pp. 157–165. Springer, Heidelberg (2008)
13. Vazirgiannis, M., Drosos, D., Senellart, P., Vlachou, A.: Web Page Rank Prediction with Markov Models. WWW poster (2008)
14. Sayyadi, H., Getoor, L.: Future Rank: Ranking Scientific Articles by Predicting their Future PageRank. In: SIAM Intern. Confer. on Data Mining, pp. 533–544 (2009)
15. Zacharouli, P., Titsias, M., Vazirgiannis, M.: Web page rank prediction with PCA and EM clustering. In: Avrachenkov, K., Donato, D., Litvak, N. (eds.) WAW 2009. LNCS, vol. 5427, pp. 104–115. Springer, Heidelberg (2009)

Towards a Semantic Calibration
of Lexical Word via EEG

Marios Poulos

Laboratory of Information Technology,
Departement of Archive and Library Science,
Ionian University, Ioannou Theotoki 72, 49100 Corfu, Greece
mpoulos@ionio.gr

Abstract. The calibration method used in this study allows for the examination of distributed, but potentially subtle, representations of semantic information between mechanistic encoding of the language and the EEG. In particular, a horizontal connection between two basic Fundamental Operations (Semantic Composition and Synchronization) is attempted. The experimental results gave significant differences, which can be considered reliable and promising for further investigation. The experiments gave helpful results. Consequently, this method will be tested along with the classification step by appropriate neural network classifiers.

Keywords: Hybrid method, EEG, Signal Processing, Semantic Processing, Lexical Word.

1 Introduction

With the blossoming of the Internet, the semantic interpretation of the word is now more imperative than ever. Two scientific approaches can lead one to achieve this aim: linguistic formalism and the neuroscience method.

Linguistic formalisms are served by semantic nets (such as ontology schema), and providing by a well-defined semantic syntax, which are also combining features of object-oriented systems, of frame-based systems, and of modal logics. However, the use of these systems creates many problems. The main problem of information extraction systems is low degree of portability due to language dependent linguistic resources and to domain-specific knowledge (ontology) [1]. Additionally, the individual differences in information needs, polysemy (multiple meanings of the same word), and synonymy (multiple words with same meaning) pose problems [2] in that a user may have to go through many irrelevant results or try several queries before finding the desired information. Although, using ontologies to support information retrieval and text document processing has lately involved more and more attention, existing ontology-based methods have not shown benefits over the outdated keywords-based Latent Semantic Indexing (LSI) technique [3]. A partial solution to the above problems uses the semantic measurement of similarity between words and terms, which plays an important role in information retrieval and information integration [4, 5]. Nowadays, this measurement is implemented by the allocation of

L. Iliadis et al. (Eds.): EANN/AIAI 2011, Part II, IFIP AICT 364, pp. 250–258, 2011.

the words in a metric space, which is called semantic map [6]. Many methods have been developed for this aim such as maps based on the representation of semantic difference of the word as geometrical distance [7–10] and the maps that depict the semantic positions of the words using the likelihood of the word which appears in a particular topic or document [6].

In neuroscience practice, the problem of data sharing in brain electromagnetic research, similar to other scientific fields, is challenged by data scale, multivariate parameterizations, and dimensionality [11]. The research about organization and localization of lexico-semantic information in the brain has been discussed in the past. Decoding methods, on the other hand, allow for a powerful multivariate analysis of multichannel neural data. A significant work about this problem showed the decoding analysis to demonstrate that the representations of words and semantic category are highly distributed both spatially and temporally [12]. In particular, many studies in the past showed that the suitable Support Vector Machines (SVMs) [14,13], which have been constructed by decoding multichannel EEG data, possess critical features in relation to the conceptual understanding of written words. These features are depicted in an acceptable time span of a 300-600 mc EEG recording and especially in spectral features (8–12 Hz) power [14, 12]. Furthermore, in recent work [15] the EEG decoding of semantic category reveals distributed representations for single concepts is implemented by applying data mining and machine learning techniques to single trials of recorded EEG signals.

However, until now, the gap between the linguistics and the neuroscience has been considered unbridgeable [16]. This is illustrated in Table 1.

Table 1. The two unordered lists enumerate some concepts canonically used to explain neurobiological or linguistic phenomena. There are principled ontology-process relationships within each domain (i.e., vertical connections) [16]

]Linguistics	Neuroscience
Fundamental Elements	
Distinctive Feature	Dendrites, spines
Syllable	Neuron
Morpheme	Cell-
Assembly/Ensemble	Assembly/Ensemble
Noun Phrase	Population
Clause	Cortical Column
Fundamental Operations on Primitives	
Concatenation	Long-Term Potentiation
Linearization	Receipt Field
Phrase-Structure	Oscillation
Semantic Composition	Synchronization

This study attempts to bridge the gap between the two methodologies. In particular, a horizontal connection between two basic Fundamental Operations (Semantic Composition and Synchronization) is proposed via a Semantic Calibration of Lexical Word via EEG. The idea is based on the following four (4) approaches:

1. The determination of any ordered sequence of k characters occurring in each word. This approach follows the Kernel learning Philosophy [17,18] and consists of an early semantic interpretation of the word "on step beyond of the word" [19]
2. The isolation of a significant feature of an EEG segment 500ms duration according to aforementioned reference is attempted[16, 14, 12]
3. A new signal generation is derived from ordered sequence of k characters and the suitable modulated EEG signal.
4. Features are extracted from the new signal and statistical testing of the semantic feature.

2   Method

The section is divided into four subsections. In the first subsection, "Numerical Encoding of Word's characters," the determination of any ordered sequence of k characters occurring in each word is considered. Preprocessing of the EEG signal and feature extraction is described in the second subsection. Data acquisition, to be used in the experimental part, is outlined in the third subsection. And the fourth subsection presents a statistical approach to the semantic feature of this calibration.

2.1   Numerical Encoding of Word's Characters

At this stage, the characters of the selected word are considered as input vector. Then, using a conversion procedure where a symbolic expression (in our case an array of characters of a word) is converted to ASCII characters in a string of arithmetic values. As a result, we obtained a numerical value vector for each. These values ranged between 1–128.

Thus, a vector $\vec{a}$ with length k is constructed, where k is the number of characters in each investigated word.

2.2   Preprocessing of the EEG Signal

Electroencephalographic (EEG) data contains changes in neuro-electrical rhythm measured over time (on a millisecond timescale), across two or more locations, using noninvasive sensors ("electrodes") that are placed on the scalp surface. The resulting measures are characterized by a sequence of positive and negative defections across time at each sensor. For example, to examine brain activity related to language processing, the EEG may be recorded during donation of the words, using 128 sensors in a time span of 500ms. In principle, activity that is not event-related will tend toward zero as the number of averaged trials increases. In this way, ERPs provide increased signal-to-noise (SNR) and thus increased sensitivity to functional (e.g., task-related) manipulations [11].

In order to model the linear component of an EEG signal $x(n)$ known to represent the major part of its power (especially in the alpha rhythm frequency band), the selected segment is submitted in alpha rhythm filtering. As it is known, the alpha rhythm is the spectral band of 8-12 Hz, extracted from the original EEG spectrum and recorded mainly from the occipital part of the brain, when the subjects are at rest with their eyes closed. Thus, the spectral values of the EEG signal are obtained and then restricted to the alpha rhythm band values only in a new signal $y(n)$ which becomes

from the time domain difference equation describing the general Mth-order IIR filter, having N feed forward stages and M feedback stages in filter cut upper (a) and lower (b) limit. The time domain expression for an Mth-order IIR filter is given by the following equation (1):

$$y(n) = b(0)x(n) + b(1)x(n-1) + b(2)x(n-2) + ... + b(N)x(n-N) \quad (1)$$
$$+ a(1)y(n-1) + a(2)y(n-1) + ...a(M)y(n-M)$$

2.3 Semantic Calibration of Lexical Words via EEG

In this stage, in order to create a new signal $z(n)$ with specific hybrid features, the vector

$$\mathbf{a} = \begin{bmatrix} a_1 \\ . \\ . \\ . \\ a_k \end{bmatrix} \quad (2)$$

and the signal $y(n)$ are combined by the following steps:

1. The length of the signal $y(n)$ is divided in k equal segments, where each has

length $l = \lfloor \dfrac{n}{k} \rfloor$ and is given by the following equation (3)

$$y(n) = \sum_{n=1+(i-1)l}^{il} y(n), \text{ where } i = 1,...,k \quad (3)$$

2. A new signal $z(n)$ is generated by the residuals between the vector $\mathbf{a}$ and signal $y(n)$. The calculation takes place for each character per segment and is depicted in the following equation:

$$z(n) = \sum_{n=1+il}^{(i+1)l} (y(n) - a_i) \quad (4)$$

2.4 The AR Model- Feature Extraction

The linear component of the signal $z(n)$ is implemented via a linear, rational model of the autoregressive type, AR [20]. This signal is treated as a superposition of a signal component (deterministic) plus additive noise (random). Noise is mainly due to imperfections in the recording process. This model can be written as

$$x_t + \sum_{i=1}^{p} b_i x_{t-1} = 0 \tag{5}$$

It is an independent, identically distributed driving noise process with zero mean and unknown variance σ_e^2; model parameters $\{b_i \ , i = 1, 2,..., p\}$ are unknown constants with respect to time.

It should be noted that the assumption of time invariance for the model of the text vector can be satisfied by restricting the signal basis of the method to a signal "window" or "horizon" of appropriate length.

The linear model can usually serve as a (more or less successful) approximation when dealing with real world data. In the light of this understanding, the linear model is the simpler among other candidate models in terms of computing spectra, covariances, etc.

In this work, a linear model of the specific form AR(p) is adopted. The choice of the order of the linear models is usually based on information theory criteria such as the Akaike Information Criterion (AIC) [21], which is given by

$$AIC(r) = (N - M) \log \sigma_e^2 + 2r \tag{6}$$

where,

$$\sigma_e^2 = \frac{1}{N - M} \sum_{t=M+1}^{N} e_t^2 \tag{7}$$

N represents the length of the data record; M is the maximal order employed in the model; (N-M) is the number of data samples used for calculating the likelihood function; and r denotes the number of independent parameters present in the model. The optimal order r* is the minimizer of AIC(r).

We have used the AIC to determine the order of the linear part of the model in i.e. the optimal order p of the AR part of the model. For each candidate order p in a range of values [pmin, pmax], the AIC(p) was computed from the residuals of each record in the ensemble of the EEG records available. This is because we deal with recordings of real world data rather than the output of an ideal linear model. We have thus seen that AIC(p) takes on its minimum values for model orders p ranging between 5 and 8, record-dependent. In view of these findings, we have set the model order of the AR part to p = 7, for parsimony purposes [22].

2.5 Identification Procedure

In this stage, the extracted sets of the 7 order AR coefficients x of the generated signal Z(n) are submitted to compute the difference between the variances for two response variables—see equation (8).

$$s = \frac{\sum_{i=1}^{n}(x - \bar{x})^2}{n-1} \tag{8}$$

For the difference of the variances, the variance is computed for each of the two samples before their difference is taken.

3 Experimental Part

As example of this study, the same simple words are used with related study [18] in order to the degree of contiguity between homonymous words to be investigated. For this reason, the words "cat," "car," "bat," "bar" are investigated, and all the algorithms according to the aforementioned method are applied. More details are depicted in figures 1 and 2. Thereafter, the AR coefficients are extracted for each word (an example of this is presented on Table 2). Finally, in Table 3 the differences between each of the pairs of AR coefficients are isolated. Specifically, the isolation of a significant feature of an EEG segment 500ms duration according to aforementioned reference and the appropriate filtering is depicted in fig 1.

The calibration of the characters of each word on the new filtering signal is presented in figure 2

The determination of any ordered sequence of k characters occurring in each word is depicted in table 2 as well as the difference between the variances of the tested words which are presented in table 3.

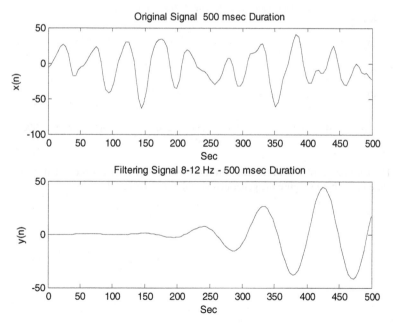

Fig. 1. The first two steps (EEG Selection and Filtering) of the proposed method are applied

256   M. Poulos

Table 2. The AR coefficients of the word "car" from the generated signal z(n)

AR coefficients "car"
0.8008
-0.2886
0.0460
0.2848
0.2151
-0.0251
-0.2828
0.8008

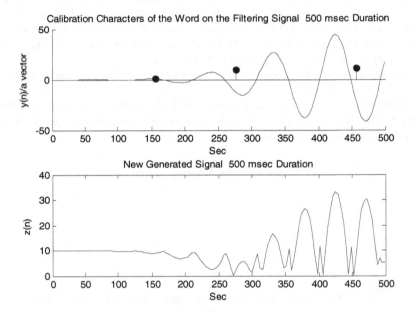

Fig. 2. The calibration of the word "car" on the filtered signal y(n) is presented in the upper figure, while in the below figure the generated signal z(n) is depicted

Table 2. The two unordered lists enumerate some concepts canonically used to explain neurobiological or linguistic phenomena. Principled ontology-process relationships connect words in each domain (i.e., vertical connections) [16]

Identification Procedure (Difference of Variance)

	cat	car	bat	bar
cat	0	0.0036	6.3972e-004	0.0030
car	-0.0036	0	-0.0042	-6.4958e-004
bat	-6.3972e-004	0.0042	0	-0.0036
bar	-0.0030	6.4958e-004	0.0036	0

4 Results-Conclusions

The calibration method used in this study allows for the examination of distributed, but potentially subtle, differences in representations of semantic information between mechanistic encoding of the language and the EEG.

It was noted that all comparisons in table 2 gave significant differences, which outcome can be considered reliable and promising for further investigations. It should be noted that the words with the same suffix as bat-cat and bar-car showed more consistency. This observation is in agreement with research in the field of neuroscience, which indicates that it is "the syntactically relevant word category information in the suffix, available only after the word stems which carried the semantic information" [23].

These multivariate techniques offer advantages over traditional statistical methodologies in linguistics and neuroscience. The proposed method creates a new basis in the measurements of writing because, for the first time, a code of the digital lexical-word, such as ASCI code, is attempted to calibrate based on a biological signal. The experiments gave helpful results. Consequently, this method will be tested along with the classification step by appropriate neural network classifiers. The proposed metrics have been implemented in the Matlab Language.

In conclusion, the proposed method differs from all existing methods of semantic decoding EEG because it aims to build a model that explains how an acoustic signal lexical content may be shaped so that it can form the basis of linguistic education of the brain. In other words, the proposed model is based on a different logic in relation to aforementioned studies because it creates a combination of two scientific areas, which are neuroscience and linguistics.

References

1. Todirascu, A., Romary, L., Bekhouche, D.: Vulcain—an ontology-based information extraction system. In: Andersson, B., Bergholtz, M., Johannesson, P. (eds.) NLDB 2002. LNCS, vol. 2553, pp. 64–75. Springer, Heidelberg (2002)
2. Deerwester, S., Dumais, S.T., Furnas, G.W., Landauer, T.K., Harshman, R.: Indexing by latent semantic analysis. Journal of the American Society for Information Science 41, 391–407 (1990)
3. Wang, J.Z., Taylor, W.: Concept forest: A new ontology-assisted text document similarity measurement method (2007)
4. Rodríguez, M.A., Egenhofer, M.J.: Determining semantic similarity among entity classes from different ontologies. IEEE Transactions on Knowledge and Data Engineering 15(2), 442–456 (2003)
5. Corley, C.: Mihalcea, and R.: Measuring the semantic similarity of texts. In: Proceedings of the ACL Workshop on Empirical Modeling of Semantic Equivalence and Entailment, pp. 13–18 (2005)
6. Samsonovic, A.V., Ascoli, G.A., Krichmar, J.: Principal Semantic Components of Language and the Measurement of Meaning. PloS one 5(6), e10921 (2010)
7. Tversky, A., Gati, I.: Similarity, separability, and the triangle inequality. Psychological Review 89, 123–154 (1982)
8. Fauconnier, G.: Mental Spaces. Cambridge University Press, Cambridge (1994)
9. Gardenfors, P.: Conceptual spaces: The geometry of thought. MIT Press, Cambridge (2004)
10. Landauer, T.K., McNamara, D.S., Dennis, S.: Kintsch,: Handbook of Latent Semantic Analysis. Lawrence Erlbaum Associates, Mahwah (2007)
11. LePendu, P., Dou, D., Frishkoff, G., Rong, J.: Ontology database: A new method for semantic modeling and an application to brainwave data. Scientific and Statistical Database Management, 313–330 (2008)
12. Chan, A.M., Halgren, E., Marinkovic, K., Cash, S.S.: Decoding word and category-specific spatiotemporal representations from MEG and EEG. Neuroimage (2010)
13. Indefrey, P., Levelt, W.J.M.: The spatial and temporal signatures of word production components. Cognition 92, 101–144 (2004)
14. Canolty, R.T., et al.: Spatiotemporal dynamics of word processing in the human brain. Frontiers in Neuroscience 1, 185 (2007)
15. Murphy, B., et al.: EEG decoding of semantic category reveals distributed representations for single concepts. Brain and Language (2011)
16. Poeppel, D., Embick, D.: Defining the relation between linguistics and neuroscience. In: Twenty-first Century Psycholinguistics: Four Cornerstones, pp. 103–118 (2005)
17. Vapnik, V.N.: The nature of statistical learning theory. Springer, Heidelberg (2000)
18. Cristianini, N., Shawe-Taylor, J.: An introduction to support Vector Machines: and other kernel-based learning methods. Cambridge University Press, Cambridge (2004)
19. Lodhi, H., Saunders, C., Shawe-Taylor, J., Cristianini, N., Watkins, C.: Text classification using string kernels. The Journal of Machine Learning Research 2, 419–444 (2002)
20. Box, G.E.P., Jenkins, G.M., Reinsel, G.C.: Time Series Analysis Forecasting and control. Wiley John Wiley & Sons, Inc., Chichester (1970)
21. Stone, M.: An asymptotic equivalence of choice of model by cross-validation and Akaike's criterion. Journal of the Royal Statistical Society. Series B (Methodological) 39, 44–47 (1977)
22. Poulos, M., Rangoussi, M., Alexandris, N., Evangelou, A.: Person identification from the EEG using nonlinear signal classification. Methods of Information in Medicine 41, 64–75 (2002)
23. Friederici, A.D., Gunter, T.C., Hahne, A., Mauth, K.: The relative timing of syntactic and semantic processes in sentence comprehension. NeuroReport 15, 165 (2004)

Data Mining Tools Used in Deep Brain Stimulation – Analysis Results

Oana Geman

Faculty of Electrical Engineering and Computer Science,
"Stefan cel Mare" University,
Suceava, Romania
geman@eed.usv.ro

Abstract. Parkinson's disease is associated with motor symptoms, including tremor. The DBS (Deep Brain Stimulation) involves electrode implantation into sub-cortical structures for long-term stimulation at frequencies greater than 100Hz. We performed linear and nonlinear analysis of the tremor signals to determine a set of parameters and rules for recognizing the behavior of the investigated patient and to characterize the typical responses for several forms of DBS. We found patterns for homogeneous group for data reduction. We used Data Mining and Knowledge discovery techniques to reduce the number of data. To support such predictions, we develop a model of the tremor, to perform tests determining the DBS reducing the tremor or inducing tolerance and lesion if the stimulation is chronic.

Keywords: Parkinson's disease, Deep Brain Stimulation, Data Mining.

1 Parkinson's Disease, Tremor and Deep Brain Stimulation

Parkinson's disease (PD) is a serious neurological disorder with a large spectrum of symptoms (rest tremor, bradykinesia, muscular rigidity and postural instability). The neurons do not produce dopamine anymore or produce very low level of this chemical mediator, necessary on movement coordination. Parkinson's disease seems to occur in about 100 - 250 cases on 100 000 individuals. In Europe were reported about 1.2 million Parkinson patients [1]. The missing of a good clinical test, combined with the patient's reticence to attend a physician, make the diagnostic to be established very often too late. The accuracy of the diagnosis of PD varies from 73 to 92 percent depending on the clinical criteria used.

Parkinsonian tremor is a rhythmic, involuntary muscular contraction characterized by oscillation of a part of the body. Initial symptoms include resting tremor beginning distally in one arm at a 4 – 6 Hz frequency. Deep Brain Stimulation (DBS) is an electric therapy approved by FDA (Food and Drug Administration) for the treatment of Parkinson's disease (PD) in 2002 and used now to treat the motor symptoms like essential tremor. It consists of a regular high frequency stimulation of specific subcortical sites involved in the movement-related neural patterns [1].

The exact neurobiological mechanism by which DBS exerts modulator effects on brain tissue are not yet full understood. It is unknown which part of the neuronal

L. Iliadis et al. (Eds.): EANN/AIAI 2011, Part II, IFIP AICT 364, pp. 259–264, 2011.

structure (cell body, axon) is primarily modulated by DBS. Since the causes are confused, the treatment does not provide cure but slows down the illness evolution and attenuate the invalidating symptoms – tremor.

2 Methods

The current study was motivated by the wish to develop a better understanding of potential, mechanisms of the effects of Deep Brain Stimulation on Parkinsonian tremor by studying the tremor dynamics that occur during the on/off of high frequency DBS in subjects with Parkinson's disease.

Currently there are insufficient data in the literature to include data and specific knowledge of nonlinear dynamics, which would eventually lead to the improvement of Parkinson's disease database, to get an accurate a medical diagnosis, with rare exceptions [2], [3], [4]. The field of nonlinear dynamics introduced idea that a random behavior of tremor time series might have been generated by a low-dimensional chaotic deterministic dynamical system. Parkinsonian tremor exhibits a nonlinear oscillation that is not strictly periodic. Titcombe et al. [2] applied various methods from linear and nonlinear time series analysis to tremor time series.

First, we made a linear, nonlinear and statistical analysis of the tremor signals (available on the internet at [5]) to determine a set of parameters and rules for recognizing the behavior of the investigated patient and to characterize the typical responses for several forms of DBS. Second, we found representatives for homogeneous groups in order to perform data reduction. Then we found "clusters" and describe their unknown properties. We used k-means and k-medoids clustering.

Therefore, it is desirable to predict, before we pursuit with this type of operation, if the DBS will give good results in a specific case. To make such predictions, we need a model of the tremor, on which to perform tests to determine if the DBS procedure will reduce the tremor. An accurate, general model of a system can allow the investigation of a wide variety of inputs and behaviors in sillico, without the time or resources required to perform the same investigations in vivo.

3 Clinical Databases Structure

Parkinsonian tremor time series are available on internet [5]. These data were recorded using a low-intensity velocity laser in subjects with Parkinson's disease receiving chronic, high frequency DBS at a sampling rate of 100 Hz. Tremor signals were recorded to the subject's index finger in a resting position continuously throughout switching the DBS on and off.

We analyzed tremor signals from 16 subjects with Parkinson disease, ages between 37 and 71 years, 11 men and 5 women. The recordings of this database are of rest tremor velocity in the index finger of 16 subjects with Parkinson's disease (PD) who receive chronic high frequency electrical Deep Brain Stimulation (DBS) either uni- or bi-laterally within one of three targets: Vim = the ventro-intermediate nucleus of the thalamus, GPi = the internal Globus pallidus, STN = the subthalamic nucleus.

4 Data Analysis Results

4.1 Linear and Nonlinear Time Series Analysis of Parkinsonian Tremor

Linear analysis of the signal is mainly using Fourier analysis and reporting made comparison of the amplitude frequency bands. Based on the Fourier spectrum in the range of 0-25 Hz and amplitude-time representation, a number of parameters are used to characterize the tremor signal [6], [7]. Many reports suggest that Parkinsonian tremor is typically in the range of 4-6 Hz, and the essential tremor is in the range of 4-12 Hz.

A very first phase on non-linear analysis is to draw the phase diagram. In the signal period, the phase diagram is a closed curve. If the signal is chaotic, the diagram is a closed curve called *strange attractor*. The positive Lyapunov exponent is the main chaotic dynamic indicator. In case at least one Lyapunov exponent is bigger that 0, the system is chaotic. Using CDA software solution to analyze the tremor signals from our database, we found the Lyapunov exponent varies from 0.02 to 0.28, depending on the signal type (Fig.1 shows the Lyapunov exponent (a), fractal dimension (b), and correlation dimension (c)).

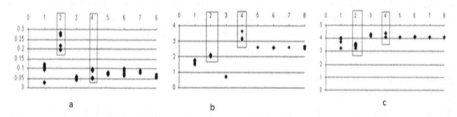

Fig. 1. The values of Lyapunov exponent (a), fractal dimension (b), and correlation dimension (c) (class 2 and class 4 are marked because results will be analyzed)

We have 8 preset classes of tremor: class1 - DBS on/medication on; **class 2 - DBS on/medication off;** class 3 - DBS off/medication on; **class 4 - DBS off/medication off (Parkinsonian tremor);** class 5...8 – DBS off/medication off (the signals recorded after 15, 30, 45 and, respectively, 60 minutes after ceasing the DBS action).

Analyzing the data, we observed that the fractal dimension allows us to recognize with good accuracy the high-amplitude tremor classes. Using the other parameters/rules, the classification becomes more accurate.

Based on the information (for example the values of Lyapunov exponent – Fig. 1 (a)) we may conclude: using the values of the Lyapunov exponent, we may distinguish class 2 (DBS on/medication off) from all the other classes. In this case the values of Lyapunov exponent are between 0.19 and 0.28 **(1)**.

Using the values of the fractal dimension, we observe that in case of class 4 (DBS off/medication off – simple Parkinsonian tremor) there are important differences relative to other classes (for class 4, values between 3.2 and 3.8) **(2)**. Using the correlation dimension, we notice that for class 2 (DBS on/medication off), its values are between 3.3 and 3.7 and for class 4 (DBS off /medication off) the correlation dimension can reach a value of 4.5 (this value does not exist for any other class) **(3)**.

In conclusion, for the patients with Parkinson's disease, using the correlation dimension and calculation of Lyapunov exponents, we found evidence that the dynamics are generated by a nonlinear deterministic and chaotic process. We argue that nonlinear dynamic parameters of Parkinsonian tremor have certain peculiarities and can be used in knowledge-discovery. This data and new knowledge will be integrated in a Knowledge-based System aimed to identify each class of tremor, to worn on atypical responses at DBS and to recommend the stimulation targets (thalamus nuclei).

4.2 Data Mining Tools

Data Mining has been defined as the automatic analysis of a large and complex data sets in order to discover meaningful patterns. Recent advances in Data Mining research has led to developement of efficient methods for finding patterns and important knowledge in large volumes of data, efficient method of classification, clustering, frequent pattern analysis, sequential, structured [8]. This raises the acute need to link the two areas of Arificial Intelligence: Bioinformatics and Data mining for efective analysis of biomedical data, with aim of discovering new knowledge and data in the medical field.

Cluster analysis tools based on k-means, k-medoids, and several other methods have also been buit into many statistical analysis software systems. Given a data set of n objects, and k = the number of clusters to form, a partitioning algorithm organizes the objects into k partitions, where each partition represents a cluster.

We adopt two of a few popular heuristic methods, such as:

1. the k-means algorithm, where each cluster is represented by the mean value of the objects in the cluster;
2. the k-medoids algorithm, where each cluster is represented by one of the objects located near the center of the cluster.

Because it is easier to illustrate, but also because, classes 2 and 4 are the most significant (class 2 - signals acquired from patients tremor with DBS without medication, class 4 - patients without medication and without DBS, Parkinsonian Tremor), we will continue this k-means algorithm and k-medoids algorithm only for these classes.

Using STATISTICA software version 9 for classes 2 and 4 we obtained the following results:

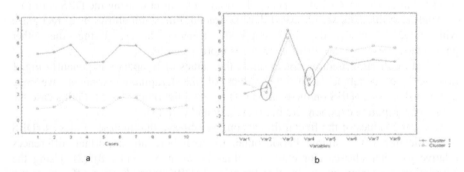

a   b

Fig. 2. K-means Clustering Algorithm (a) and K-medoids Clustering Algorithm (b) for class 2 and 4 (cluster 1 (a)/var2 (b) is class 2 – DBS on/medication off and cluster 2 (a)/var4 (b) is class 4 – DBS off, medication off)

We have shown that applied k-means clustering algorithm and k-medoids clustering algorithm solve the well known clustering problem and DBS treatments results identify. Being an invasive method it is important to predict patient's response to DBS and therefore the next step is to model the Parkinsonian tremor.

4.3 A Nodel of the Parkinsonian Tremor

Modeling of biological systems is difficult due to nonlinearities, time dependence and internal interactions common in physiological and neural systems. To get a true understanding of the behavioral of a complex biological systems technique of nonlinear dynamical systems modeling can be used. Modeling complex biological systems is very difficult, but the simplest method for predicting output from input is the linear regression. We proposed the usage of a nonlinear iterated function in modeling simple biological processes. Controlling the parameters of the nonlinear function, we control the dynamic of the generated process [9].

We notice that between the original Parkinson tremor and the model there are visible similarities up to the time scale. The model based on adaptive systems searching the matching Lyapunov exponent, is a preliminary model, which can be further improved [9]. Developing an accurate nonlinear dynamical model of a patient's PD symptoms as a function of DBS could both increase the effectiveness of DBS and reduce the clinical time needed to program the implanted stimulator [10], [11], [12]. If the model were able to accurately predict the level of symptoms as a result of past stimulation, the model could be used to test all possible parameter settings very quickly in simulation to identify the globally optimal stimulus parameters.

5   Conclusions and Future Works

As it has been demonstrated, classes 2 and 4 are easy to identify, which reinforces the following conclusions: only ordinary statistical parameters are not enough for identifying the tremor classes (**1**); the nonlinear dynamic parameters are not sufficient in distinguishing between different tremor classes (**2**); a more careful analysis of elements from point 1 and 2 helps to a more easy class identification (**3**); refinement of current indication for DBS procedure is possible, regarding the risks and benefits of STN versus Gpi DBS for PD (**4**).

To determine if the DBS procedure will reduce the tremor, we need a model of the tremor, on which to perform tests. The results obtained are encouraging and we will continue the research on this issue, by modeling other known nonlinear processes. We also demonstrated that, by iterating a linear function, we could model a Parkinsonian tremor, and we developed an adaptive system to build this model.

Acknowledgments. This paper was supported by the project "Progress and development through post-doctoral research and innovation in engineering and applied sciences – PRiDE – Contract no. POSDRU/89/1.5/S/57083", project co-funded from European Social Fund through Sectorial Operational Program Human Resources 2007-2013.

References

1. National Parkinson Foundation, http://www.parkinson.org
2. Titcombe, M.S., Glass, l., Guehl, D., Beuter, A.: Dynamics of Parkinsonian tremor during Deep Brain Stimulation. Chaos 11(4), 201–216 (2001)
3. Teodorescu, H.N., Kandel, A.: Nonlinear Analysis of Tremor and Applications. JJME Japanese Journal of Medical Electronics and Biological Engineering 13(5), 11–19 (2004)
4. Teodorescu, H.N., Chelaru, M., Kandel, A., Tofan, I., Irimia, M.: Fuzzy methods in tremor assessment, prediction, and rehabilitation. Artificial Intelligence in Medicine 21(1-3), 107–130 (2001)
5. PhysioNet, http://www.physionet.org/database
6. Schlapfer, T.E., Bewernick, B.H.: Deep brain stimulation for psychiatric disorders – state of the art. In: Advances and Technical Standards in Neurosurgery, vol. 34, pp. 115–211. Springer, Heidelberg (2002)
7. Rocon, E., Belda-Lois, J.M., Sancez-Lacuesta, J.J., Pons, J.L.: Pathological tremor management: Modelling, compensatory technology and evaluation. Technology and Disability 16, 3–18 (2004) ISSN 1055-4181/04
8. Danubianu, M.: Advanced Databases. In: Modern Paradigms in Computer Science and Applied Mathematics, p. 30. AVM, Munchen (2011) ISBN: 978-3-86306-757-1
9. Geman (Voroneanu), Teodorescu, H.N., Zamfir, C.: Nonlinear Analysis and Selection of Relevant Parameters in Assessing the Treatment Results of reducing Tremor, using DBS procedure. In: Proceedings of International Joint Conference on Neural Networks, Paper 1435 (2004)
10. Tuomo, E.: Deep Brain Stimulation of the Subthalamic Nucleus in Parkinson Disease, a Clinical Study, faculty of Medicine, University of Oulu, Finland (2006)
11. Van Battum, E.Y.: The Role of Astrocytes in Deep Brain Stimulation, Master thesis, Master Neuroscience and Cognition (2010)
12. Elias, J.W., Lozano, A.M.: Deep Brain Stimulation: The Spectrum of Application, Neurosurg. Focus, 29(2), American Association of Neurological Surgeon (2010)

Reliable Probabilistic Prediction for Medical Decision Support

Harris Papadopoulos

Computer Science and Engineering Department, Frederick University,
7 Y. Frederickou St., Palouriotisa, Nicosia 1036, Cyprus
h.papadopoulos@frederick.ac.cy

Abstract. A major drawback of most existing medical decision support systems is that they do not provide any indication about the uncertainty of each of their predictions. This paper addresses this problem with the use of a new machine learning framework for producing valid probabilistic predictions, called Venn Prediction (VP). More specifically, VP is combined with Neural Networks (NNs), which is one of the most widely used machine learning algorithms. The obtained experimental results on two medical datasets demonstrate empirically the validity of the VP outputs and their superiority over the outputs of the original NN classifier in terms of reliability.

Keywords: Venn Prediction, Probabilistic Classification, Multiprobability Prediction, Medical Decision Support.

1 Introduction

Medical decision support is an area in which the machine learning community has conducted extensive research that resulted in the development of several diagnostic and prognostic systems [8,11]. These systems learn to predict the most likely diagnosis of a new patient based on a past history of patients with known diagnoses. The most likely diagnosis however, is the only output most such systems produce. They do not provide any further information about how much one can trust the provided diagnosis. This is a significant disadvantage in a medical setting where some indication about the likelihood of each diagnosis is of paramount importance [7].

A solution to this problem was given by a recently developed machine learning theory called *Conformal Prediction* (CP) [24]. CP can be used for extending traditional machine learning algorithms and developing methods (called Conformal Predictors) whose predictions are guaranteed to satisfy a given level of confidence without assuming anything more than that the data are independently and identically distributed (i.i.d.). More specifically, CPs produce as their predictions a set containing all the possible classifications needed to satisfy the required confidence level. To date many different CPs have been developed, see e.g. [14,15,17,18,19,21,22], and have been applied successfully to a variety of important medical problems such as [3,6,9,10,16].

L. Iliadis et al. (Eds.): EANN/AIAI 2011, Part II, IFIP AICT 364, pp. 265–274, 2011.
© IFIP International Federation for Information Processing 2011

This paper focuses on an extension of the original CP framework, called Venn Prediction (VP), which can be used for making *multiprobability predictions*. In particular multiprobability predictions are a set of probability distributions for the true classification of the new example. In effect this set defines lower and upper bounds for the conditional probability of the new example belonging to each one of the possible classes. These bounds are guaranteed (up to statistical fluctuations) to contain the corresponding true conditional probabilities. Again, like with CPs, the only assumption made for obtaining this guaranty is that the data are i.i.d.

The main aim of this paper is to propose a Venn Predictor based on Neural Networks (NNs) and evaluate its performance on medical tasks. The choice of NNs as basis for the proposed method was made due to their successful application to many medical problems, see e.g. [1,2,12,20], as well as their popularity among machine learning techniques for almost any type of application. The experiments performed examine on one hand the empirical validity of the probability bounds produced by the proposed method and on the other hand compare them with the probabilistic outputs of the original NN classifier.

The rest of this paper starts with an overview of the Venn Prediction framework in the next section, while in Section 3 it details the proposed Neural Network Venn Predictor algorithm. Section 4 presents the experiments performed on two medical datasets and reports the obtained results. Finally, Section 5 gives the conclusions and future directions of this work.

2   The Venn Prediction Framework

This section gives a brief description of the Venn prediction framework; for more details the interested reader is referred to [24]. We are given a training set $\{(x_1, y_1), \ldots, (x_l, y_l)\}$ of examples, where each $x_i \in \mathbb{R}^d$ is the vector of attributes for example i and $y_i \in \{Y_1, \ldots, Y_c\}$ is the classification of that example. We are also given a new unclassified example x_{l+1} and our task is to predict the probability of this new example belonging to each class $Y_j \in \{Y_1, \ldots, Y_c\}$ based only on the assumption that all $(x_i, y_i), i = 1, 2, \ldots$ are generated independently by the same probability distribution (i.i.d.).

The main idea behind Venn prediction is to divide all examples into a number of categories and calculate the probability of x_{l+1} belonging to each class $Y_j \in \{Y_1, \ldots, Y_c\}$ as the frequency of Y_j in the category that contains it. However, as we don't know the true class of x_{l+1}, we assign each one of the possible classes to it in turn and for each assigned classification Y_k we calculate a probability distribution for the true class of x_{l+1} based on the examples

$$\{(x_1, y_1), \ldots, (x_l, y_l), (x_{l+1}, Y_k)\}. \tag{1}$$

To divide each set (1) into categories we use what we call a *Venn taxonomy*. A Venn taxonomy is a finite measurable partition of the space of examples. Typically each taxonomy is based on a traditional machine learning algorithm, called the *underlying algorithm* of the Venn predictor. The output of this algorithm for each attribute vector $x_i, i = 1, \ldots, l+1$ after being trained either on the whole set (1), or on the set resulting after removing the pair (x_i, y_i) from (1), is used to assign (x_i, y_i) to one of a predefined set of categories. For example, a Venn taxonomy that can be used with every traditional algorithm puts in the same category all examples that are assigned the same classification by the underlying algorithm. The Venn taxonomy used in this work is defined in the next section.

After partitioning (1) into categories using a Venn taxonomy, the category T_{new} containing the new example (x_{l+1}, Y_k) will be nonempty as it will contain at least this one example. Then the empirical probability of each label Y_j in this category will be

$$p^{Y_k}(Y_j) = \frac{|\{(x^*, y^*) \in T_{new} : y^* = Y_j\}|}{|T_{new}|}. \tag{2}$$

This is a probability distribution for the label of x_{l+1}. After assigning all possible labels to x_{l+1} we get a set of probability distributions that compose the multi-probability prediction of the Venn predictor $P_{l+1} = \{p^{Y_k} : Y_k \in \{Y_1, \ldots, Y_c\}\}$. As proved in [24] the predictions produced by any Venn predictor are automatically valid multiprobability predictions. This is true regardless of the taxonomy of the Venn predictor. Of course the taxonomy used is still very important as it determines how efficient, or informative, the resulting predictions are. We want the diameter of multiprobability predictions and therefore their uncertainty to be small and we also want the predictions to be as close as possible to zero or one.

The maximum and minimum probabilities obtained for each class Y_j define the interval for the probability of the new example belonging to Y_j:

$$\left[\min_{k=1,\ldots,c} p^{Y_k}(Y_j), \max_{k=1,\ldots,c} p^{Y_k}(Y_j) \right]. \tag{3}$$

If the lower bound of this interval is denoted as $L(Y_j)$ and the upper bound is denoted as $U(Y_j)$, the Venn predictor finds

$$j_{best} = \arg \max_{j=1,\ldots,c} L(Y_j) \tag{4}$$

and outputs the class $\hat{y} = Y_{j_{best}}$ as its prediction together with the interval

$$[L(\hat{y}), U(\hat{y})] \tag{5}$$

as the probability interval that this prediction is correct. The complementary interval

$$[1 - U(\hat{y}), 1 - L(\hat{y})] \tag{6}$$

gives the probability that $\hat{y}$ is not the true classification of the new example and it is called the *error probability interval.*

3   Venn Prediction with Neural Networks

This section describes the proposed Neural Network (NN) based Venn Prediction algorithm. In this work we are interested in binary classification problems ($Y_j \in \{0,1\}$) and therefore this algorithm is designed for this type of problems. The NNs used were 2-layer fully connected feed-forward networks with tangent sigmoid hidden units and a single logistic sigmoid output unit. They were trained with the scaled conjugate gradient algorithm minimizing cross-entropy error (log loss). As a result their outputs can be interpreted as probabilities for class 1 and they can be compared with those produced by the Venn predictor.

The outputs produced by a binary classification NN can be used to define an appropriate Venn taxonomy. After assigning each classification $Y_k \in \{0,1\}$ to the new example x_{l+1} we train the underlying NN on the set (1) and then input the attribute vector of each example in (1) as a test pattern to the trained NN to obtain the outputs $o_1, \ldots, o_{l+1}$. These output values can now be used to divide the examples into categories. In particular, we expect that the examples for which the NN gives similar output will have a similar likelihood of belonging to class 1. We therefore split the range of the NN output $[0,1]$ to a number of equally sized regions λ and assign the examples whose output falls in the same region to the same category. In other words each one of these λ regions defines one category of the taxonomy.

Using this taxonomy we divide the examples into categories for each assumed classification $Y_k \in \{0,1\}$ of x_{l+1} and follow the process described in Section 2 to calculate the outputs of the Neural Network Venn Predictor (NN-VP). Algorithm 1 presents the complete NN-VP algorithm.

Algorithm 1. Neural Networks Venn Predictor

Input: training set $\{(x_1, y_1), \ldots, (x_l, y_l)\}$, new example x_{l+1}, number of categories λ.
for $k = 0$ **to** 1 **do**
    Train the NN on the extended set $\{(x_1, y_1), \ldots, (x_l, y_l), (x_{l+1}, k)\}$;
    Supply the input patterns $x_1, \ldots, x_{l+1}$ to the trained NN to obtain the outputs $o_1, \ldots, o_{l+1}$;
    for $i = 1$ **to** λ **do**
        Find all examples with NN output between $(i-1)/\lambda$ and i/λ and assign them to category T_i;
    end
    Find the category $T_{new} \in \{T_1, \ldots, T_\lambda\}$ that contains (x_{l+1}, k);
    $p^k(1) := \frac{|\{(x^*, y^*) \in T_{new} : y^* = 1\}|}{|T_{new}|}$;
    $p^k(0) := 1 - p^k(1)$;
end
$L(0) := \min_{k=0,1} p^k(0)$; and $L(1) := \min_{k=0,1} p^k(1)$;
Output:
    Prediction $\hat{y} = \arg\max_{j=0,1} L(j)$;
    The probability interval for $\hat{y}$: $[\min_{k=0,1} p^k(\hat{y}), \max_{k=0,1} p^k(\hat{y})]$.

4   Experiments and Results

Experiments were performed on two medical datasets from the UCI Machine Learning Repository [5]:

- **Mammographic Mass**, which is concerned with the discrimination between benign and malignant mammographic masses based on 3 BI-RADS attributes (mass shape, margin and density) and the patient's age [4]. It consists of 961 cases of which 516 are benign and 445 are malignant.
- **Pima Indians Diabetes**, which is concerned with forecasting the onset of diabetes mellitus in a high-risk population of Pima Indians [23]. It consists of 768 cases of which 500 tested positive for diabetes. Each case is described by 8 attributes.

In the case of the Mammographic Mass data the mass density attribute was not used as it did not seem to have any positive impact on the results. Furthermore all cases with missing attribute values were removed and the 2 nominal attributes (mass shape and margin) were converted to a set of binary attributes, one for each nominal value; for each case the binary attribute corresponding to the nominal value of the attribute was set to 1 while all others were set to 0. The resulting dataset consisted of 830 examples described by 10 attributes each.

The NNs used consisted of 4 hidden units for the Mammographic Mass data, as this was the number of units used in [4], and 10 for the Pima Indians Diabetes data, as this seemed to have the best performance with the original NN classifier. All NNs were trained with the scaled conjugate gradient algorithm minimizing cross-entropy error and early stopping based on a validation set consisting of 20% of the corresponding training set. In an effort to avoid local minima each NN was trained 5 times with different random initial weight values and the one that performed best on the validation set was selected for being applied to the test examples. Before each training session all attributes were normalised setting their mean value to 0 and their standard deviation to 1. The number of categories λ of NN-VP was set to 6, which seems to be the best choice for small to moderate size datasets.

4.1   On-Line Experiments

This subsection demonstrates the empirical validity of the Neural Networks Venn Predictor (NN-VP) by applying it to the two datasets in the on-line mode. More specifically, starting with an initial training set consisting of 50 examples, each subsequent example is predicted in turn and then its true classification is revealed and it is added to the training set for predicting the next example. Figure 1 shows the following three curves for each dataset:

- the cumulative error curve

$$E_n = \sum_{i=1}^{n} err_i, \tag{7}$$

where $err_i = 1$ if the prediction $\hat{y}_i$ is wrong and $err_i = 0$ otherwise,

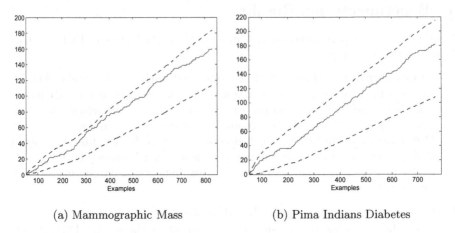

(a) Mammographic Mass   (b) Pima Indians Diabetes

Fig. 1. On-line performance of NN-VP on the two datasets. Each plot shows the cumulative number of errors E_n with a solid line and the cumulative lower and upper error probability curves LEP_n and UEP_n with dashed lines

- the cumulative lower error probability curve (see (6))

$$LEP_n = \sum_{i=1}^{n} 1 - U(\hat{y}_i) \tag{8}$$

- and the cumulative upper error probability curve

$$UEP_n = \sum_{i=1}^{n} 1 - L(\hat{y}_i). \tag{9}$$

Both plots confirm that the probability intervals produced by NN-VP are well-calibrated. The cumulative errors are always included inside the cumulative upper and lower error probability curves produced by the NN-VP.

Two analogous plots generated by applying the original NN classifier to the two datasets are shown in Figure 2. In this case the cumulative error curve (7) for each NN is plotted together with the cumulative error probability curve

$$EP_n = \sum_{i=1}^{n} |\hat{y}_i - \hat{p}_i|, \tag{10}$$

where $\hat{y}_i \in \{0, 1\}$ is the NN prediction for example i and $\hat{p}_i$ is the probability given by NN for example i belonging to class 1. In effect this curve is a sum of the probabilities of the less likely classes for each example according to the NN. One would expect that this curve would be very near the cumulative error curve if the probabilities produced by the NN were well-calibrated. The two plots of Figure 2 show that this is not the case. The NNs underestimate the true error probability in both cases since the cumulative error curve is much higher than the

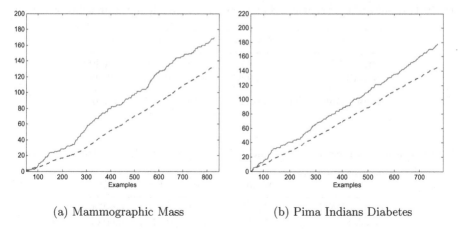

(a) Mammographic Mass   (b) Pima Indians Diabetes

Fig. 2. On-line performance of the original NN classifier on the two datasets. Each plot shows the cumulative number of errors E_n with a solid line and the cumulative error probability curve EP_n with a dashed line.

cumulative error probability curve. To confirm this, the p-value of obtaining the resulting total number of errors E_N by a Poisson binomial distribution with the probabilities produced by the NN was calculated for each dataset. The resulting p-values were 0.000179 and 0.000548 respectively. This demonstrates the need for probability intervals as opposed to single probability values as well as that the probabilities produced by NNs can be very misleading.

4.2   Batch Experiments

This subsection examines the performance of NN-VP in the batch setting and compares its results with those of the direct predictions made by the original NN classifier. For these experiments both datasets were divided randomly into a training set consisting of 200 examples and a test set with all the remaining examples. In order for the results not to depend on a particular division into training and test sets, 10 different random divisions were performed and all results reported here are over all 10 test sets.

Since NNs output a single probabilistic output for the true class of each example being 1, in order to compare this output with that of NN-VP the latter was converted to the mean of $L(1)$ and $U(1)$; corresponding to the estimate of NN-VP about the probability of each test example belonging to class 1. For reporting these results four quality metrics are used. The first is the accuracy of each classifier, which does not take into account the probabilistic outputs produced, but it is a typical metric for assessing the quality of classifiers. The second is cross-entropy error (log loss):

$$CE = - \sum_{i=1}^{N} y_i \log(\hat{p}_i) + (1 - y_i) \log(1 - \hat{p}_i), \qquad (11)$$

Table 1. Results of the original NN and NN-VP on the Mammographic Mass dataset

	Accuracy	CE	BS	REL
Original NN	78.83%	3298	0.1596	0.0040
NN-VP	78.92%	3054	0.1555	0.0023
Improvement (%)	0.11	7.40	2.57	42.50

Table 2. Results of the original NN and NN-VP on the Pima Indians Diabetes dataset

	Accuracy	CE	BS	REL
Original NN	74.56%	3084	0.1760	0.0074
NN-VP	74.26%	3014	0.1753	0.0035
Improvement (%)	-0.40	2.27	0.40	52.70

where N is the number of examples and $\hat{p}_i$ is the probability produced by the algorithm for class 1; this is the error minimized by the training algorithm of the NNs on the training set. The third metric is the Brier score:

$$BS = \frac{1}{N} \sum_{i=1}^{N} (\hat{p}_i - y_i)^2. \tag{12}$$

The Brier score can be decomposed into three terms interpreted as the uncertainty, reliability and resolution of the probabilities, by dividing the range of probability values into a number of intervals K and representing each interval $k = 1, \ldots, K$ by a 'typical' probability value r_k [13]. The fourth metric used here is the reliability term of this decomposition:

$$REL = \frac{1}{N} \sum_{k=1}^{K} n_k (r_k - \phi_k)^2, \tag{13}$$

where n_k is the number of examples with output probability in the interval k and ϕ_k is the percentage of these examples that belong to class 1. Here the number of categories K was set to 20.

Tables 1 and 2 present the results of the original NN and NN-VP on each dataset respectively. With the exception of the accuracy on the Pima Indians Diabetes dataset the NN-VP performs better in all other cases. Although the difference between the two methods on the first three metrics is relatively small, the improvement achieved by the VP in terms of reliability is significant. Reliability is the main concern of this work, as if the probabilities produced by some algorithm are not reliable, they are not really useful.

5   Conclusions

This paper presented a Venn Predictor based on Neural Networks. Unlike the original NN classifiers VP produces probability intervals for each of its predictions, which are valid under the general i.i.d. assumption. The experiments performed in the online setting demonstrated the validity of the probability intervals produced by the proposed method and their superiority over the single probabilities produced by NN, which can be significantly different from the observed frequencies. Moreover, the comparison performed in the batch setting showed that even when one discards the interval information produced by NN-VP by taking the mean of its multiprobability predictions these are still much more reliable than the probabilities produced by NNs.

An immediate future direction of this work is the definition of a Venn taxonomy based on NNs for multilabel classification problems and experimentation with the resulting VP. Furthermore, the application of VP to other challenging problems and evaluation of the results is also of great interest.

Acknowledgments. The author is grateful to Professors V. Vovk and A. Gammerman for useful discussions. This work was supported by the Cyprus Research Promotion Foundation through research contract ORIZO/0609(BIE)/24 ("Development of New Venn Prediction Methods for Osteoporosis Risk Assessment").

References

1. Anagnostou, T., Remzi, M., Djavan, B.: Artificial neural networks for decision-making in urologic oncology. Review in Urology 5(1), 15–21 (2003)
2. Anastassopoulos, G.C., Iliadis, L.S.: Ann for prognosis of abdominal pain in childhood: Use of fuzzy modelling for convergence estimation. In: Proceedings of the 1st International Workshop on Combinations of Intelligent Methods and Applications, pp. 1–5 (2008)
3. Bellotti, T., Luo, Z., Gammerman, A., Delft, F.W.V., Saha, V.: Qualified predictions for microarray and proteomics pattern diagnostics with confidence machines. International Journal of Neural Systems 15(4), 247–258 (2005)
4. Elter, M., Schulz-Wendtland, R., Wittenberg, T.: The prediction of breast cancer biopsy outcomes using two CAD approaches that both emphasize an intelligible decision process. Medical Physics 34(11), 4164–4172 (2007)
5. Frank, A., Asuncion, A.: UCI machine learning repository (2010), http://archive.ics.uci.edu/ml
6. Gammerman, A., Vovk, V., Burford, B., Nouretdinov, I., Luo, Z., Chervonenkis, A., Waterfield, M., Cramer, R., Tempst, P., Villanueva, J., Kabir, M., Camuzeaux, S., Timms, J., Menon, U., Jacobs, I.: Serum proteomic abnormality predating screen detection of ovarian cancer. The Computer Journal 52(3), 326–333 (2009)
7. Holst, H., Ohlsson, M., Peterson, C., Edenbrandt, L.: Intelligent computer reporting 'lack of experience': a confidence measure for decision support systems. Clinical Physiology 18(2), 139–147 (1998)
8. Kononenko, I.: Machine learning for medical diagnosis: History, state of the art and perspective. Artificial Intelligence in Medicine 23(1), 89–109 (2001)

9. Lambrou, A., Papadopoulos, H., Gammerman, A.: Reliable confidence measures for medical diagnosis with evolutionary algorithms. IEEE Transactions on Information Technology in Biomedicine 15(1), 93–99 (2011)

10. Lambrou, A., Papadopoulos, H., Kyriacou, E., Pattichis, C.S., Pattichis, M.S., Gammerman, A., Nicolaides, A.: Assessment of stroke risk based on morphological ultrasound image analysis with conformal prediction. In: Papadopoulos, H., Andreou, A.S., Bramer, M. (eds.) AIAI 2010. IFIP Advances in Information and Communication Technology, vol. 339, pp. 146–153. Springer, Heidelberg (2010)

11. Lisboa, P.J.G.: A review of evidence of health benefit from artificial neural networks in medical intervention. Neural Networks 15(1), 11–39 (2002)

12. Mantzaris, D., Anastassopoulos, G., Iliadis, L., Kazakos, K., Papadopoulos, H.: A soft computing approach for osteoporosis risk factor estimation. In: Papadopoulos, H., Andreou, A.S., Bramer, M. (eds.) AIAI 2010. IFIP Advances in Information and Communication Technology, vol. 339, pp. 120–127. Springer, Heidelberg (2010)

13. Murphy, A.H.: A new vector partition of the probability score. Journal of Applied Meteorology 12(4), 595–600 (1973)

14. Nouretdinov, I., Melluish, T., Vovk, V.: Ridge regression confidence machine. In: Proceedings of the 18th International Conference on Machine Learning (ICML 2001), pp. 385–392. Morgan Kaufmann, San Francisco (2001)

15. Papadopoulos, H.: Inductive Conformal Prediction: Theory and application to neural networks. In: Fritzsche, P. (ed.) Tools in Artificial Intelligence, InTech, Vienna, Austria, ch. 18, pp. 315–330 (2008),
http://www.intechopen.com/download/pdf/pdfs_id/5294

16. Papadopoulos, H., Gammerman, A., Vovk, V.: Reliable diagnosis of acute abdominal pain with conformal prediction. Engineering Intelligent Systems 17(2-3), 115–126 (2009)

17. Papadopoulos, H., Haralambous, H.: Reliable prediction intervals with regression neural networks. Neural Networks (2011),
http://dx.doi.org/10.1016/j.neunet.2011.05.008

18. Papadopoulos, H., Proedrou, K., Vovk, V., Gammerman, A.: Inductive confidence machines for regression. In: Elomaa, T., Mannila, H., Toivonen, H. (eds.) ECML 2002. LNCS (LNAI), vol. 2430, pp. 345–356. Springer, Heidelberg (2002)

19. Papadopoulos, H., Vovk, V., Gammerman, A.: Regression conformal prediction with nearest neighbours. Journal of Artificial Intelligence Research 40, 815–840 (2011), http://dx.doi.org/10.1613/jair.3198

20. Pattichis, C.S., Christodoulou, C., Kyriacou, E., Pattichis, M.S.: Artificial neural networks in medical imaging systems. In: Proceedings of the 1st MEDINF International Conference on Medical Informatics and Engineering, pp. 83–91 (2003)

21. Proedrou, K., Nouretdinov, I., Vovk, V., Gammerman, A.: Transductive confidence machines for pattern recognition. In: Elomaa, T., Mannila, H., Toivonen, H. (eds.) ECML 2002. LNCS (LNAI), vol. 2430, pp. 381–390. Springer, Heidelberg (2002)

22. Saunders, C., Gammerman, A., Vovk, V.: Transduction with confidence and credibility. In: Proceedings of the 16th International Joint Conference on Artificial Intelligence, vol. 2, pp. 722–726. Morgan Kaufmann, Los Altos (1999)

23. Smith, J.W., Everhart, J.E., Dickson, W.C., Knowler, W.C., Johannes, R.S.: Using the ADAP learning algorithm to forecast the onset of diabetes mellitus. In: Proceedings of the Annual Symposium on Computer Applications and Medical Care, pp. 261–265. IEEE Computer Society Press, Los Alamitos (1988)

24. Vovk, V., Gammerman, A., Shafer, G.: Algorithmic Learning in a Random World. Springer, New York (2005)

Cascaded Window Memoization for Medical Imaging

Farzad Khalvati, Mehdi Kianpour, and Hamid R. Tizhoosh

Department of Systems Design Engineering,
University of Waterloo,
Waterloo, Ontario, Canada
{farzad.khalvati,m2kianpo,tizhoosh}@uwaterloo.ca

Abstract. *Window Memoization* is a performance improvement technique for image processing algorithms. It is based on removing computational redundancy in an algorithm applied to a single image, which is inherited from data redundancy in the image. The technique employs a fuzzy reuse mechanism to eliminate unnecessary computations. This paper extends the window memoization technique such that in addition to exploiting the data redundancy in a single image, the data redundancy in a sequence of images of a volume data is also exploited. The detection of the additional data redundancy leads to higher speedups. The cascaded window memoization technique was applied to Canny edge detection algorithm where the volume data of prostate MR images were used. The typical speedup factor achieved by cascaded window memoization is 4.35x which is 0.93x higher than that of window memoization.

Keywords: Fuzzy memoization, Inter-frame redundancy, Performance optimization.

1 Introduction

While high volume data processing in medical imaging is a common practice, the high processing time of such data significantly increases computational cost. The urgent need for optimizing the medical imaging algorithms have been frequently reported from both industry and academia. As an example, the processing times of 1 minute [1], 7 minutes [2], and even 60 minutes [3] are quite common for processing the volume data in medical imaging. Considering the high volume of data to be processed, these reported long processing times introduce significant patient throughput bottlenecks that have a direct negative impact on access to timely quality medical care. Therefore, accelerating medical imaging algorithms will certainly have a significant impact on the quality of medical care by shortening the patients' access time.

Currently, the medical image processing end-product market is dominated by real-time applications with both soft and hard time constraints where it is imperative to meet the performance requirements. In hard real-time medical systems,

L. Iliadis et al. (Eds.): EANN/AIAI 2011, Part II, IFIP AICT 364, pp. 275–284, 2011.

such as image guided surgery, it is crucial to have a high-performance system. In soft real-time systems, such as many diagnostic and treatment-planning tasks in medical imaging, although it is not fatal to not have a high-performance system, it is important to increase the speed of the computations to enable greater patient throughput. The combination of complex algorithms, high volume of data, and high performance requirements leads to an urgent need for optimized solutions. As a result, the necessity and importance of optimization of medical image processing is a hard fact for the related industry.

In general, the problem of optimizing computer programs are tackled with the following solutions:

- Fast Processors (e.g. Intel Core i7): The increasing speed of processors is mainly due to more transistors per chip, fast CPU clock speeds, and optimized architectures (e.g. memory management, out of order superscalar pipeline).
- Algorithmic Optimization (e.g. Intel IPP: Integrated Performance Primitives [4]): The optimization is done by decreasing the algorithm complexity by reducing the number of computations.
- Parallel Processing: It takes advantage of the parallel nature of hardware design in which the operations can be done in parallel. This can be exploited in both hardware (Graphics Processing Unit: e.g. NVIDIA [5]) or software (RapidMind Multi-core Development Platform, now integrated with Intel called as Intel Array Building Blocks [6]).

Among these optimization solutions, using fast processors are the most obvious and simplest one. In academia, in particular, this has been an immediate response in dismissing the need for optimization methods for image processing and medical imaging. Although using fast processors can always be an option to tackle the performance requirements of a medical imaging solution, it suffers from two major shortcomings; first, it is not always possible to use high-end processors. One example is embedded systems where the low-end embedded processors are used. Second, in the past 20 years, the processors' performance has increased by 55% per year [7]. For a medical imaging solution that currently takes 10 minutes to run on a high-performance computer, in order to reduce the processing time down to 1 minute, it will take more than 6 years for the processors to catch up with this performance requirement. Obviously, this is not a viable solution for many performance-related problems.

Algorithmic optimization is a fundamental method which aims at reducing the number of operations used in an algorithm by decreasing the number of steps that the algorithm requires to take. Parallel processing breaks down an algorithm into processes that can be run in parallel while preserving the data dependencies among the operations. In both schemes, the building blocks of the optimization methods are the operations where the former reduces the number of operations and the latter runs them in parallel. In both cases, the input data is somewhat ignored. In other words, the algorithmic optimization will be done in the same way regardless of the input data type. Similarly, running the operations in parallel is independent of the input data. Window memoization, introduced in [8], is a new optimization technique which can be used in conjunction with either of these two optimization methods.

Window memoization is based on fuzzy *reuse* or *memoization* concept. Memoization is an optimization method for computer programs which avoids repeating the function calls for previously seen input data. Window memoization applies the general concept of memoization to image processing algorithms by exploiting the image data redundancy and increases the performance of image processing. The reuse method used by window memoization is a simple fuzzy scheme; it allows for the reuse of the results of similar but non-identical inputs. Thus far, window memoization has been applied to single frame images and therefore, it has been able to exploit the intra-frame data redundancy of image data to speed up the computations significantly [8]. In this work, we extend the window memoization technique to be applicable to sets of volume data where in addition to intra-frame data redundancy, the inter-frame data redundancy is exploited as well. This extra data redundancy (i.e. inter-frame) leads to higher speedups compared to intra-frame redundancy based window memoization. We call this new window memoization technique that exploits both inter-and intra-frame data redundancy *Cascaded Window Memoization.*

The organization of the remaining of this paper is as follows. Section 2 provides a brief background review on the window memoization technique. Section 3 presents the cascaded window memoization technique. Sections 4 and 5 present the results and conclusion of the paper, respectively.

2   Background

In this section, a background review of the window memoization technique and its fuzzy reuse mechanism is given.

2.1   Window Memoization

Introduced in [8], window memoization is an optimization technique for image processing algorithms that exploits the data redundancy of images to reduce the processing time. The general concept of memoization in computer programs has been around for a long time where the idea is to speed up the computer programs by avoiding repetitive/redundant function calls [9] [10] [11]. Nevertheless, in practice, the general notion of memoization has not gained success due to the following reasons: 1) the proposed techniques usually require detailed profiling information about the runtime behaviour of the program which makes it difficult to implement [12], 2) the techniques are usually generic methods which do not concentrate on any particular class of input data or algorithms [13], and 3) the memoization engine used by these techniques are based on non-fuzzy comparison where two input data is considered equal (in which case one's result can be reused for another) if they are identical. In contrast, in designing the window memoization technique in [8], the unique characteristics of image processing algorithms have been carefully taken into account to enable window memoization to both be easy to implement and improve the performance significantly. One important property of image data is the fact that it is tolerant to small changes that a fuzzy reuse mechanism would produce. This has been exploited by window memoization to increase the performance gain.

The window memoization technique minimizes the number of redundant computations by identifying similar groups of pixels (i.e. window) in the image using a memory array, reuse table, to store the results of previously performed computations. When a set of computations has to be performed for the first time, they are performed and the corresponding results are stored in the reuse table. When the same set of computations has to be performed again, the previously calculated result is reused and the actual computations are skipped. This eliminates the redundant computations and leads to high speedups.

2.2 Fuzzy Memoization

Although window memoization can be applied to any image processing algorithm, local algorithms, in particular, are the first to adopt the technique. Local processing algorithms mainly deal with extracting local features in image (e.g. edges, corners, blobs) where the input to the algorithm is a window of pixels and the output is a single pixel. When a new window of pixels arrives, it is compared against the ones stored in the reuse table. In order for a window to find the exact match in the reuse table (i.e. hit), all pixels should match:

$$\forall pix \in win_{new}, \forall pix' \in win_{reuse},\ pix = pix' \implies win_{new} = win_{reuse} \qquad (1)$$

where pix and pix' are corresponding pixels of the new window win_{new} and the one already stored in the reuse table win_{reuse}, respectively. In the above definition, a hit occurs if all the corresponding pixels of the windows are identical. This limits the *hit rate*: the percentage of windows that match the one stored in the reuse table. In order to increase the hit rate, we use fuzzy memoization. The human vision system cannot distinguish small amounts of error in an image, with respect to a reference image. Therefore, small errors can usually be tolerated in an image processing system. The idea of using this characteristic of image data to improve performance was introduced by [14] where a fuzzy reuse scheme was proposed for microprocessor design in digital hardware.

By using the fuzzy window memoization, those windows that are similar but not identical are assumed to provide the same output for a given algorithm.

$$\forall pix \in win_{new}, \forall pix' \in win_{reuse},\ MSB(d, pix) = MSB(d, pix') \implies$$
$$win_{new} = win_{reuse} \qquad (2)$$

where $MSB(d, pix)$ and $MSB(d, pix')$ represent d most significant bits of pixels pix and pix' in windows win_{new} and win_{reuse}, respectively. By reducing d, similar but not necessarily identical windows are assumed to have identical results for a given algorithm. This means that the response of one window may be assigned to a similar but not necessarily identical window. As d decreases, more windows with minor differences are assumed equal and thus, the hit rate of window memoization increases drastically. Assigning the response of a window to a similar but not necessarily identical window introduces inaccuracy in the result of the algorithm to which window memoization is applied. However, in practice, the accuracy loss in responses is usually negligible.

We perform an experiment, using our memoization mechanism, to pick an optimal d that gives high hit rates with small inaccuracy in results. For our experiments, we use an ideal algorithm which outputs the central pixel of the input 3×3 window as its response. We use different values for d from 1 to 8. As the input images, we use a set of 40 natural images of 512×512 pixels. The accuracy of the results is calculated as SNR[1]. In calculating SNR, we replace an infinite SNR with 100, in order to calculate the average of SNRs of all images.

Figure 1 shows the average hit rate and SNR of the result for each value of d. It is seen that as d decreases, hit rate increases and at the same time, SNR decreases. The error in an image with SNR of $30dB$ is nearly indistinguishable by the observer [15]. Therefore, we pick d to be 4 because it gives an average SNR of $29.68dB$, which is slightly below the value $30dB$. Reducing d from 8 to 4 increases the average hit rate from 10% to 66%. For our experiments in the remaining of this chapter, we will choose d to be 4.

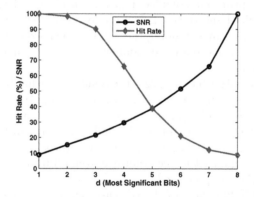

Fig. 1. Average hit rate and SNR versus the number of the most significant bits used for assigning windows to symbols. Infinite SNRs have been replaced by SNR of 100.

In the previous work [8], an optimized software architecture for window memoization has been presented where the typical speedups range from 1.2x to 7.9x with a maximum factor of 40x.

---

[1] The error in an image (Img) with respect to a reference image (R_{Img}) is usually measured by *signal-to-noise ratio (SNR)* as $SNR = 20log_{10}(\frac{A_{signal}}{A_{noise}})$ where A_{signal} is the RMS (root mean squared) amplitude. A_{noise}^2 is defined as: $A_{noise}^2 = \frac{1}{rc}\sum_{i=0}^{r-1}\sum_{j=0}^{c-1}(Img(i,j) - R_{Img}(i,j))^2$ where $r \times c$ is the size of Img and R_{Img}.

3   Cascaded Window Memoization

In this paper, we improve the the window memoization technique by extending it from exploiting only intra-frame data redundancy to inter-frame data redundancy as well. The main objective is that for a given set of images and an image processing algorithm, to further speed up the computations. The actual speedup achieved by the window memoization technique is calculated by equation 3 [8].

$$speedup = \frac{t_{mask}}{t_{memo} + (1 - HR) \times t_{mask}} \qquad (3)$$

where:

- t_{mask} is the time required for the actual mask operations of the algorithm at hand.
- t_{memo} or the memoization overhead cost is the extra time that is required by the memoization mechanism to actually reuse a result.
- HR or hit rate determines the percentage of the cases where a result can be reused.

In the equation above, for a given processor, t_{mask} depends on the complexity of the algorithm to be optimized. For a fixed t_{mask}, to achieve high speedups, the hit rate must be maximized and the memoization overhead cost must be minimized. We increase the hit rate by extending the window memoization technique from exploiting only intra-frame data redundancy to utilize both intra- and inter-frame data redundancy.

While window memoization minimizes the number of redundant computations by identifying similar groups of pixels in a single image (i.e. intra-frame data redundancy), cascaded window memoization discovers the data redundancy not only in a single frame, but also among the multiple frames of the same volume data (i.e. inter-frame data redundancy). In other words, the similar groups of pixels are compared against the ones that are from all frames of a volume data that have been previously processed.

Figure 2 shows the flowchart of the the cascaded window memoization technique. In cascaded window memoization, the reuse table is initialized only once for the entire volume data inputs; that is when the first frame arrives. After the first initialization, for the rest of the frames in the same volume data, it is assumed that the reuse table contains valid data and needs not to be reset again. As a result, not only the similar neighborhoods in a singe frame are detected and the corresponding results are reused, the same scenario occurs for the neighborhoods which belong do different frames of the same volume data. This, in fact, exploits the inter-frame data redundancy, which is in addition to the intra-frame data redundancy. Thus, the cascaded window memoization technique increases the average hit rate and hence, the average speedup for each frame increases compared to regular window memoization which is based on intra-frame redundancy only.

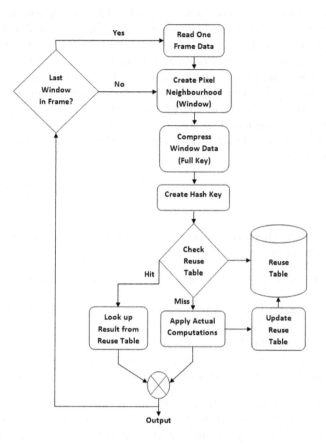

Fig. 2. Cascade Window Memoization: The flowchart

In order to translate the extra hit rates to actual speedups, it is crucial that exploiting inter-frame redundancy does not introduce extra memoization overhead time. Otherwise, the increase in hit rate will be compensated by the extra overhead time, leading to no increase (or decrease) in the speedup.

At each iteration, a window of pixels are compressed (using equation 2) to create $full\_key$. A $full\_key$ is a unique number that represents similar groups of pixels which are spread among different frames of a volume data. Compressing $full\_key$ is a necessary step; it causes that similar but not necessarily identical groups of pixel to be considered equal. This introduces a small amount of inaccuracy in the results, which is usually negligible. Moreover, the compression reduces the reuse table size which holds the $full\_keys$ and the results.

In order to map a $full\_key$ to the reuse table, a hash function is used to generate $hash\_key$. As discussed in [8], the multiplication method is used as the hash function. In order to maximize the HR and minimize t_{memo} (i.e. to minimize the collisions), a constant number is used for calculating the $hash\_key$. It is important to note that this optimal constant number is different for 32-bit and 64-bit processor architectures [16][2]. Thus, it is imperative that the window memoization technique detects (automatically or by user input) the architecture type and set the appropriate constant number. Using 64-bit processor reduces the memoization overhead time in comparison to a 32-bit processor. The reason is that for a given $full\_key$, in the 64-bit processor less number of memory accesses is required compared to that of the 32-bit processor. Less number of memory accesses means lower memoization overhead time and hence, higher speedup (equation 3)[3].

4   Results

In order to measure the performance of the cascaded window memoization technique, the following setups were used:

- Case study algorithm: Canny edge detection algorithm that uses windows of 3×3 pixels.
- Processor: Intel (R) Core (TM)2 Q8200 processor; CPU: 2.33GHz, Cache size 4MB.
- Reuse Table size: 64KB.
- Input images: Volume data for Prostate MR Images of 5 patients. The MR images are T2 weighted with endorectal coil (180 frames in total)[4].

We applied Canny edge detection algorithm to all five patients volume data. Figure 3 shows the hit rate of the frames of a sample patient volume data for regular and cascaded window memoization applied to the Canny Edge detector. As it is seen, exploiting inter-frame redundancy by cascaded window memoization increases the hit rate, on average, by 4.44%.

Figure 4 shows the actual speedup of the frames of the sample patient volume data for regular and cascaded window memoization applied to the Canny Edge detector. As it is seen, exploiting inter-frame data redundancy by cascaded window memoization increases the speedup, on average, by a factor of 1.01x. All the reported speedup factors are based on running the actual C++ code implemented for (cascaded) window memoization and measuring the CPU time. The average accuracy of the results is above 96%.

As for the entire data, on average, cascaded window memoization increased the hit rate and speedup for Canny edge detection algorithm from 84.88% and 3.42x to 90.03% and 4.35x respectively; an increase of 5.15% and 0.93x, respectively (Table 1).

---

[2] This constant number for 32-bit and 64-bit processor architecture is '2,654,435,769' and '11,400,714,819,323,198,485', respectively [16].

[3] This was verified by experimental data.

[4] Taken from online Prostate MR Image Database:
(http://prostatemrimagedatabase.com/Database/index.html).

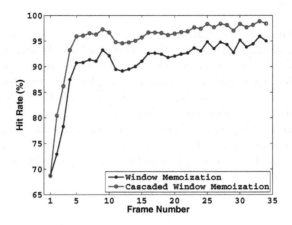

Fig. 3. Hit Rate: Cascaded Window Memoization versus Regular Window Memoization

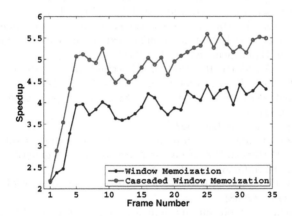

Fig. 4. Speedup: Cascaded Window Memoization versus Regular Window Memoization

Table 1. Hit Rate and Speedup Results

Patient data	Hit Rate	Hit Rate (Cascaded)	Speedup	Speedup (Cascaded)
1	84.07%	89.88%	3.24x	4.12x
2	79.96%	85.36%	3.09x	3.94x
3	82.45%	87.60%	3.29x	4.21x
4	87.20%	92.17%	3.62x	4.61x
5	90.70%	95.14%	3.85x	4.86x
Average	84.88%	90.03%	3.42x	4.35x

5   Conclusion

Window memoization is an optimization technique that exploits the intra-frame data redundancy of images to reduce the processing time. It uses a fuzzy memoization mechanism to avoid unnecessary computations. In this paper, we extended the window memoization technique to exploit the inter-frame data redundancy as well. It was shown that the elimination of the extra data redundancy leads to higher speedup factors. For a case study algorithm applied to Prostate MR Images of 5 patients volume data, on average, the original window memoization technique yielded the speedup factor of 3.42x. Cascaded window memoization increased the speedup factors by, on average, 0.93x yielding average speedup factor of 4.35x.

References

1. Haas, B., et al.: Automatic segmentation of thoracic and pelvic CT images for radiotherapy planning using implicit anatomic knowledge and organ-specific segmentation strategies. Phys. Med. Biol. 53, 1751–1771 (2008)
2. Hodgea, A.C., et al.: Prostate boundary segmentation from ultrasound images using 2D active shape models: Optimisation and extension to 3D. Computer Methods and Programs in Biomedicine 84, 99–113 (2006)
3. Gubern-Merida, A., Marti, R.: Atlas based segmentation of the prostate in MR images. In: MICCAI: Segmentation Challenge Workshop (2009)
4. Intel Integrated Performance Primitives, `http://software.intel.com/en-us/articles/intel-ipp/`
5. NVIDIA, `http://www.nvidia.com/`
6. RapidMind, software.intel.com/en-us/articles/intel-array-building-blocks/
7. Hennessy, J.L., Patterson, D.A.: Computer Architecture - A quantitative approach, 4th edn. Morgan Kaufmann, San Francisco (2007)
8. Khalvati, F.: Computational Redundancy in Image Processing, Ph.D. thesis, University of Waterloo (2008)
9. Michie, D.: Memo functions and machine learning. Nature 218, 19–22 (1968)
10. Bird, R.S.: Tabulation techniques for recursive programs. ACM Computing Surveys 12(4), 403–417 (1980)
11. Pugh, W., Teitelbaum, T.: Incremental computation via function caching. In: ACM Symposium on Principles of Programming Languages, pp. 315–328 (1989)
12. Wang, W., Raghunathan, A., Jha, N.K.: Profiling driven computation reuse: An embedded software synthesis technique for energy and performance optimization. In: IEEE VLSID 2004 Design, p. 267 (2004)
13. Huang, J., Lilja, D.J.: Extending value reuse to basic blocks with compiler support. IEEE Transactions on Computers 49, 331–347 (2000)
14. Salami, E., Alvarez, C., Corbal, J., Valero, M.: On the potential of tolerant region reuse for multimedia applications. In: International Conference on Supercomputing, pp. 218–228 (2001)
15. Alvarez, C., Corbal, J., Valero, M.: Fuzzy memoization for floating-point multimedia applications. IEEE Transactions on Computers 54(7), 922–927 (2005)
16. Preiss, B.R.: Data Structures and Algorithms with Object-Oriented Design Patterns in C++. John Wiley and Sons, Chichester (1999)

Fast Background Elimination in Fluorescence Microbiology Images: Comparison of Four Algorithms*

Shan Gong and Antonio Artés-Rodríguez

Department of Signal Theory and Communications,
Universidad de Carlos III de Madrid,
{concha,antonio}@tsc.uc3m.es

Abstract. In this work, we investigate a fast background elimination front-end of an automatic bacilli detection system. This background eliminating system consists of a feature descriptor followed by a linear-SVMs classifier. Four state-of-the-art feature extraction algorithms are analyzed and modified. Extensive experiments have been made on real sputum fluorescence images and the results reveal that 96.92% of the background content can be correctly removed from one image with an acceptable computational complexity.

Keywords: Tuberculosis bacilli, background remove, object detection, fluorescence image.

1 Introduction

Tuberculosis is a contagious disease that causes thousands of death every year around the whole word. Sputum specimens are firstly stained with fluorescent dye and fluorescence images can be obtained by scanning these specimens with a fluorescence microscope. Since conventional diagnosis require amount of human efforts, detecting Tuberculosis bacilli automatically has drove many attentions [1], [2], [3]. All of these systems use a specific segmentation method to highlight the bacilli regions from the original image.

However, all these methods are based on statistical analysis of the profile and color information of bacilli, so they have disadvantages: *Poor generalization ability*. Stained M. tuberculosis bacilli display in images with varying appearances. Thus in [1], [2], [3], both color and contour features have been considered. But, statistical knowledge about the shape and color of the labeled bacilli is limited to dataset which is sensitive to experimental conditions such as luminance and camera types. *Poor ability dealing with outliers*. The contrast between bacilli and the background is regularly manifest given that stained bacilli fluoresce in the range between green and yellow up to white. In this case, color segmentation might be able to produce desired results. Nevertheless, when one bacteria

---

* This work has been partly supported by Ministerio de Educación of Spain (projects 'DEIPRO, id. TEC2009-14504-C02-01, and 'COMONSENS', id. CSD2008-00010).

L. Iliadis et al. (Eds.): EANN/AIAI 2011, Part II, IFIP AICT 364, pp. 285–290, 2011.

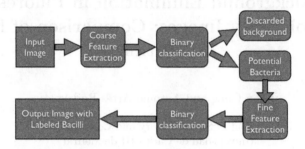

Fig. 1. The input original image passes through a front-end quick background subtraction system and drops most of background content. The filtered image (i.e., retained potential bacteria region) will be reanalyzed and reclassified with the purpose of improving the specificity of the whole system.

is overlapped by other debris objects such as cells, or when images have been taken under inapplicable luminance conditions, similar segmentation attempts may be proved futile.

Instead of getting involved in intricate segmentation algorithm investigation, we propose a quick background subtraction system as a front-end of the M. tuberculosis bacillus detection system shown in Fig.1. This background elimination system is termed as a coarse feature descriptor followed by a binary classification (i.e., bacillus or background). The essential idea of designing such a front-end is to reduce the computational complexity while maintaining elevated sensibility and compensatory specificity. As the majority of the original image has been disposed by the front-end processor, steps of locating bacilli are only required to take on the retained possible regions which will dramatically accelerate the whole detection process. In section 2, four feature extraction algorithms: SIFT, color SIFT, Haralick Features and Histograms of Colors are modified to suit our M. bacillus recognition task. Section 3 gives experimental results based on real fluorescence images.

2   Feature Extraction Algorithms

Patches are extracted by sliding a local window across the image and classified as positive (i.e., presence of object) or negative (i.e, absence of object). Unlike articulated objects (e.g., pedestrians), bacilli do not contain a lot of structural or intra-class variations, specially scale variations, thus, it is reasonable to use single fixed-size sliding window. Characters based on small sliding window are more discriminative, but computational expensive. On the contrary, big sliding window avoids expensive computations by allowing more flexible moving steps, however, as more background pixels have been included, the results are less reliable. Given that bacillus generally is of size up to 20×20, in our experiment the size of patches has been chosen as 36×36.

Patch classifier is chosen as SVMs with linear kernel [4], [5].

2.1 SIFT and Color Based SIFT

The original SIFT [6] descriptor initials with a keypoint detector followed by a local image descriptor. Different with the general image matching problem, in our case, neither multiscale invariability nor distinctive points localization should be necessarily required. Smooth the i-th color component $I(x, y, i)$ of one image with a variable-scale Gaussian mask $G(x, y, \sigma)$,

$$L(x, y, i, \sigma) = G(x, y, \sigma) * I(x, y, i), \quad 1 \le i \le c. \tag{1}$$
$$G(x, y, \sigma) = \frac{1}{2\pi\sigma^2} e^{-(x^2+y^2)/2\sigma^2},$$

where $*$ is a convolution operator.

Then the gradient of the Gaussian mask smoothed image can be written as

$$J(x, y, i) = \nabla[L(x, y, i, \sigma)] = \nabla[G(x, y, \sigma) * I(x, y, i)]$$
$$= [\nabla G(x, y, \sigma)] * I(x, y, i), \quad 1 \le i \le c. \tag{2}$$

where ∇ is the derivative operator, σ indicates the width of the Gaussian and determines the quantity of smoothing (i.e., the gradient can be obtained by convolving the image with the horizontal and vertical derivatives of the Gaussian kernel function).

SIFT features for the i-th color channel of the patch $p(u, v; c)$ are formed by computing the gradient $J(u, v, i)$ at each pixel of the patch. The gradient orientation is then weighted by its magnitude. This patch is then split into 4×4 grid, in each tile a gradient orientation histogram is formed by adding the weighted gradient value to one of the 8 orientation histogram bins.

An extension is CI-SIFT which embeds color information to conventional SIFT descriptor by applying the five color invariant sets presented in Table 3 of [7].

2.2 Histogram of Color

Color signatures are extracted by using the histograms of color method proposed in [8].

2.3 Haralick Features

Based on co-occurrence matrix, several statistics were proposed in [9], known as *Haralick features*.

Co-occurrence matrix of a n bits gray-level image is of size $2^n \times 2^n$. One way to reduce the dimension of this matrix is to quantitate the image. The quantization for images of multidimensional color space can be done with K-means clustering algorithm. In the case of images of unit color component, uniform quantization can be employed.

3   Patches Classification Experiments

3.1   Datasets Description

Images are required from a 20x fluorescence Nikon microscope with a prior automatic slide loader and a Retiga 2000R camera with acquisition time 0.6 sec, gain five and zero offset. For each patient, 200 RGB 1200 × 1600 (8 bits per color) images have been taken and stored. 4,987 bacilli of 952 images of 45 infected patients have been identified and labeled manually by experts. Patches centered at each bacteria haven been extracted. Furthermore, bacilli have been rotated 90, 180 and 270 grades so as to enrich the dataset. Finally, a positive dataset of 19,890 patches has been generated. Regarding the negative dataset, 100,031 negative patch samples haven been created by arbitrarily moving the local windows in images of a couple of healthy patients.

In the following experiments, 3,000 positive as well as negative randomly selected patches are used for training step and all the rest for test. Results are obtained by averaging 100 runs of a linear SVM. Due to space limitation, full detailed results can be found in technical report [10].

3.2   Experiments

Experiment E01: Without any feature descriptor. For the purpose of comparison, in this experiment none feature extraction processes have been taken. Patches are considered to be of size 20 × 20. PCA has been used to reduce the feature dimension. And, only the *BLUE* component has been analyzed. As long as the patch is small enough, even without feature extraction, one simple linear-SVM classifier is capable to achieve a sensibility up to 93.6%.

Experiment E02: SIFT. Both the sensibility and specificity are comparable regardless of the color spaces. For white color bacilli surrounded by yellow tissues, *BLUE* is the only discriminative in the *RGB* color space, which may explain why *BLUE* component obtains the optimal results.

Experiment E03: CI-SIFT. In [7], the author stated that the degree of invariance of the five color invariant sets have the following relation: $E_\omega < W_\omega < C_\omega < N_\omega < H_\omega$. H_ω which is the most robust to changes in imaging condition (see Table 1 in [7]) had the lowest discriminative power while E_ω which is the most sensitive to any changes achieved the best recognition precision.

Experiment E04: HF-without Quantization. It is always better not to normalize features sets obtained from the four adjacently measurements (see Fig. 2 of [10]) regardless of the implemented color spaces.

Experiment E05: HF-with Quantization. Quantization (see section 2.3) is firstly employed for each patch. 1,000 positive as well as negative patches have been picked randomly from the entire data set from which codebooks of colors are constructed with K-means clustering method. Regarding unit dimensional color spaces, uniform quantization method also has been studied. Excluding

RGB, sensibility as well as specificity obtained in all the unit dimensional color spaces in this experiment are comparable to or even higher than those obtained in the previous experiment which reveals that quantizations are highly recommended. Furthermore, uniform quantization is much more reliable than K-means quantization.

4   Conclusion and Future Line

In this paper, we have analyzed the performance of several feature extraction algorithms. In Table 1 are the optimal cases for each algorithm. The *BLUE* scale state-of-the-art SIFT-like descriptor outperforms the others. The specificity achieved by Histograms of colors method is more than 15% lower than the average level. *CI-SIFT* as well as *HF* take use both the geometrical and color signatures and are capable to obtain appealing results. Small patch may contain more discriminative signatures, but we aim to reduce the background as fast as possible which means that with the same accuracy bigger patches are always recommended. Since the computational complexity of most of attributes of the Haralick feature sets are $O(\text{number of bits of the image}^2)$, the quantization step considerably accelerate the algorithm. Another conclusion we can make is that the *BLUE* channel is the most robust compared with other color representations. By applying this background elimination process, approximated 96.92% background pixels can be subtracted from the original image. One example is given in Fig. 2.

One possible way to improve sensibility without degrading specificity is to use a different evaluation function (such as F-score, AUC) of the linear SVM classifier or add other simple implemented characters into the feature set.

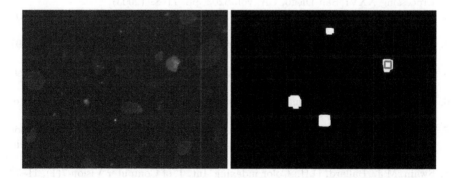

Fig. 2. The left figure is a fluorescence image with one bacteria. The right figure is the filtered image in which the white regions are detected potential bacilli locations, with the real bacteria bounded by the red rectangular for purpose of reference. Parameters used here are: SIFT descriptor on green color channel with step of the moving window equal to 1. Most of the pixels of the image have been classified as background (i.e., the black regions).

Table 1. Comparison of all the algorithms

method	parameters	sensibility	specificity	computational complexity (approximated number of executions per patch)		
				+	×	others
SIFT	Blue	95.199(0.187)	97.234(0.151)	$11l^2$	$12l^2$	$2l^2$
HoColor	RGB, NumBin = 8	93.418(0.357)	79.672(0.464)	0	0	$3l^2$
	BLUE, NumBin = 8	91.097(0.484)	75.931(0.525)	0	0	l^2
CI-SIFT	E_ω	94.000(0.272)	96.949(0.167)	$30l^2$	$31l^2$	$2l^2$
HF	RGB, no normalization	95.260(0.312)	93.704(0.256)	$75k^2$	$66k^2$	$21k^2$
	BLUE, no normalization	93.792(0.232)	93.409(0.218)	$25k^2$	$22k^2$	$7k^2$
HF	BLUE, codebook = 8, uniform Quantization, no normalization	94.845(0.256)	92.577(0.272)	$l^2 + 25k_Q^2$	$l^2 + 22k_Q^2$	$6k_Q^2$
None feature extraction	BLUE, patch size = 20×20, feature vector is of size 1 × 400	93.617(0.307)	97.046(0.189)	0	0	0

where $l = 36$, $k = 2^8$, $k_Q = 2^3$. And for HF the complexity is considered for one single adjacency measurement

References

1. Osman, M.K., Mashor, M.Y., Saad, Z., Jaafar, H.: Colour image segmentation of Tuberculosis bacillus in Ziehl-Neelsen-Stained tissue Images using moving K-Mean clustering procedure. In: 4th Asia International Conference on Mathematical/Analytical Modelling and Computer Simulation, pp. 215–220. IEEE Press, Washington, DC (2010)
2. Foreo, M.G., Gristóbal, G., Alvarez-Borrego, J.: Automatic identification techniques of tuberculosis bacteria. In: Proceedings of the Applications of Digital Image Processing XXVI, San Diego, CA, vol. 5203, pp. 71–81 (2003)
3. Foreo, M.G., Sroukeb, F., Cristóbal, G.: Identification of tuberculosis bacteria based on shape and color. J. Real-Time Imaging 10(4), 251–262 (2004)
4. Cortes, C., Vapnik, V.: Support-vector network. Machine Learning 20, 273–297 (1995)
5. Chih-Chung, Lin, C.-J.: LIBSVM: A library for Support Vector Machines, http://www.csie.ntu.edu.tw/~cjlin/papers/libsvm.pdf
6. Lowe, D.G.: Distinctive image features from scale-invariant keypoints. Int. J. of Computer Vision 60, 91–110 (2004)
7. Geusebroek, J.M., Boomgaard, R.V.D., Smeulders, A.W.M., Geerts, H.: Color invariance. J. IEEE Transactions on Pattern Analysis and Machine Intelligence 23(12) (2001)
8. Swain, M.J., Ballard, D.H.: Color indexing. Int. J. of Computer Vision 7(1), 11–32 (1991)
9. Haralick, R.M., Shanmugan, K., Dinstein, I.: Textural features for image classification. IEEE Transactions on Systems, Man, and Cybernetics SMC-3, 610–662 (1973)
10. Gong, S., Artés-Rodrigués, A.: Science report, http://www.tsc.uc3m.es/~concha/TecReportBacilos.pdf

Experimental Verification of the Effectiveness of Mammography Testing Description's Standardization

Teresa Podsiadły-Marczykowska[1] and Rafał Zawiślak[2]

[1] Instytut Biocybernetyki i Inżynierii Biomedycznej PAN,
ul.Trojdena 4, 02-109 Warszawa
tpodsiadly@ibib.waw.pl
[2] Politechnika Łódzka, Instytut Automatyki,
ul. Stefanowskiego 18/22 90-912 Łódź, Polska
rafal.zawislak@p.lodz.pl

Abstract. The article presents assumptions, and results of a test of experimental verification of a hypothesis stating that the use of the MammoEdit - a tool that uses its own ontology and the embedded knowledge of mammography – increase the diagnostic accuracy as well as the reproducibility of mammographic interpretation. The graphical user interface of the editor was similarly assessed, as well as the rules for visualization which assists the radiologist in the interpretation of the lesions' character.

Keywords: mammogram interpretation, biomedical engineering, ontology, breast scanning, cancer diagnostics.

1 Introduction

Mammography is well founded, effective imaging technique that can detect breast cancer in early and even pre-invasive stage, widely used in screening programs. However, the diagnostic value of mammography is limited by significant and up to 25% high, rate of missed breast cancers. Mammography is commonly seen as the most difficult imaging modality. Computer-aided Detection (CADe) systems aiming to reduce detection errors are now commercially available and are gaining increasing practical importance, but no serious attempts have been made to apply Computer-aided diagnosis (CADx) systems for lesion diagnosis in practical clinical situations. The paper presents an assessment of diagnostic accuracy of MammoEdit - an ontology-based editor supporting mammograms description and interpretation. In our opinion MammoEdit fills the gap among the existing solutions to the problem of supporting the interpretation of mammograms.

While working on the MammoEdit project, they used the ontology of mammography, created specifically for this need. It was used as a partial set of project requirements for the user interface and database that stores patients descriptions. This role of the ontology of mammography in the MammoEdit editor project was in line with literature indications for the use of ontology in IT systems [1].

L. Iliadis et al. (Eds.): EANN/AIAI 2011, Part II, IFIP AICT 364, pp. 291–296, 2011.
© IFIP International Federation for Information Processing 2011

2 The Ontology of Mammography

The mammographic ontology was created using Protege-2000 editor and the OWL language. Presented version of ontology includes: comprehensive, standard description patterns of basic lesion in mammography (masses and microcalcification clusters) enhanced with subtle features of malignancy, classes presenting descriptions of real lesions carried out on the basis of completed patterns and grades modeling diagnostic categories of the BI-RADS system.

Radiologists usually agree in their assessments of the diagnostic categories of BI-RADS (*Kappa=0.73*), when they have to deal with lesions presenting features typical for radiological image of breast cancer, clusters of irregular, linear or pleomorphic microcalcifications, or with spicular massess. The agreement in assessing is significantly lower when it comes to lesions from changes of the 4 BI-RADS[1] category (*Kappa=0.28*). Why does it happen? Admittedly, the BI-RADS system controlled vocabulary contains terms important with specific diagnostic value, but it lacks complete lesions definitions and does not stress the importance of early and subtle signs of cancer. Moreover BI-RADS system recommendations for estimating lesion diagnostic category are descriptive, incomplete and imprecise, they are expressed using different generality levels. The recommendations refer to the knowledge and experience of the radiologist and to typical images of the mammographic lesions. As a consequence, there is a large margin of freedom left for individual interpretation. Those conclusions are confirmed by large study based on 36 000 mammograms [2], showing wide variability in mammograms interpretation. While creating ontology, ambiguity in BI-RADS system recommendations caused by the lack of data, necessary to express restrictions on allowed values of the ontology classes modeling diagnostic categories of the BI-RADS system. The lack of these classes in the domain model clearly prevents the use of the ontology's classification mechanism to the diagnostic categories' assessment of the real lesions. The important part of this presentation is filling of the gap in imprecisely formulated BI-RADS system recommendations. This work was necessary to define through ontology classes modeling BI-RADS categories.

3 MammoEdit – A Tool Supporting Description and Interpretation of the Changes in Mammography

Analyzing reasons for errors it can be generally concluded that potential source of mammograms' interpretation mistakes is uneven level of knowledge and diagnostic abilities of radiologists and their subjectivity. Considering the above, it should be accepted that reduction of the interpretation errors can be obtained by describing mammograms using standardized protocols which include complete, standardized definitions of the lesions taking into account initial, subtle signs of malignancy and indication of feature values qualifying the nature of the changes during the description (potentially malignant or benign).

---

[1] These are suspicious changes with ambiguous configuration of feature values, without evidence of malignancy.

Suggested, by the authors of the article, method of errors' reduction is to support radiologists with specialized IT tool editor – MammoEdit. In the case of the MammoEdit, created ontology of mammography was used as a set of partial project needs for the user's interface and for database storing patients descriptions. The main project assumption of the interface was that the description of the mammogram should be done in the graphical mode, using such components as menu and button, while the text of the final description should be generated automatically. The choice of the graphical mode seemed to be obvious, because radiologists are used to images. Pictographic way of recording information enables both fast entry and immediate visualization. Over 300 pictograms illustrating the most important features of the changes in a vivid and explicit way have been created for the need of the editor. The collection of graphical objects (projects of buttons, icons, pictograms etc.) was gathered after the ontological model had been created and it reflects its content. The system of mutual dependence of entering/representing data is also based on the ontology. It let to obtain legibility and explicitness of the interface and blocked entering conflict data (exclusive). Detailed description of the editor can be found in [1].

4 The Evaluation of the Mammogram's Interpretation Supporting Tool

The basic aim of this testing was experimental verification of the hypothesis which states that: *"the mammogram's interpretation, using the editor that systemizes the description process and indicates the values of the features that are significant for the accurate interpretation of their nature to the radiologist, increases diagnostic efficiency of the radiologist"*.

Receiver Operating Characteristic analysis (ROC) has been used to assess testing. Quantitative assessment of diagnostic test performance involves taking into account the variability of the cut-off criteria in the whole variation of the parameter. ROC curves, which are created as a result of drawing the correlation of test's sensitiveness (Y axis) in Specificity function (X axis), are served for this purpose. Sensitiveness (SE) is a test ability to detect disease among sick patients, specificity (SP) is a test ability to exclude disease for healthy patients. The whole field contained under the ROC curve (AUC – *Area Under Curve*) is interpreted as a measure of test's effectiveness [3-4]. It has been decided to use *Multireader Receiver Operating Characteristic analysis* for testing MammoEdit. The mathematic model of the method takes into account typical for radiology sources of variability in image testing assessment and possible correlations among doctor's assessments within the same technique and also possible correlations among assessed techniques. Comparing the effectiveness of different image testing interpretation techniques involves carrying out an experiment, where trial testing consists of:

- Assessing techniques – imaging, interpretation or medical image processing algorithms;
- Assessing cases (the collection must include the study of pathological states and negatives);
- Radiologists who interpret testing cases using testing techniques.

These kinds of experiments, marked briefly as „technique x radiologist x case" are based on, so-called, factorial design [5]. The choice of the scale, which was used for assessing test results reflects substantive criteria of the medical problem that is being considered. [4]. Diagnostic category of the BI-RADS system matched by radiologists was chosen to represent the assessment of test results. The test which was performed included 80 mammograms that were difficult to interpret (20 negatives and 60 pathologies) The mammograms were described by three different groups of radiologists representing different levels of competence (a trainee, a specialist, an expert). They used two interpretation methods with and without MammoEdit editor. The mammograms came from a public data base DDSM[2] (*Digital Database for Screening Mammography*). Taking under consideration different origin of testing, doctors, and widely varied, uneven level of professional competence, model: random testing/random viewer was used.

The conditions of mammograms interpretation mimicked clinical environment as closely as possible. The mammograms were presented to the radiologists on the medical screens with the MammoViewer application, which function for testing had been reduced to an advanced medical browser, in a dark room. The assessment of mammograms was held during many sessions that lasted from 1 to 3 hours. The speed and time of sessions were under doctors control, the breaks or suspensions were taken on their requests. During one session the doctor performed only one type of mammogram's assessment – with or without the assisting tool. Changes were assessed on the basis of breast images in two basic projections, without any additional diagnostic projections, clinical trials, patient's medical records or previous images to compare. Before the first mammograms reading, the meaning of scales was explained and, during testing, the scales description was available at the workplace. The minimum time interval between the interpretation of the same testing with and without MammoEdit was estimated on the basis of literature. It is assumed that it should be minimum 4-5 weeks. In the experiment it ranged from 7 to 12 weeks.

The DBM method of experiments assessment MRMC ROC [6] was used to calculate indicators of diagnostics performance, while the PropRoc method [7], which provides reliable indicators estimation for the analysis of the small trial testing and discrete assessment scale, was used to estimate ROC curves indicators.

The use of MammoEdit to support mammograms interpretation has raised the average diagnostic performance (statistically significant increase, significance level less than 0.05). Detailed results are presented in Table 1.

Table 1. The result of MammoEdit editor's influence on mammograms interpretation performance with 3 groups of radiologists

lesion type	interpretation methods	AUC	SL	AC%	SE%	SP%
masses	without MammoEdit	0,610	0,016	55,0	20,0	94,2
	with MammoEdit	0,853		79,3	63,3	97,1
microcalcifications	without MammoEdit	0,677	0,728	71,3	36,1	96,1
clusters	with MammoEdit	0,730		75,9	44,4	98,0

---

[2] The material comes from four medical centers in the USA (Massachusetts General Hospital, University of South Florida, Sandia National Laboratories and Washington University School of Medicine).

The rise of indicators mentioned above was accompanied by the reduction of the deviation/variation of the standard field under the ROC curve.

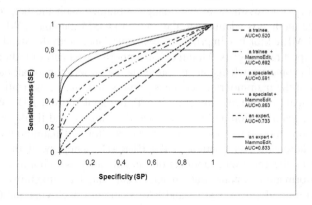

Fig. 1. MammoEdit's influence on the diagnostic effectiveness of the radiologists

The biggest increase of the diagnostic effectiveness was observed in case of a specialist radiologist, the smallest – in case of an expert radiologist. It proves bad influence of routine on the quality of mammograms assessment. The influence of MammoEdit on the variation of radiologists diagnostic opinions was assessed comparing the value of Kappa statistics achieved in the I and II testing period. The variability of opinions was assessed for the pair of radiologists and for the group of all radiologists, separately for microcalsifications, tumors and for both types of pathologies.

The use of MammoEdit for the mammograms interpretations has increased diagnostics opinions consistency for the group of radiologists for both types of changes masses and microcalcification clusters and for each of the pathologies; the biggest increase in consistency has been observed for microcalsifications clusters. MammoEdit has also influenced the increase of opinions consistency in all pairs of viewers and in all groups of changes.

5 Conclusion

The use of MammoEdit for mammograms interpretation support has raised the average diagnostic performance for the group of radiologists, other ROC analysis indicators have also increased. It brings the conclusion, that initially, in the trial testing, the hypothesis of MammoEdit usefulness as a tool for mammograms interpretation support has been confirmed. The level of diagnostic knowledge representation in the ontology of mammography has also been estimated as a satisfactory one. Thorough analysis of radiologists performance, in distinction of the types changes, has proved a difference in supporting tumors and microcalsifications interpretation effectiveness (to the detriment of tumors).

The next result of the experiment is the initial assessment of the logical correctness of the class construction method. It consisted of comparing BI-RADS category of selected testing cases in DDSM data base and the assessment of the grades of the same changes performed using the MammoEdit and automatic evaluation of the changes grade on the basis of ontology classification. The experiment has proved that BI-RADS category assessment of the changes obtained with the help of the ontology's inference module is adequate to the ones in DDSM data base. What is interesting, the use of ME editor has improved asses compliance with the standard, even for the group of expert radiologists. MammoEdit supports just a part of radiologists work, we need to gain some knowledge about mathematical descriptors of all basic types for feature values of the mammograms changes, tumors and microcalsification clusters to apply it in clinical practice and integrate it with the application CAD. We also need to connect changes detection and their verified description, and finally widen the application with the cooperative function with the ontology of mammography in order to receive an automatic verification of the BI-RADS category of the change. The complement of the integrated application functionality is: administrative features (Radiology Information System), data management (data base functions) and security, finally integration with a workstation and PACS (Picture Archiving and Communication System).

References

1. Podsiadły-Marczykowska, T., Zawiślak, R.: The role of domain ontology in the design of editor for mammography reports. Dziedzinowa ontologia mammografii w projektowaniu edytora raportów, Bazy Danych: Struktury, Algorytmy, Metody, ch. 37 (2006)
2. Miglioretti, D.L., Smith-Bindman, R., Abraham, L., Brenner, R.J., Carney, P.A., et al.: Radiologist Characteristics Associated With Interpretive Performance of Diagnostic Mammography. J. Natl. Cancer Inst. 99, 1854–1863 (2007)
3. Swets, J.A.: ROC analysis applied to the evaluation of medical imaging tests. Invest. Radiol. 14, 109–121 (1979)
4. Zhou, X.H., Obuchowski, N.A., Obuchowski, D.M.: Statistical Methods in Diagnostic Medicine. John Wiley and Sons, New York (2002)
5. Obuchowski, N.A., Beiden, S.V., Berbaum, K.S., Stephen, L., Hillis, S.L.: Multireader, Multicase Receiver Operating Characteristic Analysis: An Empirical Comparison of Five Methods. Acad. Radiol. 11, 980–995 (2004)
6. Dorfman, D.D., Berbaum, K.S., Lenth, R.V., Chen, Y.F., Donaghy, B.A.: Monte Carlo validation of a multireader method for receiver operating characteristic discrete rating data: factorial experimental design. Acad. Radiol. 5, 591–602 (1998)
7. Pesce, L.L., Metz, C.E.: Reliable and computationally efficient maximum-likelihood estimation of "proper" binormal ROC curves. Acad. Radiol. 14(7), 814–882 (2007)

Ethical Issues of Artificial Biomedical Applications

Athanasios Alexiou, Maria Psixa, and Panagiotis Vlamos

Department of Informatics, Ionian University, Plateia Tsirigoti 7, 49100 Corfu, Greece
{alexiou,vlamos}@ionio.gr, marypsiha@hotmail.com

Abstract. While the plethora of artificial biomedical applications is enriched and combined with the possibilities of artificial intelligence, bioinformatics and nanotechnology, the variability in the ideological use of such concepts is associated with bioethical issues and several legal aspects. The convergence of bioethics and computer ethics, attempts to illustrate and approach problems, occurring by the fusion of human and machine or even through the replacement of human determination by super intelligence. Several issues concerning the effects of artificial biomedical applications will be discussed, considering the upcoming post humanism period.

Keywords: Bioethics, Artificial Intelligence in Biomedicine, Bioinformatics, Nanotechnology, Post humanism.

1 Introduction

Aldo Leopold, states that 'a thing is right when it tends to preserve the integrity, stability and beauty of the biotic community while it is wrong when it tends otherwise' [1]. The converging of science and technology in several levels through their realizations is used to profit individuals, under the condition that humans serve always-high values and humanitarian ideal. The world of moral debt is, as it should be, the world of science, serving fields sensitive to the human biological identity and uniqueness. The basic characterization of an individual is to have freedom to make decisions, to have emotions and consciousness. Obviously this is not the consciousness of a super-intelligent machine or significant to the morality of Artificial Intelligence (AI). Additionally, nanotechnology is having a great impact on the fields of biology, biotechnology and medicine. This area of nanotechnology is generally referred to as nanomedicine, and sometimes widely called bio-nanotechnology [2]-[3]. Undoubtedly several issues are also related to advanced topics and applications of nanotechnology such as human immortality and democracy, artificial intelligence on nanomachines, therapeutic limitations etc.

Several authors have also argued that there is a substantial chance that super intelligence may be created within a few decades, perhaps as a result of growing hardware performance and increased ability to implement algorithms and architectures similar to those used by human brains [4]. While super intelligence is any intellect that is vastly outperforms the best human brains in practically every field, including scientific creativity, general wisdom, and social skills [5], we must be very hesitant before we convict the artificial moral way of thinking. AI can adopt our

L. Iliadis et al. (Eds.): EANN/AIAI 2011, Part II, IFIP AICT 364, pp. 297–302, 2011.
© IFIP International Federation for Information Processing 2011

culture, but it is not possible to identify and become a part of our evolution. Obviously super intelligence can produce new knowledge and solutions to hard or NP-complete problems. A smart nanomachine can recognize cancer cells using the tunnelling phenomenon. A machine learning algorithm could give right decisions in the forecast of neurogenerative diseases like epilepsy.

Of course, it is not ethical to clone humans, but is it unethical to copy the products of artificial intelligence? While scientists are not allowed to 'duplicate' themselves, artificial products can easily be distributed in millions of copies. Emotional disturbance or motives are totally absent from the definitions of artificial intellects. Therefore it is important to control artificial process, but is it ethical any effort for the humanization of super-intelligence? In what way can we criticize or punished a serious error of an intelligent robot during a surgical process? There is no consciousness' risk on prognosis or medical treatment using AI, while the seriousness of illness or death is emotional absent in these cases.

2   Emerging Issues in Biomedicine

While Bioinformatics deal with biotechnology, computer technology and also life sciences, the ethics emerging from this scientific field has to be an amalgam of the two major strands of applied ethics: computer ethic and bioethics [6]. On the other hand the parallel innovating structure of the so called 'convergent technologies', referring to the NBIC tools and including nanoscience and nanotechnology, biotechnology, biomedicine and genetic engineering, information technology and cognitive science, seems to remove any barrier in scientific and technological achievement [7]. The nanodevices which can repair cells, promise great improvements in longevity and quality of life, involving radical modifications of the human genome and leading to the old but diachronic issue of human immorality [8].

Biology itself provides a fully worked out example of a functioning nanotechnology, with its molecular machines and precise, molecule by molecule chemical syntheses. What is a bacterium if not a self-replicating, robot? Already, there are a few nanomedicine products on the market with numerous other potential applications under consideration and development [2],[3],[9]. In vivo disease detection and monitoring using micro-electromechanical systems (MEMS) also appears to be making applications for creating "lab-on-a-chip" devices to detect cells, fluids or even molecules that predict or indicate disease even more probable [10]. The use of MEMS chips and other devices for the purpose of diagnosing or monitoring healthy or diseased conditions is likely to raise grave questions about health information systems, privacy and confidentiality in our healthcare system [11]. The manufacturing of devices able to provide real time processing of several blood levels, leads to a strong cost benefit for people with chronic diseases or organ transplant. Artificial diagnosis could possible prevent illnesses or the impact of a disease and reduce the cost of drug discovery and development using nanodevices, unlike the traditional medicine.

Super intelligence therapeutics can be expected within the next decade for the specific delivery of drugs and genetic agents. This will provide a degree of specificity in the action of therapeutics that could prevent side-effects and improve efficacy [12].

Long-term therapeutic developments would include nanosystems that totally replace, repair, and regenerate diseased tissue. This could involve the correction of developmental defects or the resolution of problems from disease and trauma [13].

Let's assume a Case-Based Reasoning (CBR) system for decision making in reproductive technologies when having children. The system will have the ability to identify the best sperm-eggs, offering potential on individual choice and human rights. Therefore, we have to distinguish the ethics of the reproductive technologies as a human's risk, from the ethics concerning the scientific progress and the AI's progress. The use of AI techniques is totally separated from religious sentiments, human dignity or essential emotional states. But what about the pseudo-philosophical question for the upcoming era of the posthumans?

3   Substituting Human Consciousness with Artificial Intelligence

Unlike mankind, AI applications can use and manipulate the storage knowledge through scientific or social networks in a more efficient way. Therefore one of the best ways to ensure that these super intelligent creations will have a beneficial impact on the world is to endow it with philanthropic values and friendliness [14]. Additionally, a super intelligence could give us indefinite lifespan, either by stopping and reversing the aging process through the use of nanomedicine [15]. Especially in the field of medical care and biomedicine, the convergence of nanotechnology and biotechnology with cognitive science began to produce new materials for improving human performance. Researchers looked toward biology as a guide to assemble nanostructures into functional devices, where only a small amount of subunits are required to produce a rich and diverse group of functional systems [16]. Bioinspired materials may create difficulties for biologic systems and ecosystems, as their small size allows them to be easily internalized in organisms. These materials can mimic biologic molecules and disrupt their function although toxic effects on cells and animals had been recognized [17].

In most of the cases, the relation of applied artificial intelligence with several environmental and social threats, where humans seems to 'play the God' with natural processes, cause questions of social and environmental nature to arise and wake up fears of the past about who patents and controls this new technology [18]. How immoral and harmful for the human freedom can be the effort to force biology to do a better job than nature has done?

The use of certain kinds of nanomaterials, nanomedicines or nanodevices also raises fundamental questions about human enhancement and human nature, about what are living and non-living and the definition of normal and human entity, against the possibility of post humanism. Nevertheless the extreme possibility of nanomachines going out of control by using their power of reconstruction and self-replication might more likely to happen in terms of a terrorist attack, despite than a machinery revolution. Of course these scenarios concerning the unethical cases of advanced bio-terroring weapons are totally hypothetical against the applications on artificial biomedicine, genetics and human enhancement technologies.

4 Ethical and Legal Aspects

The prospect of post humanity is feared for at least two reasons. One is that the state of being post human might in itself be degrading, so that by becoming post human we might be harming ourselves. Another is that post humans might pose a threat to ordinary humans [19]. While the beneficial of technological progress seems to be subjective, homogenization, mediocrity, pacification, drug-induced contentment, debasement of taste, souls without loves and longings – these are the inevitable results of making the essence of human nature the last project of technical mastery [20].

The new species, or post human, will likely view the old "normal" humans as inferior, even savages, and fit for slavery or slaughter. The normal's, on the other hand, may see the post humans as a threat and if they can, may engage in a preemptive strike by killing the post humans before they themselves are killed or enslaved by them. It is ultimately this predictable potential for genocide that makes species-altering experiments potential weapons of mass destruction, and makes the unaccountable genetic engineer a potential bioterrorist [21].

Is it possible to define the degree of influence of human conscience, dignity, rights and fundamental freedom by merging human and machine? Is it possible to achieve and control confidentiality and privacy on genetic data, without of course increasing tremendously the high quality treatment cost? Who can develop and participate in such scientific experiments, who will be the subject of the experiment and how can we make provision for individuals with special needs? It is important to note, that such social and ethical issues are not specific to nanotechnology alone; any modern technology is the product of a complex interplay between its designers and the larger society in which it develops [22]. The development of nanotechnology is moving very quickly, and without any clear public guidance or leadership as to the moral tenor of its purposes, directions and outcomes; where nanotechnology is leading and what impact it might have on humanity is anyone's guess [23]. What appears to be missing at the present time is a clearly articulated prognosis of the potential global social benefits and harms that may develop from further scientific and technological advances in all of these areas [24].

According to Humanism, human beings have the right and responsibility to give meaning and shape to their own lives, building a more humane society through an ethic based on human and other natural values in the spirit of reason and free inquiry through human capabilities. The moral person guided from his evolving social behaviour, can easily comprehend and be committed to laws and principles that a scientific field, such as AI or Nanoscience set as a precondition, in order to improve the structural elements of man's biological existence. What is mainly the problem therefore? Are there any laws and ethical aspects for the consequences of artificial implants in humans, or mainly the consequences from their non-application?

Super Intelligence should be comprehensible to the public, including activities that benefit society and environment, guided by the principles of free participation to all decision-making processes; AI techniques in Biomedicine, should respect the right of access to information, having the best scientific standards and encouraging creativity, flexibility and innovation with accountability to all the possible social, environmental and human health impacts.

Another aspect of the applied research in biomedicine is the economical cost. The cost of care and also the expectations of citizens are increasing rapidly. Even if research is directed towards solutions economically accessible, the more effective solutions can possibly lead to the unequal distribution of medical care and prompting assurances companies to stop their attendance in the social health system [25].

It is obvious that these ethical codes differ from AI applications in Biomedicine to traditional medical applications due to the influence of other ethical frameworks and perspectives on their basic research and development. Novel Biomedicine methods will allow us to understand down to the atomic and single-cell level how our bodies are performing at any given moment. For some, this information could be helpful, empowering or enlightening and may enhance human health. For others, it is likely that such information could result in fear, anxiety and other mental health issues. Nevertheless, the development of novel therapies based on the convergent of AI, Bioinformatics and Nanotechnology will arise several ethics principles about human rights which have to be followed: moral, political and religious issues but also individual privacy, human dignity, justice, and fair access to the knowledge of the diseases but further more to any possible beneficial therapy. Therefore, it is ethically essential that researchers inform potential research subjects in clinical trials of all details pertaining to the study [26].

5 Conclusion

The new realizations of science put the modern person in front of dilemmas, as far as the moral and the legal, the socially acceptable and the pioneering dangerous, the naturally valorized and the technologically innovative [27].

The ethical considerations of innovating technologies have to be announced and explained to the social target groups. Therefore, it is ethically desirable to determine whether new artificial types of prognosis or treatment will be more effective and safe for humans when compared to conventional ones. Undoubtedly artificial bio-technologies will deliver a variety of improvements or a technological and healthcare revolution. The main problem is to study at early stage any social side effects.

References

1. Leopold, A.: A Sand County Almanac: And Sketches here and there. Oxford University Press, New York (1987)
2. Vo-Dinh, T.: Nanotechnology in Biology and Medicine: Methods, Devices, and Applications. CRC Press, Boca Raton (2007)
3. Niemeyer, C.M., Mirkin, C.A.: Nanobiotechnology: Concepts, Applications and Perspectives. Wiley-VCH, Weinheim (2004)
4. Bostrom, N.: Existential Risks: Analyzing Human Extinction Scenarios and Related Hazards. Journal of Evolution and Technology 9 (2002)
5. Bostrom, N.: How Long Before Superintelligence? International Journal of Futures Studies 2 (1998)
6. Hongladarom, S.: Ethics of Bioinformatics: A convergence between Bioethics and Computer Ethics. Asian Biotechnology and Development Review 9(1), 37–44 (2006)

7. Roco, M.C., Bainbridge, W.S.: Converging technologies for improving human performance. Journal of Nanoparticle Research 4, 281–295 (2002)
8. Drexler, E.: Engines of Creation. Bantam, New York (1986)
9. Kubik, T., et al.: Nanotechnology on duty in medical applications. Current Pharmaceutical Biotechnology 6, 17–33 (2005)
10. Craighead, H.: Future lab-on-a-chip technologies for interrogating individual molecules. Nature 442, 387–393 (2006)
11. Allhoff, F., Lin, P.: What's so special about nanotechnology and nanoethics? International Journal of Applied Philosophy 20(2), 179–190 (2006)
12. Patri, A., Majoros, I., Baker, J.: Dendritic polymer macromolecular carriers for drug delivery; using nanotechnology for drug development and delivery. Current Opinion in Chemical Biology 6(4), 466–471 (2003)
13. Bainbridge, W.S., Roco, M.C.: Managing Nano-Bio-Info-Cogno Innovations, pp. 119–132. Converging Technologies in Society, Springer (2006)
14. Kurzweil, R.: The Age of Spiritual Machines: When Computers Exceed Human Inteligence. Viking, New York (1999)
15. Moravec, H.: Robot: Mere Machine to Transcendent Mind. Oxford University Press, New York (1999)
16. Mardyani, S., Jiang, W., Lai, J., Zhang, J., Chan, C.W.: Biological Nanostructures and Applications of Nanostructures in Biology. Springer, US (2004)
17. Oberdörster, G., et al.: Nanotoxicity: An emerging discipline evolving from studies of ultrafine particles. Environmental Health Perspective 113, 823–839 (2005)
18. Preston, C.J.: The Promise and Threat of Nanotechnology. Can Environmental Ethics Guide US? International Journal for Philosophy of Chemistry 11(1), 19–44 (2005)
19. Bostrom, N.: In Defence of Posthuman Dignity. Bioethics 19(3), 202–214 (2005)
20. Kass, L.: Life, Liberty, and Defense of Dignity: The Challenge for Bioethics, p. 48. Encounter Books, San Francisco (2002)
21. Annas, G., Andrews, L., Isasi, R.: Protecting the Endangered Human: Toward an International Treaty Prohibiting Cloning and Inheritable Alterations. American Journal of Law and Medicine 28(2&3), 162 (2002)
22. Pool, R.: How society shapes technology. In: Teich, A.H. (ed.) Technology and the Future, 9th edn., pp. 13–22. Wadsworth/Thomson Learning, Belmont, CA (2003)
23. Berne, R.W.: Towards the conscientious development of ethical nanotechnology. Science and Engineering Ethics 10, 627–638 (2004)
24. Sweeney, E.A.: Social and Ethical Dimensions of Nanoscale Research. Science and Engineering Ethics 12(3), 435–464 (2006)
25. Solodoukhina, D.: Bioethics and legal aspects of potential health and environmental risks of nanotechnology. In: Nanotechnology-Toxicological Issues and Environmental Safety, pp. 167–184. Springer, Heidelberg (2007)
26. Donaldson, K.: Resolving the nanoparticles paradox. Nanomedicine 1, 229–234 (2006)
27. Alexiou, A., Vlamos, P.: Ethics at the Crossroads of Bioinformatics and Nanotechnology. In: 8th International Conference of Computer Ethics and Philosophical Enquiry (2009)

ECOTRUCK: An Agent System for Paper Recycling

Nikolaos Bezirgiannis[1] and Ilias Sakellariou[2]

[1] Dept. of Inf. and Comp. Sciences, Utrecht University,
P.O. Box 80.089, 3508 TB Utrecht, The Netherlands
[2] Dept. of Applied Informatics, University of Macedonia,
156 Egnatia Str. 54124 Thessaloniki, Greece
n.bezirgiannis@students.uu.nl iliass@uom.gr
http://eos.uom.gr/~iliass

Abstract. Recycling has been gaining ground, thanks to the recent progress made in the related technology. However, a limiting factor to its wide adoption, is the lack of modern tools for managing the collection of recyclable resources. In this paper, we present EcoTruck, a management system for the collection of recyclable paper products. EcoTruck is modelled as a multi-agent system and its implementation employs Erlang, a distribution-oriented declarative language. The system aims to automate communication and cooperation of parties involved in the collection process, as well as optimise vehicle routing. The latter have the effect of minimising vehicle travel distances and subsequently lowering transportation costs. By speeding up the overall recycling process, the system could increase the service throughput, eventually introducing recycling methods to a larger audience.

Keywords: Agent Systems, Contract-Net, Functional Logic Programming, Erlang.

1 Introduction

The term *Recycling* refers to the reinsertion of used materials to the production cycle, the initial phase of which is collection of the recycled materials from the consumers. In a sense, this is a transportation problem i.e. collecting from various end-point users items in a highly dynamic manner. Due to this dynamic nature, we consider it to be an excellent area of application for multi-agent technology [1]. Although the latter has been around for many years, still presents a number of excellent opportunities for its application in new domains. In the light of new technological advances in the areas of telecommunications and portable computing devices, such applications can really "go" mainstream.

Paper recycling has gain significant attention due to a) the large quantities of paper used in everyday tasks, (offices, packaging, etc.) and b) possible reductions in energy consumption and landfill waste. Currently, municipal authorities employ a rather outdated procedure for the collection of recyclable large packaging cartons and other paper waste, that is based on telephone communication

L. Iliadis et al. (Eds.): EANN/AIAI 2011, Part II, IFIP AICT 364, pp. 303–312, 2011.

and manual truck assignment for pick up. The drawbacks of such a procedure are rather obvious: delays in the collection process, unoptimised use of resources (trucks, etc.), which can potentially lead to a low acceptance of recycling practise and thus to failure of the whole process.

EcoTruck aims to address the above issues, replacing the current "manual" collection process by a multi-agent system. This offers a number of advantages, such as distributed coordination of collection services, speed, robustness, absence of a central point of control, etc. The system is implemented entirely in Erlang, a concurrency-oriented declarative language. While Erlang is not widely adopted by the agent community, it is quite heavily used in industry to develop mission critical server products and soft real-time systems. We found that the language provides the necessary distributed infrastructure and carries enough expressiveness to allow the development of our multi-agent system.

Thus, the aim of this paper is twofold: firstly, the paper supports that the application of multi-agent technology can improve the recyclable paper collection process; secondly, we argue that the development of such multi-agent systems can be easily done in Erlang and prove the latter by presenting the implementation of our system in the language.

The rest of the paper is organised as follows. We introduce in more detail the problem we address in section 2. Section 3 presents the system's architecture and cooperation protocol, i.e. the Contract-net protocol. Section 4 describes the implementation of the system in Erlang while section 5 presents related work on the problem. Finally, section 6 concludes the paper and presents potential extensions of the current system.

2   Collecting Paper for Recycling

While recycling technology has made significant progress, the collection of the recyclable materials still relies on old-fashion practises. Currently, municipal offices responsible for the paper collection and management operations act as mediators between companies (i.e. large companies, shopping centres, supermarkets) and dedicated truck vehicles for transporting paper from the former to the materials recovery facilities (MRFs). Typically, offices rely on telephone communication: a) the company contacts the office and places a request, b) the municipal office checks the current truck schedules and finds an appropriate truck for the task c) the truck is immediately informed of the new task and adds it to its current schedule. Of course the procedure described is an ideal "manual" operation; usually, offices collect requests and construct the next-day trucks' schedule.

There are quite a few problems in this procedure, mainly due to the centralised approach followed, that imposes a serious bottleneck to the system. One can easily see that:

- Efficient resource allocation (trucks to tasks) cannot be easily achieved, since the municipal office is unaware of the current status or location of the trucks.
- Communication is slow, resulting to a low number of handled requests.

- Trucks do not coordinate their actions and thus resources are underexploited in most of the cases.
- Finally, customers have limited information on the progress of their request and this has an impact on the adoption of recycling by a wider community of professionals.

The approach proposed in this paper attempts to resolve most of the above problems. Imposing a distributed cooperation model between interested parties alleviates the need for a central coordination office, allows trucks to form their daily schedule dynamically based on their current state (e.g. load, distance from client, etc.), and increases the number of requests serviced. Finally, since the system's status is available, clients have access to information regarding the progress of their request.

3   EcoTruck Agents

A natural modelling of the problem is to map each interested party in the process to an agent. Thus, EcoTruck is conceived as a multi-agent system that consists of two types of agents: *Customer Agents*, each representing a company in the system and *Truck Agents*, each representing a truck, responsible to manage the latter's everyday schedule.

The *Customer agent* has the goal to satisfy "its" user's recycling request by allocating the latter to an available truck. Thus, a customer agent after receiving input, i.e. the paper amount the company wishes to recycle, initiates a cooperation protocol to find the best possible truck to handle the request. The best truck in this case is the one that can service the request in the *shortest attainable amount of time*. Furthermore, the agent is responsible for monitoring the entire progress and provide a user friendly display, as well as take action in the light of possible failures.

A *Truck agent* is modelled as a software agent, mounted on truck vehicles that operates as an assistant to the driver. The agent has the goal of collecting as much recyclable material as possible, thus tries to maximise the number of Customer agent requests it can service. Each incoming request is evaluated based on the current available capacity of the truck, its geographical position and the tasks (paper collections) it has already committed to, i.e. its plan. Once a new collection task is assigned to the agent, it is added to the truck's plan. Each Truck agent maintains its own plan, that is a queue of jobs it intends to execute. Additionally, the truck agent proactively adds a "paper unloading" job to the plan upon detecting current low capacity. Finally, by processing real-time routing information, the agent "guides" the driver inside the city, much like a GPS navigation system, reducing travel distances and response time.

The operation of the system is *cooperative*, in the sense that all involved parties work together in order to fulfil the overall goal of increasing the amount of recyclable material collected during a working day. Consequently, the interaction protocol that was selected for EcoTruck system was the Contract-Net protocol, as discussed in the subsection that follows.

3.1   Agent Cooperation

The Contract-Net protocol (CNP) [2,3], is probably the most well-known and mostly implemented task sharing protocol in distributed problem solving. It offers an elegant yet simple way of task sharing within a group of agents and we considered it to be suitable for the EcoTruck system. In the latter, Customer agents play the role of *managers* while Truck agents are *contractors* according to the protocol's terminology. Thus, the overall process is the following:

1. A Customer agent initiates a CNP protocol by announcing the task to truck agents. This "call for proposals" (CFP) contains information regarding the geographical location of company and the paper quantity for collection. The latter plays the role of the *eligibility criterion* in the protocol.
2. Trucks evaluate the CFP and decide on their eligibility for the task. If they they can carry the paper load, they answer with a *bid* that contains the *estimated time of servicing* (ETS) the request; the latter is the sole criterion on which Customer agents decide on who to assign the task. Naturally, Trucks can also refuse a request, if they cannot handle it, are currently full or for any other reason, by issuing an appropriate message.
3. The Customer agent receives bids, processes them, decides on the best truck to assign the task, and broadcasts the corresponding messages (accept-proposal/refuse-proposal).
4. The *winner* truck adds the task to its current plan. Upon arriving at the designated location, it will collect the paper, mark the job as finished and inform the corresponding Customer agent.

Obviously [2], there is always the risk that the Customer agent receives no bids. This can occur when the paper quantity in the CFP exceeds either a) the current capacity of any truck in the system, or b) the maximum capacity of even the largest truck. It was decided that both these cases, although different in nature, are going to be treated uniformly: the Customer Agent decides to decompose the original request to two smaller ones of equal size and initiate the protocol on each one once more. While the decision is obvious in the case that the CFP quantity exceeds the maximum capacity of all trucks, it requires some explanation in case (a). Since for trucks to regain their capacity they need to unload their cargo in Material Recovery Facility (MRF), a rather time consuming process, it was considered better in terms of service time reduction to allow the decomposition of the task, instead of the Customer agent waiting for some truck with the appropriate capacity to appear in the community.

Another issue, that is also present in the original CNP specification, regards whether to allow contractors to participate in multiple call for proposals [4], [5]. Since, in the EcoTruck system such cases would naturally occur (multiple companies exist that might have simultaneous requests), it was decided that when a "winner" truck is awarded a contract (accept-proposal) then it will check it against its current schedule (note that the latter can be different than the one the agent had during the bidding phase). If there are no differences in the ETS then it simply adds it to the schedule; if there are then it sends back a *failure* message to the customer agent and the latter re-initiates a CNP protocol once more. Although, this approach is not very sophisticated like the one described in [4], and certainly leaves plenty of room for improvements, it was considered to be adequate for the specific case.

4   Implementing EcoTruck

4.1   Platform of Choice

Erlang [6], is a concurrent declarative language aiming at the development of large distributed systems with soft real-time constraints [7]. The language offers dynamic typing, a single assignment variable binding scheme and automatic garbage collection so as to facilitate program development. Support for concurrency is achieved through process based programming and asynchronous message passing communication. Erlang follows the now popular "execution through virtual machine" model and a distributed system is composed of a number of different Erlang nodes running in one or more computers. Each node being a unique instance of a virtual machine.

A useful extension to the standard Erlang language is the Open Telecom Platform (OTP), that is a set of libraries and tools to facilitate the design and development of large distributed systems. OTP includes a set of the most common design patterns (*OTP behaviours*) that occur in a distributed setting.

EcoTruck employs *Server behaviour* processes, for asynchronous client-server communication, *FSM behaviour* processes, for interactions that require state transitions, and *Supervisor behaviour* processes, that monitor other OTP processes and restart them if they crash. These processes are nicely packaged into three distinct applications, using another OTP behaviour, called *Application behaviour*. Thus, our agents in Erlang consist of a number of processes, each being an instance of a specific OTP behaviour. This approach greatly facilitated the design and implementation of the EcoTruck system, since it provides all the tools necessary to built the agents.

4.2   Directory Facilitator

Since our implementation did not rely on any agent specific platforms, in order to provide directory services for the agent community, a *Directory Facilitator* (DF) was introduced in the system. This entity stores the roles and addresses of every active agent.

EcoTruck agents subscribe to the system by passing relevant information to the DF; the DF saves this information in an Erlang Term Storage (ETS) table, i.e. a fast in-memory data-store, part of the standard Erlang distribution. The DF is implemented as a server OTP behaviour, that monitors all subscribed agent processes and if any of them becomes unresponsive, due to network failure or abnormal termination, automatically unsubscribes them from the system. The server process itself is monitored by a supervisor behaviour.

4.3 Agent Structure

A set of different Erlang processes constitute the customer agent. The *customer Server*, an OTP Server instance monitored by a *customer Supervisor*, is the master process of the agent, that stores agent's preferences and controls the user interface. When the user places a request, the server spawns a new *Manager process* and passes the relevant information.

The *customer Manager* process will acquire the list of truck agents from the DF and initiate a CNP interaction as described above. The latter is implemented as an OTP Finite State Machine (FSM) behaviour, for the reason that CNP requires successive steps of interaction. These communication steps are essentially the transition states of the Finite State Machine. Finally, the Manager will link itself with the "winner" truck and specifically with its Contractor process. This link can be perceived as a "bidirectional monitor"; if one of the linked processes exits abnormally, the other process becomes aware of it.This ensures that if a winner Contractor crashes, the Manager will notice it and start a new CNP interaction. When the requested job is completed, the Manager reaches its final state and gracefully dies.

A Manager process also gracefully exits in the case of a request decomposition: it spawns two new Manager processes, passing to each one a request divided in two, and terminates its execution. Compared to the customer Server which lives as long as the application is running, the Manager's lifetime span covers only the period of time from request announcement to collection. The described behaviours together with a GUI are packaged into an OTP application.

The truck agent is also composed of a number of supervised processes; the *truck Server*, that is responsible for maintaining the plan (list of jobs) of the agent. For every incoming CFP request, the Server will spawn a new *Contractor process* (FSM behaviour) and pass to it the request. An eligible Contractor places a bid and in the case that it is the "winning" bidder, it will instruct the truck Server to schedule the job for execution by appending it to the plan. Upon completing the job, the Contractor process will notify the customer Manager process and terminate.

A conflict that commonly arises in a multi-CFP setting is when two or more Contractors try to place a recycling job in the exact same place in the plan queue. EcoTruck resolves this, by setting the truck Server to accept only one of the Contractors and its job, while rejecting the rest. The rejected Contractors will signal a failure to the corresponding Manager, who will in turn restart a CNP interaction.

Finally, the *truck driver process* is an extra Server, which simulates the motion of the truck vehicle inside the city roads. Of course, this is not necessary in a production environment. These three behaviours constitute a truck application.

It should be noted that since Erlang follows the virtual machine model, the applications developed can execute both in desktop and mobile environments. This is particularly interesting in the case of the truck application, since it allows easily executing it on a mobile device mounted on truck vehicles. Figure 1 presents the agent processes and message exchange of EcoTruck agents.

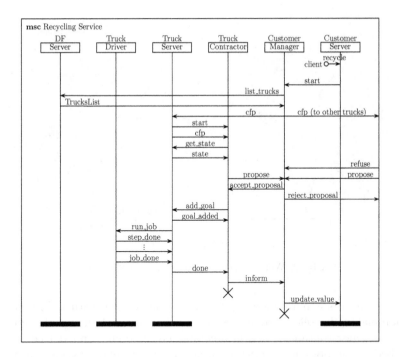

Fig. 1. Agent processes and message exchange in EcoTruck

4.4 Google Maps Integration

The Truck application uses Google Maps Directions Service to provide the truck driver with routing information as well as live traffic data where applicable. Additionally, both Customer and Truck applications have a GMAPS based web monitoring tool. The service responses are parsed with the *xmerl* Erlang parsing library. Each application relies on Google Maps Javascript API to display to the user a live view of the system's state. The Google Maps JSON encoded content along with any HTML files are displayed to the user by Misultin, a lightweight Erlang web server. The reason for not having, instead, a native GUI, is that a web interface can be more portable, thus the EcoTruck software will run on any hardware an Erlang VM exists for. Figure 2 shows the GUI of EcoTruck.

5 Related Work

The MAS approach has been applied to recycling/environmental processes in various contexts. For instance, the EU funded project E-MULT [8] aimed at developing a dynamic network of SMEs based on multi-agent solutions for the recycling of end-of-life vehicles, a process that is reported to be particularly complex. In [9] a simulation of an animal waste management system between producers and potential consumers (farms) is presented that allows to test different management scenarios.

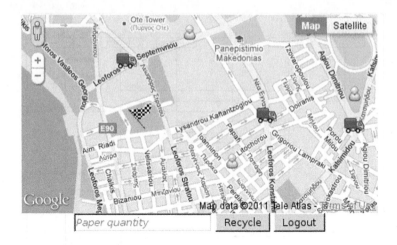

Fig. 2. Web Interface of EcoTruck Application

In [10] authors employ multi-agent simulation coupled with a GIS to assist decision making in solid waste collection in urban areas. Although the approach concerned truck routing, it was in a completely different context than the one described in the present paper: the former aimed at validating alternative scenarios by multi-agent modelling in the waste collection process, while EcoTruck aims at dynamically creating truck paths based on real time user requests.

The problem of efficiently collecting recyclable material from customers is in essence a dynamic transportation problem, although simpler in the sense that the destination location is fixed. A number of approaches in the literature have attacked the same problem, such as [5], where an extended contract net protocol (ECNP) is used to solve a transportation logistics problem. The ECNP protocol introduces new steps in the original contract net (temporal grant/reject and definitive grant/reject) and allows to dynamically decompose large demands (contracts). In [11] authors propose a new protocol the "provisional agreement protocol", where agents under the protocol are committed to bids sent only when they are (provisionally) granted the award. Thus, agents are allowed to participate simultaneously in multiple task announcements. The PAP protocol has been applied together with open distributed planning to the simulation of a real logistics data of a European company. Since the problem EcoTruck is dealing with is less complex than the dynamic transportation problem, EcoTruck follows a simpler approach: it allows agents to place multiple bids and allows multiple task announcements for an unserviced request if the bidder has already committed to another "contract" by the time it receives the award.

Finally, the Erlang language has been used to implement software agents with quite promising results. In [12] authors proposed the agent platform eXAT and support that it offers a number of benefits compared to the well-known JADE [13] platform, such as robustness and fault tolerance, more natural integration of the necessary components to implement such systems (FSM support, production rules components, etc). Although eXAT was not used in the implementation of the EcoTruck agents, we tend to agree with the arguments stated in [14] for the suitability of the Erlang language compared to JAVA based platforms.

6   Conclusions and Future Work

EcoTruck is a multi-agent system for the management of recyclable paper collection process. We believe that the specific approach can greatly facilitate and optimise the process, thus allow its wider adoption by the parties involved.

As discussed, the system's implementation is based on concurrency and distribution mechanisms of the Erlang language. We strongly believe that robustness and fault-tolerance are important qualities that a multi-agent system should meet. Although, the Erlang language is not a MAS platform it appears to have the necessary features to facilitate simple and elegant implementations of multi-agent applications.

There are quite a few features that could be incorporated in the current system among which the most interesting ones include:

- *Dynamic re-planning and scheduling.* In the present system, each truck agent maintains its own plan that has a static ordering of jobs, to which new jobs are inserted in a FIFO manner. A more intelligent planning process would include more dynamic features, such as prioritisation of jobs based on a number of criteria such as proximity to the current position and estimated time of arrival, and could help minimise the total travel path of the truck.
- *Smarter Truck Positioning.* Based on past data, truck agents can identify geographic areas where most requests appear in, place themselves closer to those areas and consequently increase system performance and success rate.
- *Better Decomposition of CFP's.* A more informed manner of splitting large requests could involve taking advantage of information attached in "refuse" messages of the CNP, and decomposing them in smaller ones of appropriate size, based on the current truck availability. Thus, the overall system's performance would increase, since fewer agent interactions would occur.

There are also a few improvements that could be done on the implementation level. For instance, the Erlang Term Storage (ETS), can be replaced by a Mnesia database, allowing exploitation of the fault-tolerance and distribution advantages of the latter. This change would allow to have multiple replicated instances of the DF database, and thus achieve robustness through redundancy.

Deployment of the application in a real-world environment, would lead to fine-tuning the system and examine possible patterns and procedures emerging in real-life situations. Since the system is based on a extensively tested, industrial strength platform (Erlang), we believe that the transition to a full-fledged real-world application can be accomplished with relative ease.

Finally, we should note that an earlier version of EcoTruck with a different structure and implementation by Nikolaos Bezirgiannis and Katerina Sidiropoulou, won the 1st prize on the Panhellenic Student Contest of Innovative Software (Xirafia.gr).

References

1. Jennings, N., Sycara, K., Wooldridge, M.: A roadmap of agent research and development. Journal of Autonomous Agents and Multi-Agent Systems 1, 275–306 (1998)
2. Smith, R.: The contract net protocol: High-level communication and control in a distributed problem solver. IEEE Transactions on Computers C-29, 1104–1113 (1980)
3. Smith, R.G., Davis, R.: Frameworks for cooperation in distributed problem solving. IEEE Transactions on Systems, Man, and Cybernetics 11, 61–70 (1981)
4. Aknine, S., Pinson, S., Shakun, M.F.: An extended multi-agent negotiation protocol. Autonomous Agents and Multi-Agent Systems 8, 5–45 (2004)
5. Fischer, K., Mller, J.P., Pischel, M.: Cooperative transportation scheduling: an application domain for dai. Journal of Applied Artificial Intelligence 10, 1–33 (1995)
6. Armstrong, J.: Programming Erlang: Software for a Concurrent World. Pragmatic Bookshelf (2007)
7. Armstrong, J.: The development of Erlang. In: Proceedings of the Second ACM SIGPLAN International Conference on Functional Programming, ICFP 1997, pp. 196–203. ACM, New York (1997)
8. Kovacs, G., Haidegger, G.: Agent-based solutions to support car recycling. In: 2006 IEEE International Conference on Mechatronics, pp. 282–287 (2006)
9. Courdier, R., Guerrin, F., Andriamasinoro, F., Paillat, J.M.: Agent-based simulation of complex systems: Application to collective management of animal wastes. Journal of Artificial Societies and Social Simulation 5 (2002)
10. Karadimas, N.V., Rigopoulos, G., Bardis, N.: Coupling multiagent simulation and GIS - an application in waste management. WSEAS Transactions on Systems 5, 2367–2371 (2006)
11. Perugini, D., Lambert, D., Sterling, L., Pearce, A.: Provisional agreement protocol for global transportation scheduling. In: Calisti, M., Walliser, M., Brantschen, S., Herbstritt, M., Klgl, F., Bazzan, A., Ossowski, S. (eds.) Applications of Agent Technology in Traffic and Transportation. Whitestein Series in Software Agent Technologies and Autonomic Computing, pp. 17–32. Birkhauser, Basel (2005)
12. Stefano, A.D., Santoro, C.: eXAT: An experimental tool for programming multi-agent systems in Erlang. In: IN AI*IA/TABOO Joint Workshop On Objects And Agents (WOA 2003), Villasimius (2003)
13. Di Stefano, A., Santoro, C.: Designing collaborative agents with eXAT. In: 13th IEEE International Workshops on Enabling Technologies: Infrastructure for Collaborative Enterprises, WET ICE 2004, pp. 15–20 (2004)
14. Varela, C., Abalde, C., Castro, L., Gulías, J.: On modelling agent systems with Erlang. In: Proceedings of the 2004 ACM SIGPLAN Workshop on Erlang, ERLANG 2004, pp. 65–70. ACM, New York (2004)

Prediction of CO and NO$_x$ Levels in Mexico City Using Associative Models

Amadeo Argüelles, Cornelio Yáñez, Itzamá López, and Oscar Camacho

Centro de Investigación en Computación, Instituto Politécnico Nacional.
Av. Juan de Dios Bátiz s/n casi esq. Miguel Othón de Mendizábal,
Unidad Profesional "Adolfo López Mateos", Edificio CIC.
Col. Nueva Industrial Vallejo, C. P. 07738, México, D.F. México
Tel.: (+52) 5557296000-56593;
Fax: (+52) 5557296000-56607
{jamadeo,cyanez,oscarc}@cic.ipn.mx, ilopezyb05@ipn.mx

Abstract. Artificial Intelligence has been present since more than two decades ago, in the treatment of data concerning the protection of the environment; in particular, various groups of researchers have used genetic algorithms and artificial neural networks in the analysis of data related to the atmospheric sciences and the environment. However, in this kind of applications has been conspicuously absent from the associative models, by virtue of which the classic associative techniques exhibit very low yields. This article presents the results of applying Alpha-Beta associative models in the analysis and prediction of the levels of Carbon Monoxide (CO) and Nitrogen Oxides (NO$_x$) in Mexico City.

Keywords: Associative memories, pollution prediction, atmospheric monitoring.

1 Introduction

In recent years, the care and protection of the environment have become priorities of the majority of the world's governments [1-4] and actively through non-governmental organizations and civil society [5, 6]. The length and breadth of the globe there are specialized agencies on the recording of data corresponding to various environmental variables, whose study and analysis is useful in many cases, in the decision-making related to the preservation of the environment in the local and global. During the 1990s of the 20th century, was established the importance of the paradigm of artificial intelligence, as a valuable assistant, in the tasks of analysis of data related to the atmospheric sciences and the environment [7]. It is noticeable the use of artificial neural networks in the assessment of ecosystems, in the regression of functions of high non-linearity and the prediction of values associated with the variables inherent to the environment [8-11]. Neural networks have evolved over time, and in the year 2002 were created, in the Center for Research in Computing of the National Polytechnic Institute of Mexico, the Alpha-Beta associative models [12, 13] whose efficiency has been shown through different applications in actual databases in different areas of human knowledge [14-40]. In this article, concepts and

L. Iliadis et al. (Eds.): EANN/AIAI 2011, Part II, IFIP AICT 364, pp. 313–322, 2011.

experimental results obtained by the members of the alpha-beta research group [42] are shown, when applying Alpha-Beta associative models in both CO and NO_x levels included in the databases of the atmospheric monitoring system used in Mexico City (SIMAT) [41]. The rest of the article is organized as follows: sections 2 and 3 describe concisely the SIMAT and the alpha-beta associative models, respectively. Section 4 contains the main proposal of this work, and in section 5 we discuss the experiments and results obtained.

2 SIMAT

Atmospheric Monitoring System of Mexico City (SIMAT, Sistema de Monitoreo ATmosférico) [41] was used to develop this section. Their principal purpose is the measurement of pollutants and meteorological parameters to provide information for the government's decision making related with environment conditions. It is composed of subsystems that capture the information about several pollutants presented in local environment. Below are mentioned the different parts of the SIMAT:

RAMA (Red Automática de Monitoreo Atmosférico, automatic atmospheric monitoring network) makes continuous and permanent ozone measurements (O_3), sulfur dioxide (SO_2), nitrogen oxides (NO_x), carbon monoxide (CO), particulate matter smaller than 10 microns (PM10), particles less than 2.5 micrometers (PM2.5) and hydrogen sulphide (H_2S). All the data is taken hourly from January 1986 to present.

REDMA (Red Manual de Monitoreo Atmosférico, air quality monitoring network) monitors particles suspended and determines the concentration of some elements and components contained in the air. The structure of the database is outlined in table 1, with hourly data taken since January 1986.

REDMET (Red Meteorológica, meteorological network) provides information regarding meteorological parameters in the forecast meteorological and dispersion models. Their main purpose is to analyze the movement of contaminants through the time and allow, in addition, inform the population the UV index, aimed at promoting a healthy exposure to the sun's rays. Table 2 shows some of the parameters provided by REDMET. There are weekly samplings since 1989.

Table 1. Information provided by REDMA stations

Pollutant	Abbreviation	Units
Total Suspended Particles	PST	$\mu g/m^3$
Particles smaller than 10 micrometers	*PM	$\mu g/m^3$
Particles smaller than 2.5 micrometers	PM2.5	$\mu g/m^3$
Total Lead suspended particles	PbPS	$\mu g/m^3$
Lead particles smaller than 10 micrometers	PbPM	$\mu g/m^3$

Table 2. Information provided by REDMET stations

Meteorological Parameters	Abbreviation	Units	Stations
Temperature	TMP	°C	15
Relative Humidity	RH	%	15
Wind Direction	WDR	azimuth	15
Wind Velocity (1986 - March 1995)	WSP	miles/hr	15
Wind Velocity (April 1995 - actual)	WSP	m/seg	15

REDDA (Red de Depósito Atmosférico, atmospheric warehouse network) takes samples from wet and dry deposits, whose analysis allows knowing the flow of toxic substances in the atmosphere to the earth's surface and its involvement in the alteration of typical elements of the soil and chemical properties of rain water. REDDA takes samplings 24 hours every six days since 1989.

3 Associative Models

The associative models used in this paper, associative memories and neural networks, are based on the Alpha-Beta models [12, 13]. In the learning and recovery phases, minimum and maximum are used. At the same time, there are two binary operations, α and β [12,13]. To be defined, α and β must specify two numerical sets: $A = \{0, 1\}$ and $B = \{0, 1, 2\}$. Table 3 and 4 shows the result of both operations over the A and B sets.

Table 3. α Binary operation

$\alpha : A \times A \to B$		
x	Y	$\alpha(x, y)$
0	0	1
0	1	0
1	0	2
1	1	1

Table 4. β Binary operation

$\beta : B \times A \to A$		
x	y	$\beta(x, y)$
0	0	0
0	1	0
1	0	0
1	1	1
2	0	1
2	1	1

There are 4 matrix operations:

$$\alpha_{max} : P_{m \times r} \nabla_\alpha Q_{r \times n} = \left\lfloor f_{ij}^\alpha \right\rfloor_{m \times n}, \text{ where } f_{ij}^\alpha = \bigvee_{k=1}^{r} \alpha\left(p_{ik}, q_{kj} \right) \tag{3.1}$$

$$\beta_{max} : P_{m \times r} \nabla_\beta Q_{r \times n} = \left\lfloor f_{ij}^\beta \right\rfloor_{m \times n}, \text{ where } f_{ij}^\beta = \bigvee_{k=1}^{r} \beta\left(p_{ik}, q_{kj} \right) \tag{3.2}$$

$$\alpha_{\min} : P_{m \times r} \Delta_\alpha Q_{r \times n} = \left\lfloor h_{ij}^\alpha \right\rfloor_{m \times n}, \text{ where } h_{ij}^\alpha = \overset{r}{\underset{k=1}{\wedge}} \alpha\left(p_{ik}, q_{kj}\right) \tag{3.3}$$

$$\beta_{\max} : P_{m \times r} \Delta_\beta Q_{r \times n} = \left\lfloor h_{ij}^\beta \right\rfloor_{m \times n}, \text{ where } h_{ij}^\beta = \overset{r}{\underset{k=1}{\wedge}} \beta\left(p_{ik}, q_{kj}\right) \tag{3.4}$$

Lemma 3.1. Let $\mathbf{x} \in A^n$ and $\mathbf{y} \in A^m$; then $\mathbf{y} \nabla_\alpha \mathbf{x}^t$ is a matrix of dimension $m \times n$, and will also fulfill that: $\mathbf{y} \nabla_\alpha \mathbf{x}^t = \mathbf{y} \Delta_\alpha \mathbf{x}^t$.

Simbol $\boxplus$ is used to represent the operations ∇_α y Δ_α when operates a column vector of dimension n: $\mathbf{y} \nabla_\alpha \mathbf{x}^t = \mathbf{y} \boxplus \mathbf{x}^t = \mathbf{y} \Delta_\alpha \mathbf{x}^t$

The ij-nth component of the matrix is $\mathbf{y} \boxplus \mathbf{x}^t$ given by: $[\mathbf{y} \boxplus \mathbf{x}^t]_{ij} = \alpha(y_i, x_j)$

Given an association index μ, the above expression indicates that the ij-nth component of the matrix $\mathbf{y}^\mu \boxplus (\mathbf{x}^\mu)^t$ is expressed in the following manner:

$$[\mathbf{y}^\mu \boxplus (\mathbf{x}^\mu)^t]_{ij} = \alpha\left(y_i^\mu, x_j^\mu\right) \tag{3.5}$$

Lemma 3.2. Let $\mathbf{x} \in A^n$ y $\mathbf{P}$ an array of dimension $m \times n$. The operation $\mathbf{P}_{mxn} \nabla_\beta \mathbf{x}$ gives the result of a column vector of dimension m, whose i-nth component tales the following form:

$$\left(\mathbf{P}_{mxn} \nabla_\beta \mathbf{x}\right)_i = \overset{n}{\underset{j=1}{\vee}} \beta\left(p_{ij}, x_j\right) \tag{3.6}$$

There are two types of Alpha-Beta heteroassociative models: type $\mathbf{V}$ and type $\boldsymbol{\Lambda}$. Let look at the type $\mathbf{V}$.

Learning phase

Step 1. For each $\mu = 1, 2, ..., p$, from the couple $(\mathbf{x}^\mu, \mathbf{y}^\mu)$ builds the array:

$$[\mathbf{y}^\mu \boxtimes (\mathbf{x}^\mu)^t]_{mxn} \tag{3.7}$$

Step 2. Applying the binary maximum operator $\vee$ to the arrays obtained in step 1:

$$\mathbf{V} = \overset{p}{\underset{\mu=1}{\vee}} \left[\mathbf{y}^\mu \, ?(\mathbf{x}^\mu)^t\right] \tag{3.8}$$

The ij-nth entry is given by the following expression:

$$v_{ij} = \overset{p}{\underset{\mu=1}{\vee}} \alpha(y_i^\mu, x_j^\mu) \tag{3.9}$$

Recovery phase

During this phase, a set of patterns $\mathbf{x}^\omega$ are presented, with $\omega \in \{1, 2, ..., p\}$ and the operation Δ_β: $\mathbf{V}\,\Delta_\beta\,\mathbf{x}^\omega$ is performed. Given that the dimensions of the matrix $\mathbf{V}$ are m x n and $\mathbf{x}^\omega$ is a column vector of dimension n, the result of the previous operation must be a column vector of dimension m, whose i-nth component is:

$$(V\Delta_\beta x^\omega)_i = \overset{n}{\underset{j=1}{\wedge}} \beta(v_{ij}, x_j^\omega) \qquad (3.10)$$

In the alpha-beta heteroasociative models type Λ, maximums and minimums are exchanged in the expressions above. And for the autoassociative models, the fundamental set is $\{(\mathbf{x}^\mu, \mathbf{x}^\mu) \mid \mu = 1, 2, ..., p\}$.

In the last five years the applications of the associative models Alpha-Beta in databases of real problems have been heavy and constant. They have been implemented in representative topics of the areas of current knowledge on the border between science and technology, namely: memory architectures [14], mobile robotics [15], software engineering [16], classification algorithms [17-20, 26, 29, 38], BAM [21-24, 32, 34, 40], equalization of industrial colors [25], feature selection [27, 39], image compression [28], Hopfield model [30], binary decision diagrams [31], images in gray levels [33], color images [35], Parkinson's disease [36] and cryptography [37], among others. This work is one of the first incursions of the Alfa-Beta algorithm in environmental issues.

4 Proposal

It was proposed the implementation of the Alpha-Beta associative models in the foundations of data from SIMAT subsystems, specifically in the value of CO and NO$_x$ concentrations reported in the database created in the 15 stations. Both CO and NO$_x$ were chosen due to the importance of their impact to the environment. The government of the Federal District of Mexico said that the results of epidemiological studies in Mexico City and other cities with similar problems of pollution indicate that their people are conducive to the early development of chronic respiratory disease because of such type of pollutants.

5 Experiments and Results

At the stage of experimentation CO and NO$_x$ data levels were used (both included in the RAMA database). To carry out the experiments, it took the whole of measurement in ppm units (parts per million) of both pollutants obtained in Iztacalco station, sampled every hour for the year 2010. The patterns taken from RAMA database are used in vector form, which were provided to the Alpha-Beta associative model as inputs, being the output data pattern the one that follows 10 input data patterns provided. Thus, the fundamental set was composed of 8749 associations with input patterns of dimension 10 and output patterns of dimension 1. As a set of test took the data obtained by the same monitoring station during the month of February 2011, with an integrated set of 673 associations. The importance of the results presented

Table 5. Performance measurement values used in the prediction of pollutants

Pollutant	RMSE	Bias
CO	0.00414	-1.8
NO_x	0.000887	-0.546

here lies in the fact that the Alpha-Beta models learned the data generated in the year 2010, and they are able to predict, automatically, the data that would be transferred to some time in one day of 2011. For example, consider the use of Iztacalco station. There was a measurement where the concentration of CO for the February 26th, 2010 was 1.6 ppm, while the Alfa-Beta associative model proposed here predicted by the same day in 2011 at the same time a value of 1.6 ppm too. i.e. the prediction coincided with the real value. On the other hand, on January 25th, 2010, it was recorded 0.022 ppm concentration of NO_x, while our system predicted a concentration of 0.018 ppm, which means a difference ppm of NO_x. In some cases during the same month, differences were in excess. For example, on February 23rd, 2011, the recorded concentration was of 0,040 ppm of NO_x, while our system predicted a concentration of 0,053 ppm, which means a difference of +0.013 ppm. As numerical metric performance, the ingrain square root mean square error (RMSE) is used, which is one of the performance measures more used in forecasting intelligent, and is according to the equation 5.1. On the other hand, to describe how much underestimates or overestimated the situation the model, bias was used, which is calculated in accordance with the equation 5.2. RMSE and bias values are presented in Table 5. For both equations, P_i is the i-nth value predicted and O_i is the i-th original value.

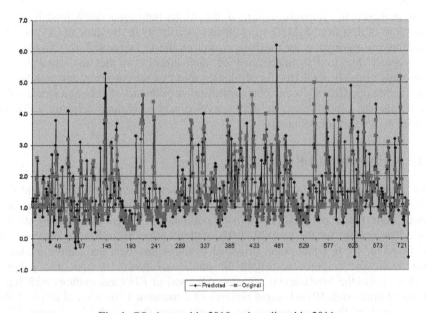

Fig. 1. CO observed in 2010 and predicted in 2011

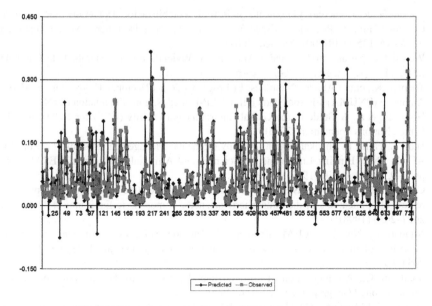

Fig. 2. NO$_x$ emissions observed in 2010 and predicted in 2011

$$RMSE = \sqrt{\frac{1}{n}\sum_{i=1}^{n}(P_i - O_i)^2} \quad (5.1) \qquad\qquad Bias = \frac{1}{n}\sum_{i=1}^{n}(P_i - O_i) \quad (5.2)$$

Figure 1 and 2 shows the graphs containing the results derived from the application of the associative models reported in section 3, applied to the information provided by SIMAT. A very close follow between the curves in both figures can be observed, which indicates the prediction performance reached by the application of the associative models.

Acknowledgments. The authors gratefully acknowledge the Instituto Politécnico Nacional (Secretaría Académica, COFAA, SIP, and CIC), CONACyT, SNI, and ICyTDF (grants PIUTE10-77 and PICSO10-85) for their support to develop this work.

References

1. Secretaría de Comercio y Fomento Industrial. Protección Al Ambiente - Contaminación Del Suelo - Residuos Sólidos Municipales - Determinación De Azufre, Norma Mexicana NMX-AA-092-1984, México (1984)
2. Secretaría de Comercio y Fomento Industrial. Protección Al Ambiente - Contaminación Atmosférica - Determinación De Neblina De Acido Fosfórico En Los Gases Que Fluyen Por Un Conducto, Norma Mexicana NMX-AA-090-1986, México (1986)

3. Secretaría de Comercio y Fomento Industrial. Potabilizacion Del Agua Para Uso Y Consumo Humano-Poliaminas-Especificaciones Y Metodos De Prueba, Norma Mexicana NMX-AA-135-SCFI-2007, México (2007)

4. Web del Departamento de Medio Ambiente y Vivienda de la Generalitat de Cataluña (2007), http://mediambient.gencat.net/cat

5. Toepfer, K., et.al.: Aliados Naturales: El Programa de las Naciones Unidas para el Medio Ambiente (PNUMA) y la sociedad civil, UNEP-United Nations Foundation (2004)

6. Hisas, L., et.al. A Guide to the Global Environmental Facility (GEF) for NGOs, UNEP-United Nations Foundation (2005)

7. Hart, J., Hunt, I., Shankararaman, V.: Environmental Management Systems - a Role for AI?. In: Proc. Binding Environmental Sciences and Artificial Intelligence, BESAI 1998, pp. 1–9, Brighton, UK (1998)

8. Gardner, M.W., Dorling, S.R.: Artificial neural networks (the multilayer perceptron)–a review of applications in the atmospheric sciences. Atmospheric Environment 32(14/15), 2627–2636 (1998)

9. Nunnari, G., Nucifora, A.F.M., Randieri, C.: The application of neural techniques to the modelling of time-series of atmospheric pollution data. Ecological Modelling 111, 187–205 (1998)

10. Spellman, G.: An application of artificial neural networks to the prediction of surface ozone. Applied Geography 19, 123–136 (1999)

11. Corchado, J.M., Fyfe, C.: Unsupervised Neural Network for Temperature Forecasting. Artificial Intelligence in Engineering 13(4), 351–357 (1999)

12. Yáñez Márquez, C.: Memorias Asociativas basadas en Relaciones de Orden y Operadores Binarios. Tesis Doctoral del Doctorado en Ciencias de la Computación, Centro de Investigación en Computación del Instituto Politécnico Nacional, México (2002)

13. Yáñez Márquez, C., Díaz-de-León Santiago, J.L.: Memorias Asociativas Basadas en Relaciones de Orden y Operaciones Binarias. Computación y Sistemas 6(4), 300–311 (2003) ISSN 1405-5546

14. Camacho Nieto, O., Villa Vargas, L.A., Díaz de León Santiago, J.L., Yáñez Márquez, C.: Diseño de Sistemas de Memoria Cache de Alto Rendimiento aplicando Algoritmos de Acceso Seudo-Especulativo. Computación y Sistemas 7(2), 130–147 (2003) ISSN 1405-5546

15. Yáñez-Márquez, C., Díaz de León-S., Juan, L., Camacho-N., Oscar.: Un sistema inteligente para telepresencia de robots móviles. In: Proc. de la Décimoquinta Reunión de Otoño de Comunicaciones, Computación, Electrónica y Exposición Industrial "La Convergencia de Voz, Datos y Video", ROC&C 2004, IEEE Sección México, Acapulco, Guerrero, México (2004)

16. López-Martín, C., Leboeuf-Pasquier, J., Yáñez-Márquez, C., Gutiérrez-Tornés, A.: Software Development Effort Estimation Using Fuzzy Logic: A Case Study, en IEEE Computer Society. In: Proc. Sixth Mexican International Conference on Computer Science, pp. 113–120 (2005) ISBN: 0-7695-2454-0, ISSN: 1550 4069

17. Flores Carapia, R.,Yáñez Márquez, C.: Minkowski's Metrics-Based Classifier Algorithm: A Comparative Study, en Memoria del XIV Congreso Internacional de Computación CIC 2005, celebrado en las instalaciones del Instituto Politécnico Nacional, México, del 7 al 9 de septiembre de 2005, pp. 304–315 (2005) ISBN: 970-36-0267-3

18. Muñoz Torija, J.M., Yáñez Márquez, C.: Un Estudio Comparativo del Perceptron y el Clasificador Euclideano, en Memoria del XIV Congreso Internacional de Computación CIC 2005, celebrado en las instalaciones del Instituto Politécnico Nacional, México, del 7 al 9 de septiembre de 2005, pp. 316–326 (2005) ISBN: 970-36-0267-3

19. Sánchez-Garfias, F.A., Díaz-de-León Santiago, J.L., Yáñez-Márquez, C.: A new theoretical framework for the Steinbuch's Lernmatrix. In: Proc. Optics & Photonics 2005, Conference 5916 Mathematical Methods in Pattern and Image Analysis, organizado por la SPIE (International Society for Optical Engineering), San Diego, CA., del 31 de julio al 4 de agosto de 2005, pp. (59160)N1-N9 (2005) ISBN: 0-8194-5921-6, ISSN: 0277-786X

20. Argüelles, A.J., Yáñez, C., Díaz-de-León Santiago, J.L, Camacho, O.: Pattern recognition and classification using weightless neural networks and Steinbuch Lernmatrix. In: Proc. Optics & Photonics 2005, Conference 5916 Mathematical Methods in Pattern and Image Analysis, organizado por la SPIE (International Society for Optical Engineering), San Diego, CA., del 31 de julio al 4 de agosto de 2005, pp. (59160)P1-P8 (2005) ISBN: 0-8194-5921-6, ISSN: 0277-786X

21. Acevedo-Mosqueda, M.E., Yáñez-Márquez, C.: Alpha-Beta Bidirectional Associative Memories. Computación y Sistemas (Revista Iberoamericana de Computación incluida en el Índice de CONACyT) 10(1), 82–90 (2006) ISSN: 1405-5546

22. Acevedo-Mosqueda, M.E., Yáñez-Márquez, C., López-Yáñez, I.: Alpha-Beta Bidirectional Associative Memories. IJCIR International Journal of Computational Intelligence Research 3(1), 105–110 (2006) ISSN: 0973-1873

23. Acevedo-Mosqueda, M.E., Yáñez-Márquez, C., López-Yáñez, I.: Complexity of alpha-beta bidirectional associative memories. In: Gelbukh, A., Reyes-Garcia, C.A. (eds.) MICAI 2006. LNCS (LNAI), vol. 4293, pp. 357–366. Springer, Heidelberg (2006) ISSN: 0302-9743

24. Acevedo-Mosqueda, M.E., Yáñez-Márquez, C., López-Yáñez, I.: Alpha-Beta Bidirectional Associative Memories Based Translator. IJCSNS International Journal of Computer Science and Network Security 6(5A), 190–194 (2006) ISSN: 1738-7906

25. Yáñez, C., Felipe-Riveron, E., López-Yáñez, I., Flores-Carapia, R.: A novel approach to automatic color matching. In: Martínez-Trinidad, J.F., Carrasco Ochoa, J.A., Kittler, J. (eds.) CIARP 2006. LNCS, vol. 4225, pp. 529–538. Springer, Heidelberg (2006)

26. Román-Godínez, I., López-Yáñez, I., Yáñez-Márquez, C.: A New Classifier Based on Associative Memories, IEEE Computer Society. In: Proc. 15th International Conference on Computing, CIC 2006, pp. 55–59 (2006) ISBN: 0-7695-2708-6

27. Aldape-Pérez, M., Yáñez-Márquez, C., López -Leyva, L.O.: Feature Selection using a Hybrid Associative Classifier with Masking Technique, en IEEE Computer Society. In: Proc. Fifth Mexican International Conference on Artificial Intelligence, MICAI 2006, pp. 151–160 (2006) ISBN: 0-7695-2722-1

28. Guzmán, E., Pogrebnyak, O., Yáñez, C., Moreno, J.A.: Image compression algorithm based on morphological associative memories. In: Martínez-Trinidad, J.F., Carrasco Ochoa, J.A., Kittler, J. (eds.) CIARP 2006. LNCS, vol. 4225, pp. 519–528. Springer, Heidelberg (2006)

29. Aldape-Pérez, M., Yáñez-Márquez, C., López -Leyva, L.O.: Optimized Implementation of a Pattern Classifier using Feature Set Reduction. Research in Computing Science 24, 11–20 (2006) Special issue: Control, Virtual Instrumentation and Digital Systems, IPN México, ISSN 1870-4069

30. Catalán-Salgado, E.A., Yáñez-Márquez, C.: Non-Iterative Hopfield Model, en IEEE Computer Society. In: Proc. Electronics, Robotics, and Automotive Mechanics Conference, CERMA 2006, Vol. II, pp. 137–144 (2006) ISBN: 0-7695-2569-5, ISSN/Library of Congress Number 2006921349

31. López-Yáñez, I., Yáñez-Márquez, C. Using Binary Decision Diagrams to Efficiently Represent Alpha-Beta Associative Memories, en IEEE Computer Society. In: Proc. Electronics, Robotics, and Automotive Mechanics Conference, CERMA 2006, Vol. I, pp. 172–177 (2006) ISBN: 0-7695-2569-5, ISSN/Library of Congress Number 2006921349

32. Acevedo-Mosqueda, M.E., Yáñez-Márquez, C., López-Yáñez, I.: A new model of BAM: Alpha-beta bidirectional associative memories. In: Levi, A., Savaş, E., Yenigün, H., Balcısoy, S., Saygın, Y. (eds.) ISCIS 2006. LNCS, vol. 4263, pp. 286–295. Springer, Heidelberg (2006) ISSN: 0302-9743

33. Yáñez-Márquez, C., Sánchez-Fernández, L.P., López-Yáñez, I.: Alpha-beta associative memories for gray level patterns. In: Wang, J., Yi, Z., Żurada, J.M., Lu, B.-L., Yin, H. (eds.) ISNN 2006. LNCS, vol. 3971, pp. 818–823. Springer, Heidelberg (2006) ISSN: 0302-9743

34. Acevedo-Mosqueda, M.E., Yáñez-Márquez, C., López-Yáñez, I.: A New Model of BAM: Alpha-Beta Bidirectional Associative Memories. Journal of Computers 2(4), 9–56, Academy Publisher; ISSN: 1796-203X

35. Yáñez-Márquez, C., Cruz-Meza, M.E., Sánchez-Garfias, F.A., López-Yáñez, I.: Using alpha-beta associative memories to learn and recall RGB images. In: Liu, D., Fei, S., Hou, Z., Zhang, H., Sun, C. (eds.) ISNN 2007. LNCS, vol. 4493, pp. 828–833. Springer, Heidelberg (2007) ISBN: 978-3-540-72394-3

36. Ortiz-Flores, V.H., Yáñez-Márquez, C., Kuri, A., Miranda, R., Cabrera, A., Chairez, I.: Non parametric identifier for Parkinson's disease dynamics by fuzzy-genetic controller. In: Proc.of the 18th International Conference Modelling And Simulation, Montreal, Quebec, Canada, May 30–June 1, pp. 422–427 (2007) ISBN: 978-0-88986-663-8

37. Silva-García, V.M., Yáñez-Márquez, C., Díaz de León-Santiago, J.L.: Bijective Function with Domain in N and Image in the Set of Permutations: An Application to Cryptography. IJCSNS International Journal of Computer Science and Network Security 7(4), 117–124 (2007); ISSN: 1738-7906

38. Román-Godínez, I., López-Yáñez, I., Yáñez-Márquez, C.: Perfect recall on the lernmatrix. In: Liu, D., Fei, S., Hou, Z., Zhang, H., Sun, C. (eds.) ISNN 2007. LNCS, vol. 4492, pp. 835–841. Springer, Heidelberg (2007) ISBN: 978-3-540-72392-9

39. Aldape-Pérez, M., Yáñez-Márquez, C., Argüelles-Cruz, A.J.: Optimized Associative Memories for Feature Selection. In: Martí, J., Benedí, J.M., Mendonça, A.M., Serrat, J. (eds.) IbPRIA 2007. LNCS, vol. 4477, pp. 435–442. Springer, Heidelberg (2007) ISBN: 978-3-540-72846-7

40. Acevedo-Mosqueda, M.E., Yáñez-Márquez, C., López-Yáñez, I.: Alpha-Beta Bidirectional Associative Memories: Theory and Applications. In: Neural Processing Letters, August 2007, vol. 26(1), pp. 1–40. Springer, Heidelberg (2007) ISSN: 1370-4621

41. Sistema de monitoreo atmosférico de la Ciudad de México, http://www.sma.df.gob.mx/simat2/

42. http://www.cornelio.org.mx

Neural Network Approach to Water-Stressed Crops Detection Using Multispectral WorldView-2 Satellite Imagery

Dubravko Ćulibrk, Predrag Lugonja, Vladan Minić, and Vladimir Crnojević

Faculty of Technical Sciences, University of Novi Sad,
Trg Dositeja Obradovica 6, 21000 Novi Sad, Serbia
{dculibrk,lugonja,minic,crnojevic}@uns.ac.rs
http://www.biosense.uns.ac.rs/

Abstract. The paper presents a method for automatic detection and monitoring of small waterlogged areas in farmland, using multispectral satellite images and neural network classifiers. In the waterlogged areas, excess water significantly damages or completely destroys the plants, thus reducing the average crop yield. Automatic detection of (waterlogged) crops damaged by rising underground water is an important tool for government agencies dealing with yield assessment and disaster control.

The paper describes the application of two different neural network algorithms to the problem of identifying crops that have been affected by rising underground water levels in WorldView-2 satellite imagery. A satellite image of central European region (North Serbia), taken in May 2010, with spatial resolution of $0.5m$ and 8 spectral bands was used to train the classifiers and test their performance when it comes to identifying the water-stressed crops. WorldView-2 provides 4 new bands potentially useful in agricultural applications: coastal-blue, red-edge, yellow and near-infrared 2. The results presented show that a Multilayer Perceptron is able to identify the damaged crops with 99.4% accuracy. Surpassing previously published methods.

Keywords: Water stress, Agriculture, Satellite imagery, Neural networks, Waterlogged farmland, Remote sensing.

1 Introduction

The advances in the satellite imaging technology provide researchers and practitioners with ever more data that needs to be processed to extract meaningful and useful information.

In agriculture, the applications of satellite imagery typically deal with crop monitoring [23][5][19][15][2]. A general goal is to be able to predict the yield of specific crops and monitor their health. Such information can be used directly by government agencies to ensure that the subsidies given to farmers are allocated correctly and that they are compensated for damage occurring in their fields.

L. Iliadis et al. (Eds.): EANN/AIAI 2011, Part II, IFIP AICT 364, pp. 323–331, 2011.

While satellite imagery for crop monitoring is a practice with more than 30 years of tradition [5][19], the sensors have only recently gained resolution that allows for precision monitoring in the case of agriculture practices based on small land parcels [15][2][8][23].

In the work presented here we address the problem of detecting small wetland areas in the farmland. Particularly, we are interested to detect these areas in the plains of Northern Serbia (Central Europe), where the small areas of wetland appear in the arable land as the consequence of the rise of a groundwater. This phenomenon emerges usually after the period of heavy rains, affects small mutually isolated areas, and disappears in a couple of weeks. Although temporary, it influences crop yields significantly and therefore it is necessary to discover it in order to perform damage assessment. In addition, farmers in Serbia are entitled to compensation if more than 30% of their crop is affected. Since crops are still present in the wetland areas, their stress due to excess water can be used to detect the stretches of waterlogged area within arable land. In addition, the presence of the crops in the water occludes the water surface, rendering the methods which wold rely on reflection from water inapplicable.

High-resolution 8-band images, provided by the remote sensing equipment mounted on the recently launched WorldView-2 satellite are used to detect crops damaged by excess water. Besides the standard red, green, blue and near-infra red band, WorldView-2 it offers new bands: costal blue, yellow, red-edge and an additional near-infra red band [11]. The new channels carry significant information one one is concerned with watelogged areas detection [18]. In addition, multispectral images are available in high spatial resolution of 50 cm, making them suitable for applications in areas with small-land-parcel-based agriculture.

Human-annotated ground truth was used to train 2 types of neural network classifiers to distinguish between pixels pertinent to waterlogged land and other pixels: Multi Layer Perceptron (MLP) and Radial Basis Function (RBF) neural nets. The experimental results presented show that the MLP approach achieves superior performance both in terms of accuracy and detection speed, when compared to previously published results [18]. RBF however, proved unsuitable for the task at hand, since it achieves accuracy below the simplest (logistic regression) classifier.

The rest of the paper is organized as follows: Section 2 provides an overview of the relevant published work. Section 3 provides the details of our approach. Section 4 describes the experiments conducted and the results achieved. Finally, Section 5 provides our conclusions.

2   Related Work

The problem of waterlogged areas detection falls into the scope of land-cover classification. Lu and Weng [10] provide a comprehensive and fairly recent survey of the different approaches used to distinguish between different types of cover. Within their taxonomy the proposed approach is supervised, non-parametric, per-pixel, spectral and makes hard decisions. This means that we use ground truth data created by human experts to train the classifier, make no assumptions about the data itself, consider single pixels with no notion of spatial dependencies and derive unambiguous classification decisions.

As far as precision agriculture applications are concerned, using satellite imagery for crop identification and crop-covered area estimation is a practice with more than 30 years of tradition [5][19]. Initially low-resolution Landsat MSS and TM data was used [5][19][21][14]. As new satellites became available SPOT imagery [13][1], Indian Remote Sensing (IRS) satellite data [4][16] and moderate-resolution Moderate Resolution Imaging Spectroradiometer (MODIS) [9][20] data has been used. More recently several studies have looked at using high-resolution satellite imagery from QuickBird [15][2], IKONOS [8] and SPOT 5 [23].

Using neural networks for classification of multispectral image pixels was considered as early as 1992 [7]. At the time, the authors of the study were concerned with the feasibility of training the network classifying the pixels of a satellite image in reasonable time, using the hardware available. The considered a single hidden layer feed-forward neural network trained and proposed an adaptive back-propagation algorithm, which reduced the training time from 3 weeks to 3 days, for a network with 24, 24 and 5 nodes in the input, hidden and output layer. At the time, a single input neuron was used for each bit in the RGB pixel value of a Landsat image. The authors concluded that the neural network is able to generalize well across different images, both real and synthetic. Since this early work there have been numerous applications of neural networks to classification of remotely sensed data. Mas and Flores provide an overview [12].

A recent study by Marchisio et al. [11] evaluated the capability of new WorldView-2 data, when it comes to general land-cover classification. They considered the relative predictive value of the new spectral bands in the WorldWiew-2 sensor. Four types of classifiers were employed: logistic regression, decision trees, random forest and neural network classifiers. The study showed that new bands lead to an improvement of 5-20% in classification accuracy on 6 different scenes. In addition, the authors proposed an approach to evaluate the relative contribution of each spectral band to the learned models and, thus, provide a list of bands important for the identification of each specific land cover class. Although the neural network classifiers used by Marchisio et al. [11] have not been described in detail, they state that they used a topology with two hidden layers, containing $10 - 20$ neurons. The neural network used was most likely a Multilayer Perceptron trained by back propagation.

Our work builds up on the study of Petrovic et al. [18]. The authors used WorldView-2 data to address the problem of water-logged agricultural land detection. They evaluated logistic regression, Support Vector Machines (SVMs) and Genetic Programming (GP) classifiers. SVMs and GP achieved best classification results when cross validated (98,80% and 98,87% accurate respectively). The GP solution was deemed the best by the authors, since it was able to achieve best classification and do it significantly faster than the SVM classifier that contained 6000 support vectors. Neural network classifiers were not considered in that study, and to the best of our knowledge were never applied to the problem of waterlogged area detection nor the problem of plant water-stress detection.

3   Detecting Waterlogged Crops Using WorldView-2 Data

3.1   WorldView-2 Data

The data used in this work is provided by commercial satellite WorldView-2. The image was made during July 2010 and covers the area of arable land from in the northern part of Serbia. Figure 1 shows visible light components of a part of the scene. The discolored irregular shaped blotches within the arable land correspond to waterlogged parts.

WorldView-2 is the first commercial satellite which provides eight high-resolution spectral bands. Its sensors are sensitive to visible and near-infrared spectral bands and producce images of 50 cm resolution. This satellite detects radiation in following spectral bands: red (630-690 nm), green (510-580 nm), blue (450-510 nm), near-infrared 1 (NIR1 - 770-895 nm), coastal blue (400-450 nm), yellow (585-625 nm), red-edge (705-745 nm) and near-infrared 2 (NIR2 - 860-1040 nm). The last four components provide additional value when compared to the data available from other commercial satellites and each of them has been designed with specific applications in mind. The costal blue is least absorbed by water, but it is absorbed by chlorophyll in healthy plants. NIR2 component to some extent resembles the characteristics of NIR1 component, initially aimed at vegetation detection and classification tasks, but is less influenced by atmospheric interference. Red-edge is especially designed for maximal reflection from vegetation and is intended for the measurement of plant health and classification of vegetation [3].

The data set used for training and testing of classifiers was derived based on ground truth, manually annotated on a part of a single satellite image covering an area of $10km^2$. It contains environ 200,000 data samples, corresponding to the values of 8 bands for both waterlogged and non-waterlogged pixels, selected out some 25,000,000 pixels in the single scene. The values for each of the 8 bands are 16 bits wide. All values were normalized in our experiments.

3.2   Neural Network Classification

Two types of feed forward neural networks were considered for the classification of multispectral pixel values: Multilayer Perceptron (MLP)[6] and Radial Basis Function (RBF)[17].

Both architectures represent feed-forward networks the signals are propagated from the input to output neurons, with no backward connections. The MLP contained a single hidden layer and was trained using a back-propagation algorithms as detailed in [22].

The parameters of the RBF network used were determined using k-means clustering. Both MLP and RBF implementation is available within the open-source data mining and artificial intelligence suite Wakaito Environment for

Fig. 1. Part of the satellite image used to derive the training data set and test the classifiers. Only red, green and blue bands are shown.

Knowledge Discovery (WEKA) [22]. This is the suite that has been used to conduct experiments detailed in Section 4.

Once the neural networks are trained they are used to classify each pixel in the image separately.

4   Experiments and Results

To evaluate the performance of the two neural network models 10-fold cross validation was used. The process involves holding out 10 percent of data in the training data set and using the rest to train the classifier. The 10% is used to test the model. The process is repeated 10 times and average performance measures are reported for each type of classifier.

The MLP contained 100 neurons in a single hidden layer, learning rate was set to 0.3, momentum to 0.2 and 500 training epochs were used. The time taken to build the model on an Intel Core2Duo processor running at 2.93 GHz was 6251 seconds. Table 1 provides the performance statistics for the MLP model: true-positive (TP) rate, false-positive (FP) rate, precision, recall, f-measure and Response Operating Characteristic (ROC) Area. The accuracy achieved by the MLP is 99.4043%.

The RBF model was built much faster. It took only 43.69 seconds to train the network. Minimum standard deviation for the clusters was set to 0.1, while the

Table 1. Classification results for Multilayer Perceptron

	TP Rate	FP Rate	Precision	Recall	F-Measure	ROC Area
Not waterlogged	0.998	0.037	0.995	0.998	0.997	0.995
Waterlogged	0.963	0.002	0.987	0.963	0.975	0.995
Weighted Avg.	0.994	0.033	0.994	0.994	0.994	0.995

Table 2. Classification results for RBF

	TP Rate	FP Rate	Precision	Recall	F-Measure	ROC Area
Not waterlogged	0.992	0.410	0.947	0.992	0.969	0.925
Waterlogged	0.590	0.008	0.904	0.590	0.714	0.925
Weighted Avg.	0.944	0.362	0.942	0.944	0.938	0.925

Fig. 2. MLP classification result for the image shown in Fig. 1

ridge parameter for the logistic regression was set to 10^{-8}. Table 1 provides the performance statistics for the RBF model. The accuracy achieved by the RBF is significantly lover than that of the MLP (94.3538%).

Figures 2 and 3 show the classification results for a part of the satellite image shown in Figure 1. The parts of the scene classified as waterlogged are indicated by white pixels. Since the waterlogged areas represent a relatively small part of the overall area, the classification result shown in two images differs significantly, although the difference in accuracy between two methods is just 5%.

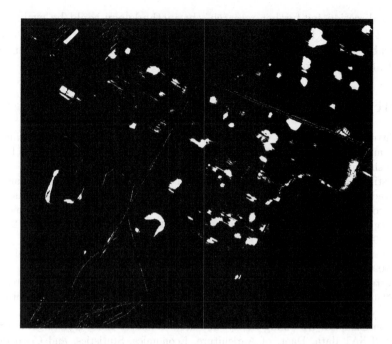

Fig. 3. MLP classification result for the image shown in Fig. 1

When compared to the results of other classifiers as reported in [18], the MLP achieved results superior to any of the classifiers reported there. The best classifier reported in the previous study was achieved by Genetic Programming and its accuracy was 98.87% on the same data set. Unfortunately the RBF classifier performed worse even than the simple logistic regression, which achieved 97.58% accuracy, as reported previously. This can only be attributed to the errors in the k-means clustering used to position the kernels of the RBF.

5   Conclusion

The application of two different neural network classifiers to the problem of detecting waterlogged farmland using multispectral satellite imagery, has been evaluated in the study presented.

Using imagery acquired by the new WorldView-2 satellite, we showed that Multilayer Perceptron can achieve accuracy superior to that of other published approaches and classify 99.4% of the pixels in the scene accurately.

RBFs have not proved successful when applied to the problem at hand, since the k-means procedure used to position the kernel functions was unable to cope. Other ways of doing this, such as random sampling from the training data or orthogonal least square learning should be considered if RBFs are to be used in this scenario.

330 D. Ćulibrk et al.

As a direction for further research, the methodology could be evaluated in terms of detecting less pronounced effects of water stress, where plants are not irreparably damaged. It could also be possible to design crop yield estimation models, akin to the methodology presented in [15].

References

1. Büttner, G., Csillag, F.: Comparative study of crop and soil mapping using multitemporal and multispectral SPOT and Landsat Thematic Mapper data. Remote Sensing of Environment 29(3), 241–249 (1989)
2. Castillejo-González, I., López-Granados, F., García-Ferrer, A., Peña-Barragán, J., Jurado-Expósito, M., de la Orden, M., González-Audicana, M.: Object-and pixel-based analysis for mapping crops and their agro-environmental associated measures using QuickBird imagery. Computers and Electronics in Agriculture 68(2), 207–215 (2009)
3. DigitalGlobe: The Benefits of the 8 Spectral Bands of WorldView-2. whitepaper (2009)
4. Dutta, S., Sharma, S., Khera, A., et al.: Accuracy assessment in cotton acreage estimation using Indian remote sensing satellite data. ISPRS Journal of Photogrammetry and Remote Sensing 49(6), 21–26 (1994)
5. Hanuschak, G.: Obtaining timely crop area estimates using ground-gathered and LANDSAT data. Dept. of Agriculture, Economics, Statistics, and Cooperatives Service: for sale by the Supt. of Docs., US Govt. Print. Off. (1979)
6. Haykin, S.: Neural Networks: A Comprehensive Foundation. Macmillan, New York (1994)
7. Heermann, P., Khazenie, N.: Classification of multispectral remote sensing data using a back-propagation neural network. IEEE Transactions on Geoscience and Remote Sensing 30(1), 81–88 (1992)
8. Helmholz, P., Rottensteiner, F., Heipke, C.: Automatic qualtiy control of cropland and grasland GIS objects using IKONOS Satellite Imagery. IntArchPhRS (38) Part 7, 275–280 (2010)
9. Lobell, D., Asner, G.: Cropland distributions from temporal unmixing of MODIS data. Remote Sensing of Environment 93(3), 412–422 (2004)
10. Lu, D., Weng, Q.: A survey of image classification methods and techniques for improving classification performance. International Journal of Remote Sensing 28(5), 823–870 (2007)
11. Marchisio, G., Pacifici, F., Padwick, C.: On the relative predictive value of the new spectral bands in the WorldWiew-2 sensor. In: 2010 IEEE International Geoscience and Remote Sensing Symposium (IGARSS), pp. 2723–2726. IEEE, Los Alamitos (2010)
12. Mas, J., Flores, J.: The application of artificial neural networks to the analysis of remotely sensed data. International Journal of Remote Sensing 29(3), 617–663 (2008)
13. Murakami, T., Ogawa, S., Ishitsuka, N., Kumagai, K., Saito, G.: Crop discrimination with multitemporal SPOT/HRV data in the Saga Plains, Japan. International Journal of Remote Sensing 22(7), 1335–1348 (2001)
14. Oetter, D., Cohen, W., Berterretche, M., Maiersperger, T., Kennedy, R.: Land cover mapping in an agricultural setting using multiseasonal Thematic Mapper data. Remote Sensing of Environment 76(2), 139–155 (2001)

15. Pan, G., Sun, G., Li, F.: Using QuickBird imagery and a production efficiency model to improve crop yield estimation in the semi-arid hilly Loess Plateau, China. Environmental Modelling & Software 24(4), 510–516 (2009)
16. Panigrahy, S., Sharma, S.: Mapping of crop rotation using multidate Indian Remote Sensing Satellite digital data. ISPRS Journal of Photogrammetry and Remote Sensing 52(2), 85–91 (1997)
17. Park, J., Sandberg, I.: Universal approximation using radial-basis-function networks. Neural Computation 3(2), 246–257 (1991)
18. Petrovic, N., Lugonja, P., Culibrk, D., Crnojevic, V.: Detection of Wet Areas in Multispectral Images of Farmland. In: The Seventh Conference on Image Information Mining: Geospatial Intelligence from Earth Observation, pp. 21–24 (April 2011)
19. Ryerson, R., Dobbins, R., Thibault, C.: Timely crop area estimates from Landsat. Photogrammetric Engineering and Remote Sensing 51(11), 1735–1743 (1985)
20. Wardlow, B., Egbert, S.: Large-area crop mapping using time-series MODIS 250 m NDVI data: An assessment for the US Central Great Plains. Remote Sensing of Environment 112(3), 1096–1116 (2008)
21. Williams, V., Philipson, W., Philpot, W.: Identifying vegetable crops with Landsat Thematic Mapper data. Photogrammetric Engineering and Remote Sensing 53, 187–191 (1987)
22. Witten, I.H., Frank, E.: Data Mining: Practical machine learning tools and techniques, 2nd edn. Morgan Kaufmann, San Francisco (2005)
23. Yang, C., Everitt, J., Murden, D.: Evaluating high resolution SPOT 5 satellite imagery for crop identification. Computers and Electronics in Agriculture (2011)

A Generalized Fuzzy-Rough Set Application for Forest Fire Risk Estimation Feature Reduction

T. Tsataltzinos[1], L. Iliadis[2], and S. Spartalis[3]

[1] PhD candidate Democritus University of Thrace, Greece
tsataltzinos@yahoo.gr
[2] Associate Professor Democritus University of Thrace, Greece
liliadis@fmenr.duth.gr
[3] Professor Democritus University of Thrace, Greece
sspart@pme.duth.gr

Abstract. This paper aims in the reduction of data attributes of a fuzzy-set based system for the estimation of forest fire risk in Greece, with the use of rough-set theory. The aim is to get as good results as possible with the use of the minimum amount of data attributes possible. Data manipulation for this project is done in MS-Access. The resulting data table is inserted into Matlab in order to be fuzzified. The final result of this clustering is inserted into Rossetta, which is a Rough set exploration software, in order to estimate the reducts. The risk estimation is recalculated with the use of the reduct set in order to measure the accuracy of the final minimum attribute set. Nine forest fire risk factors were taken into consideration for the purpose of this paper and the Greek terrain was separated into smaller areas, each concerning a different Greek forest department.

Keywords: Fuzzy sets, rough sets, forest, fire, risk, attribute reduction.

1 Introduction

Forest fire risk estimation is one of the major assessments of present and future forestry. The number of fire instances is growing annually and one of the most basic concerns of every modern country is to find a way to estimate where and when the next forest fire incident will occur. Fire risk evaluation is an important part of fire prevention, since the planning that needs to be done before a fire instance occurs, requires tools to be able to monitor where and when a fire is more prone to ignite, or when it will cause the most negative effects. The most commonly used definition of Fire risk (danger) is the one that is expressed as an index of both constant and variable risk factors which affect the ignition, spread and difficulty of control of fires and the damage they cause [15].

A lot of theories have been applied but all of them luck in precision. Traditional fire danger systems rely on meteorological indices, based on atmospheric conditions which are not the only factors that compose the overall risk of fire. Human caused factors, fuel loads, and any other non atmospherical factor needs to be taken under consideration. Modeling fuel trends have been proposed to analyze spatial and

L. Iliadis et al. (Eds.): EANN/AIAI 2011, Part II, IFIP AICT 364, pp. 332–341, 2011.

temporal changes in fire risk conditions [14]. Many methods from various countries such as USA, Canada, Australia and France take into consideration weather factors, soil water and fuel loads. Most of those methods use statistical analysis of past and present data in order to evaluate a short term fire risk.

Many other methods were used to evaluate forest fire risk. Each one of these methods is adapted to the data available, or even needs specific data in order to produce results. The kind of data indices used in every one of these occasions is different, according to the objectives of the forest management planning. Many countries have greatly invested in fire risk evaluation and data collection because risk estimation produces better results, when a great amount of data is available. On the other hand, when data entries are missing, imprecise, cannot be measured or when specific data is not available at all, risk estimation becomes a very difficult task.

One of the most interesting methods that helps in overcoming the problem of luck of accurate data, concerns the use of fuzzy sets theory, introduced by Lotfi A. Zadeh in 1965, in order to present vagueness in everyday life as a mathematical way. This theory is widely used for everyday appliances and helps into depicting real life situations and problems, such as uncertainty and luck of precise information into computer data, available for use in order to estimate the risk. Despite its positive points, fuzzy set theory has a basic backdrop. In order to increase the accuracy of each fuzzy set system, the user has to increase the available data, which results in problems like difficulty in obtaining this amount of data and the increase in rules that need to be inserted to the system. With the use of bigger data sets the total number of rules becomes the basic problem of the user.

This paper concentrates in the application of a rough-set based method in order to reduce the total number of data entries needed. Rough set theory was developed by Zdzislaw Pawlak in the early 80's, it uses data that can be acquired from measurements or human experts and it deals with the classification analysis of data tables [16]. The concepts in rough set theory can be used to define the necessity of features, which in this case are the factors that influence the forest fire risk index. The basic concept is to produce as good results as possible, with a minimum data set. Data is not always easy to find in time and sometimes might not be available. The final goal is to be able to produce a flexible system that will be able to use more than one data set, from any kind of available data, in order to produce results that will be helpful to the expert, in order to deploy human and vehicle power in the right place at the right time.

2   Research Area and Software

The research area of the DSS used for the assessment of the risk index is the Greek terrain. Data has been gathered from the Greek public services that are responsible for the meteorological, morphological and statistical data. The initial data table have been processed through MS Access so that the results become more meaningful, but due to the lack of resent detailed information from the public services, the studies concerns a past period of time between 1983 and 1993. The processed data were inserted into Matlab in order to apply the fuzzy rule based decision support system. Reducts were calculated with the use of rough set theory in the Rosetta software [1], [2], [3], [4], [5]. The final results were imported into MS Excel, so that comparisons with reality are possible.

The Greek terrain has been separated into smaller areas, each one concerning a different forest department. The final results do not produce a specific risk index, but a ranking between each and every forest department under study, according to their overall forest fire risk.

3 Fuzzy Set and Rough Set Theory

The use of computers for assigning risk degrees has its positives due to ease of computation but comes with a prize. The huge amount of data that is needed cannot always be described with classical mathematics and crisp sets. At this point, fuzzy set theory can be used to describe almost every situation. The fact that in fuzzy sets there exists a degree of membership (DOM) $\mu s(X)$ that is mapped on [0,1] for every data entry helps in the production of rational and sensible clustering [9] and any element of the real world can belong to each set with a different degree of membership.

Fuzzy Sets use membership functions $\mu s(x)$ which are mappings of the universe of discourse X on the closed interval [0,1]. That is, the degree of belonging of some element x to the universe X, can be any number $\mu s(x)$, where $0 \leq \mu_s(x) \leq 1$. Below are the five major steps that should be followed when designing Fuzzy Models.

Step 1: Define the Universe of discourse (e.g. domains, input and output parameters)
Step 2: Specify the Fuzzy Membership Functions
Step 3: Define the Fuzzy Rules
Step 4: Perform the numerical part, like T-Norms, S-Norms or other custom inference
Step 5: If not-Normal Fuzzy sets exist, perform Defuzzification

The 3rd step in which the fuzzy rules must be defined, is actually the basic backdrop of any fuzzy – rule based system. The more the data attributes increase, the more accurate the system is, but the complexity of creating the rule set becomes very difficult. In an effort to reduce the data attributes as much possible various methods can be used varying from simple statistical ones, to more complex neural network systems.

One of the most promising methods, introduced by Zdzisław Pawlak in 1991 [13], which can be used for attribute reduction, is the Rough-set theory. Rough set theory is still another approach to vagueness. Similarly to fuzzy set theory it is not an alternative to classical set theory but it is embedded in it. Rough set theory can be viewed as a specific implementation of Frege's idea of vagueness, i.e., imprecision in this approach is expressed by a boundary region of a set, and not by a partial membership, like in fuzzy set theory. Rough set concept can be defined quite generally by means of topological operations, interior and closure, called approximations. Given a set of objects U called the universe and an indiscernibility relation $R \subseteq U \times U$, representing our lack of knowledge about elements of U, we assume that R is an equivalence relation. Let X be a subset of U. We want to characterize the set X with respect to R. To this end we will need the basic concepts of rough set theory given below.

- The *lower approximation* of a set X with respect to R is the set of all objects, which can be for certain classified as X with respect to R (are certainly X with respect to R).
- The *upper approximation* of a set X with respect to R is the set of all objects which can be possibly classified as X with respect to R (are possibly X in view of R).
- The *boundary region* of a set X with respect to R is the set of all objects, which can be classified neither as X nor as not-X with respect to R.

Now we are ready to give the definition of rough sets.

Set X is crisp (exact with respect to R), if the boundary region of X is empty.
Set X is rough (inexact with respect to R), if the boundary region of X is nonempty.

4   The Case of Forest Fire Risk Assessment

The first step in assigning forest fire risk is related to the determination of the main n risk factors (RF) affecting this specific risk. The risk factors are distinguished in two basic categories; Human factors and Natural ones [11]. Another issue is that forest fire risk is a general concept. In order to be more precise, the expert needs to concentrate in a more specific kind of risk. In this case, there exist to basic risks:

1. A risk concerning the speed with which each fire is spread

2. A risk concerning the total amount of fire incidents that occur in a specific area during a given period of time

Of course these two different risk types are influenced in a different way by numerous factors that most of them are difficult to measure and also need to be measured for a long period of time. This project is going to concentrate on the first kind of risk, concerning the speed with which each fire incident is spread.

Forest fires can have various factors that can influence both of the above risk indicators, either long term or short term. The long term forest fire risk factors are distinguished in two basic categories; Human factors and Natural ones [11]. Each of these general risk types consists of several other sub-factors that influence in their own way the overall risk degree (ORD). Some of the major factors can be seen in the table below.

Table 1. Factors used by the fuzzy rule based system

Level 1	Natural factors		Human caused factors	
Level 2	Weather factors	Landscape factors	Measurable	Immeasurable
Level 3	Temperature	Forest cover	Tourism	Land value
	Humidity	Altitude	Population	Other
	Wind speed			

Of course there are multiple other factors that can influence the forest fire risk but it is not this paper's purpose to explore them in depth. Further separation of the "Natural" factors into two more sub-groups, the "Weather" and the "Landscape" factors is being made, so that it becomes easier to apply the rule set of the fuzzy decision support system. The same applies for the "Human caused" factors that are being separated into "Measurable" and "Immeasurable".

5   The Decision Support System

The system was designed and developed under the Matlab platform. With the use of fuzzy logic and fuzzy sets, the system assigns a degree of the long term forest fire risk degree to each area by using Matlab's fuzzy toolbox. The functions used to define the DOM are called fuzzy membership functions (FMF) and in this project the triangular FMF (TRIAMF) and the semi-triangular FMF (semi-TRIAMF) were applied [10]. Functions 1 and 2 below represent the TRIAMF and semi-TRIAMF which are implemented in Matlab.

$$\mu_s(X) = \begin{cases} 0 \text{ if } X < a \\ (X-a)/(c-a) \text{ if } X \in [a,c] \\ (b-X)/(b-c) \text{ if } X \in [c,b] \\ 0 \text{ if } X > b \end{cases} \tag{1}$$

$$\mu_s(X) = \begin{cases} 0 \text{ if } X < a \\ (X-a)/(b-a) \text{ if } X \in [a,b] \end{cases} \tag{2}$$

The system assigns each forest department three Partial Risk Indices (PRI), for every one of the nine factors that are taken under consideration, as it is shown below:

- Low Danger due to each factor
- Medium Danger due to each factor
- High Danger due to each factor

This method allows the use of any kind of data and does not need specific metrics for every factor. Due to this fact, there was no need to do any changes in the row data provided by the Greek national services. The above steps resulted in having 27 different PRIs. The more detailed the linguistics become the greater the number of PRIs.

In order to unify those partial risk indices in one Unified Risk Index (URI), the system has to take into consideration the human experience and to apply the rules that a human expert would use. These rules are distinct for each risk factor and most of them are specified in bibliography [11]. The total amount of rules that has to be applied is $3^9=19683$, but with the use of sub groups as shown in the table above, this number reduces to 33+6*32=81. This is accomplished by creating partial rule sets in order to combine the factors as shown in figure 1 below

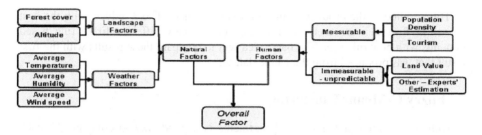

Fig. 1. Categorization of risk factors

Previous studies [7], [8] have shown that this way the system retains its simplicity, while still producing the same results. This way, the system is able to provide a ranking of the Greek forest departments and the comparison of these results to reality gives the user a degree of "compatibility" with the actual ranking of these forest departments according to each year's total burnt area.

6   Attribute Reduction with the Use of Rough Sets

Despite the fact that the accuracy of the system was good, the fact that 9 factors had to be combined, resulted in a rather complicated amount of data sets. Another issue was also the fact that in order to find the optimal data set that could produce the best result would need a huge amount of tests. Each test would require the built of a new rule set. Building 511rule sets, which is the total number of combination of 9 attributes and testing each one of them for every year of the data available would be time consuming.

Rough set theory is a good tool for attribute reduction and with the help of the Rosetta software, the process of finding the reducts is much easier. The whole dataset, including the FFR-FIS' final risk degree index, was inserted into Rosetta and the reducts were calculated with the help of RSES dynamic reducer. The algorithm used was the generic RSES. The values used for the rest of the parameters are shown in the Screen 1 below.

Subtable sampling

Number of sampling levels: 50

Number of subtables to sample per level: 100

☑ Include whole table as a sample

Dimensions (in % of whole table)

Smallest subtable size (lowest level): 10

Largest subtable size (highest level): 100

Seed to RNG:

54321

Screen 1. Values for the RSES dynamic reducer parameters

This resulted in the production of 85 possible reducts. The 3 reducts with the best results in Rosetta were used in order to create 3 new different fuzzy rule based systems, with new rule sets. The basic idea was to compare these results with the first result that was produced from the whole dataset.

7   Fuzzy C-Means Clustering

Providing a ranking of the forest departments is a good way to compare the each other, but the actual ranking of those departments according to their burnt areas is not possible to be accurately found. This is due to the fact that the fire incidents are influenced by a great number of factors that are unpredictable. The speed of expansion on the other hand is influenced by more "predictable" factors but still the ranking does not provide a sharp image of reality. For this reason it is more helpful to cluster the forest departments and compare the results afterwards.

Table 2. Example of actual data set

Forest Department	Temperature	Humidity	Burnt area	Risk degree
A	14	0,1	1500	0.7785
B	17	0,4	1100	0.3722
C	22	0,7	200	0.2162

Table 3. Example of clustered data set

Forest Department	Temperature	Humidity	Burnt area	Risk Degree
A	Low	Low	Vast	High
B	Average	Low	Vast	Average
C	High	High	Small	Small

In this paper, the whole data set that is under study has been clustered with the use of fuzzy c-means clustering method. Each factor, the results of the forest fire risk fuzzy inference system and the "Burnt area" attribute have been clustered into three different clusters. This way each data entry can be described as shown in the example of tables 2 and 3 above.

This makes the comparison between results and reality easier. The user can compare whether a forest department that has a "vast" burnt area, like forest departments A and B, has been indexed as a "high" or "average" risk department. Comparison between these two clustered attributes gives an index of quality, about the final results of the fuzzy inference system.

8   Results and Discussion

Fuzzy rule based systems can be helpful in many various occasions and forest fire risk assessment is one of these occasions. The built of a system that could utilize as many attributes as possible was the first stage. Results from this system varied from 25% to 70%. During the built of this system, it became obvious that the luck of data is a frequent problem. The accuracy gets smaller, when more risk attributes are missing and reaches its peak with a full data set. The creation of smaller systems that could produce acceptable risk indices was necessary. The utilization of a rough set attribute reduction system in order to find possible reducts, gave some interesting results. The 10 best reducts are shown in the table 4 below.

Table 4. The reducts that produce the best results

	Reduct	Support
1	{Altitude, Forest Cover}	5093
2	{Humidity, Altitude}	5093
3	{Temperature, Altitude}	5093
4	{Humidity, Wind Speed, Population}	4497
5	{Temperature, Forest Cover, Wind Speed, Other (expert's evaluation)}	3430
6	{Temperature, Forest Cover, Wind Speed, Tourism}	3264
7	{Altitude, LandValue}	2813
8	{Altitude, Population}	2813
9	{Humidity, Wind Speed, Tourism}	2728
10	{Temperature, Wind speed, Land Value}	2688

Human factors are known to be the most possible cause for forest fires, but they are not the basic cause for each forest fire incidents expansion. According to these results, only in 5 out of 10 reducts include human caused factors.

Careful study of these results gives some interesting discovering about forest fire risk in Greece. It is obvious it possible to study forest fire risk with just the use of long term factors alone. This shows that forest fires, during the period of time that is under study, which expand the most rapidly and cause the most damage are most of the times (almost 100%) in the same forest departments. 5093 cases out of the 5100 can be described with Average Altitude and Forest Cover as only knowledge about the terrain. It wouldn't require an "expert" to prove that almost all cases of big fire incidents were near the sea, or near cities that are by the sea, that also had great forest cover and unprotected forests.

Another interesting point of view is that the attribute "Altitude" is found in 6 of those 10 reducts. This also proves that the altitude of a forest department is one of the most basic factors that influence the risk indices. All the forest departments that are far away from the sea have greater altitude in Greece and none of them seems to have major fire incidents. Knowing the altitude of a forest department and with the use of just the humidity or the temperature, it is safe to say that this department is a risky one or not.

An interesting aspect of this study is the fact that the best reducts that occur are small, composed by only 2 or 3 attributes. For the first 3 reducts, a fuzzy rule based system was built and new indices were extracted. The results were not that accurate, reaching a maximum similarity with the real ranking of 43%. On the other side the application fuzzy c-means clustering on the results and the total burnt area of each department, and further comparison between those two, showed that it is possible to cluster the departments into "Dangerous" and "non-Dangerous" with great accuracy, reaching even 90% in some cases, with the use of just a couple of attributes.

The target of future studies is to find the optimum attributes set for the Greek terrain, that will provide results, as close to reality as possible. A basic categorization of the Greek forest departments due to their long term factors that influence the forest fires, in order to provide a standard map of the dangerous areas, with the combination of short term factors in order to rank the dangerous areas in daily basis is a future thought.

References

1. Øhrn, A., Komorowski, J.: ROSETTA: A Rough Set Toolkit for Analysis of Data. In: Proc. Third International Joint Conference on Information Sciences, Fifth International Workshop on Rough Sets and Soft Computing (RSSC 1997), Durham, NC, USA, March 1-5, vol. 3, pp. 403–407 (1997)
2. Øhrn, A., Komorowski, J., Skowron, A., Synak, P.: The Design and Implementation of a Knowledge Discovery Toolkit Based on Rough Sets: The ROSETTA System. In: Polkowski, L., Skowron, A. (eds.) Rough Sets in Knowledge Discovery 1: Methodology and Applications. STUDFUZZ, vol. 18, ch. 19, pp. 376–399. Physica-Verlag, Heidelberg (1998) ISBN 3-7908-1119-X
3. Øhrn, A., Komorowski, J., Skowron, A., Synak, P.: The ROSETTA Software System. In: Polkowski, L., Skowron, A. (eds.) Rough Sets in Knowledge Discovery 2: Applications, Case Studies and Software Systems. STUDFUZZ, vol. 19, pp. 572–576. Physica-Verlag, Heidelberg (1998) ISBN 3-7908-1130-3
4. Øhrn, A.: ROSETTA Technical Reference Manual, Department of Computer and Information Science, Norwegian University of Science and Technology (NTNU), Trondheim, Norway (2000)
5. Øhrn, A.: The ROSETTA C++ Library: Overview of Files and Classes, Department of Computer and Information Science, Norwegian University of Science and Technology (NTNU), Trondheim, Norway (2000)
6. Komorowski, J., Øhrn, A., Skowron, A.: The ROSETTA Rough Set Software System. In: Klösgen, W., Zytkow, J. (eds.) Handbook of Data Mining and Knowledge Discovery, ch. D.2.3. Oxford University Press, Oxford (2002) ISBN 0-19-511831-6
7. Tsataltzinos, T.: A fuzzy decision support system evaluating qualitative attributes towards forest fire risk estimation. In: 10th International Conference on Engineering Applications of Neural Networks, Thessaloniki, Hellas, August 29-31 (2007)
8. Tsataltzinos, T., Iliadis, L., Spartalis, S.: An intelligent Fuzzy Inference System for risk estimation using Matlab platform: The case of forest fires in Greece. In: Tsataltzinos, L., Iliadis, S. (eds.): (Proceedings): "Artificial Intelligence Applications and Innovations III", 5th IFIP Conference (AIAI 2009), pp. 303–310. Springer, Heidelberg (2009) ISBN 978-1-4419-0220-7
9. Kandel, A.: Fuzzy Expert Systems. CRC Press, Boca Raton (1992)

10. Iliadis, L., Spartalis, S., Maris, F., Marinos, D.: A Decision Support System Unifying Trapezoidal Function Membership Values using T-Norms. In: Proceedings of ICNAAM (International Conference in Numerical Analysis and Applied Mathematics). J. Wiley-VCH Verlag, GmbH Publishing co., Weinheim, Germany (2004)
11. Kailidis, D.: Forest Fires (1990)
12. Johnson, E.A., Miyanishi, K.: Forest Fire: Behavior and Ecological Effects (2001)
13. Zdzisław, P.: Rough sets: theoretical aspects of reasoning about data (1991)
14. He, H.S., Shang, B.Z., Crow, T.R., Gustafson, E.J., Shifley, S.R.: Simulating forest fuel and fire risk dynamics across landscapes – LANDIS fuel module design. Ecological Modeling 180, 135–151 (2004)
15. Chandler, C.C.: Fire in forestry / Craig Chandler... [et al.]. Wiley, Chichester (1983)
16. Komorowski, J., Pawlak, Z., Polkowski, L., Skowron, A.: Rough sets: A tutorial (1998)

Pollen Classification Based on Geometrical, Descriptors and Colour Features Using Decorrelation Stretching Method

Jaime R. Ticay-Rivas[1], Marcos del Pozo-Baños[1], Carlos M. Travieso[1],
Jorge Arroyo-Hernández[2], Santiago T. Pérez[1], Jesús B. Alonso[1],
and Federico Mora-Mora[2]

[1] Signals and Communications Department,
Institute for Technological Development and Innovation in Communications,
University of Las Palmas de Gran Canaria,
Campus University of Tafira, 35017, Las Palmas de Gran Canaria, Las Palmas, Spain
{jrticay,mpozo,ctravieso,sperez,jalonso}@idetic.eu
[2] Escuela de Matemáticas, Universidad Nacional, Costa Rica
{fmoram,jarroy}@una.ac.cr

Abstract. Saving earth's biodiversity for future generations is an important global task, where automatic recognition of pollen species by means of computer vision represents a highly prioritized issue. This work focuses on analysis and classification stages. A combination of geometrical measures, Fourier descriptors of morphological details using Discrete Cosine Transform (DCT) in order to select their most significant values, and colour information over decorrelated stretched images are proposed as pollen grains discriminative features. A Multi-Layer neural network was used as classifier applying scores fusion techniques. 17 tropical honey plant species have been classified achieving a mean of 96.49% $\pm$ 1.16 of success.

Keywords: Pollen grains, pollen classification, colour features, geometrical features, neural networks, decorrelation stretch.

1 Introduction

Over 20% of all the world's plants are already at the edge of becoming extinct [1]. Saving earth's biodiversity for future generations is an important global task [2] and many methods must be combined to achieve this goal. Saving flora biodiversity involves mapping plant distribution by collecting pollen and identifying them in a laboratory environment. Pollen grain classification is a qualitative process, involving observation and discrimination of features.

The manual method, heavily depending on experts, is still the most accurate and effective, but takes considerable amounts of time and resources, limiting research progress [3]. Therefore, automatic recognition of pollen species by means of computer vision is a highly prioritized subject in palynology.

A lot of progress has been made in the field of automatic pollen species classification. In [4], it showed one of the first works in which texture features and

L. Iliadis et al. (Eds.): EANN/AIAI 2011, Part II, IFIP AICT 364, pp. 342–349, 2011.

neural networks were used for pollen grains identification task. In [5] detection and classification of pollen grains were identified as the main and difficult tasks. In both cases neural networks were used. The characteristics of pollen grains for their correct classification were studied in different studies. In [6], the shape and ornamentation of the grains were analyzed, using simple geometric measures. Concurrence matrices were applied for the measurement of texture in [7]. Both techniques were combined in [8] and [9]. In [10] brightness and shape descriptors were used as vector features.

Another interesting approach is found in [11]. Here, a combination of statistical reasoning, feature learning and expertise knowledge from the application was used. Feature extraction was applied alternating 2D and 3D representations. Iterative refinement of hypotheses was used during the classification process. A full 3D volume database of pollen grain was recorded in [12] using a confocal laser scanning microscope. 14 invariant gray-scale features based on integration over the 3D Euclidian transformation were extracted and used for classification. A more recent work with an ambitious approach is shown in [13]. It describes an automatic optical recognition and classification of pollen grains system. This is able to locate pollen grains on slides, focus and photograph them, and finally identify the species applying a trained neural network.

There is no doubt that significant progress has been made, but more knowledge is needed in the combination of features that have been studied and colour features in order to increase the number of species that systems are capable to recognize and classify with a high rate of success.

This work introduces the use of colour features over decorrelated images. A novel combination of these features with geometrical measures, and Fourier descriptors of morphological details are proposed as pollen grains discriminative features, being this our first work in this area. The remainder of this paper is organized as follows. In Section 2 the process of pollen grains detection is explained. Section 3 introduces the techniques used for feature extraction. The classification block is presented in Section 4. Experimental methodology and its results are exposed in Section 5. Finally in Section 6, discussions and conclusions are shown.

2   Pollen Grains Detection

Since there is not a uniform focus and conglomerates may appear in the images, a simple semi-automatic algorithm was developed to capture the pollen grains, indicating and cropping the pollen image. Thus, only one grain appears in each pollen grain sub-image. An example of each pollen class and its Scientist name (family, gender and specie by column) can be seen in figure 1.

Once sub-images of the pollen grains have been selected, automatic detection is performed. First, the colour contrast of the image is increased in order to locate the pollen grain in the image. Then a decorrelation stretching (DS) of the image is done. This technique is explained in detail in [15] and [16]. The main idea is to achieve a variance-equalizing enhancement in order to uncorrelate and to fill the colour space.

The pollen grain image is represented by a matrix with dimensions $M{\times}N{\times}3$ where $M{\times}N$ represents the image dimensions of each colour channel.

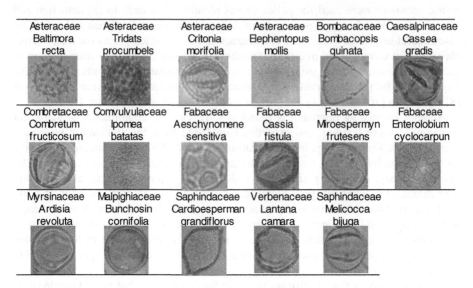

Asteraceae Baltimora recta	Asteraceae Tridats procumbels	Asteraceae Critonia morifolia	Asteraceae Elephentopus mollis	Bombacaceae Bombacopsis quinata	Caesalpinaceae Cassea gradis
Combretaceae Combretum fructicosum	Comvulvulaceae Ipomea batatas	Fabaceae Aeschynomene sensitiva	Fabaceae Cassia fistula	Fabaceae Miroespermyn frutesens	Fabaceae Enterolobium cyclocarpun
Myrsinaceae Ardisia revoluta	Malpighiaceae Bunchosin cornifolia	Saphindaceae Cardioesperman grandiflorus	Verbenaceae Lantana camara	Saphindaceae Melicocca bijuga	

Fig. 1. Examples of each pollen class

Thus, the pixels of the image are represented by the vector $x \in \mathfrak{R}^l$. Let V express the covariance or correlation matrix of x and the eigenanalysis associated [16] with the principal component analysis described as:

$$V = UEU^t \tag{1}$$

Where the columns of U denote the eigenvectors and the diagonal elements of the diagonal matrix E denote the corresponding eigenvalues. Using the transform U^t the vector x is transformed in a new vector space where the new data y is given by:

$$y = U^t x \tag{2}$$

The scaling, applied in order to equalize the variance, is achieved in this space dividing each value of y by its corresponding standard deviation $e_i^{1/2}$, thus the new w matrix is

$$w = E^{-1/2} U^t x \tag{3}$$

where

$$E = diag(e_1, e_2, e_3) \tag{4}$$

The transformation to the original space is given by

$$z = UE^{-1/2} U^t x \tag{5}$$

The resulting image is transformed to the HSV colour model (Hue, Saturation and Value). The S channel is used in order to take advantage of the difference of brightness between the background and the pollen grain. Histogram equalized image is necessary before the binarization of the image. The next factor is multiplied to the histogram equalization:

$$f = 1/e^{((m+0.1)^{*}1.5)} + 0.8 \qquad (6)$$

where m is the mean of the S channel insuring that the equalization is made in a correct range.

The binarization is applied setting the threshold at 0.45. Morphological methods like close, erosion, dilation, etc. are used in order to consider only the area where the pollen grain is found. Figure 2 shows the obtained images:

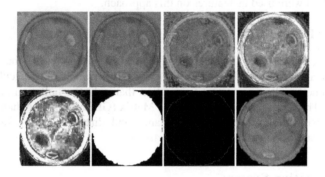

Fig. 2. Images obtained in the pollen grains detection

3 Discriminative Feature Extractions

Discriminative feature extractions were divided in three steps and are described in the next sections:

3.1 Geometrical Features

The geometrical features have been used in almost all works related to grain classification. The area, convex area, and perimeter are used in this work as they have been proven to provide a fair amount of the basic information of the pollen grain. Area is the number of pixels in the considered region. Convex area is the number of pixels of the binary image that specifies the convex hull. Perimeter is calculated as the distance between each adjoining pair of pixels around the border of the pollen grain.

3.2 Fourier Descriptors

The Fourier Descriptor represents the shape of the pollen grain in the frequency domain [10], [16]. Each boundary point is represented as a complex number. In order to eliminate the bias effect, the centroid of the shape of the pollen grain is introduced in the complex representation:

$$s = [x - x_c] + i[y - y_c] \qquad (7)$$

The contour is sampled each 2 degrees using the Cartesian to radian coordinates transformation. The Fourier Descriptor is defined as the power spectrum of the discrete Fourier Transform of s. In this work the Discrete Cosine Transformation (DCT) is applied to the resulting power spectrum in order to select the most significant values of the Fourier Descriptor. Finally, the first three coefficients of the transformation were used as features on this approach.

3.3 Colour Features

In section 2, decorrelation stretching method (DS) is applied to the pollen grain images. The averages of R (Red), B (Blue), H (Hue) channels were calculated as three colour parameters. It is important to mention that only the pollen grain region was considered. The Gray Level Co-occurrence Matrix (GLCM) from H channel was used in order to obtain its contrast. The distance and direction setting in the GLCM calculation was 1 pixel and 0 degree respectively.

4   Classification System

Artificial neural networks (NN) have been used in several studies [4], [5], [7], [8], [13] demonstrating good results. Multi-Layer Perceptron NN trained by Back Propagation algorithm (MLP-BP) were used. Neurons implemented the hyperbolic tangent sigmoid transfer function. The number of neurons in the input layer is the same as the number of selected features. Following experimentation, it was determined to use one hidden layer. The number of neurons that showed better results ranged between 20 and 80. The best results were found using 50 neurons. The numbers of neurons in the output layer was 17, that is, the number of classes. 5000 iterations were performed in the training phase.

Since weights of neurons in the input layer were initialized randomly, 30 NNs were trained in order to determine the classes. Two decision techniques (DT) were applied: class majority voting (CMV) and adding scores (AS). The former selects the most repeated result along all 30 NNs. These results at every NN are computed as the index of the maximum scoring neuron. The AS adds the output scores of corresponding neurons from each 30 NNs, before selecting the index of the maximum value.

5 Experiments and Results

5.1 Database

The database contains a total of 426 pollen grain sub-images as those presented in figure 1. These sub-images correspond to 17 genders and species from 11 different families of tropical honey plants from Costa Rica, Central America.

5.2 Experimental Methodology and Results

Parameters introduced in section 3 were tested for both raw images and DS processed images. For each of them, apart from the original dataset of parameters (ORIG), a second dataset of parameters (ORIGPCA) was obtained by applying PCA to the original dataset. About the classification, CMV and AS decision techniques were applied separately. Moreover, both parameter sets were fused at the score level using again the AS technique. This makes a total of 8 combinations of raw/DS-processed images, original/PCA data set, and CVM/AS decision technique.

In order to obtain valid results, a series of cross-validation methods were performed on every experiment: 30 iterations of 50% hold out (50-HO), and k-folds with k equal 3, 4, and 5 (3-, 4-, and 5-folds). A second set of features were obtained applying the DS method to the original images. Tables 1 and 2 show all obtained results for both original and DS processed set of images, showing results as mean (%) and standard desviation (±) from 30 iterations.

Table 1. Results obtained for the original set of parameters and after applying PCA, for both CMV and AS score fusion methods

DS	PCA	DT	50-HO	3-folds	4-folds	5-folds
No	No	CMV	86.25% ± 2.67	87.98% ± 3.13	90.01% ± 3.77	90.44% ± 3.53
No	Yes	CMV	86.13% ± 3.00	88.98% ± 7.54	91.79% ± 7.01	92.62% ± 7.17
No	No	AS	86.48% ± 2.75	88.87% ± 2.70	90.30% ± 3.45	90.19% ± 3.84
No	Yes	AS	86.55% ± 3.08	90.25% ± 3.36	93.34% ± 4.74	93.80% ± 4.25
Yes	No	CMV	88.08% ± 1.73	90.91% ± 1.11	90.98% ± 2.45	91.37% ± 3.34
Yes	Yes	CMV	93.99% ± 1.63	95.22% ± 0.24	94.22% ± 2.76	95.73% ± 1.84
Yes	No	AS	88.20 % ± 2.19	90.91% ± 0.62	90.86% ± 2.12	91.76% ± 3.27
Yes	Yes	AS	94.18 % ± 1.71	95.55% ± 0.15	95.36% ± 1.88	95.73% ± 1.84

Table 2. Results obtained merging ORIG and ORIGPCA parameters using AS score fusion

Param.	DS	DT	50-HO	3-folds	4-folds	5-folds
ORIG + ORIGPCA	No	AS	91.19% ± 2.07	92.96% ± 1.74	94.61% ± 3.05	95.78% ± 4.45
ORIG + ORIGPCA	Yes	AS	95.33% ± 1.31	96.53% ± 0.21	96.63% ± 1.17	97.48% ± 1.97

6 Discussion and Conclusions

This work has introduced the use of colour features over decorrelated stretched images, for enhance pollen grain classification. More standard characteristics like geometrical features and Fourier descriptors have been added to the pollen grain descriptions. Over this multiple feature vector, PCA has proven to increase the system's performance.

Experimental results shown in Table 1 indicate how the use of DS technique significantly improves the recognition of pollen grain species. Moreover, this improvement can be seen in terms of success rate and stability (standard deviation). This can be explained by the fact that colours channel variances are equalized and uncorrelated, reducing intra-class variation and differentiating inter-class samples. Furthermore, the use of PCA on the original database increased the success rate. This is consistent with the fact that PCA transformation projects samples over more meaningful dimensions, maximizing variance.

Finally, both data sets (ORIG and ORIGPCA) were fused using the AS score fusion, obtaining a maximum success rate of 96.4%, and a standard deviation of 1.16 with 17 pollen species. It is worth to mention that these results improve those achieved by other authors [7] [8] [9] [10], even though the number of classified species was significantly larger.

Acknowledgements. This work has been supported by Spanish Government, in particular by "Agencia Española de Cooperación Internacional para el Desarrollo" under funds from D/027406/09 for 2010, and D/033858/10 for 2011.

To M.Sc. Luis A. Sánchez Chaves, Tropical Bee Research Center (CINAT) at Universidad Nacional de Costa Rica, for provides the database images and palynology knowledge.

References

1. Plants under pressure: a global assessment. The first report of the IUCN Sampled Red List Index for Plants. Royal Botanic Gardens, Kew, UK (2010)
2. Sytnik, K.M.: Preservation of biological diversity: Top-priority tasks of society and state. Ukrainian Journal of Physical Optics 11(suppl. 1), S2–S10 (2010)
3. Stillman, E.C., Flenley, J.R.: The Needs and Prospects for Automation in Palynology. Quaternary Science Reviews 15, 1–5 (1996)
4. Li, P., Flenley, J.: Pollen texture identification using neural networks. Grana 38(1), 59–64 (1999)
5. France, I., Duller, A., Duller, G., Lamb, H.: A new approach to automated pollen analysis. Quaternary Science Reviews 18, 536–537 (2000)
6. Treloar, W.J., Taylor, G.E., Flenley, J.R.: Towards Automation of Palynology 1: Analysis of Pollen Shape and Ornamentation using Simple Geometric Measures, Derived from Scanning Electron Microscope Images. Journal of Quaternary Science 19(8), 745–754 (2004)
7. Li, P., Treloar, W.J., Flenley, J.R., Empson, L.: Towards Automation of Palynology 2: The Use of Texture Measures and Neural Network Analysis for Automated Identification of Optical Images of Pollen Grains. Journal of Quaternary Science 19(8), 755–762 (2004)

8. Zhang, Y., Fountain, D.W., Hodgson, R.M., Flenley, J.R., Gunetileke, S.: Towards Automation of Palynology 3: Pollen Pattern Recognition using Gabor Transforms and Digital Moments. Journal of Quaternary Science 19(8), 763–768 (2004)
9. Rodriguez-Damian, M., Cernadas, E., Formella, A., Fernandez-Delgado, M., De Sa-Otero, P.: Automatic detection and classification of grains of pollen based on shape and texture. IEEE Trans. on Systems, Man, and Cybernetics, Part C: Applications and Reviews 36(4), 531–542 (2006)
10. Rodriguez-Damian, M., Cernadas, E., Formella, A., Sa-Otero, R.: Pollen classification using brightness-based and shape-based descriptors. In: Proceedings of the 17th International Conference on Pattern Recognition, ICPR 2004, August 23-26, vol. 2, pp. 23–26 (2004)
11. Boucher, A., Thonnat, M.: Object recognition from 3D blurred images. In: 16th International Conference on Pattern Recognition, vol. 1, pp. 800–803 (2002)
12. Ronneberger, O., Burkhardt, H., Schultz, E.: General-purpose object recognition in 3D volume data sets using gray-scale invariants - classification of airborne pollen-grains recorded with a confocal laser scanning microscope. In: 16th International Conference on Pattern Recognition, vol. 2, pp. 290–295 (2002)
13. Allen, G.P., Hodgson, R.M., Marsland, S.R., Flenley, J.R.: Machine vision for automated optical recognition and classification of pollen grains or other singulated microscopic objects. In: 15th International Conference on Mechatronics and Machine Vision in Practice, December 2-4, pp. 221–226 (2008)
14. Karvelis, P.S., Fotiadis, D.I.: A region based decorrelation stretching method: Application to multispectral chromosome image classification. In: 15th IEEE International Conference on Image Processing, ICIP 2008, October 12-15, pp. 1456–1459 (2008)
15. Cambell, N.: The decorrelation stretch transformation. Int. J. Remote Sensing 17, 1939–1949 (1996)
16. Amanatiadis, A., Kaburlasos, V.G., Gasteratos, A., Papadakis, S.E.: A comparative study of invariant descriptors for shape retrieval. In: IEEE International Workshop on Imaging Systems and Techniques, May 11-12, pp. 391–394 (2009)

Global Optimization of Analogy-Based Software Cost Estimation with Genetic Algorithms

Dimitrios Milios[1], Ioannis Stamelos[2], and Christos Chatzibagias[2]

[1] School of Informatics, University Of Edinburgh, Edinburgh UK EH8 9AB
[2] Department Of Informatics, Aristotle University Of Thessaloniki, Thessaloniki Greece 54124

Abstract. Estimation by Analogy is a popular method in the field of software cost estimation. A number of research approaches focus on optimizing the parameters of the method. This paper proposes an optimal global setup for determining empirically the best method parameter configuration based on genetic algorithms. We describe how such search can be performed, and in particular how spaces whose dimensions are of different type can be explored. We report results on two datasets and compare with approaches that explore partially the search space. Results provide evidence that our method produces similar or better accuracy figures with respect to other approaches.

1 Introduction

One of the most challenging, yet essential, tasks in software industry is cost estimation. The term cost refers to the human effort spent in every phase of the development life cycle, including analysis, design, coding and testing. The most widely researched estimation methods are expert judgement, algorithmic/parametric estimation and analogy based estimation [4]. The first one involves consulting one or more experts who provide estimates using their own methods and experience. Although this practice produces estimates easily, the dependency on human factor has certain drawbacks, like subjectivity or availability of the expert. Algorithmic estimation models, such as COCOMO [4] and Function Points Analysis [1], have been proposed in order to generate estimates in a more repeatable and controllable manner. They are based on mathematical models that produce a cost estimate as a function of a number of variables. Several researchers have investigated the accuracy of such models. For example, Kemerer has suggested in [14] that the efficiency of such models is strongly dependent on the calibration of these models to local conditions. Finnie et al [9] conducted a comparison between algorithmic cost models, Artificial Neural Networks and Case-Based Reasoning for software cost estimation. It was verified that algorithmic models do not effectively model the complexities involved in software development projects. Both CBR and ANN produced more accurate estimations for the effort, however, ANN do not adequately explain their predictions. More recent research has reinforced such findings, providing further

L. Iliadis et al. (Eds.): EANN/AIAI 2011, Part II, IFIP AICT 364, pp. 350–359, 2011.

evidence in support of calibration. For example, in [16] it was found that estimates based on *within-company* projects are more accurate w.r.t. estimates based on *cross-company* historical data. In addition, research efforts have been dedicated to approaches that aim to explain better cost estimates to users and managers [3].

Estimation by Analogy (or EbA) is a case-based reasoning technique, in which cost estimation is produced by comparing the characteristics of the target project with the ones of old cases. These old cases are projects with known costs, also called *analogies*. It is assumed that the cost of the project to be estimated will be similar to that of the most similar projects. The similarity between two projects can be specified using the distance between them in the N-dimensional space defined by their characteristics. In most of EbA recent implementations [12][15][17], Euclidean distance is adopted for measuring project dissimilarity. In this work, the following distance metrics are also considered: *Euclidean, Manhattan, Minkowski, Canberra, Czekanowski, Chebychev* and *Kaufman-Rousseeuw*.

Projects are classified differently for each one of the metrics above, thus method's performance is affected. The choice of attributes that describe the projects also affect this classification, since the attributes are not of the same importance [21]. Attribute importance can be expressed either as the subset of attributes that best describes the given dataset, or as a vector of weights that are applied to the attributes. Furthermore, it has been already stated that the prediction accuracy is sensitive to the number N of analogies [21]. One more issue is how to combine their values in order to produce an estimate. In this paper the following statistics are used for the adjustment of the efforts of the most similar projects: *Mean, Median, Size Adjusted Mean* and *Size Adjusted Median*.

The objective of the current paper is to propose an automated strategy to select a combination of the parameters described above that result in the best performance for EbA. What we have is an optimization problem featuring a huge search space. The width of the possible solutions as well as the lack of an underlying mathematical model imposes the use of approximate search algorithms. In this domain, a popular optimization method is Genetic Algorithms (or GA).

A brief overview of related work is presented in Sect. 2. Section 3 describes how global optimization can be implemented with GA. Section 4 provides some experimental results. Finally, Section 5 concludes and provides future research ideas.

2   Related Work

Estimation by Analogy, was considered as valid alternative to conventional methods long ago [4], although it was not presented as estimation methodology until 1997 by Shepperd and Schofield [21]. The efficiency of EbA is dependent on numerous parameters that affect the way of retrieval of similar projects and producing estimations. In [21], it was found that the exclusion of the least important attributes results in improvement of prediction accuracy. In the same paper, the number of analogies was marked as another parameter that has to be carefully determined, since it has a strong effect on final estimates. In [2], different distance metrics were also investigated and it was proposed that interval estimates be produced for EbA.

The optimization of certain parameters in applying analogy was among the goals of AQUA method [18]. More specifically, a learning phase was applied in order to find the number of analogies and the threshold for the degree of similarity that granted the maximum prediction accuracy. Exhaustive search was used, by applying jackknife [8] for each number of analogies and similarity threshold. It was observed that although, in general, high numbers of analogies tend to produce increased values of MMRE, best performance is achieved for a different number of analogies for each database. Furthermore, the optimum number of analogies varied for different values of the threshold.

The importance of the project attributes was also discussed in various papers. For example in [15], a stepwise procedure based on statistical analysis was used, in order to select the most important attributes. However, it is more common that weighting factors are used to adjust the influence of each attribute in correspondence with its own significance. In [12], a GA approach was proposed as a search method for identifying a combination of weights that maximizes the accuracy of the EbA. Further improvement was possible after applying both linear and nonlinear equations as weighting factors. The identification of the optimum weights was the objective of a second version of AQUA that was published in 2008 [17], establishing a way of attribute weighting based on rough set analysis.

In 2007, Chiu and Hung [5] proposed an adjustment of the efforts of the closest analogies, based on similarity distances. According to experimental results, this way of adjustment outperformed the typical adjustment methods. It is also important that more than one distance metrics were used in these experiments, including Euclidean, Manhattan and Minkowski. It is interesting, however, that there was no distance metric found that was globally superior to others. In fact, performance of each distance metric varied both for different ways of adjustment and different datasets.

From what has been discussed so far, there is no optimization proposal that considers all parameters involved in EbA. Consequently, although genetic algorithms have already been adopted in the field of cost estimation [12][19], the current paper focuses on a fitness function and a problem representation that enable the search to be expanded to any application parameter available. The challenge is to combine the variables that represent the alternatives in applying EbA with the N-dimensional space derived from the weighting factors of project attributes.

3   Global Optimization by Genetic Algorithms

GA [11] is a search technique inspired from the genetic evolution in nature. Given an optimization problem, populations of possible solutions are considered, from which the fittest ones are going to survive and get combined so as to form a probably better population. The new set of solutions, or else the next generation, goes through the same process. It has been observed that the average fitness of each generation is increased. So, it is quite possible that the population will converge to one of the local optima.

In the case of EbA optimization, the individuals of the population are different combinations of the alternatives for applying the EbA method. In general, most GAs consist of the following components [10]:

- Encoding refers to the assignment of a unique string to each solution. The characteristics of a solution are translated into *genes*. Genes are parts of a larger string, the chromosome, in which the entire solution is encoded.
- Genetic operators are applied to the chromosomes. Their purpose is to produce the genes of the chromosomes of the next generation.
- Fitness evaluation function assigns to each solution-individual a value that corresponds to its relative fitness.
- Selection algorithm is applied to the evaluated solutions. The selection favors the fittest among the individuals, which is basic principal of evolution theory.

However, only two of these components are problem dependent: the encoding and the evaluation function.

3.1 Selection and Reproduction

Starting from the selection algorithm, the roulette-wheel selection is a typical choice in GA textbooks [10]. The probability of selecting an individual to participate in the process of reproduction is proportional to its fitness. This feature captures the main principle of evolution, the survival of the fittest.

As we will see in Sect. 3.3, the chromosomes are encoded in a binary form, so we can use some of the simplest, yet effective reproduction operators that can be found in the literature [10]. One-point binary crossover randomly combines parents' chromosomes, producing children that inherit most of the genes of their parents, while there is still a probability to create random genes with unknown effect. In order to prevent premature convergence, a mutation operator is used with much less probability e.g. about 5%. Binary mutation involves the random inversion of any bit of a chromosome, destroying its properties.

3.2 Evaluation

As already mentioned, individuals are combinations of EbA parameters. The fitness of any combination can be expressed as an accuracy measure of its application on the known projects of an available dataset. Accuracy measures, as defined in [6], make use of the Magnitude Relative Error:

$$MRE = \frac{|act - est|}{act} \tag{1}$$

where *act* is the project's actual effort, and *est* is the estimated value produced by applying EbA.

A widely accepted accuracy measure is the Mean MRE (or MMRE). The purpose of the algorithm when using this criterion is to minimize the error. Another accuracy measure that is also considered is Pred(0.25), which computes the number of projects that have been estimated with an MRE of less than 25%. The fitness of an individual is directly proportional to the second criterion. Moreover, the median of MRE values, MdMRE, is also used as it is less likely to be affected by extreme values. Eventually, in our approach these criteria are combined using the following ad-hoc fixed formula:

$$fitness = -0.4MMRE - 0.3MdMRE + 0.3Pred(0.25) \qquad (2)$$

The purpose of the GA is to maximize Equation (2) for the projects of some training dataset. A jackknife method is used consisting of the following steps:

1. For each project with known effort that belongs to the database
2. Remove the current project from the database
3. Apply EbA with the selected configuration, in order to produce an estimate
4. Push the project back to its original dataset

The procedure above will be repeated for each one of the individuals of a population, which correspond to different configurations of EbA.

3.3   Encoding

We have chosen to encode EbA configuration parameters into binary strings. This is a convenient choice, since we can then apply the simple genetic operators described in Sect. 3.1. Moreover, this choice allows to easily explore a massive search space whose dimensions are essentially of different type.

Application Parameters. The use of one or another parameter for applying EbA should be translated into binary words (i.e. genes). Provided that genes have fixed size, each one should be long enough to store all the possible options for a parameter. The number of bits needed to encode a parameter will be:

$$bits = \lceil \log_2 S \rceil \qquad (3)$$

where S is the number of the possible values of the parameter encoded to the specified gene. In this paper, we have considered three parameters for EbA: the distance metric, the number of analogies, and the analogies' adjustment method. It is straightforward to allocate an appropriate number of bits to represent the possible values for these parameters in a binary string. This policy of encoding is quite flexible in the case we want to examine the effect of more parameters.

Project Attributes. Project attributes can be either numerical or categorical. In the first case, they are normalized between 0 and 1, in order to neutralize the side-effects of different units. This normalization is described by the type below:

$$normalized_{xy} = \frac{attr_{xy} - min_x}{max_x - min_x} \qquad (4)$$

Where $attr_{xy}$ is the value of the x-th numerical attribute for the y-th project of the dataset, min_x is the minimum value observed for the x-th attribute in the current dataset, while max_x is respectively the maximum value observed. In the case of categorical attributes, only values selected from a limited set can be assigned. As there is no order imposed among possible values, a '0' indicates that two compared values are the same, while '1' is used when they are different. In this way, the definition of similarity is compatible with that of numerical values.

Encoding of the Attribute Subsets. In the simplest case, an attribute will either participate in distance calculation or not. Just one bit is enough for each attribute indicating its presence in calculations. Finally, we have as many one-bit genes as the maximum of attributes available in the dataset, with each gene corresponding to a certain attribute.

Non-Normalized Weights Encoding (NNWE). Any real number could be applied as weighting factor to an attribute. However, we do not have to consider negative weights, since there is no use in counting the degree of unsuitability. It is also reasonable to allow values only in the range $[0, 1)$.

The next step is to choose an appropriate level of accuracy for the weighting factors. Two decimal digits are considered enough to represent all significant variations in attributes' importance. So, the length of each attribute gene will be determined by Equation (3). This Non-Normalized Weights Encoding (or NNWE) is used in the majority of the experiments run in Sect. 4. Although it performs well, it is not considered as optimum, due to the existence of weight vectors that have same effect.

An alternative encoding is also proposed, that allows weighting vectors that are linearly independent only. However, NNWE is still preferred. In the next paragraph, there is a detailed description about the need for excluding linearly dependent vectors and new problems that arise.

Normalized Weighting Vector Encoding (NWVE). In general, when applying a weighting vector to a set of attributes, the N-dimensional space defined by these attributes is modified. Weighting vectors that are linearly dependent produce the same space, in a different scale however. This means that distance between any two points will be different in each one of the modified spaces, but the relative position of these points will be the same. Thus, we should use only normalized vectors as candidate solutions, which means that a new representation has to be specified. It is difficult to generate binary words from normalized vectors, as any bit inversion or interchange would probably destroy the unary length property.

Instead, we could take advantage of the fact that all these vectors define an $(N-1)$-sphere in an N-dimensional space. Attribute weights are normally expressed in Cartesian coordinates. Equations (5), which are adapted from [20], show the relationship between hyperspherical and Cartesian coordinates in an N dimensional space, such as the one defined by the weights of N attributes.

$$x_1 = r \cos \phi_1$$
$$x_2 = r \sin \phi_1 \cos \phi_2$$
$$\cdots \tag{5}$$
$$x_{n-1} = r \sin \phi_1 \ldots \sin \phi_{n-2} \cos \phi_{n-1}$$
$$x_n = r \sin \phi_1 \ldots \sin \phi_{n-2} \sin \phi_{n-1}$$

Vectors in the hyperspherical system include a radius r and N-1 angles ϕ_i of the vector with each one of the N-1 axes. Since we are interested in the portion of space where all weights are non-negative, we can restrict the range of all $N-1$ angles to $[0, \pi/2]$. Moreover, as far as only normalized vectors are used, radius will always be equal to 1, so it can be considered as a constant. In this way, any combination of N-1 ϕ_i angles represents a vector of unary norm. Each one of them has unique influence in EbA, since linear independence among different individuals is guaranteed. The vector of angles can easily be encoded into binary strings that can be recombined without destroying the vector's unary norm. Each angle corresponds to a gene that consists of a numbers of bits. Therefore, the hyperspherical system is used to encode the attribute weights into genes, while Equation (5) is used to transform to Cartesian coordinates to calculate distances.

In this way, there is a unique bit string representation for any weighting vector. However, there is a critical drawback that is derived from the way of computation of Cartesian coordinates. In Equations (5), we can see that a weighting factor x_k, where $1 \leq k \leq n-1$, is dependent on k sines or cosines, while x_{n-1} is dependent on $n - 1$ sines. It is evident that for large values of k, x_k tends to be zero, as it is a product of a large set of numbers whose absolute values are smaller than 1.

On the other hand, the existence of few attributes could probably diminish that bias. So, given a dataset described by a small number of attributes, it is worth exploring the possibility of further improvement in estimation accuracy, when using Normalized Weighting Vector Encoding (or NWVE). Hence, we are going to perform experiments for this encoding too. In any case however, NWVE needs to be revised so as to be generally applicable.

4   Experiments

The experiments are conducted in two phases. In the first phase, the optimization method is applied in Desharnais dataset [7]. This is sort of a sanity check of the method, trying to measure the improvement, if any, of EbA after applying a global optimization with GA. In the second phase, the goal is to produce comparative results with other EbA approaches [12][17]. ISBSG is a popular database for project estimation and benchmarking [13], and there is already evidence of the performance of various methods using this dataset. In both cases, the datasets were randomly split into training and validation sets. The 2/3 of each dataset are used for the training phase, which involves applying the GA.

The results presented correspond to the remaining 1/3 of the projects. It is noted that the datasets have been split several times and the results presented are the averages of the different dataset instances. Is it also noted that any project with one or more missing values is excluded from EbA optimization.

4.1 Results on Desharnais Dataset

The global optimization method is applied using three different encodings for project attributes, as described in Sect. 3.3. Table 1 summarizes the results of the simple EbA and the three different encodings. The MMRE, MdMRE and Pred(0.25) values presented are the average values out of ten different trials.

The first thing we observe is a significant improvement in all of the evaluation criteria, when comparing the simple EbA with the GA optimized ones. Moreover, the MMRE for NNWE has been further reduced to 44.33%, compared to the 46.18% of Subset Encoding. This is due to the fact that weighting factors provide a more refined way of regulating the influence of each attribute in distance calculations.

Table 1. Results of Global Optimization approaches in Desharnais dataset

	EbA	Global Optimization with GA		
		Subset Encoding	NNWE	NWVE
MMRE %	73.02	46.18	44.43	40.67
MdMRE %	48.5	37.14	38.11	36.8
Pred(0.25) %	24.8	39.2	38.4	38.8

Finally, NWVE produced the best results in terms of MMRE and MdMRE, 40.67% and 36.8% respectively. This seems to confirm that attribute weights should be handled as normalized vectors. However, as reported in Sect. 3.3, a large number of project attributes would introduce serious bias towards certain attributes. That is the reason why we use NNWE to measure methods efficiency at the sections that follow, as ISBSG dataset makes use of a large number of attributes.

4.2 ISBSG Results and Comparison

In this test case, we have adopted the assumption made in AQUA+, including projects with effort between 10000 and 20000, as representatives of medium size projects [17]. Such an assumption is realistic, since it is possible for an experienced manager to define a priori the size of a project in general terms. After eliminating projects with missing values, a total of 65 projects remain in effort range [10000 − 20000]. Table 2 presents the results of AQUA+ for the same estimation range and the best case obtained by attribute weighting by genetic algorithms [12].

It is clear that our method outperformed Attribute Weighting GA. Actually, this was expected to happen, since the search has been expanded to parameters that had not been taken into account in the past. It also outperforms by a

Table 2. Results of EbA approaches in ISBSG dataset

	AQUA+	Attribute Weighting GA	Global Optimization GA
MMRE%	26	54	23
MdMRE%	-	34	23.4
Pred(0.25)%	72	45	59.4

limited amount AQUA+ MMRE. However, it is important to note that AQUA+ had a higher value for Pred(0.25). It should be noted that AQUA+ used a larger collection of projects as a result of tolerating missing values. In any case, method's efficiency is promising, however the elimination of a number of projects due to missing values seems to be an issue.

5   Conclusions and Future Work

The purpose of the current work was to explore the unified search space that emerges from the combination of many parameters which are incompatible at first sight. Genetic Algorithms were selected to conduct this search, due to their ability to approach global optima, no matter what the properties of the search space are. However, the key element was the encapsulation of variables that are different in nature, such as the distance metric to be used and the attribute weights, into a single encoding.

Summarizing the results for Desharnais dataset, optimization based on attribute weights outperformed attribute selection. Two different kinds of encoding were used for attribute weights. Both of them resulted in considerable improvement in prediction accuracy, while NWVE seems to be superior to NNWE. This fact supports the hypothesis that there should be a single representation for each weighting vector. However, because of some computational issues, as reported in Sect. 3.3, NWVE is not used in the rest of the experiments in the current paper. Further adjustment for dealing with this problem is subject of future research. Finally, the results in ISBSG dataset seem to support the concept of broadening the search space. By comparing with optimization attempts in earlier researches [12][17], evidence was found that the optimum configuration is obtained with combinations of multiple parameters.

The generation of the current experimental data involved distance metrics, analogy numbers, attribute weights and a small set of adjustment methods only. Of course, there are more options than the ones reported. Nevertheless, the encoding adopted can be easily expanded to contain more alternatives for any parameter. Such an expansion demands no change in the rationale of the encoding. An even more extensive search is an interesting subject of future work. Another issue to be looked into is to deal with missing values. Moreover, Equation (2), which was adopted for fitness evaluation, is ad-hoc and needs further investigation.

References

1. Albrecht, A.: Software function, source lines of code, and development effort prediction: a software science validation. IEEE Trans. Softw. Eng. 9(6), 639–648 (1983)
2. Angelis, L., Stamelos, I.: A simulation tool for efficient analogy based cost estimation. Empirical Softw. Engg. 5(1), 35–68 (2000)
3. Bibi, S., Stamelos, I., Angelis, L.: Combining probabilistic models for explanatory productivity estimation. Inf. Softw. Technol. 50(7-8), 656–669 (2008)
4. Boehm, B.: Software engineering economics, vol. 1. Prentice-Hall, Englewood Cliffs (1981)
5. Chiu, N., Huang, S.: The adjusted analogy-based software effort estimation based on similarity distances. J. Syst. Softw. 80(4), 628–640 (2007)
6. Conte, S.D., Dunsmore, H.E., Shen, V.Y.: Software Engineering Metrics and Models. Benjamin-Cummings Publishing Co., Inc. (1986)
7. Desharnais, J.M.: Analyse statistique de la productivitie des projets informatique a partie de la technique des point des fonction (1989)
8. Efron, B., Gong, G.: A leisurely look at the bootstrap, the jackknife, and cross-validation. The American Statistician 37(1), 36–48 (1983)
9. Finnie, G.R., Wittig, G.E., Desharnais, J.M.: A comparison of software effort estimation techniques: using function points with neural networks, case-based reasoning and regression models. J. Syst. Softw. 39(3), 281–289 (1997)
10. Goldberg, D.: Genetic algorithms in search, optimization, and machine learning. Addison-Wesley, Reading (1989)
11. Holland, J.: Adaptation in natural and artificial systems. The University of Michigan Press, Ann Arbor (1975)
12. Huang, S., Chiu, N.: Optimization of analogy weights by genetic algorithm for software effort estimation. Inf. Softw. Technol. 48(11), 1034–1045 (2006)
13. International Software Benchmark and Standards Group. ISBSG Data Release 10
14. Kemerer, C.: An empirical validation of software cost estimation models. Communications of the ACM 30(5), 416–429 (1987)
15. Keung, J., Kitchenham, B.: Analogy-X: Providing statistical inference to analogy-based software cost estimation. IEEE Trans. Softw. Eng. 34(4), 471–484 (2008)
16. Kitchenham, B., Mendes, E., Travassos, G.: Cross versus within-company cost estimation studies: a systematic review. IEEE Trans. Softw. Eng. 33(5), 316–329 (2007)
17. Li, J., Ruhe, G.: Analysis of attribute weighting heuristics for analogy-based software effort estimation method AQUA+. Empirical Softw. Engg. 13(1), 63–96 (2008)
18. Li, J., Ruhe, G., Al-Emran, A., Richter, M.: A flexible method for software effort estimation by analogy. Empirical Softw. Engg. 12(1), 65–106 (2007)
19. Oliveira, A., Braga, P., Lima, R., Cornélio, M.: GA-based method for feature selection and parameters optimization for machine learning regression applied to software effort estimation. Inf. Softw. Technol. 52(11), 1155–1166 (2010)
20. Schweizer, W.: Numerical quantum dynamics. Kluwer Academic Publishers, Dordrecht (2001)
21. Shepperd, M., Schofield, C.: Estimating software project effort using analogies. IEEE Trans. Softw. Eng. 23(11), 736–743 (1997)

The Impact of Sampling and Rule Set Size on Generated Fuzzy Inference System Predictive Accuracy: Analysis of a Software Engineering Data Set

Stephen G. MacDonell

SERL, School of Computing and Mathematical Sciences, AUT University,
Private Bag 92006, Auckland 1142, New Zealand
smacdone@aut.ac.nz

Abstract. Software project management makes extensive use of predictive modeling to estimate product size, defect proneness and development effort. Although uncertainty is acknowledged in these tasks, fuzzy inference systems, designed to cope well with uncertainty, have received only limited attention in the software engineering domain. In this study we empirically investigate the impact of two choices on the predictive accuracy of generated fuzzy inference systems when applied to a software engineering data set: sampling of observations for training and testing; and the size of the rule set generated using fuzzy c-means clustering. Over ten samples we found no consistent pattern of predictive performance given certain rule set size. We did find, however, that a rule set compiled from multiple samples generally resulted in more accurate predictions than single sample rule sets. More generally, the results provide further evidence of the sensitivity of empirical analysis outcomes to specific model-building decisions.

Keywords: Fuzzy inference, prediction, software size, source code, sampling, sensitivity analysis.

1 Introduction

Accurate and robust prediction of software process and product attributes is needed if we are to consistently deliver against project management objectives. In recent years fuzzy inference systems (FIS) have gained a degree of traction (with empirical software engineering researchers if not practitioners) as an alternative or complementary method that can be used in the prediction of these attributes [1-4].

Empirical software engineering research that has focused on prediction is normally performed using a standard sample-based approach in which a data set is split into two sub-samples, the first, larger set for building a predictive model and the second for non-biased evaluation of predictive performance (for instance, see [1]). Any reasonable splitting strategy can be used. Typically, two-thirds of the total sample is randomly allocated to the build subset, leaving one third of the observations to comprise the test subset. Given the incidence of outlier observations and skewed distributions in software engineering data sets [5], stratified sampling may be useful, but it has not been widely employed in software engineering analyses. Similarly,

L. Iliadis et al. (Eds.): EANN/AIAI 2011, Part II, IFIP AICT 364, pp. 360–369, 2011.

where information on the ordering of observations is available this could be used to further inform the sampling strategy. Again, however, such an approach has not been used to any great extent in empirical software engineering research [6].

In this paper we analyze the sensitivity of prediction outcomes to systematic changes in two parameters relevant to fuzzy inference model building and testing: data sampling and rule set size. In essence, then, this is a paper focused on the *infrastructure* of predictive modeling – our intent is to highlight the potential pitfalls that can arise if care is not taken in the building and testing of models. In the next section we describe the empirical work we have undertaken using a software engineering data set. Section 3 reports the results of our analysis. A discussion of the key outcomes is provided in Section 4, after which we briefly address related work. We conclude the paper and highlight opportunities for future research in Section 6.

2 Empirical Analysis

In this section we describe the empirical analysis undertaken to assess the impact of changes in both data sampling and rule set size on the predictive accuracy of fuzzy inference systems. After providing a description of the data set employed in the analysis we explain the three approaches used in constructing our prediction systems.

2.1 The Data Set

The data set we have used here is a simple one in that it involves a small number of variables. This does not detract, however, from its effectiveness in illustrating the impact of sample selection and rule set size, the principal issues of concern in this study. The data were collected in order to build prediction systems for software product size. The independent variables characterized various aspects of software specifications, including the number of entities and attributes in the system data model, the number of data entry and edit screens the system was to include, and the number of batch processes to be executed. As each implemented product was to be built using a 4GL (PowerHouse) our dependent variable was the number of non-comment lines of 4GL source code. The systems, which provided business transaction processing and reporting functionality, were all built by groups of final-year undergraduate students completing a computing degree.

A total of 70 observations were available for analysis. Although this is not a large number, it did represent the entire population of systems built in that environment over a period of five years. It is in fact quite a homogeneous data set – although the systems were built by different personnel over the five-year period they were all constructed using the same methodology and tool support by groups of four people of similar ability. Furthermore, it is quite common in software engineering to have access to relatively small numbers of observations (for example, see [7]), so the data set used here is not atypical of those encountered in the domain of interest.

In a previous investigation [2] we identified that two of the 8 independent variables were significantly correlated to source code size and were also not correlated to one another. These were the total number of attributes in the system data model (Attrib) and the number of non-menu functions (i.e. data entry and reporting modules rather than menu selection modules) in the system's functional hierarchy (Nonmenu). We therefore used these two variables as our predictors of size.

2.2 FIS Development

We adopted three model-building approaches in order to investigate the impact of sampling and fuzzy rule set size on the coverage and accuracy of source code size predictions. We first built membership functions and rule sets (via the algorithms below) using the entire data set of 70 observations (Full Approach); we also randomly generated ten separate build subsets of 50 cases that we used to build fuzzy prediction systems (Sampled Approach); finally we analyzed the ten rule sets generated by the sampling process to create a 'mega' rule set comprising the 50 most frequently generated rules (Top 50 Approach). For each of the three approaches we generated fuzzy inference systems comprising from one to fifty rules. Each FIS was tested against the ten test subsets of 20 observations. While performance was assessed using a variety of accuracy indicators, we report here the results in terms of the sum of the absolute residuals and the average and median residual values, as these are preferred over other candidate measures [8]. We also assessed the coverage of each FIS i.e. the proportion of the test subsets for which predictions were generated (meaning the rule set contained rules relevant to the test subset observations that were then fired to produce a predicted size value).

Previous research using this data set [2] had indicated that the three variables of interest were most effectively represented by seven triangular membership functions, mapping to the concepts <VerySmall, Small, SmallMedium, Medium, MediumLarge, Large, VeryLarge>. The membership function extraction algorithm is as follows:

1. select an appropriate mathematically defined function for the membership functions of the variable of interest (i), say $f_i(x)$

2. select the number of membership functions that are desired for that particular variable, m_i functions for variable i (m_i may be set by the user or may be found automatically, to desired levels of granularity and interpretability)

3. call each of the m_i functions $f_{ij}([x])$ where $j = 1 \ldots m_i$ and $[x]$ is an array of parameters defining that particular function (usually a center and width parameter are defined, either explicitly or implicitly)

4. using one-dimensional fuzzy c-means clustering on the data set find the m_i cluster centers, c_{ij} from the available data (m_i may be set by the user or may be found automatically, to desired levels of granularity and interpretability)

5. sort the cluster centers c_{ij} into monotonic (generally ascending) order for the given i

6. set the membership function center for f_{ij}, generally represented as one of the parameters in the array $[x]$, to the cluster center c_{ij}

7. set the membership function widths for f_{ij} in $[x]$ such that $\sum_{n=1}^{mi} f_{in}$ ($[c_{in}, \ldots]$) = 1, or as close as possible for the chosen $f(x)$ where this cannot be achieved exactly (for example for triangular membership functions each function can be defined using three points, a, b, and c where a is the center of the next smaller function and c is the center of the next larger function).

Rules were extracted using the same clustering process with multiple dimensions (matching the number of antecedents plus the single consequent):

1. start with known membership functions $f_{ij}([x])$ for all variables, both input and output, where j represents the number of functions for variable i and $[x]$ is the set of parameters for the particular family of function curves

2. select the number of clusters k (which represents the number of rules involving the k-1 independent variables to estimate the single output variable)

3. perform fuzzy c-means clustering to find the centers (i dimensional) for each of the k clusters

4. for each cluster k with center c_k

    (a) determine the kth rule to have the antecedents and consequent f_{ij} for each variable i where f_{ij} (c_k) is maximized over all j

    (b) weight the rule, possibly as $\prod_{n=1}^{i} f_{ij}\,(c_k)$ or $\sum_{n=1}^{i} f_{ij}\,(c_k)$

5. combine rules with same antecedents and consequents, either summing, multiplying, or bounded summing rule weights together

6. (optionally) ratio scale all weights so that the mean weight is equal to 1.0 to aid interpretability.

3 Results

We now consider the outcomes of each of the three approaches in turn – Full, Sampled and Top 50.

3.1 Full Approach

Development of an FIS using the entire set of observations is the most optimistic of the three approaches, and in fact represents more of a model-fitting approach rather than one of unbiased prediction where a hold-out sample is used. We employed this approach, however, to provide a benchmark for comparative model performance. The clustering approaches described above were used to build membership functions and rule sets based on all 70 observations. Essentially this meant that there were fifty distinct FIS produced, comprising from one to fifty rules. These FIS were then tested against each of the ten test subsets produced from the Sampled Approach (described below). Accuracy and coverage were assessed for each FIS, illustrated by the example shown for the sample 4 test subset in Figures 1 and 2 and Table 1.

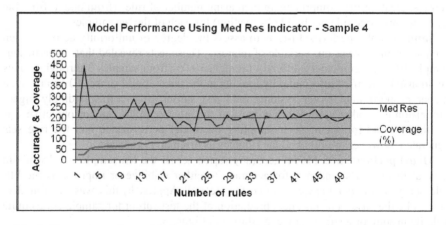

Fig. 1. Change in median residual and coverage with increasing rule set size for sample 4

Table 1. Overall coverage for sample 4

Coverage	Sample 4			
100%	22	44%	Mean	86%
80%-99%	16	32%	Med	95%
60%-79%	9	18%		
40%-59%	1	2%		
20%-39%	2	4%		
0%-19%	0	0%		

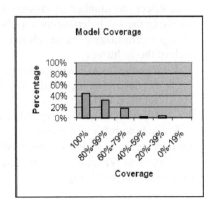

Fig. 2. Graphical depiction of coverage

For sample 4 we can see in Figure 1 that predictive accuracy measured using the median residual value taken over the predictions made shows some volatility but is generally stable within the range 150-250 SLOC, once the rule set size reaches a value of around 18. At this point also we can see that coverage becomes stable at close to 100% i.e. rules were fired and predictions made for close to all 20 test subset observations. Further evidence of the generally high degree of test observation coverage is provided in Table 1 and Figure 2. This latter result is not unexpected as in the Full Approach all 70 observations were used to build the FIS, and so the test observations had been 'seen' and accommodated during model-building.

3.2   Sampled Approach

As stated above, in this approach we randomly allocated 50 of the 70 observations to a build subset and left the remaining 20 observations for testing, repeating this process ten times to create ten samples. FIS development then used only the build subsets, creating FIS comprising an increasing number of rules (from one to fifty, the latter meaning one rule per build observation) for each of the ten samples.

Sampling. We performed t-tests to assess the degree of uniformity across the ten samples, using Size as our variable of interest. These tests revealed that, in particular, sample 10 was significantly different (at alpha=0.05) from samples 3 through 8. This could also be verified informally by looking at the data – build subset 10 comprised a higher proportion of larger systems (and test subset 10 a correspondingly higher proportion of smaller systems) than most of the other samples. As our objective was to illustrate the impact of such sampling on model performance and the rule sets generated, however, we retained the samples as originally produced.

Model performance. Model accuracy varied across the ten samples, and while in general there was a tendency for accuracy to either stabilize or improve as the FIS rule set grew (as per Figure 1 above for the Full Approach) this was not found to always be the case. For instance, prediction of the test subset for sample 2 was quite volatile in spite of good coverage, as shown in Figure 3.

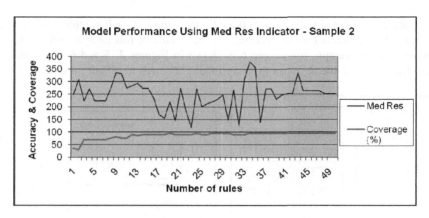

Fig. 3. Change in median residual and coverage with increasing rule set size for sample 2

Table 2. Best models for each sample in terms of accuracy measures (when coverage ignored)

Sample	1	2	3	4	5	6	7	8	9	10
Coverage	80%	90%	90%	100%	90%	95%	80%	90%	85%	95%
Minimum rules?	19	23	31	34	18	14	12	35	29	26
Abs Res	3100	4340	5430	6156	4077	4623	5257	5996	6823	5000
Ave Res	194	241	302	308	227	243	329	333	401	263
Med Res	111	118	163	177	159	200	266	288	261	176

Table 3. Best models for each sample in terms of accuracy measures (maximum coverage)

Sample	1	2	3	4	5	6	7	8	9	10
Coverage	95%	95%	100%	100%	100%	100%	95%	100%	100%	100%
Minimum rules?	44	36	40	34	50	15	38	49	45	36
Abs Res	5825	5017	6394	6156	7193	4807	5970	5842	6789	5028
Ave Res	307	264	320	308	360	240	314	292	339	251
Med Res	187	139	208	177	274	203	177	296	269	173

The best models achieved for each sample are shown in Tables 2 and 3. The first table depicts the results irrespective of coverage whereas the second considers only those FIS that achieved maximum coverage for each test subset. What is evident in both tables is the variation in values for all three accuracy measures – for instance, the median residual value in the best maximum coverage models (Table 3) varies from a low of 139 SLOC (for sample 2) up to 296 SLOC (for sample 8).

Rule Distribution. For each of the ten samples we also considered the composition of the generated rule sets, comprising from one to fifty rules each at increments of one, meaning a total of 1275 rules per sample. Each rule was a triple made up of two antecedents and a single consequent e.g. IF <Attrib> IS [Small] AND <Nonmenu> IS [SmallMedium] THEN <Size> IS [Small]. Our analysis considered the frequency of particular memberships in the rule sets for each variable (e.g. count of '<Attrib> IS [VerySmall]'), as well as the frequency of each complete rule across the sets. Space precludes full reporting of the results of this analysis across all ten samples but an example for sample 6 is shown in Figures 4, 5 and 6. (Note that in these figures Set Value '1' maps to 'VerySmall', '2' maps to 'Small', and so on.)

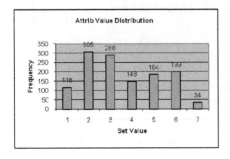

Fig. 4. Membership frequency for Attrib

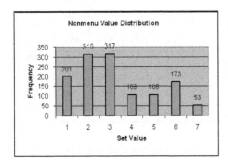

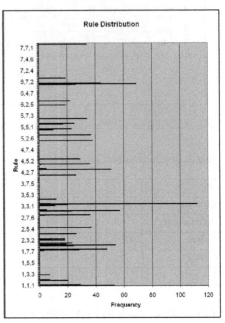

Fig. 5. Membership frequency for Nonmenu   **Fig. 6.** Rule frequency for sample 6

For this sample 47 distinct rules of the 343 possible (the latter representing each possible triple from '1,1,1' through to '7,7,7') made up the 1275 rules in the fifty sets, the most frequently occurring rule for this sample being '3,3,4': IF <Attrib> IS [SmallMedium] AND <Nonmenu> IS [SmallMedium] THEN <Size> IS [Medium], with 112 instances over the fifty rule sets (shown as the longest bar in Figure 6). Across the ten samples, the number of distinct rules varied from 45 for sample 8 up to 60 for sample 1, and these sets were distinct in part, meaning that over the ten samples 131 distinct rules were generated from the 343 possible candidates.

3.3 Top 50 Approach

The above analysis of the ten samples allowed us to also identify the Top 50 distinct rules *across* the sampled data sets i.e. the 50 rules that were generated most frequently (where 50 was an arbitrary choice). The most common rule (at 506 instances) was: IF <Attrib> IS [MediumLarge] AND <Nonmenu> IS [MediumLarge] THEN <Size> IS [MediumLarge], whereas the 50th most common rule, with 89 instances, was IF <Attrib> IS [SmallMedium] AND <Nonmenu> IS [Small] THEN <Size> IS [VerySmall]. We then applied this set of 50 rules to the ten test subsets, in the expectation that over the entire set they may outperform the potentially over-fitted Sampled rule sets generated from each specific build subset. We compare the results obtained from this approach to those derived from the originally Sampled approach.

3.4 Comparison of Approaches

As might be expected, the Full approach models performed well across the ten samples. When considered in terms of the sum of absolute residual and average and median residual measures, the Full model was the most accurate in 14 of the 30 cases (30 being 10 samples by 3 error measures). As noted above, however, this does not represent an unbiased test of predictive accuracy as the Full models were built using the entire data set. Our comparison is therefore focused on the Sampled and Top 50 approaches. Figure 7 illustrates one such comparison, based on the median residual measure for sample 3. We can see that in this case the model created from the Top 50 approach performed better than those from the Sampled approach for FIS comprising up to 29 rules, but beyond that the Sampled approach was generally superior.

The pattern exhibited for sample 3 was not a common one across the ten samples, however. The data presented in Tables 4 and 5 indicate that, in general, the Top 50 model ('Top') outperformed the specific FIS developed with the Sampled approach ('Sam.'). Of the 30 comparisons, 9 favored the Sampled approach. (Note that totals can exceed 100% for a sample because the two methods may achieve equivalent performance for a given number of rules, and so are both considered 'Best'.)

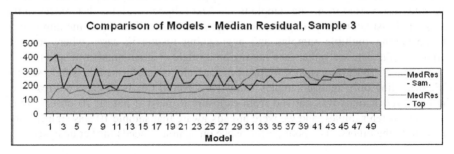

Fig. 7. Sample 3 comparison in terms of median residual (Sam.: Sampled, Top: Top 50)

Table 4. Comparison of model accuracy across the ten samples (best shown in bold type)

Sample	1		2		3		4		5	
Approach	Sam.	Top	Sam.	Top	Sam.	Top	Sam.	Top	Sam.	Top
Abs Res Best	**52%**	50%	50%	**52%**	**52%**	48%	14%	**86%**	14%	**86%**
Ave Res Best	38%	**64%**	24%	**78%**	36%	**64%**	10%	**90%**	8%	**92%**
Med Res Best	42%	**60%**	**56%**	46%	40%	**60%**	30%	**72%**	0%	**100%**
Sample	6		7		8		9		10	
Approach	Sam.	Top	Sam.	Top	Sam.	Top	Sam.	Top	Sam.	Top
Abs Res Best	16%	**84%**	**82%**	20%	26%	**74%**	30%	**72%**	**70%**	30%
Ave Res Best	24%	**76%**	**74%**	28%	6%	**94%**	20%	**82%**	**62%**	38%
Med Res Best	20%	**82%**	**58%**	46%	12%	**88%**	22%	**80%**	**72%**	28%

Table 5. Comparison of model accuracy – summary statistics

Summary Statistic	Mean		Median	
Approach	Sam.	Top	Sam.	Top
Abs Res Best	41%	60%	40%	62%
Ave Res Best	30%	71%	24%	77%
Med Res Best	35%	66%	35%	66%

4 Discussion

The impact of sampling was evident in the t-tests of difference in Size across the ten samples. The mean value for Size varied from 1043 to 1211 SLOC in the build subsets, and from 843 to 1264 in the test subsets, with a significant difference evident in relation to sample ten. Similarly, FIS model performance also varied over the samples in terms of both the residual values and the coverage of the test observations. For example, the outright best model for sample 1 comprised just 19 rules, had an absolute error value of 3100 SLOC and achieved 80% coverage, whereas the sample 4 best model achieved 100% coverage using 34 rules and an absolute residual of 6156 SLOC. No consistent pattern was evident in relation to rule set size even for FIS achieving maximum coverage – across the ten samples the most accurate FIS was obtained from rule sets varying in size from 15 to 49 rules (out of a possible 50 rules). The rules also varied from one sample to the next (aligned with the differences in the data) – 131 distinct rules were generated. These outcomes all reinforce the importance of considering multiple samples and aggregating results in cases where there is variance in the data set, rather than relying on a single or small number of splits.

The Top 50 rules (in terms of frequency) generally outperformed the single-sample FIS, struggling only against samples 7 and 10. Overall this is an expected result, as the mega-set of rules was derived from exposure to most if not all 70 observations, albeit across ten samples, and we were already aware that sample 10 was significantly different to the others. The poor performance on sample 7 needs further analysis.

5 Related Work

As far as we are aware there have been no prior *systematic* assessments of the impact of these parameters on analysis outcomes in the empirical software engineering community. While several studies have used FIS in software engineering (e.g. [3]) only single splits of the data set have been used, and there has been no specific assessment of the impact of rule set size. That said, [4] describes the selection of rules based on error thresholds. Related work has also been undertaken in other domains. For instance, [9] considered the trade-off between a smaller rule set and the accuracy of decisions made in regard to credit scoring, concluding that substantially fewer rules did not lead to proportional reductions in accuracy. The author notes, however, that "[t]he cost in accuracy loss one is willing to pay for the benefit of a smaller rule-base is entirely domain and organization dependent." [9, p.2787]. [10] investigated the impact of stratification of training set data in data mining (employing evolutionary algorithms) noting significant differences in classification outcomes. Within the software engineering research community the influence of sampling on data analysis has received some attention, but as this does not relate to fuzzy inference the interested reader is referred to [11] for an example of this work.

6 Conclusions and Future Work

We have conducted an analysis of the sensitivity of predictive modeling accuracy to changes in sampling and rule set size in the building and testing of fuzzy inference systems for software source code size. Our results indicate that both aspects influence analysis outcomes – in particular, the splits of data across build and test sets lead to significant differences in both predictive accuracy and test set coverage.

We are continuing to work on aspects of prediction infrastructure – in particular, our current work is focused on considering the impact of temporal structure on modeling outcomes as well as the development of more suitable error measures for accuracy assessment. Further research should also address the degree to which stability in predictor variables over the life of a project affects predictions of size, quality and other aspects of project management interest.

References

1. Azzeh, M., Neagu, D., Cowling, P.: Improving Analogy Software Effort Estimation Using Fuzzy Feature Subset Selection Algorithm. In: 4th Intl. Workshop on Predictive Models in Softw. Eng., pp. 71–78. ACM Press, New York (2008)
2. MacDonell, S.G.: Software Source Code Sizing Using Fuzzy Logic Modeling. Info. and Softw. Tech. 45(7), 389–404 (2003)
3. Huang, X., Ho, D., Ren, J., Capretz, L.F.: A Soft Computing Framework for Software Effort Estimation. Soft Computing 10, 170–177 (2006)
4. Ahmed, M.A., Saliu, M.O., AlGhambi, J.: Adaptive Fuzzy Logic-based Framework for Software Development Effort Prediction. Info. and Softw. Tech. 47, 31–48 (2005)
5. Kitchenham, B., Mendes, E.: Why Comparative Effort Prediction Studies may be Invalid. In: 5th Intl. Conf. on Predictor Models in Softw. Eng., ACM DL. ACM Press, New York (2009)
6. MacDonell, S.G., Shepperd, M.: Data Accumulation and Software Effort Prediction. In: 4th Intl Symp. on Empirical Softw. Eng. and Measurement, ACM DL. ACM Press, New York (2010)
7. Song, Q., Shepperd, M.: Predicting Software Project Effort: A Grey Relational Analysis Based Method. Expert Systems with Applications 38, 7302–7316 (2011)
8. Kitchenham, B.A., Pickard, L.M., MacDonell, S.G., Shepperd, M.J.: What Accuracy Statistics Really Measure. IEE Proc. - Software 148(3), 81–85 (2001)
9. Ben-David, A.: Rule Effectiveness in Rule-based Systems: A Credit Scoring Case Study. Expert Systems with Applications 34, 2783–2788 (2008)
10. Cano, J.R., Herrera, F., Lozano, M.: On the Combination of Evolutionary Algorithms and Stratified Strategies for Training Set Selection in Data Mining. Applied Soft Computing 6, 323–332 (2006)
11. Shepperd, M., Kadoda, G.: Using Simulation to Evaluate Prediction Techniques. In: 7th Intl. Symp. on Softw. Metrics, pp. 349–359. IEEE CS Press, Los Alamitos (2001)

Intelligent Risk Identification and Analysis in IT Network Systems

Masoud Mohammadian

University of Canberra,
Faculty of Information Sciences and Engineering,
Canberra, ACT 2616, Australia
masoud.mohammadian@canberra.edu.au

Abstract. With ever increasing application of information technologies in every day activities, organizations face the need for applications that provides better security. The existence of complex IT systems with multiple interdependencies creates great difficulties for Chief Security Officers to comprehend and be aware of all potential risks in such systems. Intelligent decision making for IT security is a crucial element of an organization's success and its competitive position in the marketplace. This paper considers the implementation of an integrated attack graph and a Fuzzy Cognitive Maps (FCM) to provide facilities to capture and represent complex relationships in IT systems. By using FCMs the security of IT systems can regularly be reviewed and improved. What-if analysis can be performed to better understand vulnerabilities of a designed system. Finally an integrated system consisting of FCM, Attack graphs and Genetic Algorithms (GA) is used to identify vulnerabilities of IT systems that may not be apparent to Chief Security Officers.

Keywords: Network Security, Risks Analysis, Intelligent Systems, Attack Graphs.

1 Introduction

Current IT systems are accessed anytime from any locations using the Internet. This approach increases the security risks in ever increasing IT applications and networks. IT systems are complicated with large number of interdependent networks and IT facilities. In such systems there is a need for proactive and continuous security risk assessment, identification, verification and monitoring.

A graphical representation of an IT system can improve the understanding of the designer of a system and mitigate risks of attack to designed systems. Such a graphical representation can assist in documenting security risks and identifying possible paths attackers may consider to attack a system for their undesirable goals.

Attack graphs [1, 8] are designed after analyzing an IT system purpose, its components and any set of potential attacker undesirable goals. These goals may include system's disruptions, intrusion and misuse by an attacker [2, 3]. During design, implementation and verification of an attack graph possible attacks and undesirable goals of diverse attackers are considered. The skill, access, and goals of

L. Iliadis et al. (Eds.): EANN/AIAI 2011, Part II, IFIP AICT 364, pp. 370–377, 2011.

attackers are also considered. Attack graphs are created based on several notations. These notations are attacker's goals, trust boundaries, sub-goals and paths in an attack graph to reach attacker's goals from a part of a system.

Sub-goals may be used by attackers to move through a system to reach their goals. Paths through an attack graph are identified to reach attacker's goals. With complexity of existing systems drawing attack graphs are becoming increasingly difficult. Due to the complexity and difficulty of designing attack graphs an attack graph may be flawed. Attack graphs provides a graphical representation of an IT system which make it easier to understand however an attack graph does not provide any facilities to analyze and assess different risks and possible attacks that may exist in a systematic way. An attack graphs documents the risks known at the time the system is designed. However an attack graph does not provide facilities to perform concrete risk analysis such as what-if and scenario analysis to test the designed system for possible risk of attacks. In this paper, a Fuzzy Cognitive Map (FCM) is used with graph attacks to provide facilities that will enable the system architects to perform what-if scenario analysis to better understand vulnerabilities of their designed system. Fuzzy Cognitive Maps (FCM) [4, 5, 6] are graph structures that provide a method of capturing and representing complex relationships in a system. Application of FCM has been popular in modeling problems with low or no past data set or historical information [6]. This paper proposes also a Genetic Algorithm (GA) to automatically create attack scenarios based on the given IT system. These scenarios are then evaluated using a FCM and the results are provided for analysis and mitigation of security risks. The novelty of the proposed systems based on attack graph, FCM and GA is to provide the automatic facilities to identify security vulnerabilities in complex IT systems. Our proposed approach comprises of:

(i)   a formal graphical representation of the IT systems using attack graphs,
(ii)   conversion of an attack graph into Fuzzy Cognitive Maps (FCM) to allow calculation and what-if scenario analysis,
(iii)   a GA which creates attack scenarios to be passed into a fuzzy cognitive maps to assess the results of such attacks.

Our novel approach proposes the use of attach graphs, FCM and GA to represent a give IT systems and to score each attack. In particular, a security risk value is assigned to every given attack. Such a measure of security risk provides the ability to re-assess such IT systems and modify such systems to reduce security risks.

2   Attack Graphs, Fuzzy Cognitive Maps and Genetic Algorithms

A graphical representation of a system can improve the understanding of the designer of a system and mitigate risks of attack to designed systems. Attack graphs [1, 8] are designed after analyzing a system purpose, its components and any set of potential attacker undesirable goals. Attack graphs are created based on several notations. These notations are attacker's goals, trust boundaries, sub-goals and paths in an attack graph to reach attacker's goals from a part of a system. Attacker's goals are identified on an attack graph using octagons placed at the bottom of an attack graph. Trust boundaries separate components of a system that are of different trust levels.

Sub-goals are represented using AND and OR nodes. An AND node is represented by a circle and an OR node is represented by a triangle. Sub-goals may be used by attackers to move through a system to reach their goals. Paths through an attack graph are identified to reach attacker's goals [1]. With complexity of existing systems creating attack graphs are becoming increasingly difficult. Due to the complexity and difficulty of designing attack graphs an attack graph may be flawed.

Attack graphs provides a graphical representation of a system which make it easier to understand however an attack graph does not provide any facilities to analyze and assess different risks and possible attacks that may exist in a systematic way. Fuzzy Cognitive Maps (FCM) [4, 5, 6] is graph structures that provide a method of capturing and representing complex relationships in a system. A FCM provides the facilities to capture and represent complex relationships in a system to improve the understanding of a system. A FCM uses scenario analysis by considering several alternative solutions to a given situation [9, 10]. Concepts sometimes called nodes or events represent the system behavior in a FCM. The concepts are connected using a directed arrow showing causal relations between concepts. The graph's edges are the casual influences between the concepts. The development of the FCM is based on the utilization of domain experts' knowledge. Expert knowledge is used to identify concepts and the degree of influence between them. A FCM can be used in conjunction with attack graph to provide the system architects with possibilities for what-if analysis to understand the vulnerability of the system and perform risk analysis and identification. Using FCM it is possible to identify paths through an attack graph to reach attacker's goals. Concepts in a FCM represent events and concepts are connected together with edges that describe relationships between concepts. These relationships increase or decrease of likelihood of a concept (i.e. event) to occur when other concept/s (events) occurs. Values on each edge in a FCM represent strengths or weakness of the relationship between concepts. The values on each edge are in the interval range of [−1, 1] which indicate the degree of influence of a concept to another concept. A positive value represents an increase in strength of a concept to another concept while a negative value indicates decrease in its influence to another concept. Each FCM has an activation threshold. This activation threshold provides the minimum strength required in a relationship to trigger and activate a concept. Drawing a FCM for a system requires knowledge of system's architecture. The activation level of concepts participating in a FCM can be calculated using specific updating equations in a series of iterations. A FCM can reach equilibrium or it can exhibit limit cycle behaviour. Once the system reaches equilibrium the decision-makers can use this information to make decisions about the system.

If limit cycle is reached decision-making is impossible. When limit cycle reached the experts are asked to consider the FCM and provide advice in changing the weights or one or more concepts of the FCM. The mathematical model behind the graphical representation of the FCM consists of a $1 x n$ state vector I. This state vector represents the values of the n concepts and $n x n$ weight matrix W_{IJ} represents value of weights between concepts of C_i and C_j. For each concept in a FCM a value one or zero is assigned. One represent the existence of that concept at a given time and zero represent none-exist of the respective concept. A threshold function is used in FCM. The threshold function used in this paper is sigmoid function [6] as shown below:

$$C_i(t_{n+1}) = S\left[\sum_{K=1}^{N} e_{KI}(t_n)C_k(t_n)\right] \qquad (1)$$

Having constructed an attack graph, analysts can create a FCM based on the developed attack graph and allocate values to each edge and perform analysis of various risk and vulnerabilities in a given system using What-If analysis [4, 5]. An attack graph contains several notations as explained earlier. For an analyst to be able to perform What-If analysis an attack graphs must be converted into a FCM. A FCM have some representational limits. A FCM does not represent AND and OR operators. Therefore to convert an attack graph to a FCM graph, AND and OR operators needs to be removed without changing the design of the system. This needs careful consideration. Sub-goals on an attack graph are joined using AND/OR operators. Therefore to remove AND/OR operators each sub-goal using AND operator needs to be represented as a concept by joining all concepts that are joined using AND operator on a FCM. OR operators are removed and concepts joined by OR operator are directly connected to the subsequent concept of the OR node. Paths through an attack graph are represented using edges on a FCM with weight attached to them. Paths connect concepts in FCM. The weights on the edges are then assigned by the system architects accordingly. Genetic Algorithms [7] are powerful search algorithms based on the mechanism of natural selection and use operations of reproduction, crossover, and mutation on a population of strings. A set (population) of possible solutions, in this case, a coding of the attack scenarios for an IT system, represented as a string of zero and ones. New strings are produced every generation by the repetition of a two-step cycle. First each individual string is decoded and its ability to solve the problem is assessed. Each string is assigned a fitness value, depending on how well it performed. In the second stage the fittest strings are preferentially chosen for recombination to form the next generation. Recombination involves the selection of two strings, the choice of a crossover point in the string, and the switching of the segments to the right of this point, between the two strings (the cross-over operation). A string encoded this way can be represented as:

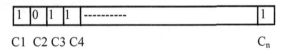

C1 C2 C3 C4 C_n

Each individual string represents an attack scenario. Each scenario is evaluated by a FCM representing a given IT system based upon a fitness value which is specific to the system. At the end of each generation, (two or more) copies of the best performing string from the parent generation is included in the next generation to ensure that the best performing strings are not lost. GA performs then the process of selection, crossover and mutation on the rest of the individual strings. The process of selection, crossover and mutation are repeated for a number of generations till a satisfactory value is obtained. We define a satisfactory value as one whose fitness value differs from the desired output of the system by a very small value. In our case we would like to find out the string that represents the most undesirable attack on our given IT system.

3 Simulation

In this paper a case study based on an IT system by S. Gupta and J. Winstead [1] is adapted. This case study will be used to represent how an attack graph can be converted into a FCM. The FCM graph created then is used to perform What-If analysis using GA. Figure 1 displays an attack graph presented by [1]. In this attack graph designer's goal was to create an IT system to protect sensitive data at a distributed set of sites with variety of constraints on development time, hardware availability, and limited business process changes [1]. The designers used encryption technology to encrypt financial information on customer sites using a set of symmetric keys distributed globally throughout the enterprise. All systems in this architecture are required to be able to read encrypted messages. The issue of physical protection of the system on customer sites was also considered. An important requirement of the systems was the availability of system in case of failure and lack of connectivity [1]. The designers with assistance of external reviewers of the system identified attack goals [1]. The attack goals identified were:

• unauthorized access to a sensitive file on the system for direct financial gain,
• unauthorized access to the encryption keys to gain access to sensitive data on other systems and
• unauthorized control of the host itself for use as a launching point for other attacks [1].

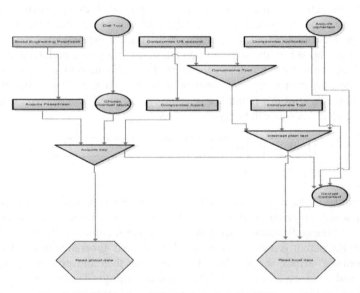

Fig. 1. Attack Graph for case study [1]

The attack graph in Figure 1 can be converted to a FCM (as shown in Figure 2) to provide the system architects with possibilities for what-if analysis and to understand the vulnerabilities and risks associated with this system. Using FCM it is now possible to identify and evaluate each path through FCM for each attacker's goals. The relationships details among all concepts in Figure 2 can be displayed using the following matrix E. The opinion of the experts and system designer is required to determine the weights of the different causal links and the initial activation level for each concept. In this scenario the author has carefully considered the system and provided the weights for the FCM shown in Figure 2. To simplify further and ease FCM in Figure 2 the following abbreviations are used for each concept: C1 = Social Engineering Passphrase, C2 = Call Tool and Chosen plaintext attack, C3 = Call Tool, C4 = Compromise OS account, C5 = Compromise Application, C6 = Acquire ciphertext and Decrypt Ciphertext, C7 = Acquire Passphrase, C8 = Compromise Agent, C9 = Compromise Tool, C10 = Impersonate Tool, C11 = Acquire Key, C12 = Read global data, C13 = Read local data. Now what-If analysis can proceed by using the matrix E. In this scenario the threshold is set to be 0.5.

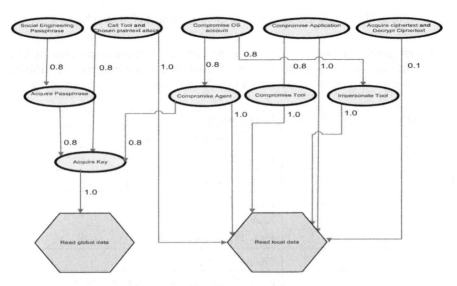

Fig. 2. A FCM displaying the routes an attacker could take to compromise the system with weights on each route (based on the attack graph in Figure 1)

For example consider the following scenario. What happens if the event C1 (i.e. Social Engineering Passphrase occurs) occurs? This scenario can be presented using vector I_0 representing this situation by $I_0 = [1, 0, 0, 0, 0, 0, 0, 0, 0, 0, 0, 0, 0]$

In vector I_0 the concept C1 is represented as the first element in the vector and it is set to 1 and all other elements are set to be zero representing other events that has not happened. It is assumed that C1 happens and no other event has happened. Now I_0*E can provide the solution for this situation as follows: $I*E =[0, 0, 0, 0, 0, 0, 0.8, 0, 0, 0, 0, 0, 0]= I_1$

which conclude that if C1 happens then it will increase the possibility of C7 (i.e. Acquire Passphrase) to occur by 0.8. This process continues: $I_1*E =[0, 0, 0, 0, 0, 0, 0, 0, 0, 0.8, 0, 0, 0]=I_2$ which concludes that if C7 happens then it will increase the possibility of C11 (i.e Acquire Key) by 0.8. Now $I_2*E = [0, 0, 0, 0, 0, 0, 0, 0, 0, 0, 0, 0, 1]$ $= I_3$ which conclude that if C10 happens then it will increase the possibility of C11 (i.e. Read global data) by 1 (or 100%). This means that the attacker will be able to read the global data.

$$E = \begin{bmatrix}
0 & 0 & 0 & 0 & 0 & 0.8 & 0 & 0 & 0 & 0 & 0 & 0 & 0 \\
0 & 0 & 0 & 0 & 0 & 0 & 0 & 0 & 0.8 & 0 & 1 & 0 \\
0 & 0 & 0 & 0 & 0 & 0 & 0 & 0 & 0 & 0 & 0 & 0 \\
0 & 0 & 0 & 0 & 0 & 0 & 0.8 & 0 & 0.8 & 0 & 0 & 0 & 0 \\
0 & 0 & 0 & 0 & 0 & 0 & 0.8 & 0 & 0 & 0 & 1 & 0 \\
0 & 0 & 0 & 0 & 0 & 0 & 0 & 0 & 0 & 0 & 1 & 0 \\
0 & 0 & 0 & 0 & 0 & 0 & 0 & 0 & 0 & 0.8 & 0 & 0 & 0 \\
0 & 0 & 0 & 0 & 0 & 0 & 0 & 0 & 0.8 & 0 & 0 & 1 \\
0 & 0 & 0 & 0 & 0 & 0 & 0 & 0 & 0 & 0 & 0 & 1 \\
0 & 0 & 0 & 0 & 0 & 0 & 0 & 0 & 0 & 0 & 0 & 1 \\
0 & 0 & 0 & 0 & 0 & 0 & 0 & 0 & 0 & 0 & 1 & 0 \\
0 & 0 & 0 & 0 & 0 & 0 & 0 & 0 & 0 & 0 & 0 & 0 \\
0 & 0 & 0 & 0 & 0 & 0 & 0 & 0 & 0 & 0 & 0 & 0
\end{bmatrix}$$

Fig. 3. Matrix representing value of connecting edges of FCM from Figure 2

Several simulations were performed using different scenarios generated by GA. The details are shown in Table 1. Table 1 displays the consequences of different scenarios. For GA mutation rate were set to 0.06 and crossover rate was set to 0.5.

What if the following event occurs	Consequences
C2	C2 $\xrightarrow{80\%}$ C11 $\xrightarrow{100\%}$ C12
C3	C3 $\xrightarrow{100\%}$ C13
C4	C4 $\xrightarrow{80\%}$ C8 $\xrightarrow{100\%}$ C13
C5	C5 $\xrightarrow{100\%}$ C13 Also C5 $\xrightarrow{80\%}$ C9 $\xrightarrow{100\%}$ C13
C6	C6 $\xrightarrow{100\%}$ C13
C7	C7 $\xrightarrow{80\%}$ C11 $\xrightarrow{100\%}$ C12
C8	C8 $\xrightarrow{80\%}$ C11 $\xrightarrow{100\%}$ C13
C9	C9 $\xrightarrow{100\%}$ C13
C10	C10 $\xrightarrow{100\%}$ C13
C11	C11 $\xrightarrow{100\%}$ C12

4 Conclusion

Attack graphs are designed to provide a graph of paths that attackers may take to reach their undesirable goals and to attack a system. Attacker's goals may include system's disruptions, intrusion and misuse. With complexity of existing systems drawing attack graphs are becoming increasingly difficult and as such an attack graphs may be flawed. Attack graphs do not provide any facilities to analyze and assess different risks and possible attacks that may exist in attack graphs in a systematic way. Fuzzy Cognitive Maps (FCM) is employed in this paper to provide the facilities to capture and represent complex relationships in a system and to improve the understanding of a system designer to analyze risks. Using a FCM different scenarios are considered. The proposed FCM is used in conjunction with attack graph to provide system architects with possibilities for what-if analysis to understand the vulnerability of their designed system. What-if scenarios were generated using a GA.

From simulation results it was found that the FCM is capable of making accurate predictions attack for the given system. The GA provided many different attack scenarios for the FCM. Using such Scenarios an expert can inspect and make any modifications if necessary to the given IT system based on analysis of each attack scenario. The research work performed in this paper is unique in the way the attack graphs, FCM and GA are integrated. The application of this method to several other IT network security analyses is currently under consideration.

References

1. Gupta, I.S., Winstead, J.: Using Attack Graphs to Design Systems. In: IEEE Security and Privacy. IEEE Computer Society Publishing, Los Alamitos (2007)
2. Peterson, G., Steven, J.: Defining Misuse within the Development Process. IEEE Security & Privacy 4(6), 81–84 (2006)
3. Peeters, J., Dyson, P.: Cost- Effective Security. IEEE Security & Privacy 5(3), 85–87 (2007)
4. Kosko, B.: Fuzzy Engineering. Prentice Hall, Upper Saddle River (1997)
5. Kosko, B.: Fuzzy Cognitive Maps. International Journal of Man-Machine Studies 24, 65–75 (1986)
6. Aguilar, J.: A Survey about Fuzzy Cognitive Maps Papers. International Journal of Computational Cognition 3(2), 27–33 (2005)
7. Goldberg, D.: Genetic Algorithms in Search, Optimisation and Machine Learning. Addison Wesley, Reading (1989)
8. Swiler, L.P., Phillips, C., Ellis, D.: Chakerian. S.: Computer-attack graph generation tool. In: DISCEX II 2001: DARPA Information Survivability Conference and Exposition Conference and Exposition, vol. 2, pp. 307–321 (2001)

Benchmark Generator for Software Testers

Javier Ferrer, Francisco Chicano, and Enrique Alba

Departamento de Lenguajes y Ciencias de la Computación
University of Málaga, Spain
{ferrer,chicano,eat} @lcc.uma.es

Abstract. In the field of search based software engineering, evolutionary testing is a very popular domain in which test cases are automatically generated for a given piece of code using evolutionary algorithms. The techniques used in this domain usually are hard to compare since there is no standard testbed. In this paper we propose an automatic program generator to solve this situation. The program generator is able to create Java programs with the desired features. In addition, we can ensure that all the branches in the programs are reachable, i.e. a 100% branch coverage is always possible. Thanks to this feature the research community can test and enhance their algorithms until a total coverage is achieved. The potential of the program generator is illustrated with an experimental study on a benchmark of 800 generated programs. We highlight the correlations between some static measures computed on the program and the code coverage when an evolutionary test case generator is used. In particular, we compare three techniques as the search engine for the test case generator: an Evolutionary Strategy, a Genetic Algorithm and a Random Search.

Keywords: Software Testing, Evolutionary Algorithm, Search Based Software Engineering, Benchmarks.

1 Introduction

Automatic software testing is one of the most studied topics in the field of Search-Based Software Engineering (SBSE) [5,6]. From the first works [10] to nowadays many approaches have been proposed for solving the automatic test case generation problem. It is estimated that half the time spent on software project development, and more than half its cost, is devoted to testing the product [3]. This explains why the Software Industry and Academia are interested in automatic tools for testing.

Evolutionary algorithms (EAs) have been the most popular search algorithms for generating test cases [9]. In fact, the term *evolutionary testing* is used to refer to this approach. In the paradigm of *structural testing* a lot of research has been performed using EAs. The objective of an automatic test case generator used for structural testing is to find a test case suite that is able to cover all the software elements. These elements can be statements, branches, atomic conditions, and so on. The performance of an automatic test case generator is usually measured

L. Iliadis et al. (Eds.): EANN/AIAI 2011, Part II, IFIP AICT 364, pp. 378–388, 2011.

as the percentage of elements that the generated test suite is able to cover in the test program. This measure is called *coverage*. The coverage obtained depends not only on the test case generator, but also on the program being tested.

Once a test case generating tool is developed, we must face the problem of evaluating and enhancing the tool. Thus, a set of programs with a known maximum coverage would be valuable for comparison purposes. It is desirable to test the tool with programs for which a total coverage can be reached. This kind of benchmark programs could help the tool designer to identify the scenarios in which the tool works well or either has problems in the search. With this knowledge, the designer can improve her/his tool. The problem is that, as far as we know, there is no standard benchmark of programs to be used for comparing test case generating tools, and the software industry is usually reticent to share their source code. Furthermore, it is not usual to find programs in which total coverage is possible. Researches working on automatic test case generators usually provide their own benchmark composed of programs they programmed or belonging to a company with which they are working. This makes hard the comparison between different tools.

In order to alleviate the previous situation we propose, design, and develop a program generator that is able to create programs with certain features defined by the user. The program generator allows us to create a great amount of programs with different features, with the aim of analyzing the behavior of our test case generator in several scenarios. In this way, researches can face their tool with programs containing, for example, a high nesting degree or a high density of conditions. In addition, the program generator is able to write programs in which 100% branch coverage is possible. This way, the branch coverage obtained by different test case generators can be used as a measure of performance of the test case generator on a given program. This automatic program generator is the main contribution of this work. But we also include a study that illustrates how such a tool can be used to identify the program features that more influence have on the performance of a given test case generator. This study could help the researches to propose test case generators that selects the most appropriate algorithm for the search of test cases depending on the value of some static measures taken from the program.

The rest of the paper is organized as follows. In the next section we present the measures that the program generator takes into account and are used in our experimental study. Then, in Section 3, we describe the automatic program generator, the main contribution of the paper. Later, Section 4 describes the empirical study performed and discusses the obtained results and Section 5 outlines some conclusions and future work.

2 Measures

Quantitative models are frequently used in different engineering disciplines for predicting situations, due dates, required cost, and so on. These quantitative models are based on some kinds of measure made on project data or items.

Software Engineering is not an exception. A lot of measures are defined in Software Engineering in order to predict software quality [14], task effort [4], etc. We are interested here in measures made on source code pieces. We distinguish two kinds of measures: *dynamic*, which requires the execution of the program, and *static*, which does not require this execution.

The static measures used in this study are eight: number of statements, number of atomic conditions per decision, number of total conditions, number of equalities, number of inequalities, nesting degree, coverage, and McCabe's cyclomatic complexity. The three first measures are easy to understand. The number of (in)equalities is the number of times that the operator $==$ ($!=$) is found in atomic conditions of a program. The nesting degree is the maximum number of conditional statements that are nested one inside another. The cyclomatic complexity is a complexity measure related to the number of ways there exist to traverse a piece of code. This measure determines the minimum number of test cases needed to test all the paths using linearly independent circuits [8].

In order to define a coverage measure, we first need to determine which kind of element is going to be "covered". Different coverage measures can be defined depending on the kind of element to cover. *Statement coverage*, for example, is defined as the percentage of statements that are executed. In this work we use *branch coverage*, which is the percentage of branches of the program that are traversed. This coverage measure is used in most of the related papers in the literature.

3   Automatic Program Generator

We have designed a novel automatic program generator able to generate programs with values for the static measures that are similar to the ones of the real-world software, but the generated programs do not solve any concrete problem. Our program generator is able to create programs for which total branch coverage is possible. We propose this generator with the aim of generating a big benchmark of programs with certain characteristics chosen by the user.

In a first approximation we could create a program using a representation based on a general tree and a table of variables. The tree stores the statements that are generated and the table of variables stores basic information about the variables declared and their possible use. With these structures, we are able to generate programs, but we can not ensure that all the branches of the generated programs are reachable. The unreachability of all the branches is a quite common feature of real-world programs, so we could stop the design for the generator at this stage. However, another objective of the program generator is to be able of creating programs that can be used to compare the performance of different algorithms, programs for which total coverage is reachable are desirable. With this goal in mind we introduce logic predicates in the program generation process.

The program generator is parameterizable, the user can set several parameters of the program under construction (PUC). Thus, we can assign through several probability distributions the number of statements of the PUC, the number of variables, the maximum number of atomic conditions per decision, and the maximum nesting degree by setting these parameters. The user can define the structure of the PUC and, thus, its complexity. Another parameter the user can tune is the percentage of control structures or assignment statements that will appear in the code. By tuning this parameter the program will contain the desired density of decisions.

Once the parameters are set, the program generator builds the general scheme of the PUC. It stores in the used data structure (a general tree) the program structure, the visibility, the modifiers of the program, and creates a main method where the local variables are first declared. Then, the program is built through a sequence of basic blocks of statements where, according to a probability, the program generator decides which statement will be added to the program. The creation of the entire program is done in a recursive way. The user can decide whether all the branches of the generated program can be traversed (using logic predicates) or this characteristic is not ensured.

If total reachability is desired, logic predicates are used to represent the set of possible values that the variables can take at a given point of the PUC. Using these predicates we can know which is the range of values that a variable can take. This range of values is useful to build a new condition that can be true or false. For example, if at a given point of the program we have the predicate $x \leq 3$ we know that a forthcoming condition $x \leq 100$ will be always true and if this condition appears in an `if` statement, the `else` branch will not be reachable. Thus, the predicates are used to guide the program construction to obtain a 100% coverable program.

In general, at each point of the program the predicate is different. During the program construction, when a statement is added to the program, we need to compute the predicate at the point after the new statement. For this computation we distinguish two cases. First, if the new statement is an assignment then the new predicate CP' is computed after the previous one CP by updating the values that the assigned variable can take. For example, if the new statement is $x = x + 7$ and $CP \equiv x \leq 3$, then we have $CP' \equiv x \leq 10$.

Second, if the new statement is a control statement, an `if` statement for example, then the program generator creates two new predicates called True-predicate (TP) and False-predicate (FP). The TP is obtained as the result of the AND operation between CP and the generated condition related to the control statement. The FP is obtained as the result of the AND operation between the CP and the negated condition. In order to ensure that all the branches can be traversed, we check that both, TP and FP are not equivalent to $false$. If any of them were false, this new predicate is not valid and a new control structure would be generated.

Once these predicates are checked, the last control statement is correct and new statements are generated for the two branches and the predicates are computed inside the branches in the same way. After the control structure is completed, the last predicates of the two branches are combined using the OR operator and the result is the predicate after the control structure.

4   Experimental Section

In this section we present the experiments performed on a benchmark of programs created by the program generator. First, we describe how our test case generator works. Then, we explain how the benchmark of test programs was generated. In the remaining sections we show the empirical results and the conclusions obtained. Particularly, in Subsection 4.3 we study the correlations between some static measures and branch coverage when three different automatic test data generators are used.

4.1   Test Case Generator

Our test case generator breaks down the global objective (to cover all the branches) into several partial objectives consisting of dealing with only one branch of the program. Then, each partial objective can be treated as a separate optimization problem in which the function to be minimized is a distance between the current test case and one satisfying the partial objective. In order to solve such minimization problem Evolutionary Algorithms (EAs) are used.

In a loop, the test case generator selects a partial objective (a branch) and uses the optimization algorithm to search for test cases exercising that branch. When a test case covers a branch, the test case is stored in a set associated to that branch. The structure composed of the sets associated to all the branches is called *coverage table*. After the optimization algorithm stops, the main loop starts again and the test case generator selects a different branch. This scheme is repeated until total branch coverage is obtained or a maximum number of consecutive failures of the optimization algorithm is reached. When this happens the test data generator exits the main loop and returns the sets of test cases associated to all the branches. In the following section we describe two important issues related to the test case generator: the objective function to minimize and the parameters of the optimization algorithms used.

Objective Function. Following on from the discussion in the previous section, we have to solve several minimization problems: one for each branch. Now we need to define an objective function (for each branch) to be minimized. This function will be used for evaluating each test case, and its definition depends on the desired branch and whether the program flow reaches the branching condition associated to the target branch or not. If the condition is reached we can define the objective function on the basis of the logical expression of the branching condition and the values of the program variables when the condition is reached. The resulting expression is called *branch distance* and can be defined recursively on the structure of the logical expression. That is, for an expression composed of other expressions joined by logical operators the branch distance is computed as an aggregation of the branch distance applied to the component logical expressions.

When a test case does not reach the branching condition of the target branch we cannot use the branch distance as objective function. In this case, we identify the branching condition c whose value must first change in order to cover the target branch (critical branching condition) and we define the objective function as the branch distance of this branching condition plus the *approximation level*. The approximation level, denoted here with $ap(c, b)$, is defined as the number of branching nodes lying between the critical one (c) and the target branch (b) [15].

In this paper we also add a real valued penalty in the objective function to those test cases that do not reach the branching condition of the target branch. With this penalty, denoted by p, the objective value of any test case that does not reach the target branching condition is higher than the one of any test case that reaches the target branching condition. The exact value of the penalty depends on the target branching condition and it is always an upper bound of the target branch distance. Finally, the expression for the objective function is as follows:

$$f_b(x) = \begin{cases} bd_b(x) & \text{if } b \text{ is reached by } x \\ bd_c(x) + ap(c, b) + p & \text{otherwise} \end{cases} \tag{1}$$

where c is the critical branching condition, and bd_b, bd_c are the branch distances of branching conditions b and c.

To finish this section, we show in Table 1 a summary of the parameters used by the two EAs in the experimental section.

Table 1. Parameters of the two EAs used in the experimental section

	ES	GA
Population	25 indivs.	25 indivs.
Selection	Random, 5 indivs.	Random, 5 indivs.
Mutation	Gaussian	Add $U(-500, 500)$
Crossover	discrete (bias = 0.6) + arith. + arith.	Uniform
Replacement	Elitist	Elitist
Stopping cond.	5000 evals.	5000 evals.

4.2 Benchmark of Test Programs

The program generator can create programs having the same value for the static measures, as well as programs having different values for the measures. In addition, the generated programs are characterized by having a 100% coverage, thus all possible branches are reachable. The main advantage of these programs is that algorithms can be tested and analyzed in a fair way. This kind of programs are not easy to find in the literature.

Our program generator takes into account the desired values for the number of atomic conditions, the nesting degree, the number of statements and the number of variables. With these parameters the program generator creates a program with a defined control flow graph containing several conditions. The main features of generated programs are: they deal with integer input parameters, their conditions are joined by whichever logical operator, they are randomly generated and all their branches are reachable.

The methodology applied for the program generation was the following. First, we analyzed a set of Java source files from the JDK 1.5 (java.util.*, java.io.*, java.sql.*, etc.) and we computed the static measures on these files. Next, we used the ranges of the most interesting values (size, nesting degree, complexity, number of decisions and atomic conditions per decision), obtained in this previous analysis as a guide to generate Java source files having values in the same range for the static measures. This way, we generated programs with the values in these ranges, e.g., nesting degree in 1-4, atomic conditions per decisions in 1-4, and statements in 25, 50, 75, 100. The previous values are realistic with respect to the static measures, making the following study meaningful.

Finally, we generated a total of 800 Java programs using our program generator and we applied our test case generator using an ES and a GA as optimization algorithms. We also add to the study the results of a random test case generator (RND). This last test case generator proceeds by randomly generating test cases until total coverage is obtained or a maximum of 100,000 test cases are generated. Since we are working with stochastic algorithms, we perform in all the cases 30 independent runs of the algorithms to obtain a very stable average of the branch coverage. The experimental study requires a total of $800 \times 30 \times 3 = 72,000$ independent runs of the test case generators.

4.3   Correlation between Coverage and Static Measures

After the execution of all the independent runs for the three algorithms in the 800 programs, in this section we analyze the correlation between the static measures and the coverage. We use the Spearman's rank correlation coefficient ρ to study the degree of correlation between two variables.

First, we study the correlation between the number of statements and the branch coverage. We obtain a correlation of 0.173, 0.006, and 0.038 for these two variables using the ES, GA and RND, respectively. In Table 2 we show the average coverage against the number of statements for all the programs and algorithms. It can be observed that the number of statements is not a significant parameter for GA and RND, however, if we use ES, the influence on coverage exists. Thus, we can state that the test case generator obtains better coverage on average with ES in large programs.

Second, we analyze the nesting degree. In Table 3, we summarize the coverage obtained in programs with different nesting degree. If the nesting degree is increased, the branch coverage decreases and vice versa. It is clear that there is an inverse correlation between these variables. The correlation coefficients are -0.526 for ES, -0.314 for GA, and -0.434 for RND, which confirms the

Table 2. Relationship between the number of statements and the average coverage for all the algorithms. The standard deviation is shown in subscript

# Statements	ES	GA	RND
25	$77.31_{13.31}$	$87.59_{12.66}$	$76.85_{16.17}$
50	$78.40_{12.15}$	$88.58_{11.36}$	$74.82_{16.10}$
75	$80.92_{10.21}$	$89.51_{9.54}$	$78.56_{13.75}$
100	$82.93_{10.36}$	$89.95_{8.75}$	$77.90_{14.02}$
ρ	0.173	0.006	0.038

Table 3. Relationship between the nesting degree and the average coverage for all the algorithms. The standard deviation is shown in subscript

Nesting degree	ES	GA	RND
1	$88.53_{6.85}$	$94.05_{5.84}$	$86.13_{10.81}$
2	$82.38_{8.36}$	$90.28_{8.22}$	$79.87_{13.07}$
3	$76.92_{10.12}$	$87.72_{11.01}$	$73.46_{13.99}$
4	$71.74_{13.41}$	$83.57_{13.38}$	$68.67_{16.03}$
ρ	-0.526	-0.314	-0.434

observations. These correlation values are the highest ones obtained in the study of the different static measures, so we can state that the nesting degree is the parameter with the highest influence on the coverage that evolutionary testing techniques can achieve.

Now we study the influence on coverage of the number of equalities and inequalities found in the programs. It is well-known that equalities and inequalities are a challenge for automatic software testing. This fact is confirmed in the results. The correlation coefficients are -0.184, -0.180, and -0.207 for equalities and -0.069, -0.138, and -0.095 for inequalities using ES, GA, and RND, respectively. We conclude that the coverage decreases as the number of equalities increases for all the algorithms. However, in the case of the number of inequalities only the GA is affected by them.

Let us analyze the rest of static measures of a program. We studied the correlation between the number of atomic conditions per decision and coverage. After applying Spearman's rank correlation we obtained low values of correlation for all the algorithms $(0.031, -0.012, 0.035)$. From the results we conclude that there is no correlation between these two variables. This could seem counterintuitive, but a large condition with a sequence of logical operators can be easily satisfied due to OR operators. Otherwise, a short condition composed of AND operators can be more difficult to satisfy.

Now we analyze the influence on the coverage of the number of total conditions of a program. The correlation coefficients are -0.004 for ES, -0.084 for GA, and -0.082 for RND. At a first glance, it seems that the number of total conditions has no influence on the coverage, nevertheless, calculating the density of conditions (number of total conditions / number of statements), one can realize that the influence exists. The correlation coefficients between the coverage and the density of conditions are -0.363 for ES, -0.233 for GA, and -0.299 for RND. In Figure 1.a) the tendency is clear: a program with a high density of conditions is more difficult to test, especially for the ES.

Finally, we study the relationship between the McCabe's cyclomatic complexity and coverage. In Figure 1.b), we plot the average coverage against the cyclomatic complexity for GA in all the programs. In general we can observe that there is no clear correlation between both parameters. The correlation coefficients are 0, -0.085, and -0.074 for ES, GA, and RND, respectively. These values are low, and confirm the observations: McCabe's cyclomatic complexity and branch coverage are not correlated.

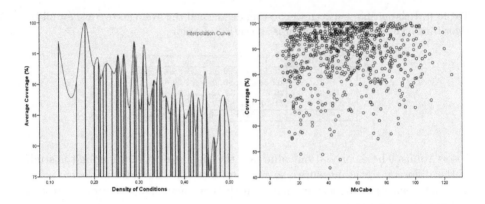

Fig. 1. Average branch coverage against the Density of conditions for GA and all the programs (a) and average branch coverage against the McCabe's cyclomatic complexity for GA and all the programs (b)

Furthermore, the correlation coefficients are lower than the coefficients we have obtained with other static measures like the nesting degree, the density of conditions, the number of equalities, and the number of inequalities. This is somewhat surprising, because it would be expected that a higher complexity implies also a higher difficulty to find an adequate test case suite. The correlation between the cyclomatic complexity and the number of software faults has been studied in some research articles [2,7]. Most such studies find a strong positive correlation between the cyclomatic complexity and the defects: the higher the complexity the larger the number of faults. However, we cannot say the same with respect to the difficulty in automatically finding a test case suite. McCabe's cyclomatic complexity cannot be used as a measure of the difficulty of getting an adequate test suite.

We can go one step forward and try to justify this unexpected behavior. We have seen in the previous paragraphs that the nesting degree is the static measure with the highest influence on the coverage. The nesting degree has no influence on the computation of the cyclomatic complexity. Thus, the cyclomatic complexity is not taking into account the information related to the nested code, it is based on some other static information that has a lower influence on the coverage. This explanation is related to one of the main criticisms that McCabe complexity has received since it was proposed, namely: Piwowarski [11] noticed that cyclomatic complexity is the same for N nested if statements and N sequential if statements.

5   Conclusions

In this work we have developed a program generator that is able to create programs for which total branch coverage is reachable. To achieve this desirable characteristic, we have used logic predicates to represent the set of possible values that the variables can get at a given point of the PUC. We illustrate the potential of the generator developing a study that could hardly be done in any other way. We created a benchmark of 800 programs and analyzed the correlations between the static measures taken on the programs and the branch coverage obtained by three different test case generators. In the empirical study we included eight static measures: number of statements, number of atomic conditions per decision, number of total conditions, density of conditions, nesting degree, McCabe's cyclomatic complexity, number of equalities, and number of inequalities. The results show that the nesting degree, the density of conditions and the number of equalities are the static measures with a higher influence on the branch coverage obtained by automatic test case generators like the ones used in the experimental section. This information can help a test engineer to decide which test case generator s/he should use for a particular test program.

As future work we plan to advance in the flexibility of the program generator. We should add more parameters to the program generator with the aim of giving total control of the structure of the program under construction to the user. We should also modify the program generator to be able to create Object Oriented programs in order to broaden the scope of applicability.

References

1. Bäck, T., Fogel, D.B., Michalewicz, Z.: Handbook of Evolutionary Computation. Oxford University Press, New York (1997)
2. Basili, V., Perricone, B.: Software errors and complexity: an empirical investigation. ACM Commun. 27(1), 42–52 (1984)
3. Beizer, B.: Software testing techniques, 2nd edn. Van Nostrand Reinhold Co., New York (1990)
4. Boehm, B., Abts, C., Brown, A.W., Chulani, S., Clark, B.K., Horowitz, E., Madachy, R., Reifer, D.J., Steece, B.: Software cost estimation with COCOMO II. Prentice-Hall, Englewood Cliffs (2000)
5. Harman, M.: The current state and future of search based software engineering. In: Proceedings of ICSE/FOSE 2007, Minneapolis, Minnesota, USA, May 20-26, pp. 342–357 (2007)
6. Harman, M., Jones, B.F.: Search-based software engineering. Information & Software Technology 43(14), 833–839 (2001)
7. Khoshgoftaar, T., Munson, J.: Predicting software development errors using software complexity metrics. Jnl. on Selected Areas in Communications (1990)
8. McCabe, T.J.: A complexity measure. IEEE Trans. on Software Engineering 2(4), 308–320 (1976)
9. McMinn, P.: Search-based software test data generation: a survey. Software Testing, Verification and Reliability 14(2), 105–156 (2004)
10. Miller, W., Spooner, D.L.: Automatic generation of floating-point test data. IEEE Trans. Software Eng. 2(3), 223–226 (1976)
11. Piwarski, P.: A nesting level complexity measure. SIGPLAN 17(9), 44–50 (1982)
12. Rechenberg, I.: Evolutionsstrategie: Optimierung technischer Systeme nach Prinzipien der biologischen Evolution. Fromman-Holzboog Verlag, Stuttgart (1973)

13. Rudolph, G.: Evolutionary Computation 1. Basic Algorithms and Operators. Evolution Strategies, vol. 1, ch. 9, pp. 81–88. IOP Publishing Lt. (2000)
14. Samoladas, I., Gousios, G., Spinellis, D., Stamelos, I.: The SQO-OSS Quality Model. In: Open Source Development, Communities and Quality, vol. 275, pp. 237–248 (2008)
15. Wegener, J., Baresel, A., Sthamer, H.: Evolutionary test environment for automatic structural testing. Information and Software Technology 43(14), 841–854 (2001)

Automated Classification of Medical-Billing Data

R. Crandall[1], K.J. Lynagh[2], T. Mehoke[1], and N. Pepper[1,2,*]

[1] Center for Advanced Computation, Reed College, Portland, OR
[2] Qmedtrix Systems Inc., Portland OR
peppern@reed.edu

Abstract. When building a data pipeline to process medical claims there are many instances where automated classification schemes could be used to improve speed and efficiency. Medical bills can be classified by the statutory environment which determines appropriate adjudication of payment disputes. We refer to this classification result as the *adjudication type* of a bill. This classification can be used to determine appropriate payment for medical services.

Using a set of 182,811 medical bills, we develop a procedure to quickly and accurately determine the correct adjudication type. A simple naïve Bayes classifier based on training set class occurrences gives 92.8% accuracy, which can be remarkably improved by instead presenting these probabilities to an artificial neural network, yielding 96.8 ± 0.5 % accuracy.

1 Introduction

In this paper we discuss a medical bill classification methodology. Our exposition focuses on the bill adjudication type, which is but one of many possible classifications that may be desired. This paper starts with a discussion of the motivation for creating this machine learning component of a larger information processing pipeline, followed by a description of the problem and data fields used for prediction. The paper then continues on to an exposition of our bayesian and neural net solutions. Finally we discuss our actual training and testing data and present results and conclusions.

A bill's adjudication type is an important factor in determining proper compensation for medical services. In some domains such as workers' compensation, there are legally mandated fee schedules that dictate the price of medical procedures. These fee schedules vary by state and facility type; a procedure performed by an outpatient hospital may have a different fee schedule rate than the same procedure provided at an ambulatory surgical center. We categorize bills into seven basic adjudication types (table 1).

The reader may be wondering why such a classification scheme is needed when the tax identification number (tin), and thus the corporate identity of the service

---

* Corresponding author.

L. Iliadis et al. (Eds.): EANN/AIAI 2011, Part II, IFIP AICT 364, pp. 389–399, 2011.

provider, is already known. While the tin does help narrow down the possible adjudication type it is by no means conclusive. For example a hospital with one tin is likely to have inpatient, outpatient and emergency room services all billed under the same tin and housed at the same street address. To make matters more confusing, many hospitals have free standing surgery centers (asc) near the actual hospital building. While the asc may be a different building and thus subject to a different set of billing rules, it will likely still be considered part of the same company and therefore bill under the same tin. In extreme cases, there are large companies which provide many types of medical services across a disparate geographic area and yet all bill as one company under one tin.

1.1 Motivation for Automation

Traditionally, in medical bill review human auditors will determine the adjudication type of a bill manually and then apply the appropriate rules or fee schedules. There are two problems with this approach which an automated solution can help solve.

First, humans are much slower than an optimized machine learning engine. While the engine described in this paper can classify around 14,000 bills per second a human takes anywhere from 10 seconds to 45 seconds to classify a single normal bill. Not only does the algorithm classify bills many orders of magnitude faster than a human, it also saves a large amount of worker time which translates to huge savings and increased efficiency. Additionally, if one were to build a fully automated system for applying fee schedules and rules to medical bills this would eliminate the human bottleneck entirely.

Second, the algorithm can be carefully tuned to reduce errors. Humans tend to make errors when trying to do many detail-oriented tasks rapidly.

While many bills may be classified using an extensive rules-based automated system, such a methodology is inferior to a machine learning approach. Rules-based systems cannot reliably handle edge cases and are labor intensive to modify or update. The genesis of this project was a desire to solve a software engineering problem: accurately route bills through a fraud detection and cost containment pipeline. Initial attempts at solving this problem using manually coded rule sets created error rates that were unacceptable. This lead to much of the classification being redone manually. The machine learning approach presented here was developed for two purposes. First the system allows for improvement of classification accuracy of manual rule sets and manual auditors by examining the output of the machine learning algorithm. Second this system can be integrated as a part of this larger information system.

2 Analog and Digital Data Fields

We have elected to split relevant medical data into two types. First, *analog* (equivalently, numerical) data are those data whose magnitude has meaning.

Such data include bill cost (in dollars), duration (in days), and entropy (in bits); that is, an analog datum carries with it a notion of physical units. On the other hand *digital* (equivalently, categorical) data are represented by strings indicating a medical procedure (such as with icd9 codes). These data have no magnitude or natural distance metric.

Now, we refer to an analog or digital datum as a *feature* of a bill. On the simplifying assumption of statistical independence, the probability of a bill with a feature set $f = \{f_1, f_2, f_3, \ldots\}$ having adjudication type A is taken as

$$p(A \mid f) = \prod_i p(A \mid f_i) \;=\; p(A \mid f_1) \, p(A \mid f_2) \, p(A \mid f_3) \cdots \qquad (1)$$

One novel aspect of our approach is that all conditional probabilities are inferred from *log-histograms* (equivalently, histograms with exponentially growing bins). Such a transformation effectively reduces the wide dynamic range of some analog data. Cost, for instance, ranges from a few dollars to more than 10^5 dollars.

Another idea—one that tends to stabilize the performance of a classifier—is to "soften". For example, say that a digital datum such as code 123ABC occurs only *once* during training, and say it occurs for an iph bill. We do not want to conclude that the 123ABC code occurs only on iph bills in post-training classification. To avoid this kind of spurious statistical inference, we estimate the conditional probabilities as

$$p(A_i \mid f) = \frac{\alpha_i + \#(A_i \mid f)}{\sum_j \left(\alpha_j + \#(A_j \mid f) \right)}, \qquad (2)$$

where α_i is a positive constant and $\#(A_i \mid f)$ is the count of bills with adjudication type A_i having feature f. (In general, the bill type $i \in [0, T-1]$, i.e. we assume T adjudication types.) The degenerate case $\alpha_i = 0$ is the usual probabilistic estimate, but we want to "soften" for the cases such as the singleton iph bill mentioned. We typically use $\forall i : \alpha_i = 1$, although any set of positive real α_i will prevent singular division. This probability transformation is equivalent to inserting a virtual entry of some certain size into every histogram, and also works effectively for analog-data histograms that have accidentally empty or near-empty bins.

A third probability transformation option applies when using a neural networks; the option *not* to normalize a length-T probability list. Because neural networks can adjust weights as a form of normalization, we can alternatively define

$$p(A_i \mid f) \;=\; \gamma \cdot \#(A_i \mid f), \qquad (3)$$

where γ is a positive constant chosen merely to keep these "unconstrained probabilities" within reasonable range. One possible advantage of this loosening of probability constraints is that, for some features, the histogram counts for all

$i \in [0, T - 1]$ might be so small as to be insignificant. By not normalizing the T vertical histogram sections, we may well reject some system noise in this way. It is a tribute to neural-network methodology that one may use this normalization, which amounts to using *a priori* adjudication frequencies, or not. Of course, the right option may be obvious after cross-validation (training/testing to assess both options).

3   Naïve Bayes

Naïve Bayes is a well known, simple, and fast classification method [4; 2] that treats bill features as statistically independent from one another. Treating item features independently is a strong assumption, but it typically holds in this domain; the codes on a bill for an organ transplant, say, are almost entirely disjoint from the codes on a bill for a broken bone. In the Bayesian approach, a bill is classified according to the most probable adjudication type A_i that *maximizes* likelihood over all relevant feature sets, said likelihood being

$$\prod_f p(A_i \mid f).\qquad\qquad(4)$$

We estimate the conditional probabilities $p(A_i \mid f_1)$ from histograms on a training set of the data (see Analog and Digital Data Fields). Because the naïve Bayes classifier chooses between relative probabilities, the inclusion of features shared by many of the classes (a blood draw code, say) is effectively a rescaling, and thus does not affect performance. Finally, note that this procedure can incorporate a variety of features for each bill; the number of terms in equation (1) depends on the total number of codes within individual bills.

The naïve Bayes classification is simple to compute, and performs well (see table 2). The solid performance on tin alone is no surprise; 81% of the tins in the training set correspond to a single adjudication type (11% two, 8% three or more). Thus, classifying an unknown bill as having the most common adjudication type from that provider is a good rule of thumb. However, large hospitals and some hospital networks use a common tin for differently adjudicated medical services, and a good classification scheme must take into account more data. Incorporating bill procedures improves classification performance; combining all the codes gives the naïve Bayesian classifier an accuracy of 92.83%. Note that, even with the tin omitted (classifying on hcpcs, hrc, svc, icd), the accuracy is 86.28%.

4   Neural Network

Artificial neural networks are a well-studied family of classifiers [1; 5; 8]. We implemented artificial neural networks with the fann library [7], and trained the network with the conditional probabilities used in the naïve Bayes classifier. See our Appendix for network-architecture parameters.

Unlike the naïve Bayes classifier, which simply multiplies all the conditional probabilities, the neural network is capable of learning nonlinear weighted combinations that improve classification performance. Because neural networks can settle into suboptimal extrema, we perform 10 training runs and average the results. These generalized capabilities resulted in a final classification performance of $96.9 \pm 0.6\%$.

5   Data and Methods

The data used in this procedure consist of medical bills from across the United States in 2000–2009, representing more than a billion dollars of charges. The bills contain both categorical and numeric features. The sets of codes pertaining to individual line items (cpt, ..., 7,527 codes) the patient (diagnostic icd9, ... 6,302 codes), and the provider (11,603 tin) represent more than 25,000 total categorical features.

The bills also contain numeric values: the total bill cost, the duration (length of treatment or overnight stay), and the service cost entropy. This latter quantity is derived from the line item charges on bills, according to

$$E = -\sum_i \frac{c_i}{c} \log_2 \left(\frac{c_i}{c} \right), \tag{5}$$

where c_i is the line charge for line i and $c = \sum c_i$ is the total bill cost (we take $0 \cdot \log(0) = 0$). Essentially, this quantity E models the distribution of charges; a bill consisting of several same-priced weekly treatments (i.e. a uniform distribution) has a very high entropy E, whereas a hospital stay with charges for meals, medication, and surgery charged together on the same bill (i.e. a very skewed price distribution) gives a relatively low entropy.

The data were split into disjoint training/testing sets of 121,852 and 60,959 bills, respectively. This was done by randomly sampling the data into two buckets of roughly $\frac{1}{3}$ for testing and $\frac{2}{3}$ for training. All results shown reflect the outcome of predicting testing data using models formed with our training data.

Table 1. Possible adjudication types of medical bills. The proportion of bills refers to our entire sample (testing and training sets) of 182811 bills.

Adjudication type	proportion of bills
amb: Ambulance or medical transport	4.86 %
asc: Ambulatory surgery center	20.77 %
dme: Durable medical equipment	11.24 %
er: Emergency room	18.23 %
iph: Inpatient hospital	5.44 %
oph: Outpatient hospital	24.72 %
pro: Professional services	14.73 %

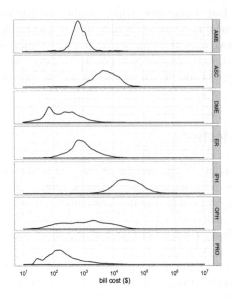

Fig. 1. Bill histograms vs. total bill cost, displayed by adjudication type. It is a key idea that the horizontal axis is nonlinear, to reduce skewness; we use logarithmic abscissa throughout our runs. See our Appendix for density-estimation parameters.

6   Results

Confusion matrices[6] are used to visualize classification results, this is a standard way to display classification accuracy and error. The true classification of a bill is given across the columns, and the predicted classification is quantified down the rows. A perfect classifier would have nonzero entries only along the diagonal, each corresponding to 100%.

Table 2. Percentage accuracy of a uniform prior naïve Bayes classifier trained on 121649 bills, tested on 61162 bills for each code type alone

code	classification accuracy
svc	71.58 %
hrc	55.08 %
ndc	4.83 %
icd	56.49 %
tin	79.03 %
hcpcs	20.26 %

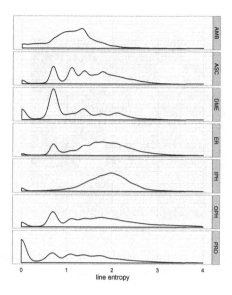

Fig. 2. Bill histograms vs. line cost entropy, displayed by adjudication type. See our Appendix for density-estimation parameters.

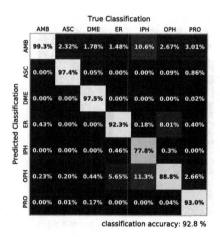

Fig. 3. Confusion matrix from analog-and-digital-data trained Bayes classifier, having 92.8% accuracy

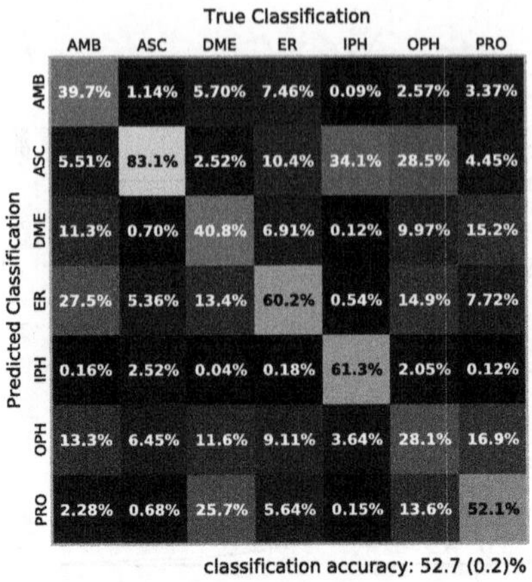

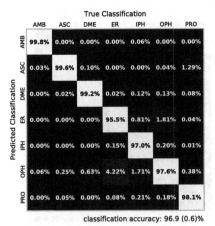

Fig. 4. Confusion matrix from analog-data trained neural network. Average classification accuracy for 10 train/test runs is $52.7 \pm 0.2\%$.

Fig. 5. Confusion matrix from digital-data trained neural network. Average classification accuracy for 10 train/test runs is $96.9 \pm 0.6\%$.

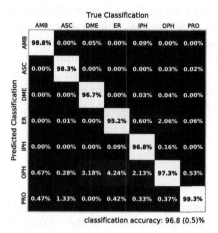

Fig. 6. Confusion matrix from analog-and-digital-data trained neural network. Average classification accuracy for 10 train/test runs is 96.8 ± 0.5%. We take this confusion matrix to be our standard, combining analog/digital data via log-histogram techniques and exploiting the power of neural networks.

7   Conclusions and Future Directions

Our method for adjudication type classification can be extended in several ways. Note that both the simple Bayesian model and the neural network return a score for each possible adjudication type, and bills are classified according to the maximum. A more nuanced scheme can be constructed with additional domain knowledge to minimize classification risk. For instance, one could classify bills when one output clearly dominates, but put a bill "on hold" for manual review if its several outputs are close to the maximum.

Also note that, though bills can be classified using the analog features, they do not provide additional discriminative power beyond the categorical features (compare figures). For the purposes of fraud detection and proper payment, this is a desirable situation–we can reliably determine a bill's adjudication type (and hence, proper reimbursement amounts) while ignoring the bill amount entirely!

In conclusion, we have theoretically motivated and technically demonstrated the viability of an automated method for medical bill classification. There are many future research directions motivated by the present work:

 - Combining various classifiers (neural networks, support vector machines, decision trees, etc.) using voting schemes.
 - Using a separate neural network for post-processing, with, say, *reduced* input count based on the analog levels of the first network; in this way, perhaps false categorization can be further reduced.
 - Using multiple classification schemes in a cascaded fashion to improve the granularity and/or accuracy of the classifications.

- Placing the classification scheme in a loop with human oracles determining correct classifications of ambiguous bills to continuously increase accuracy.
- Implement a bagging scheme to increase accuracy and mitigate the effects of possibly having incorrect labels in the training data.
- Determine whether our discovered neural-net "amplification" of first-order Bayesian classification is theoretically—or empirically—equivalent to higher-order Bayesian analysis.

Appendix

Histograms

Figures 1 and 2 were generated using the ggplot2 graphics library with the default parameters. The histograms used for the analog fields each had 10 exponentially growing bins chosen according to the domain of the training set. Analog data in the testing set outside of this range were counted in the first or last bin, as appropriate.

Neural network

Each feature maps to T (in our runs, $T := 7$ adjudication types) input neurons, giving $3T$ inputs for the analog (numeric) features (bill cost, duration, and entropy) and $5T$ for the digital (categorical) features. Note that, since the input and output vector dimensions must be fixed for a neural network, we multiplied conditional probabilities for bills with multiple features of the same kind (several icd9 codes, for instance). This gave better performance than simple heuristics like using only the most infrequently occurring code of a given type.

All neural networks were seven hidden neuron feed-forward networks, trained by standard reverse propagation. All neurons used the sigmoidal activation function proposed by Elliott [3]:

$$y(x) = \frac{xs}{2(1+|xs|)} + 0.5,$$

where the positive constant s is the neuron activation steepness.

Robustness

To show that the neural network architecture and our results are robust, we ran each combination of parameters 10 times, and report the average accuracy and its standard deviation (see figure captions). On each run, bills were split into disjoint training and testing sets (121,852 and 60,959 bills, respectively).

References

1. Bishop, C.M.: Neural Networks for Pattern Recognition. Oxford University Press Inc., Oxford (2007)
2. Cios, K., Pedrycz, W., Swiniarski, R.: Data mining methods for knowledge and discovery. Kluwer Academic Publishers, Dordrecht (1998)
3. Elliott, D.L.: A better activation function for artificial neural networks. Technical report, Institute for systems research, University of Maryland (1993)
4. Hastie, T., Tibshirani, R., Friedman, J.: The Elements of Statistical Learning. Springer, Heidelberg (2009)
5. Haykin, S.: Neural networks and learning machines, 3rd edn. Prentice Hall, Englewood Cliffs (2008)
6. Kohavi, R.: Glossary of terms. Machine Learning 30, 271–274 (1998)
7. Nissen, S.: Implementation of a fast artificial neural network library (fann). Technical report, Department of Computer Science University of Copenhagen, diku (2003)
8. Tettamanzi, A., Tomassini, M.: Soft computing: integrating evolutionary, neural, and fuzzy systems. Springer, Heidelberg (2001)

Brain White Matter Lesions Classification in Multiple Sclerosis Subjects for the Prognosis of Future Disability

Christos P. Loizou[1], Efthyvoulos C. Kyriacou[2], Ioannis Seimenis[3], Marios Pantziaris[4], Christodoulos Christodoulou[5], and Constantinos S. Pattichis[5]

[1] Department of Computer Science, Intercollege, P.O. Box 51604,
CY-3507, Limassol, Cyprus
panloicy@logosnet.cy.net
[2] Department of Computer Science and Engineering,
Frederick University, Limassol, Cyprus
[3] Medical Diagnostic Centre "Ayios Therissos",
2033 Nicosia, Cyprus
[4] Cyprus Institute of Neurology and Genetics, Nicosia, Cyprus
[5] Department of Computer Science, University of Cyprus, Nicosia, Cyprus

Abstract. This study investigates the application of classification methods for the prognosis of future disability on MRI-detectable brain white matter lesions in subjects diagnosed with clinical isolated syndrome (CIS) of multiple sclerosis (MS). For this purpose, MS lesions and normal appearing white matter (NAWM) from 30 symptomatic untreated MS subjects, as well as normal white matter (NWM) from 20 healthy volunteers, were manually segmented, by an experienced MS neurologist, on transverse T2-weighted images obtained from serial brain MR imaging scans. A support vector machines classifier (SVM) based on texture features was developed to classify MRI lesions detected at the onset of the disease into two classes, those belonging to patients with EDSS≤2 and EDSS>2 (expanded disability status scale (EDSS) that was measured at 24 months after the onset of the disease). The highest percentage of correct classification's score achieved was 77%. The findings of this study provide evidence that texture features of MRI-detectable brain white matter lesions may have an additional potential role in the clinical evaluation of MRI images in MS. However, a larger scale study is needed to establish the application of texture analysis in clinical practice.

Keywords: MRI, multiple sclerosis, texture classification.

1 Introduction

Multiple sclerosis (MS) is the most common autoimmune disease of the central nervous system, with complex pathophysiology, including inflammation, demyelination, axonal degeneration, and neuronal loss. Within individuals, the clinical manifestations are unpredictable,[1] particularly with regard to the development of disability [1].

L. Iliadis et al. (Eds.): EANN/AIAI 2011, Part II, IFIP AICT 364, pp. 400–409, 2011.
© IFIP International Federation for Information Processing 2011

Diagnostic evaluation of MS, performed by a specialized neurologist, is generally based on conventional magnetic resonance imaging (MRI) following the McDonald criteria [2], and on clinical signs and symptoms. The development of modern imaging techniques for the early detection of brain inflammation and the characterization of tissue-specific injury is an important objective in MS research. Recent MRI studies have shown that brain and focal lesion volume measures, magnetization transfer ratio and diffusion weighted imaging-derived parameters can provide new information in diagnosing MS [3]. Texture features quantify macroscopic lesions and also the microscopic abnormalities that may be undetectable using conventional measures of lesion volume and number [1], [4]. Several studies have been published, where it was documented that texture features can be used for the assessment of MS lesions in: (i) differentiating between lesions for normal white matter (NWM), and the so called normal appearing white matter (NAWM) [5-9], and (ii) monitoring the progression of the disease over longitudinal scans [10-14].

Our primary objective in this study is to develop texture classification methods that can be used to predict MS brain lesions that at a later stage are associated with advanced clinical disability and more specifically with the extended disability status scale (EDSS). Since the use of quantitative MRI analysis as a surrogate outcome is used as a surrogate measure in clinical trials, we hypothesise that there is a close relationship between the change in the extracted features and the clinical status and the rate of development of disability. We analyzed patient's images acquired at the initial stages of the disease-clinical isolated syndrome (CIS) (0 months) and we correlated texture findings with disability assessment scales. We interrelate therefore the EDSS scores [15] with standard shape and texture features.

Preliminary findings of this study, for the texture analysis of NAWM in MS subjects were also published in [8], [9]. It should be note that the same problem was also investigated in a recent study published by our group using Amplitude Modulation-Frequency Modulation (AM-FM) analysis [14]. The motivation of this study was to investigate the usefulness of classical texture analysis, as well as compare the findings with the AM-M analysis.

MRI-based texture analysis was shown to be effective in classifying MS lesions from NWM and the so called, NAWM, with an accuracy of 96%-100% 5. In [6], the authors showed that texture features can reveal discriminant features for differentiating between normal and abnormal tissue, and for image segmentation. Significant differences in texture between normal and diseased spinal cord in MS subjects were found in [7] as well a significant correlation between texture features and disability. The median value increase of these texture features suggests that the lesions texture in MS subjects is less homogeneous and more complex than the corresponding healthy tissue (NWM) in healthy subjects [7]. Similar findings were also found in [8] and [9], where it was shown that median values of lesion texture features such as standard deviation, median, sum of squares variance, contrast, sum average, sum variance, difference variance, and difference entropy increased significantly with the progression of the MS disease when compared to NWM tissue. Statistical analysis (using the spatial gray level dependence matrices) has shown that there is a significant difference between lesions and NAWM or NWM. These findings may be beneficial in the research of early diagnosis and treatment monitoring in MS.

The differentiation between active and non-active brain lesions in MS subjects from brain MRI was also investigated in [11]. Here, it was shown that active lesions could be identified without frequent gadolinium injections, using run length analysis criteria. In [12] the performance of texture analysis concerning discrimination between MS lesions, NAWM and NWM from healthy controls was investigated by using linear discriminant analysis. The results suggested that texture features can support early diagnosis in MS. When a combined set of texture features was used [12], similar findings were reported.

In [13], a pattern recognition system was developed for the discrimination of multiple sclerosis from cerebral microangiopathy lesions based on computer-assisted texture analysis of magnetic resonance images. It was shown that, MS regions were darker, of higher contrast, less homogeneous and rougher as compared to cerebral microangiopathy. Finally, in [14], the use of multi-scale Amplitude Modulation-Frequency Modulation (AM-FM) texture analysis of MS using magnetic resonance images from brain was introduced.

2   Materials and Methods

2.1   Material and MRI Acquisition

Thirty subjects (15 males, and 15 females), aged 31.4±12.6 (mean age ± standard deviation) were scanned at 1.5T within one month following initial clinical evaluation to confirm CIS diagnosis. The transverse MR images used for analysis were obtained using a T2-weighted turbo spin echo pulse sequence (repetition time=4408 ms, echo time=100 ms, echo spacing=10.8 ms). The reconstructed image had a slice thickness of 5 mm and a field of view of 230 mm with a pixel resolution of 2.226 pixels per mm. Standardized planning procedures were followed during each MRI examination. The MRI protocol and the acquisition parameters were given in detail in [8], [9].

Initial clinical evaluation was made by an experienced MS neurologist (co author M. P.) who referred subjects for a baseline MRI upon diagnosis and clinically followed all subjects for over two years. All subjects were subjected to an EDSS test two years after initial diagnosis to quantify disability [15].

A normalization algorithm was used to match the image brightness between the first (baseline) and the follow-up images (see [16] for details). For this purpose, the neurologist manually segmented cerebrovascular fluid (CSF) areas as well as areas with air (sinuses) from all MS brain scans. Similarly, ROIs representing NWM, CSF and air from the sinuses were arbitrarily segmented from the brain scans of the 20 healthy subjects. The original image histogram was stretched, and shifted in order to cover all the gray scale levels in the image.

To introduce the objective of our study, an example in Fig. 1 is presented. Here, we show a transaxial T2-weighted MR image of a female patient at the age of 32, with an EDSS [15] equal to 3. The image in Fig. 1a) corresponds to the initial diagnosis of a CIS of MS and the delineated lesion corresponding to the MS plaque is also depicted. Figure 1b) shows the magnified segmented lesion from a) (original images were acquired at a sampling rate of 2.226 pixels per mm). In what follows, texture analysis refers to the image processing of the extracted regions of interest (ROIs).

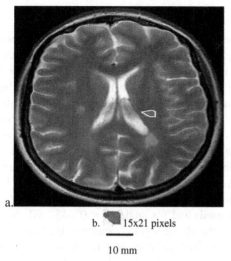

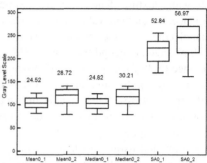

Fig. 1. a) ROI drawn on MR image of the brain obtained from a 32 year old female MS patient with an EDSS=3 at 0 months, and b) magnified segmented lesion that was acquired at a pixel resolution of 2.226 pixels per mm. The bar below the lesion shows the size of 10 mm. The grey scale median and inter-quartile range (IQR) of the segmented lesion were 108 and 9.6, respectively

Fig. 2. Box plots for the median ± inter-quartile range (IQR) values for texture features mean, median and sum average, from MS lesions at 0 months for EDSS<=2 (N=18) and EDSS>2 (N=11) corresponding to feature notation _1, and _2 respectively (see also TABLE I). Inter-quartile range (IQR) values are shown above the box plots. In each plot we display the median, lower, and upper quartiles and confidence interval around the median. Straight lines connect the nearest observations within 1.5 of the IQR of the lower and upper quartiles.

2.2   Manual Delineations and Visual Perception

All MRI-detectable brain lesions were identified and segmented by an experienced MS neurologist and confirmed by a radiologist. Only well-defined areas of hyperintensity on T2-weighted MR images were considered as MS lesions. The neurologist manually delineated (using the mouse) the brain lesions by selecting consecutive points at the visually defined borders between the lesions and the adjacent NAWM on the acquired transverse T2-weigted sections. Similar regions corresponding to NAWM were delineated contralaterally to the detected MS lesions. The manual delineations were performed using a graphical user interface implemented in MATLAB developed by our group. Manual segmentation by the MS expert was performed in a blinded manner, without the possibility of identifying the subject or the clinical findings. The selected points and delineations were saved to be used for texture analysis.

2.3 Texture Analysis

In this study texture features and shape parameters were extracted from all MS lesions detected and segmented [8], [9], while texture features were also calculated for the NAWM ROIs. The overall texture and shape features for each subject were then estimated by averaging the corresponding values for all lesions for each subject. The following texture features were extracted: (i) Statistical Features: a) mean, b) variance, c) median value, d) skewness, e) kurtosis, f) energy and g) entropy. (ii) Spatial Gray Level Dependence Matrices (SGLDM) as proposed by Haralick et al. [17]: a) angular second moment (ASM), b) contrast, c) correlation, d) sum of squares variance (SOSV), e) inverse difference moment (IDM), f) sum average (SA), g) sum variance (SV), h) sum entropy (SE), i) entropy, j) difference variance (DV), k) difference entropy (DE), and l) information measures of correlation (IMC). For a chosen distance d (in this work d=1 was used) and for angles $\theta = 0^0$, 45^0, 90^0, and 135^0, we computed four values for each of the above texture measures. Each feature was computed using a distance of one pixel. Then for each feature the mean values and the range of values were computed, and were used as two different features sets. (iii) Gray Level Difference Statistics (GLDS) [18]: a) homogeneity, b) contrast, c) energy, d) entropy, and e) mean. The above features were calculated for displacements δ=(0, 1), (1, 1), (1, 0), (1, -1), where $\delta \equiv (\Delta x, \Delta y)$, and their mean values were taken. (iv) Neighbourhood Gray Tone Difference Matrix (NGTDM) [19]: a) coarseness, b) contrast, c) busyness, d) complexity, and e) strength. (v) Statistical Feature Matrix (SFM) 20: a) coarseness, b) contrast, c) periodicity, and d) roughness. (vi) Laws Texture Energy Measures (LTEM) [18]: LL-texture energy from LL kernel, EE-texture energy from EE-kernel, SS-texture energy from SS-kernel, LE-average texture energy from LE and EL kernels, ES-average texture energy from ES and SE kernels, and LS-average texture energy from LS and SL kernels. (vii) Fractal Dimension Texture Analysis (FDTA) [21]: The Hurst coefficients for dimensions 4, 3 and 2 were computed. (viii) Fourier Power Spectrum (FPS) [18], [20]: a) radial sum, and b) angular sum. (vii) Shape Parameters: a) X—coordinate maximum length, b) Y—coordinate maximum length, c) area, d) perimeter, e) perimeter2/area, f) eccentricity, g) equivalence diameter, h) major axis length, i) minor axis length, j) centroid, k) convex area, and l) orientation.

2.4 Models Support Vector Machines Classification

Classification analysis was carried out to classify brain MS lesions delineated on the baseline MRI scans into two classes according to the EDSS score that each patient was allocated two years following initial diagnosis: (i) MS subjects with EDSS≤2, and (ii) MS subjects with an EDSS>2. Thus, the classification goal was to differentiate between lesions that were subsequently associated with mild (EDSS≤2) or advanced disability (EDSS>2). The classification was applied on 30 subjects (15 males and 15 females).

Each classifier was implemented in Matlab, using Support Vector Machines (SVM). The SVM network was investigated using Gaussian Radial Basis Function kernels [22]; this was decided as the rest of the kernel functions could not achieve satisfactory results. The Parameters for the classifier were selected using ten-fold cross validation for each set. Parameter C of SVM was 1. The leave-one-out estimate was used for validating the classification models.

The performance of the classifier models were measured using the receiver operating characteristics (ROC) curve parameters [23]. The parameters calculated for ROC curves are: (i) true positives (TP) when the system correctly classifies subjects as EDSS≤2, and EDSS >2, (ii) false positives (FP) where the system wrongly classifies subjects as EDSS>2 while they are in the group of EDSS ≤2, (iii) true negatives (TN) when the system correctly classifies subjects as EDSS ≤2. We also compute the Sensitivity (SE) which is the likelihood that a subject with EDSS >2 will be detected given that it is EDSS>2 and Specificity (SP) which is the likelihood that a subject will be classified as EDSS ≤2 given that he/she is EDSS ≤2. For the overall performance, we provide the correct classification (CC) rate which gives the percentage of correctly classified subjects.

3 Results

We present in Fig. 2 box plots of the lesion texture features mean (Mean0_1, Mean0_2) median (Median0_1, Median0_2) and sum average (SA0_1, SA0_2) for 0 months for EDSS≤2 (_1) (N=18) and EDSS>2 (_2) (N=11).

Table 1. Statistical Analysis Of The Texture Features For Lesions And NAWM Subjects With EDSS≤2 (N=18) and EDSS>2 (N=11) Based on The Mann Whitney (first row) and the Wilcoxon (2nd row) Rank Sum Tests at p<0.05. Significant Difference is Depicted With S And Non Significant Difference is Depicted With NS. The p Values Are Shown in Parentheses.

Feature	Lesions	NAWM
Mean	**S (0.03)**	NS (0.18)
Median	**S (0.03)**	NS (0.17)
STD	NS (0.83)	NS (0.07)
Contr	NS (0.16)	NS (0.36)
SOSV	NS (0.83)	NS (0.55)
Entr	NS (0.06)	NS (0.58)
IDM	**S (0.01)**	NS (0.36)
SA	**S (0.04)**	NS (0.17)
SV	NS (0.83)	NS (0.19)
DV	NS (0.44)	NS (0.18)
DE	NS (0.35)	NS (0.77)

NAWM: Normal appearing white matter, STD: Standard deviation, Contr.: Contrast, SOSV: Sum of square variance, Entr: Entropy, IDM: Inverse difference moment, SA: Sum average, SV: Sum variance, DE: Difference entropy.

Table 2. Classification Results Using The Texture And Shape Feature Set And The Support Vector Machines Classifier For Two Classes: EDSS ≤2 and EDSS>2 (N=30). The Results Are Presented Using the ROC Measures: Percentage of correct classifications (%CC), percentage of false positives (%FP), percentage of false negatives (%FN), percentage sensitivity (%SE) and percentage specificity (%SP).

Feature Group	%CC	%FP	%FN	%SE	%SP
FOS	43	47	67	33	53
SGLDM(mean)	63	53	20	80	47
SGLDM(range)	63	53	20	80	47
GLDS	67	47	20	80	53
NGTDM	67	40	27	73	60
SFM	67	40	27	73	60
TEM	67	40	27	73	60
FDTA	60	33	7	53	67
FPS	60	53	27	73	47
NGTDM & SFM	73	27	27	73	73
GLDS & NGTDM & SFM	77	27	20	80	73
SHAPE	53	53	40	60	47

Statistical analysis was carried out for all texture features given in section 2.3 for lesions and NAWM for EDSS≤2 versus EDSS>2. The results of selected features, including those that showed significant difference are tabulated in Table 1. For lesions, features mean, median, IDM and SA could differentiate between EDSS≤2 and EDSS>2 (see also Fig. 2). For the NAWM it is shown that there are no significant differences between subjects at 0 months with EDSS>2 and EDSS≤2, for all texture features.

Table 2 presents the lesion classification results. According to Table II the best correct classification rate was achieved as a combination of GLDS, NGTDM and SFM feature sets and was 77%.

4 Discussion

The objective of our study was to investigate texture classification analysis in an effort to differentiate between MS brain lesions that, in a subsequent stage, will be associated with advanced diseased disability from those lesions that will be related to a mild disability score.

Table 1 showed that there are some texture features that can be possibly used to differentiate between subjects with mild (EDSS≤2) and advanced (EDSS>2) MS disease states. These features are mean, median, IDM and SA. Table 2 showed that the combination of GLDS, NGTDM & SFM texture features gave a % of CC rate of 77% when classifying subjects according to their long-term disability status.

These results can be compared with another study carried out by our group [14], where AM-FM feature analysis was used but on a slightly larger set of subjects. The findings suggest that AM–FM characteristics succeed in differentiating (i) between

NWM and lesions, (ii) between NAWM and lesions, and (iii) between NWM and NAWM. A support vector machine (SVM) classifier succeeded in differentiating between patients that, two years after the initial MRI scan, acquired an EDSS ≤ 2 from those with EDSS > 2 (correct classification rate = 86%). The best classification results were obtained from including the combination of the low-scale instantaneous amplitude (IA) and instantaneous frequency (IF) magnitude with the medium-scale IA. Among all AM-FM features, the medium frequency IA histogram alone gave the best AUC results (AUC=0.76). Also the best % of CC achieved was 86%. The IA histograms from all frequency scales contributed to a multi-classifier system that gave 86% correct classification rate. The best classifier results also used the IF magnitude from both the low and the high frequency scales.

To the best of our knowledge, no other studies were carried out for differentiating between the aforementioned two disability scores. Several studies were carried out for differentiating and classifying NWM, and or NAWM, and lesions, as these are summarized below.

Various studies have been performed in order to establish a relationship between the various gray levels and texture features [4]-[11]. In [5] MRI texture analysis based on statistical, autoregressive model, and wavelet-derived texture analysis was performed on 30 MS subjects. The classification accuracy between MS lesions, NAWM and plaques NWM, was 96%-100%.

Likewise in [24], texture analysis was performed on MR images of MS subjects and normal controls and a combined set of texture features were explored in order to better discriminate tissues between MS lesions, NAWM and NWM. The results demonstrated that with the combined set of texture features, classification was perfect (100%) between MS lesions and NAWM (or NWM), less successful (88.89%) among the three tissue types and worst (58.33%) between NAWM and NWM. Furthermore, it was shown that compared with GLCM-based features, the combined set of texture features were better at discriminating MS lesions and NWM, equally good at discriminating MS lesions and NAWM and at all three tissue types, but less effective in classification between NAWM and NWM. The study suggested that texture analysis with the combined set of texture features may be equally good or more advantageous than the commonly used GLCM-based features alone in discriminating MS lesions and NWM/NAWM and in supporting early diagnosis of MS.

Yu et al. [11] performed textural feature analysis to discriminate active and non-active MS lesions in a study of 8 subjects with relapsing remitting MS, by using linear discriminant analysis using 42 first- and second order statistical textural features. Applications of the run-length method have been very limited compared with other methods, yet Yu et al. found that run-length matrix features actually outperformed gray-level co-occurrence matrix features in the identification of active MS lesions, with run-length matrix features distinguishing active from inactive lesions with 88% sensitivity and 96% specificity. Conversely, none of the gray-level co-occurrence features provided any discrimination between lesion subtypes.

In another study [12], it was shown that texture analysis can achieve high classification accuracy (>=90%) in tissue discrimination between MS lesions and NAWM.

In [25] the authors tried to objectively identify possible differences in the signal characteristics of benign and malignant soft tissue masses on MRI by means of texture analysis and to determine the value of these differences for computer-assisted lesion classification. Fifty-eight subjects with histologically proven soft tissue masses (benign, n=30; malignant, n=28) were included and their masses were texture analyzed. The best results of soft tissue masses classification were achieved using texture information from short tau inversion recovery images, with an accuracy of 75.0% (sensitivity, 71.4%; specificity, 78.3%) for the k-NN classifier, and an accuracy of 90.5% (sensitivity, 91.1%; specificity, 90.0%) for the artificial neural network classifier. The authors concluded that, texture analysis revealed only small differences in the signal characteristics of benign and malignant soft tissue masses on routine MRI.

5 Future Work

Further research work on a larger number of subjects is required for validating the results of this study and for finding additional shape and texture features that may provide information for longitudinal monitoring of the lesions on the initial MRI scan of the brain of patients with CIS; and improve the final classification rate. Additionally the use of other classifiers could suggest a better classification a scheme and will be examined. Finally the proposed methodology could be possibly used for the assessment of subjects at risk of developing future neurological events and disease progression as measured by increased EDSS score.

Acknowledgment. This work was supported through the project Quantitative and Qualitative Analysis of MRI Brain Images ΤΠΕ/OPIZO/0308(BIE)/15, 12/2008-12/2010, of the Program for Research and Technological Development 2007-2013, co funded by the Research Promotion Foundation of Cyprus.

References

1. Fazekas, F., Barkof, F., Filippi, M., Grossman, R.I., Li, D.K.B., McDonald, W.I., McFarland, H.F., Patty, D.W., Simon, J.H., Wolinsky, J.S., Miller, D.H.: The contribution of magnetic resonance imaging to the diagnosis of multiple sclerosis. Neur. 53, 44–456 (1999)
2. McDonald, W.I., Compston, A., Edan, G., et al.: Recommended diagnostic criteria for multiple sclerosis: guidelines from the international panel on the diagnosis of multiple sclerosis. Ann. Neurol. 50, 121–127 (2001)
3. Bakshi, R., Thompson, A.J., Rocca, M.A., et al.: MRI in multiple sclerosis: current status and future prospects. Lancet Neurol. 7, 615–625 (2008)
4. Kassner, A., Thornhill, R.E.: Texture analysis: A review of neurologic MR imaging applications. Am. J. Neuroradiol. 31, 809–816 (2010)
5. Harrison, L.C.V., Raunio, M., Holli, K.K., Luukkaala, T., Savio, S., et al.: MRI Texture analysis in multiple sclerosis: Toward a clinical analysis protocol. Acad. Radiol. 17, 696–707 (2010)
6. Herlidou-Meme, S., Constans, J.M., Carsin, B., Olivie, D., Eliat, P.A., et al.: MRI texture analysis on texture test objects, normal brain and intracranial tumors. Mag. Res. Imag. 21, 989–993 (2003)

7. Mathias, J.M., Tofts, P.S., Losseff, N.A.: Texture analysis of spinal cord pathology in multiple sclerosis. Magn., Reson. Med. 42, 929–935 (1999)
8. Loizou, C.P., Pattichis, C.S., Seimenis, I., Eracleous, E., Schizas, C.N., Pantziaris, M.: Quantitative analysis of brain white matter lesions in multiple sclerosis subjects: Preliminary findings. In: IEEE Proc., 5th Int. Conf. Inf. Techn. Appl. Biomed., ITAB, Shenzhen, China, May 30-31, pp. 58–61 (2008)
9. Loizou, C.P., Pattichis, C.S., Seimenis, I., Pantziaris, M.: Quantitative analysis of brain white matter lesions in multiple sclerosis subjects. In: 9th Int. Conf. Inform. Techn. Applic. Biomed., ITAB, Larnaca, Cyprus, November 5-7, pp. 1–4 (2009)
10. Collewet, G., Strzelecki, M., Marriette, F.: Influence of MRI acquisition protocols and image intensity normalization methods on texture classification. Magn. Reson. Imag. 22, 81–91 (2004)
11. Yu, O., Mauss, Y., Zollner, G., Namer, I.J., Chambron, J.: Distinct patterns of active and non-active plaques using texture analysis of brain NMR images in multiple sclerosis patients: Preliminary results. Magn. Reson. Imag. 17(9), 1261–1267 (1999)
12. Zhang, J., Wang, L., Tong, L.: Feature reduction and texture classification in MRI-Texture analysis of multiple sclerosis. In: IEEE/ICME Conf. Complex Med. Eng., pp. 752–757 (2007)
13. Meier, D.S., Guttman, C.R.G.: Time-series analysis of MRI intensity patterns in multiple sclerosis. NeuroImage 20, 1193–1209 (2003)
14. Loizou, C.P., Murray, V., Pattichis, M.S., Seimenis, I., Pantziaris, M., Pattichis, C.S.: Multi-scale Amplitude Modulation-Frequency Modulation (AM-FM) texture analysis of multiple sclerosis in brain MRI images. IEEE Trans. Inform. Tech. Biomed. 15(1), 119–129 (2011)
15. Thompson, A.J., Hobart, J.C.: Multiple sclerosis: assessment of disability and disability scales. J. Neur. 254(4), 189–196 (1998)
16. Loizou, C.P., Pantziaris, M., Seimenis, I., Pattichis, C.S.: MRI intensity normalization in brain multiple sclerosis subjects. In: ITAB 2009, 9th Int. Conf. on Inform. Techn. And Applic. in Biomed., Larnaca, Cyprus, November 5-7, pp. 1–5 (2009)
17. Haralick, R.M., Shanmugam, K., Dinstein, I.: Texture features for image classification. IEEE Trans. Syst., Man., and Cybernetics SMC-3, 610–621 (1973)
18. Weszka, J.S., Dyer, C.R., Rosenfield, A.: A comparative study of texture measures for terrain classification. IEEE Trans. Syst., Man. Cybern. SMC-6, 269–285 (1976)
19. Amadasun, M., King, R.: Textural features corresponding to textural properties. IEEE Trans. Syst. 19(5), 1264–1274 (1989)
20. Wu, C.M., Chen, Y.C., Hsieh, K.-S.: Texture features for classification of ultrasonic images. IEEE Trans. Med. Imag. 11, 141–152 (1992)
21. Mandelbrot, B.B.: The Fractal Geometry of Nature. Freeman, San Francisco (1982)
22. Cristianini, N., Shawe-Taylor, J.: An Introduction to Support Vector Machines and Other Kernel-based Learning Methods, 1st edn. Cambridge University Press, Cambridge (2000)
23. Ebrchart, R.C., Dobbins, R.W.: Neural Networks PC Tools A Practical Guide. Academic Pr., New York (1990)
24. Zhang, J., Tong, L., Wang, L., Lib, N.: Texture analysis of multiple sclerosis: a comparative study. Magn. Res. Imag. 26(8), 1160–1166 (2008)
25. Mayerhoefer, M.E., Breitenseher, M., Amannd, G., Dominkuse, M.: Are signal intensity and homogeneity useful parameters for distinguishing between benign and malignant soft tissue masses on MR images? Objective evaluation by means of texture analysis. Magn. Res. Imag. 26, 1316–1322 (2008)

Using Argumentation for Ambient Assisted Living

Julien Marcais[1], Nikolaos Spanoudakis[2], and Pavlos Moraitis[1]

[1] Laboratory of Informatics Paris Descartes (LIPADE),
Paris Descartes University, France
Julien.marcais@gmail.com,
pavlos@mi.parisdescartes.fr
[2] Department of Sciences,
Technical University of Crete, Greece
nikos@science.tuc.gr

Abstract. This paper aims to discuss engineering aspects for an agent-based ambient assisted living system for the home environment using argumentation for decision making. The special requirements of our system are to provide a platform with cost-effective specialized assisted living services for the elderly people having cognitive problems, which will significantly improve the quality of their home life, extend its duration and at the same time reinforce social networking. The proposed architecture is based on an agent platform with personal assistant agents that can service users with more than one type of health problems.

Keywords: Multi-Agent Systems, Ambient Assisted Living, Argumentation, Alzheimer disease.

1 Introduction

This paper is concerned with addressing the non-trivial task [9] of engineering an AAL information system addressing the needs of the elderly suffering from dementia (Alzheimer Disease) with identified risk factors as well as having cognitive problems (referred to as *end users* from now on). This application domain is about offering indoor assistance aiming to enhance the autonomy and quality of life of the user.

Agent-based computing is considered as one of the most appropriate technologies for ambient assisted living systems in an interesting survey presented by Becker [1]. Another interesting survey on agent-based intelligent decision support systems to support clinical management and research can be found in the work of Foster et al. [3]. Finally interesting information about the characteristics of residential monitoring applications designed to be used in consumers' personal living spaces and more particularly to be used by persons with dementia is presented by Mahoney et al. [7].

The HERA project[1] aims to build an AAL system to provide cost-effective specialized assisted living services for the elderly people suffering from MCI or

---

[1] The "Home sERvices for specialised elderly Assisted living" (HERA, http://aal-hera.eu) project is funded by the Ambient Assisted Living Joint Programme (AAL, http://www.aal-europe.eu)

L. Iliadis et al. (Eds.): EANN/AIAI 2011, Part II, IFIP AICT 364, pp. 410–419, 2011.

mild/moderate AD or other diseases (diabetes, cardiovascular) with identified risk factors, which will significantly improve the quality of their home life, extend its duration and at the same time reinforce social networking. In order to achieve this goal, HERA will provide to these end users the following main categories of services:

- Cognitive Exercises, the end users play cognitive reinforcement games at their TV using the remote control
- Passive communication, the end user can select to watch informative videos and news items related to his/her disease
- Pill and Exercise reminders, the end user is reminded for taking his pills or doing his exercises while he is watching TV or a video, or is playing a game
- Reality orientation, date and time is visible on screen while watching videos or the TV, also while doing their exercises
- Blood pressure and weight monitoring, the user can perform the measurements at specific intervals and he can monitor his condition along with his doctor

HERA applies technological solutions for aiding users managing their daily lives. Thus, by using the HERA system, the time to be at home, rather than in an institution, will be prolonged and relieve them from visiting the specialists often, while keeping them able to perform their daily activities and social interactions.

In [13] we presented the HERA system requirements and proposed an architecture to address the challenges of engineering such systems (see, e.g., Kleinberger et al. [6]):

- *Adaptability*. No two human beings have the same needs or everyday life habits. An AAL system must be able to adapt to a particular user. For the agents decision making we chose argumentation as it allows for decision making using conflicting knowledge, thus different experts can express their opinion that can be conflicting.
- *Natural and anticipatory Human-Computer Interaction (HCI)*. The people that need assistance very often have limitations and handicaps. In HERA the use of the TV set and remote control is ensuring a quick learning curve for our users,
- *Heterogeneity*. AAL systems are expected to be capable of being integrated with several subsystems developed by different manufacturers (e.g. sensors, etc). We use a service oriented architecture based on web services that allows the different sub-systems to be connected in a plug and play standardized way.

There are specific advantages of our approach compared with previous work (see e.g. Bravo et al. [2], or García et al. [4]). Firstly, the autonomy of the user is increased and the ambient assistance is automated. Secondly, the use of argumentation allows for decision making even in cases when a user has more than one chronic diseases situations.

The system service oriented architecture and evaluation process is presented in [11]. In this paper we focus in presenting the use of argumentation in HERA by one of the most important modules the multi-agent system.

In what follows we briefly discuss argumentation technology, then we present an overview of the agents design process and, finally, we focus in the knowledge representation and reasoning in HERA.

2 Argumentation Technology

Argumentation can be abstractly defined as the formal interaction of different conflicting arguments for and against some conclusion due to different reasons and provides the appropriate semantics for resolving such conflicts. Thus, it is very well suited for implementing decision making mechanisms dealing with the above requirements. Moreover, the dynamic nature of those conflicting decisions due to different situations or contexts needs a specific type of argumentation frameworks (such as those proposed by Kakas and Moraitis [5] or Prakken and Sartor [10]). These frameworks are based on object level arguments representing the decision policies and then they are using priority arguments expressing preferences on the object level arguments in order to resolve possible conflicts. Subsequently, additional priority arguments can be used in order to resolve potential conflicts between priority arguments of the previous level. Therefore, we are concerned with argumentation frameworks that allow for the representation of dynamic preferences under the form of dynamic priorities over arguments. In this work we are using the framework proposed by Kakas and Moraitis [5]. This framework has been applied in a successful way in different applications (see e.g. [8], [14]) involving similar scenarios of decision making and it is supported by an open source software called Gorgias[2]. The latter allows for defining dynamic priorities between arguments, which means that the priorities of rules can depend on context. Finally, the modularity of its representation allows for the easy incorporation of views of different experts.

The Gorgias Framework

Gorgias is an open source implementation of the framework presented above in the Prolog language. Gorgias defines a specific language for the representation of the object level rules and the priorities rules of the second and third levels. The language for representing the theories is given by rules with the syntax in formula (1).

$$\text{rule(Signature, Head, Body).} \tag{1}$$

In the rule presented in formula (1), Head is a literal, Body is a list of literals and Signature is a compound term composed of the rule name with selected variables from the Head and Body of the rule. The predicate prefer/2 is used to capture the higher priority relation (h_p) defined in the theoretical framework. It should only be used as the head of a rule. Using the syntax defined in (1) we can write the rule presented in formula (2).

$$\text{rule(Signature, prefer(Sig1, Sig2), Body).} \tag{2}$$

Formula (2) means that the rule with signature Sig1 has higher priority than the rule with signature Sig2, provided that the preconditions in the Body hold. If the modeler needs to express that two predicates are conflicting he can express that by using the rule presented in formula (3).

$$\text{conflict(Sig1,Sig2).} \tag{3}$$

---

[2] Gorgias is an open source general argumentation framework that combines the ideas of preference reasoning and abduction (http://www.cs.ucy.ac.cy/~nkd/gorgias)

The rule in formula (3) indicates that the rules with signatures Sig1 and Sig2 are conflicting. A literal's negation is considered by default as conflicting with the literal itself. A negative literal is a term of the form neg(L). There is also the possibility to define conflicting predicates that are used as heads of rules using the rule presented in formula (4).

$$\text{complement}(\text{Head1}, \text{Head2}). \tag{4}$$

Development Environment

The development environment involves several tools:

- Gorgias
- Logtalk
- SWI-prolog

In Prolog we use a Knowledge Base (KB), which is composed of Horn clauses. There are two types of clauses, facts and rules. Rules follow this form:

$$\text{Head :- Body.} \tag{5}$$

This means that the head of this rule will be true if and only if all the predicates which compose the body are true. The head is a single predicate. Facts are rules without body.

The developer has just to describe the world with facts and rules and then ask a question to Prolog. There are many Prolog implementations, one of which is SWI-Prolog (the free prolog environment that we used).

Logtalk is the bridge between two worlds: Logic Programming and Object-Oriented Programming. It is developed as a Prolog extension that's why Logtalk needs a Prolog compiler. A program written with the Logtalk language will first be translated by the Logtalk compiler into a full Prolog file. Then the generated Prolog file is compiled by the selected Prolog compiler. It is needed in order to be able to run simultaneously more than one prolog instances using the Java JPL interface.

3   HERA Multi-Agent System Analysis

The heart of the HERA services platform is the Multi-Agent System (MAS). For designing our agents we used the ASEME methodology [12]. In Figure 1 the System Actors Goals (SAG) model is presented. Actors are depicted as circles filled with black color, while their goals are depicted in rounded rectangles. An actor's goal can be dependent to another actor, in this case there are directed lines from the depender to the goal and then from the goal to the dependee.

A personal assistant (PA) agent serves a specific user and has access to the user's profile data. This agent is proactive in the sense that he is always active following the user's agenda and providing support and advice by taking action whenever needed. For example he is able to send a message to the user for reminding him to take his pills, to receive sensor information about the user and consult experts' knowledge on the user's behalf. Moreover, the PA uses the requests' history in order to adapt the services to the user's habitual patterns.

We also have an interface role that acts as a gateway to the MAS. Thus, while the personal assistant accesses himself all information sources that he needs, the other systems like the backoffice send information to the MAS through the personal assistant. The backoffice and notification module are external actors to the MAS and are developed by our HERA partners.

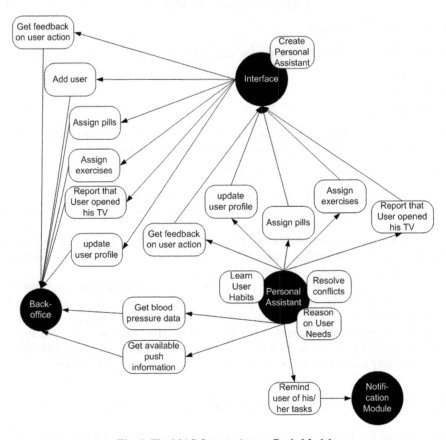

Fig. 1. The MAS System Actors Goals Model

An important analysis phase model that we define in ASEME is the System Role Model where the roles that are to be implemented as agents are refined and their specific capabilities and activities are defined. We use the Gaia operators as they were defined by Wooldridge et al. [16] for creating liveness formulas that define the process of the role. Briefly, A:B means that activity B is executed after activity A, A^{ω} means that activity A is executed forever (it restarts as soon as it finishes), A|B means that either activity A or activity B is executed, [A] means that activity A is optional and A||B means that activities A and B are executed in parallel. The left hand side term of the liveness formula is a capability, while the terms participating in the right hand side expression are activities or capabilities (if they appear on the left hand side of another liveness formula). The first formula has the role name on its left hand side.

PersonalAssistant	= initializeUserProfileStructure. initializeUserScheduleStructure. (newExercisePrescription$^{\omega}$ || newPillPrescription$^{\omega}$ || userOpenedTVSet$^{\omega}$ || serviceUser$^{\omega}$ || updateUserProfile$^{\omega}$)
userOpenedTVSet	= receiveUserOpenedTVInform. [invokeAvailableVideosInformService. [updateUserSchedule]]
updateUserSchedule	= resolveConflicts. updateUserScheduleStructure
resolveConflicts	= readUserScheduleStructure. (checkIfConflictsExist. reasonOnItemsPriorities. sortItems)+
updateUserProfile	= receiveUserProfileUpdateRequest. updateUserProfileStructure
serviceUser	= [checkUserExerciseResults]. [updateUserSchedule]. waitForUserScheduleNextItem. [reasonOnPillsQuantity]. [remindUserOfTasks]
checkUserExerciseResults	= invokeExerciseResultsService. checkIfUserDidHisExercises. [updateUserSchedule]
remindUserOfTasks	= invokeNotificationModule. [learnUserHabits]. getFeedbackOnUserAction
getFeedbackOnUserAction	= receiveUserActionInform. [updateUserSchedule]
learnUserHabits	= reasonOnUserAction. [updateUserSchedule]
newPillPrescription	= receiveNewPillPrescriptionRequest. updateUserSchedule
newExercisePrescription	= receiveNewExercisePrescriptionRequest. updateUserSchedule

Fig. 2. The PA role liveness formulas

Having defined the SRM for the PA (see Figure 2) the next activity in the ASEME methodology is to define the functionality of each activity. This is done using the Functionality Table, which is presented in Figure 3 for the PA role. In the functionality table we have on the left hand side the agent capabilities, in the middle the activities that are used by each capability and on the right hand side the functionality that is used by each activity.

Thus, for developing the personal assistant (see Figure 3) we need to implement:

- two knowledge bases (one for assigning priority to conflicting user tasks and one for reasoning on the pills quantities in specific contexts)
- five ACL message receivers
- three algorithms (one for sorting items in the user's calendar, one for checking whether new calendar items conflict with existing items, and one for determining if the user has finished his exercises based on information retrieved from the HERA database)
- two global variables (user profile and user schedule) at the agent level
- three different web service clients, and,
- a statistical learning algorithm

ASEME caters for the transformation of the liveness formulas to a statechart that allows the designer to add conditions to the statechart transitions for controlling the execution of the activities. In the rest of this paper we will focus in the knowledge bases development.

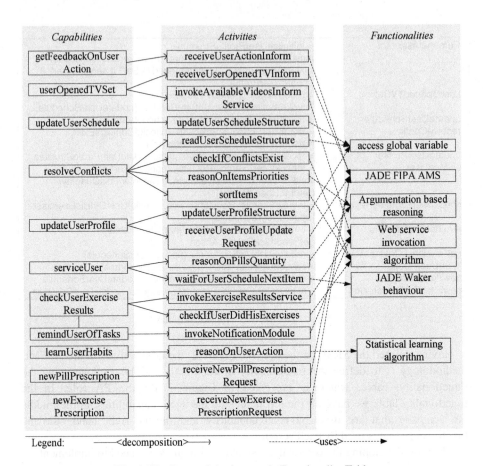

Fig. 3. The Personal Assistant role Functionality Table

4 Using Argumentation in HERA

We focus in two issues that are highly related to the assistance of people with mild or moderate Alzheimer's disease. The first suffer from frequent recent memory loss, particularly of recent conversations and events, while the second suffer from pervasive and persistent memory loss, confusion about current events, time and place [11]. Thus, it is important for the system to be able to help them in situations that they need to remember a special case for taking their pills or to schedule their tasks.

Reasoning On The Pills Quantities In Specific Contexts

When the user's scheduled time for taking pills arrives, the PA reasons on the quantity to be taken. The doctors are able to assign specific conditions when assigning pills to a patient using the HERA backoffice. For example, if the blood pressure exceeds a limit then he has to take two pills (while normally he is assigned one pill). Moreover, there can be specific contexts that define specific situations like e.g. when the temperature is cold the blood pressure fluctuates, thus the readings are not so reliable.

Pills quantities can be modified based on basic rules identified and provided by doctors. For example if we want to work on blood pressure, in order to be able to use rules we need some threshold values provided by doctors. Here we will represent them with those expressions "Lower_Threshold" and "Upper_Threshold". These upper and lower thresholds are not universal, they are defined by the doctors for each patient.

Note that we have some past measurements of the user either in a row or in the last two days. If all are up or down from a given threshold we have some rules for updating the pill quantity as prescribed by the doctor. Thus, the fact of what the new pill dosage will be if the lower threshold will be surpassed and what the pill dosage will be if an upper threshold is surpassed must be added.

If we have a specific context and precise information provided by doctors, thanks to the argumentation framework it is possible for us to transform it into an argumentation based decision making problem and find some solutions automatically. For the reader's convenience, we present a simplified extract that demonstrates the achieved functionality. We define some predicates that are instantiated by the personal assistant defaultPillDosage/2, with the pill name and the dosage as a real number, upperThresholdBreachedDosage/2, with the pill name and the dosage in the case when the upper BP threshold is breached, systolicBP/1, containing the reading of the user's blood pressure, upperSystolicThreshold/2, containing the BP threshold which must be surpassed so that a special dosage is proposed to the user, the temperature/1 characterizing the temperature of the environment, and the newPillQuantity/2 which after the reasoning phase contains the suggested dosage.

```
1.  rule(r1, newPillQuantity(Pill, Dosage1),
2.  [systolicBP(SBP), upperSystolicThreshold(Pill, Limit),
3.  upperThresholdBreachedDosage(Pill, Dosage1), SBP >
4.  Limit]).
5.  rule(r2, newPillQuantity(Pill, Dosage2),
6.  [defaultPillDosage(Pill, Dosage2)]).
7.
8.  conflict(r1,r2).
9.
10. rule(pr1, prefer(r1,r2), []).
11. rule(pr2, prefer(r2,r1), [temperature(cold)]).
12.
13. rule(c1, prefer(pr2,pr1), []).
```

Rule r2 states that the new pill dosage is equal to the default. Rule r1 says that if the upper systolic pressure threshold is breached a new pill quantity is assigned. Line 8 characterizes these rules as conflicting and line 10 assigns a default higher priority to the r1 rule. However, the rule in line 11 assigns higher priority to rule r2 in the specific context where the temperature is cold (when BP tends to fluctuate). Finally, the rule in line 13 assigns a higher priority to rule pr2 over pr1.

Assigning Priority To Conflicting User Tasks

When an item is inserted in the user's agenda or is rescheduled the agent reasons on the priority of possibly conflicting tasks. Specifically, when the user has been assigned more than one tasks for the same time (e.g. by different caregivers) or that he has specific preferences (e.g. to watch a TV series at a particular time of day the following priorities will hold (see the HERA system requirements [15]):

— Priority no1: take the assigned pills
— Priority no2: watch his favourite TV series
— Priority no3: engage with the cognitive reinforcement exercises
— Priority no4: engage with the physical reinforcement exercises

This knowledge base is the same for all users as the requirements that we got from the caretakers were such and is simpler than the previous one.

5   Conclusion

In this paper we have presented a multi-agent system part of the HERA system that is proposed for assuming Ambient Assisted Living functionalities providing at home services for people suffering from cognitive problems and more particularly from Alzheimer disease. We have presented several engineering aspects on how such a system can be designed by using the ASEME methodology along with elements concerning the reasoning mechanism used by some of the agents based on argumentation. We have thus shown that the latter allows agents to make decisions in situations where conflicting points of view, corresponding to doctors' opinions of different specialties and concerning patients having different health problems, must be taken into account. Our work has been functionally validated and is currently under user acceptance tests in the HERA trials.

The HERA system deployment will take place in two phases. In the first one it will be deployed in the medical center's premises for evaluation by the medical personnel and for controlled interaction with the patients and in the second phase it will be deployed inside the users' homes for final evaluation. The process for evaluating our system has been described in more detail in [11].

References

1. Becker, M.: Software Architecture Trends and Promising Technology for Ambient Assisted Living Systems, Dagstuhl Seminar Proceedings 07462, Assisted Living Systems-Models, Architectures and Engineering Approaches (2008)
2. Bravo, J., de Ipiña, D.L., Fuentes, C., Hervás, R., Peña, R., Vergara, M., Casero, G.: Enabling NFC Technology for Supporting Chronic Diseases: A Proposal for Alzheimer Caregivers. In: Aarts, E., Crowley, J.L., de Ruyter, B., Gerhäuser, H., Pflaum, A., Schmidt, J., Wichert, R. (eds.) AmI 2008. LNCS, vol. 5355, pp. 109–125. Springer, Heidelberg (2008)
3. Foster, D., McGregor, C., El-Masri, S.: A Survey of Agent-Based Intelligent Decision Support Systems to Support Clinical Management and Research. In: Proceedings MAS*BIOMED 2005, Ultrecht, Nederlands (2005)

4. García, O., Tapia, D.I., Saavedra, A., Alonso, R.S., García, I.: ALZ-MAS 2.0; A Distributed Approach for Alzheimer Health Care. In: 3rd Symposium of Ubiquitous Computing and Ambient Intelligence 2008, pp. 76–85. Springer, Heidelberg (2008)
5. Kakas, A., Moraitis, P.: Argumentation Based Decision Making for Autonomous Agents. In: Proc. 2nd Int'l Joint Conf. Autonomous Agents and Multi-Agent Systems (AAMAS 2003), pp. 883–890. ACM, New York (2003)
6. Kleinberger, T., Becker, M., Ras, E., Holzinger, A., Müller, P.: Ambient Intelligence in Assisted Living: Enable Elderly People to Handle Future Interfaces. In: Stephanidis, C. (ed.) UAHCI 2007 (Part II). LNCS, vol. 4555, pp. 103–112. Springer, Heidelberg (2007)
7. Mahoney, D.F., et al.: In-home monitoring old persons with dementia: Ethical quidelines for technology research and development. Alzheimer's & Dementia 3, 217–226 (2007)
8. Moraitis, P., Spanoudakis, N.: Argumentation-based Agent Interaction in an Ambient Intelligence Context. IEEE Intell. Syst. 22(6), 84–93 (2007)
9. Nehmer, J., Karshmer, A., Becker, M., Lamm, R.: Living Assistance Systems – An Ambient Intelligence Approach. In: Proc. of the Int. Conf. on Software Engineering, ICSE (2006)
10. Prakken, H., Vreeswijk, G.: Logics for Defeasible Argumentation. In: Gabbay, D., Guenthner, F. (eds.) Handbook of Philosophical Logic, vol. 4, pp. 218–319. Kluwer Academic Publishers, Dordrecht (2002)
11. Spanoudakis, N., Grabner, B., Lymperopoulou, O., Moser-Siegmeth, V., Pantelopoulos, S., Sakka, P., Moraitis, P.: A Novel Architecture and Process for Ambient Assisted Living – The HERA approach. In: Proceedings of the 10th IEEE International Conference on Information Technology and Applications in Biomedicine (ITAB 2010), Corfu, Greece, November 3-5 (2010)
12. Spanoudakis, N., Moraitis, P.: Using ASEME Methodology for Model-Driven Agent Systems Development. In: Weyns, D. (ed.) AOSE 2010. LNCS, vol. 6788, pp. 106–127. Springer, Heidelberg (2011)
13. Spanoudakis, N., Moraitis, P., Dimopoulos, Y.: Engineering an Agent-based Approach to Ambient Assisted Living. In: AmI 2009 Workshop on Interactions Techniques and Metaphors in Assistive Smart Environments (IntTech 2009), Salzburg, Austria, November 18-21 (2009)
14. Spanoudakis, N., Pendaraki, K., Beligiannis, G.: Portfolio Construction Using Argumentation and Hybrid Evolutionary Forecasting Algorithms. International Journal of Hybrid Intelligent Systems (IJHIS) 6(4), 231–243 (2009)
15. The HERA Consortium: State-of-the-art and Requirements Analysis, Deliverable D2.1, http://aal-hera.eu
16. Wooldridge, M., Jennings, N.R., Kinny, D.: The Gaia Methodology for Agent-Oriented Analysis and Design. J. Auton. Agents and Multi-Agent Syst. 3(3), 285–312 (2000)

Modelling Nonlinear Responses of Resonance Sensors in Pressure Garment Application

Timo Salpavaara and Pekka Kumpulainen

Tampere University of Technology,
Department of Automation Science and Engineering,
Korkeakoulunkatu 3, 33720,
Tampere, Finland
{timo.salpavaara,pekka.kumpulainen}@tut.fi

Abstract. Information on the applied pressure is critical to the pressure garment treatment. The use of the passive resonance sensors would be significant improvement to existing systems. These sensors have nonlinear response and thus require nonlinear regression methods. In this paper we compare three nonlinear modelling methods: Sugeno type fuzzy inference system, support vector regression and multilayer perception networks. According to the results, all the tested methods are adequate for modelling an individual sensor. The used methods also give promising results when they are used to model responses of multiple sensors.

Keywords: nonlinear regression, anfis, MLP, support vector regression, pressure sensor.

1 Introduction

The pressure garment treatment has potential to improve the healing process of burns and to reduce the swelling. The use of suitable pressure is critical to the treatment and thus, in order to ensure the proper functioning of the pressure garment, this pressure has to be measured. The typical pressure of the issued pressure garment range from 20 to 60 mmHg (millimeters of mercury). The desirable features for the sensors used in this application, besides the obvious good metrological properties, are the small size of the sensor and disposability. The size of the sensor is an issue because under the pressure garment, a thick sensor acts in a similar manner as a pressure bandage and thus increases the pressure locally which leads to a systematic error. The disposable transducers are needed to avoid contamination since the sensors are placed in contact with the skin of the patients. The pressure of the pressure garments can be measured by using an air filled pouch and tubing [1]. The tubing conducts the pressure of the clothing to a separate measurement unit. An alternative method is to use transducers based on the capacitive [2] and piezoresistive [3] principles. However, the electrical wiring is required to connect these transducers to measurement devises.

L. Iliadis et al. (Eds.): EANN/AIAI 2011, Part II, IFIP AICT 364, pp. 420–429, 2011.

A novel approach to measure the pressures under the pressure garments is the inductively read passive resonance sensors [4]. The key advantages of this approach are the wireless connection to the sensor under the garment, the simplicity and the small size of the sensor. The wireless connection makes the system more convenient to use since the electrical wiring or tubing are significant disadvantages especially when the garments are put on. The simple sensor design is an advantage since we aim at a disposable, adhesive bandage-like sensor. In addition, the simple structure of the sensor may enable the use of the new fabrication methods like printing the structures directly on supporting material. Other tested applications of the passive resonance sensors in the field of medicine are intra-ocular pressure sensing [5] and ECG-measurements [6].

The main idea of this sensing method is to inductively measure the resonance frequency of an LC-resonator with an external reader coil. The measurand affects either the inductance or the capacitance of the sensor and the inductor in the sensor serves as a link coil. One drawback of this method is the nonlinear response of the sensor. The resonance frequency of the LC-circuit is nonlinear by nature. In addition, in many applications the relation between the measurand and the capacitance or the inductance is nonlinear.

Since the responses of the resonance sensors are inherently nonlinear, methods for converting the measured resonance frequencies to the pressure are needed. The lookup tables or calculation of the pressure from electrical and mechanical models are possible but these methods do not really support the idea of disposable sensors. The needed information is impractical to measure, store and deliver. The sensor can be made less nonlinear by stiffening the mechanical structure by the cost of the sensitivity. If the nonlinearity problem can be handled, more possibilities for the structural and fabrication innovations become available. One solution for this problem is to measure the response of the sensor and to identify a nonlinear model. The model can be used to convert the measured frequencies to the measurand. For example neural network model has been used for compensating the nonlinear response of capacitive sensors [7].

In this paper we test three nonlinear regression methods to model the responses of the passive resonance sensors.

2 Passive Resonance Sensor and Used Read-Out Methods

The instrumentation in this work consists of a hand-held measurement unit and passive resonance sensors. The measurand in this application is the pressure under the clothing. It is converted to the electrical signal by using pressure dependent capacitors. The pressure on the sensor alters the impedance of the capacitors by reducing the distances between the capacitor plates. This alters the phase response of the resonance sensor. The phase response is measured wirelessly through the inductive link using the hand-held measurement unit. Afterwards, the PC post processing software calculates an estimate for the resonance frequency based on the phase response data. The post processing also includes the compensation algorithm to reduce the error caused by the unknown coupling coefficient of the inductive link.

The coil and the capacitor of the resonance circuit are created by assembling a dual-layer structure which consists of the PCB (printed circuit board) and PDMS layers (Polydimethylsiloxane). The link coil is etched on PCB. There are two rigid electrodes on the PCB layer which form two pressure dependent capacitors in a combination with the electrode layer on the PDMS. There is also a discrete capacitor on the PCB. The capacitor is needed to tune the sensor to the proper frequency range. The flexible PDMS layer is glued on top of the supporting PCB layer. It has two major features: a cavity which allows the structure to deform and a metal electrode which forms the other half of the pressure dependent capacitors. The electrical equivalent circuit of the tested sensor is similar to earlier studies [4]. The tested sensor and the parts needed for the assembly are shown in Fig. 1.

Fig. 1. The used components and the complete pressure sensor

The resonance frequency of the sensor is measured wirelessly through the inductive link with the reader device, which operates like an impedance analyzer [4]. The device sweeps over the specific frequency range and measures the phase response of the resonance sensor. The measurement distance is approximately two centimetres.

The first step of the post processing is to extract the features from the measured phase response data which contains phase values at discrete frequencies. These features are the relative frequency and the relative height of the phase dip or peak depending on the used reader coils and sensors. These features are used to calculate the compensated estimate for the resonance frequency. In the pressure garment application this frequency is compared with the frequency measured without the pressure garment. For the flowchart of the sensing method and for more detailed description see [4].

The described sensing system has a nonlinear response to a pressure stimulus. This is caused by the nonlinearity of the resonance frequency of a LC-resonance circuit

$$f = 1/2\pi\sqrt{LC}. \tag{1}$$

where L is the inductance and C is the capacitance of the circuit. Furthermore, there is nonlinearity between the capacitance of a plate capacitor and the distance d between the plates.

$$C = \epsilon_0 \, \epsilon_r \frac{A}{d}. \tag{2}$$

where ϵ_0 is permittivity of vacuum, ϵ_r is the relative permittivity of the material and A is the area of the plates. In the tested sensors, the effective capacitance of the resonance circuit is formed as a combination of the constant and variable capacitors. In addition, the relation of the distance between the capacitor plates and the applied pressure is nonlinear, especially, if the deformation of the structure is large. The mutual effect of these nonlinearities depends on the structure of the sensor and the used components and materials. The response of the sensor is also affected by imperfections in parts, materials and assembly.

The nonlinearities and the unknown imperfections of the measurement system make it practically impossible to form an exact physical model for calculation of the pressure from the measured frequency. Therefore we test nonlinear regression to model the response of the sensor based on the calibration measurements. The most important range of the measurand is from 20 to 60 mmHg, because the pressure of the proper pressure garments should be within this range. However, at the moment the mechanical properties of the used sensor allow a reliable measurement range from 0 to 45 mmHg only.

In this study we have three sensors, A1, A2 and A3. We have data from two separate measurement cycles from each sensor. Each cycle consists of 11 data points. In addition, we have a series of more dense measurement cycle from sensor A1, which consist of 21 data points. These data are divided into training and test sets in the following case studies.

3 Nonlinear Regression Methods

Regression models are used to relate two variables; the dependent variable y and the independent variable x which is typically multivariate. The goal is to find a function f to predict y from values of x: $y = f(x)$. In this application the relationship between the measurand and the sensor response is nonlinear and therefore a nonlinear function f is required. In this paper we compare three nonlinear regression methods that are presented in the following subsections.

3.1 Sugeno Fuzzy Inference System

Sugeno, or Takagi-Sugeno-Kang, is a fuzzy inference system that can be used for nonlinear regression [8]. The membership functions of the output are either constant or linear functions. The final output is a weighted average of the output functions, weighted by the membership functions of the inputs. The membership functions can be identified from the data by ANFIS (adaptive-network-based fuzzy inference system), a hybrid learning algorithm, which combines the least-squares method and the backpropagation gradient descent method [9].

In this study we use an ANFIS model with linear output functions. Examples of the system are given in section 4.

3.2 Support Vector Regression

The basic Support Vector Machine (SVM) is a classifier that uses hypothesis space of linear functions in a high-dimensional kernel-induced feature space [10]. Support Vector Regression, such as, ν-SVR [11] can be used for modelling nonlinear dependences of continuous data. Parameter ν controls the number of support vectors in the model. In this study we use Radial Basis Function (RBF) kernel: $K(\mathbf{x}_i,\mathbf{x}_j) = \exp(-\gamma\|\mathbf{x}_i - \mathbf{x}_j\|^2)$, $\gamma > 0$. It is versatile and works well in most applications [12].

In order to avoid over fitting, the optimal values for the parameters ν and γ are selected by cross validation. We use 5-fold cross validation. Parameter values that provide minimum error in the test are selected. In this study we use a software package LIBSVM [13].

3.3 Multilayer Perceptron Network

Multilayer perceptron (MLP) networks [14] are very commonly used neural networks. MLP with one hidden layer containing sufficient number of neurons acts as a universal approximator. Therefore they are excellent tools for nonlinear regression. We train the MLP network by using Levenberg-Marquardt method [15].

4 Case Study 1: Individual Sensor

The data used in this case study are acquired by measuring the compensated resonance frequencies of the sensor A1 while the pressure on sensors is altered in a test setup. The stimulus is created by funnelling pressurized air through a valve to a container. The pressure in the container is measured with a pressure calibrator (Beamex PC105). This pressure is mediated on the sensor with a rubber film which is attached to the opening of the container. The rubber film simulates the behaviour of the skin in the actual measurement event. The other side of the sensor is fixed to a rigid plate. The resonance frequency values of the data are acquired wirelessly by using the hand-held measurement unit. Each frequency value is an average of more than 100 measured values.

The training data consists of two short measurement cycles from sensor A1. The test data consist of the longer more dense measurement cycle from sensor A1.

4.1 Nonlinear Models

We train the ANFIS model with 3 membership functions in the input and output. The membership functions after training are presented in Fig. 2. The lower part depicts the 3 linear output functions. The slope of the functions steepens from function 1 to 3. The upper part presents the membership functions of the input of Gaussian bell shape. The final output is the average of the linear output functions weighted with the input functions. As the scaled frequency increases towards zero, the weight of the steepest output function also increases producing a steepening curve as seen in Fig. 4.

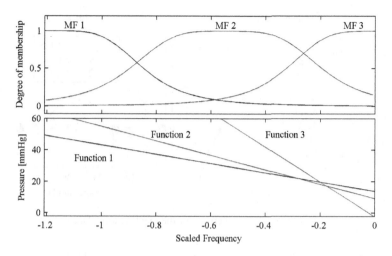

Fig. 2. Membership functions of the anfis model

We use of 5-fold cross validation to find the optimal values for v and γ parameters in v–SVR model. The cross validation error as a function of v and logarithm of γ is presented in Fig 3. The minimum is found at values $v = 0.8$ and $\gamma = 2$. These values are used to train the model using the whole training data set.

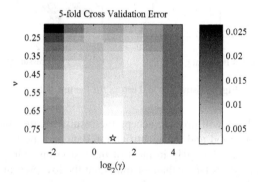

Fig. 3. Cross validation error

We use a small MLP network with one hidden layer with two neurons and sigmoid activation functions. One output neuron has a linear activation function.

4.2 Results on Training and Test Data

The trained models are used to estimate the pressure. The results are extrapolated slightly outside the range of the training data set. The results are presented in the left side of Fig. 4. The errors at the measured values in the training data are depicted on the right side. Corresponding results of the test data are presented in Fig. 5.

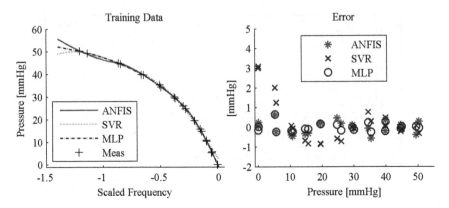

Fig. 4. Estimates and errors for the identification data

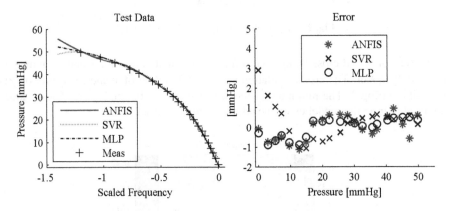

Fig. 5. Estimates and errors for the test data

All the tested methods will provide acceptable results because the repeatability of an individual sensor is good and the identification data and the test data are similar. The error is smaller than ±1 mmHg in most of the test points. Only the model made with SVR method has errors larger than ±1 mmHg at the low pressures. However, this is not significant in this the application. The error of the denser test data has smooth distribution. According to this result, the models can estimate the behaviour of the response between the identification data points. The models start to separate outside the tested range. The use of any of these models outside the tested range is not advisable. The model made with SVR is not monotonic if the shift of the resonance frequency is high. The errors of ANFIS and MLP tend to group and they differ slightly from the errors of SVR.

5 Case Study 2: Group of Three Sensors

The data of this case study were measured in a similar manner as in the case study 1. The identification data in this case consist of two short measurement cycles from sensors A1 and A2 The test data are the long measurement cycle from sensor A1 and both short cycles from sensor A3. The data of sensors A2 and A3 contain only short, less dense measurement cycles. According to the results made on case study 1, the repeatability of the sensors is adequate. The short measurement cycle is sufficient to cover the response of the sensor within the tested pressure range. Such a dense coverage of the range as used in the long cycle of A1 is not required for modelling the response.

The measured training data is presented on the left side of Fig. 6. The test data is depicted on the right side of Fig. 6. The estimates of the regression models are included in both.

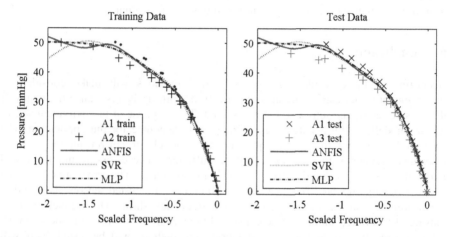

Fig. 6. Training and test data of the case study 2 with the estimates by all 3 regression models

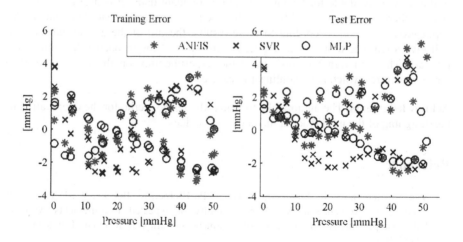

Fig. 7. Errors of the training and test cases

Sensors A2 and A3 seem to share similar behaviour. Sensor A1 differs from them especially at higher pressure levels. Because the training data contains the same number of measurements from sensors A1 and A2, the estimates of the models are in between them. The models made with ANFIS and SVR are not monotonic if the shift of frequency is high, which limits the usability of these models.

The errors of case study 2 are presented in Fig. 7. The errors are smaller than ±2 mmHg at most of the data points within the pressure range from 5 to 40 mmHg. However, especially near the edges of the tested pressure range, the error occasionally rises to over 4 mmHg. The errors seem to group according to the tested sensor rather that tested method. The errors are roughly two to three times larger than in case study 1. There is no significant difference between the errors in training and test data. The overall performance of tested regression methods is very similar.

6 Conclusion

According to these results, any of the tested methods will make considerable improvement to our earlier attempts to convert the shift of the resonance frequency of the passive resonance sensor to the measurand. The response of the individual sensor can be modelled with accuracy that is more than adequate for the application. The case study 2 shows that having just a single model for converting the shift of resonance frequency to pressure might be sufficient. The errors within the tested range of pressures are smaller than ±2 mmHg in most cases. The errors are considered to occur mainly due the differences in the responses of the tested sensors. The models are useful only at the limited range of the frequency shifts. The frequency shifts outside of this range should not be converted to pressure by using the models. In the future, more efficient and automated fabrication methods will be tested. This will diminish the the differences between sensor responses and it will increase the number of the sensors available for testing. The more abundant data set is needed in order to study if there is a significant difference between the tested regression methods in this application. However, according to these results, the use of the any tested methods will give a realistic tool for the converting the shift of the resonance frequency to a pressure which in turn will encourage the experimenting on the more innovative sensor structures which have nonlinear response.

Acknowledgments. The authors like to thank Mr. Jarmo Verho for the cooperation in the designing of the used readout methods and devices.

References

1. Van den Kerckhove, E., Fieuws, S., Massagé, P., Hierner, R., Boeckx, W., Deleuze, J.P., Laperre, J., Anthonissen, M.: Reproducibility of repeated measurements with the Kikuhime pressure sensor under pressure garments in burn scar treatment. Burns 33, 572–578 (2007)
2. Lai, C.H.Y., Li-Tsang, C.W.P.: Validation of the Pliance X System in measuring interface pressure generated by pressure garment. Burns 35, 845–851 (2009)

3. Ferguson-Pell, M., Hagisawa, S., Bain, D.: Evaluation of a sensor for low interface pressure applications. Medical Engineering & Physics 22, 657–663 (2000)
4. Salpavaara, T., Verho, J., Kumpulainen, P., Lekkala, J.: Readout methods for an inductively coupled resonance sensor used in pressure garment application. Sens. Actuators, A. (2011) (in press, corrected proof, March 5, 2011)
5. Chen, P.-J., Saati, S., Varma, R., Humayun, M.S., Tai, Y.-C.: Wireless Intraocular Pressure Sensing Using Microfabricated Minimally Invasive Flexible-Coiled LC Sensor Implant. J. Microelectromech. Syst. 19, 721–734 (2010)
6. Riistama, J., Aittokallio, E., Verho, J., Lekkala, J.: Totally passive wireless biopotential measurement sensor by utilizing inductively coupled resonance circuits. Sens. Actuators, A. 157, 313–321 (2010)
7. Patra, J.C., Kot, A.C., Panda, G.: An intelligent pressure sensor using neural networks. IEEE Trans. Instrum. And Meas. 49, 829–834 (2000)
8. Sugeno, M.: Industrial Applications of Fuzzy Control. Elsevier, New York (1985)
9. Jang, J.-S.R.: ANFIS: Adaptive-Network-based Fuzzy Inference Systems. IEEE Transactions on Systems, Man, and Cybernetics 23(3), 665–685 (1993)
10. Cristianini, N., Shawe-Taylor, J.: An Introduction to Support Vector Machines and Other Kernel-based Learning Methods. Cambridge University Press, Cambridge (2000)
11. Schölkopf, B., Smola, A., Williamson, R.C., Bartlett, P.L.: New support vector algorithms. Neural Computation 12, 1207–1245 (2000)
12. Hsu, C.-W., Chang, C.-C., Lin, C.-J.: A Practical Guide to Support Vector Classification, http://www.csie.ntu.edu.tw/~cjlin/papers/guide/guide.pdf
13. Chang, C.-C., Lin, C.-J.: LIBSVM: a library for support vector machines (2001), http://www.csie.ntu.edu.tw/~cjlin/papers/libsvm.pdf
14. Haykin, S.: Neural Networks. McMillan, New York (1994)
15. NNSYSID Toolbox - for use with MATLAB, http://www.iau.dtu.dk/research/control/nnsysid.html

An Adaptable Framework for Integrating and Querying Sensor Data

Shahina Ferdous[1], Sarantos Kapidakis[2],
Leonidas Fegaras[1], and Fillia Makedon[1]

[1] Heracleia Human Centered Computing Lab,
University of Texas at Arlington
shahina.ferdous@mavs.uta.edu,
fegaras@cse.uta.edu, makedon@uta.edu
[2] Laboratory on Digital Libraries and Electronic Publishing,
Ionian University
sarantos@ionio.gr

Abstract. Sensor data generated by pervasive applications are very diverse and are rarely described in standard or established formats. Consequently, one of the greatest challenges in pervasive systems is to integrate heterogeneous repositories of sensor data into a single view. The traditional approach to data integration, where a global schema is designed to incorporate the local schemas, may not be suitable to sensor data due to their highly transient schemas and formats. Furthermore, researchers and professionals in healthcare need to combine relevant data from various data streams and other data sources, and to be able to perform searches over all of these collectively using a single interface or query. Often, users express their search in terms of a small set of predefined fields from a single schema that is the most familiar to them, but they want their search results to include data from other compatible schemas as well. We have designed and implemented a framework for a sensor data repository that gives access to heterogeneous sensor metadata schemas in a uniform way. In our framework, the user specifies a query in an arbitrary schema and specifies the mappings from this schema to all the collections he wants to access. To ease the task of mapping specification, our system remembers metadata mappings previously used and uses them to propose other relevant mapping choices for the unmapped metadata elements. That way, users may build their own metadata mappings based on earlier mappings, each time specifying (or improving) only those components that are different. We have created a repository using data collected from various pervasive applications in a healthcare environment, such as activity monitoring, fall detection, sleep-pattern identification, and medication reminder systems, which are currently undergoing at the Heracleia Lab. We have also developed a flexible query interface to retrieve relevant records from the repository that allows users to specify their own choices of mappings and to express conditions to effectively access fine-grained data.

Keywords: Metadata, Schema Mappings, Query Interface, Digital Libraries, Pervasive Applications, Healthcare.

L. Iliadis et al. (Eds.): EANN/AIAI 2011, Part II, IFIP AICT 364, pp. 430–438, 2011.

1 Introduction

Data generated from pervasive applications can be of many different types, such as sensor readings, text, audio, video, medical records, etc. One example of a common pervasive application is the "remote activity monitoring" in an assistive environment. Continuous monitoring of activities is very important in healthcare since it helps caregivers and doctors to monitor a patient remotely and take actions in case of emergency. Such a system may also be very useful for monitoring an elderly person who lives alone in an apartment and needs occasional assistance. A pervasive application deploys various non-invasive sensors as an integral part of a persons' assistive daily living to automatically monitor his activities. It may also deploy a less-intrusive audio and video recording system, based on the needs and privacy requirements of the patient. To be effective, an assistive environment needs to store the data generated from pervasive applications into a common repository and to provide the healthcare providers a flexible and easier access to this repository.

Current advances in sensor technology allow many different sensors that use different technology to be used interchangeably to generate and record similar information about an environment. For example, a person may wear a wireless wrist watch [1] as a heart rate monitor, which may also include a 3-axis accelerometer, a pressure and a temperature monitor. On the other hand, a Sunspot sensor [2] can be used as an accelerometer and a temperature sensor as well. Although, both devices can be used to generate the same acceleration and temperature data, the format of their representation can be very different. In fact, a sensor can be programmed in many different ways to deliver data in different formats. For example, a sunspot can be programmed to transmit either 3-axis accelerometer data or the angle of acceleration. Therefore, even for a simple sensor device, such as a sunspot, the data storage may contain data in various schemas and configurations. As a result, it is very hard for a caregiver or a doctor to understand and remember each such different schema to query the data repository.

A pervasive application may sometimes refine a sensor reading specific to its settings. For example, sensor devices, such as a smoke/heat detector, only detect and transmit results when the environment reaches a predefined threshold. On the other hand, a stand-alone temperature sensor, such as a sunspot, can be configured to indirectly generate a heat alarm when the temperature exceeds some prespecified heat threshold. Moreover, such thresholds and conditions can change dynamically depending on a user's query requirements, thus making it highly infeasible for a system to cache all possible answers related to a user's query beforehand.

This paper describes a Digital Library that consists of a repository of sensor data collected from various pervasive applications and a flexible query interface for the user that requires minimum knowledge about the real metadata schemas from a user's point of view.

1.1 Framework

In this paper, we describe our framework for the sensor data repository, which contains datasets, such as C_{11}, C_{12}, ..., C_{nm}, derived from many different sensors, collected over a long period of time, possibly after years of experiments. We call each such dataset C_{ij}, collected from various applications and contexts, a *collection*. Each collection is also associated with a metadata schema, which describes the format of that collection. For example, in Fig. 1, the collection C_{11} is described by the schema S_1. A user can query over these different collections by simply using one of his preferred schemas, which we call a *virtual schema*, S_v in our framework. In our framework, given a user-query Q_v over the collections of different schemas, the user first provides the needed parts of "Map(S_i -> S_v)", the mapping for each individual schema S_i of a selected collection to the specified virtual schema, S_v. Given that the user specifies all such required mappings, the system next applies these mappings to return the final query answers, denoted by $Q_1, Q_2...Q_n$. Thus, based on our framework, a user may query on any kind and of any number of collections and obtain fine-grained results without knowing details about the real schemas. In the future, we are planning to use this framework as the building block to obtain the background knowledge for automatic metadata mappings.

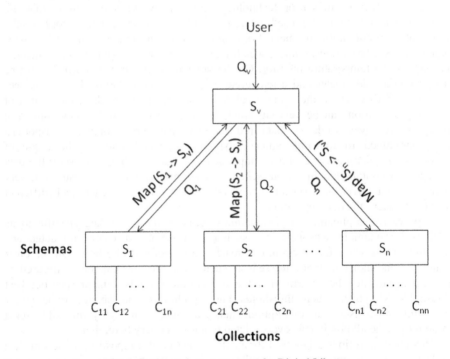

Fig. 1. Framework to query over the Digital Library

1.2 Challenges

Sensor records may not of the same format, since they could be generated from different types of sensors installed in various setups, or different versions of similar sensors, with small variations on the record formats. Besides, the values recorded from similar sensors can have different semantics too. The following example shows that even for a simple scenario, such as monitoring "whether a door of an assistive apartment is open or not", the system may require many different mappings to answer a user query.

A pervasive application can use variety of door sensors, which eventually produce door-"open/close" datasets in different formats. For example, the system may use a sensor based on door-mounted magnetic contacts, which denotes that the door is open (1) or closed (0) as an OPEN_STATUS (1/0) attribute, combined with the time of this event as an attribute named TIME_DETECTED. Any accelerometer sensor, such as a sunspot, can also be mounted on the door to obtain similar information. Since, a sunspot can be preprogrammed in different configurations, one configuration may transmit and store Cartesian (x, y, z) coordinates of the door, while the other may compute and transmit the angle of the current position of the door. Such a programmable sunspot may provide both TIME_RECEIVED and TIME_BROADCAST attributes as well.

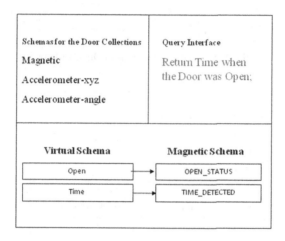

Fig. 2. Metadata Mapping between similar attributes with different names

A user may not know anything about the schema for the sensor being used to collect such door-data, but he still may ask a query such as "Give me the time when the door was open" over the collection of door datasets. Although answering such a simple query seems trivial, the mappings can be very different and complex. Fig. 2 describes the simplest mapping scenarios, where differently named attributes convey similar information, such as Open and OPEN_STATUS or Time and TIME_RECEIVED from the "Virtual" and the "Magnetic" schemas respectively.

However, a metadata field from one schema can be mapped to multiple metadata fields from the other by applying appropriate mapping function. Hence, if a user wants to select such an one-to-many mapping, he needs to specify the correct mapping function as well as to obtain meaningful results. Fig. 3 shows one example scenario, where the field "Open" in the virtual schema is mapped as a function of X_COORD, Y_COORD and Z_COORD fields of the Accelerometer-xyz schema.

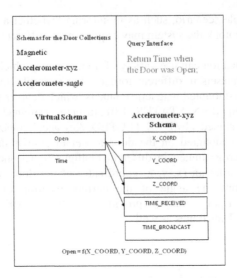

Fig. 3. Metadata Mapping from One-to-Many

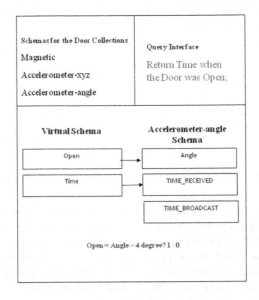

Fig. 4. Indirect Metadata Mapping

Fig. 4 shows another scenario, where a user may need to specify a condition even for a one-to-one metadata mapping. As shown in the example, a user may map the field "Open" to the field "Angle". But, since the attribute "Angle" does not directly specify the open status of the door, the user also needs to specify a condition, such as "the door is open, if the angle is greater than at least 4 degrees" and so on.

However, even if the metadata field describing an attribute is the same for two schemas, the associated units of their values can be completely different. For example, the temperature can be described in either Celsius or Fahrenheit, distances can be written either in feet or in meters, time can be a represented in milliseconds or as a general date/time expression. Therefore, in addition to field mappings, the system needs to be able to apply value mappings as well.

1.3 Contribution

In this paper, we describe a flexible query interface for searching relevant records from a Digital Library of sensor data. Our interface requires minimum background knowledge and returns results in a format chosen by the user himself. The Library also stores the history of mappings as part of a user profile. Thus, a user can re-use existing mappings to query over the same collections multiple times, thus providing minimal information. However, our system is flexible enough to allow the user to re-write some existing mappings or to add new conditions to the previously specified mappings. Our system may also suggest existing mappings from other user's profiles, which helps a new user to get some idea about the mapping between schemas. The flexibility of reusing and revising metadata mappings make the framework adaptive to different data management needs and different sensor metadata and formats.

2 Related Work

There is an increasing need for digital libraries to manage and share the vast variety of scientific data generated from different applications [3]. Even when sensor data are stored in clouds or used in grids, they need a mapping handling mechanism, like ours, to interoperate. DSpace@MIT [4] is an example repository that supports metadata creation and is built to store, share and search digital materials such as conference papers, images, scholarly articles and technical reports etc. Stanford Digital Library Project (SDLP) [5] proposes a metadata infrastructure to achieve interoperability among heterogeneous digital data, thus providing a uniform access to such datasets or collections. SDSC (San Diego Supercomputer Center) Storage Resource Broker (SRB) [6] along with a metadata catalog component, MCAT provides a uniform interface to search over heterogeneous data sources distributed over the network. Although SRB is successful enough to store and query over a federated information system, it is not sufficient to manage and share real time data collected from wireless sensor networks [7]. In general, very little work can been found in literature on building digital libraries to store and search the data collected from pervasive applications.

3 Description of the System

3.1 Design

Based on our design, a user may have one of two different roles: He can either contribute or retrieve data to/from the repository. Whenever a user adds a new collection to the repository, he must also specify the metadata schema to describe that collection. Our system stores all such metadata schemas in the repository and stores the link between each collection and its corresponding metadata schema.

Our system provides a suitable query interface for the users who want to search over the repository for relevant data. The query interface consists of a suitable window to browse for different collections and metadata schemas. The user can either select a metadata schema from the list of collections or may use a "virtual" schema, which is not used in any collection. The user may either query over the entire library by selecting all collections or select some specific collections. As soon as the user asks to execute a query, the system first checks the stored metadata mappings to see whether some mappings already exist in the system from the schemas of the selected collections to the preferred virtual schema. Our interface collects and displays all such mappings and asks the user to select any of the following mapping choices:

1. Re-use an existing mapping.
2. Select the mapping used last time, which is the default choice for the preexisting mappings.
3. Specify a new mapping. This is the default option for the mappings that have not been specified yet. Since, initially, the system will not have any stored mappings; our interface will ask the user to enter the mappings first. A user may only provide mappings for the fields that he wants to query at that moment. However, he may enter a new mapping based on similar attribute names or RDF descriptions. He may also define a function to map one schema attribute to the others.
4. Select the recommended mapping from the system. Since our system stores mapping preferences into a user profile, it can identify the most commonly used mappings by different users for a pair of schemas and can recommend such mappings to the user.

Next, as soon as the user specifies his preferred mappings, the system retrieves the resulting records or data from the collections and returns those to the user. The user may either view all such records in a separate output window or browse individual collection manually and only view the results for that particular collection.

3.2 Implementation

We have implemented a prototype interface in Java to store and view mappings and collection data from the Library. A screenshot of our interface is shown in Fig. 5. From the interface, the user may express a query using the attributes listed under a particular metadata schema. Executing such a query is straightforward, as it does not require any mapping. However, the user does not need to use any such schema to search over the collections. Instead, the user can select any type and number of collections and can query using a virtual schema. In this case, he may be asked to provide the right pair-wise mapping for each selected dataset of new metadata schema to his preferred virtual schema, before the query could be executed. The user may also specify a mapping condition using the interface and as soon as he saves it, the mappings become part of the user profile. However, whenever he is done with all the mappings, the system executes the query and returns the results to the user.

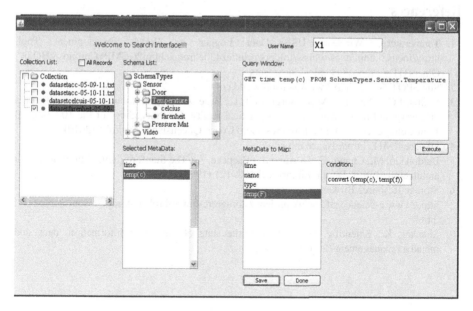

Fig. 5. A Screenshot of the Query Interface

3.3 Management of Mappings

In our framework, a mapping, M_{ij}, from schema S_i to the schema S_j is associated with a set of bindings $A_{jk}:f_{jk}(A_{i1},...A_{in})$ that derive the attribute value of A_{jk} in S_j from the attributes $A_{i1},...A_{in}$ in S_i. The expression f_{jk} may consist of simple arithmetic operations, such as the comparison of an attribute value with a constant threshold, and string manipulation operations, such as string concatenation. These expressions are represented as abstract syntax trees and are stored in a mapping repository along with the rest of the mapping information. When a query is expressed in the schema S_i and the mapping M_{ij} is selected to query the data collections that conform to the schema S_j, then the query attributes are mapped to the S_j attributes using the expression f_{jk} and the derived query is used for querying the data collections that match S_j.

4 Conclusions and Future Work

In this paper, we develop a framework for the repository of heterogeneous sensor data. We also design and implement a prototype for a flexible query interface, which allows the user to search over the collections of different metadata schemas using minimal background knowledge. Our system collects and stores possible mappings from various users incrementally, which could work as a building block to derive commonly accepted mappings. As a future work, we are planning to use our mapping repository as a knowledge base to facilitate automatic metadata mappings.

References

1. Temperature, Wireless, USB Data Logger or Watch Development Tool, http://focus.ti.com/docs/toolsw/folders/print/ez430chronos.html?DCMP=Chronos&HQS=Other+OT+chronos#technicaldocuments
2. Sun SPOT World, http://www.sunspotworld.com/
3. Wallis, J.C., Mayernik, M.S., Borgman, C.L., Pepe, A.: Digital libraries for scientific data discovery and reuse: from vision to practical reality. In: Proceedings of the 10th Annual Joint Conference on Digital Libraries, Gold Coast, Queensland, Australia (2010)
4. DSpace@MIT, http://dspace.mit.edu/
5. Baldonado, M., Chang, C., Gravano, L., Paepcke, A.: The Stanford digital library metadata architecture. Intl. J. Digital Libraries 1, 108–121 (1997)
6. Storage Resource Broker, http://www.e-science.stfc.ac.uk/projects/storage-resource-broker/storage-resource-broker-.html
7. Shankar, K.: Scientific data archiving: the state of the art in information, data, and metadata management (2003)

Feature Selection by Conformal Predictor

Meng Yang, Ilia Nouretdunov, Zhiyuan Luo, and Alex Gammerman

Computer Learning Research Centre, Royal Holloway, University of London,
Egham Hill, Egham, Surrey TW20 0EX, UK

Abstract. In this work we consider the problem of feature selection in the context of conformal prediction. Unlike many conventional machine learning methods, conformal prediction allows to supply individual predictions with valid measure of confidence. The main idea is to use confidence measures as an indicator of usefulness of different features: we check how many features are enough to reach desirable average level of confidence. The method has been applied to abdominal pain data set. The results are discussed.

Keywords: feature selection, conformal predictor, confidence estimation.

1 Introduction

When we deal with classification or regression problems, the size of the training data and noise in the data may affect the speed and accuracy of the learning system. Are all the features really important or can we use less features to achieve the same or better results? The irrelevant features will induce greater computational cost and may lead to overfitting. For example, in the domain of medical diagnosis, our purpose is to infer the relationship between the symptoms and their corresponding diagnosis. If by mistake we include the patient ID number as one input feature, an over-tuned machine learning process may come to the conclusion that the illness is determined by the ID number.

Feature selection is the process of selecting a subset of features from a given space of features with the intention of meeting one or more of the following goals.

1. Choose the feature subset that maximises the performance of the learning algorithm.
2. Minimise the size of the feature subset without reducing the performance of a learning problem significantly.
3. Reduce the requirement for storage and computational time to classify data.

The feature selection problem has been studied by the statistics and machine learning communities for many years. Many methods have been developed for feature selection and these methods can basically be classified into three groups: filter, wrappers and embedded feature selection [1,2]. The filter method employs a feature ranking function to choose the best features. The ranking function gives a relevance score based on a sequence of examples. Intuitively, the more relevant the feature, the higher its ranking. Either a fixed number of (at most) t features

L. Iliadis et al. (Eds.): EANN/AIAI 2011, Part II, IFIP AICT 364, pp. 439–448, 2011.

with the highest ranking are selected, or a variable number of features above a preset threshold are selected. Filter methods have been successful in a number of problem domains and are very efficient. Wrapper methods are general-purpose algorithms that searches the space of feature subsets, testing performance of each subset using a learning algorithm. The feature subset that gives the best performance is selected for final use. Some learning algorithms include an embedded feature selection method. Selecting features is then an implicit part of the learning process. This is the case, for example, with decision tree learners like ID3 [3] that use an information measure to choose the best features to make a decision about the class label. Other learning algorithms have been developed with embedded feature selection in mind. For example, Littlestone's WINNOW algorithm is an adaptation of the Perceptron algorithm [4] that uses multiplicative weight updates instead of additive.

The methods we mentioned above often use accuracy as criterion to select features; in our paper, we consider confidence for the feature selection. This is because in our approach we can regulate the accuracy by choosing a certain confidence level. We will use conformal predictors as a tool to perform feature selection. Conformal predictors are recently developed machine learning algorithms which supply individual predictions with valid measure of confidence [5]. Level of confidence in predictions produced by such algorithm can used as a performance measure instead of just accuracy [1].

Conformal predictors could be proceed in on-line and batch mode. In batch mode, a fixed size of training set will be used, we may get good results by chance. In on-line mode, the size of training set grows after prediction, it could consider all different sizes of the training set. So we will extend on-line conformal predictors in order to get the average confidence, make feature selection and present results of application of this approach on a medical database.

2   Conformal Predictor

Conformal predictor is a method that not just makes predictions but also provides corresponding confidences [5]. When we use this method, we predict labels for new objects and use the degree of conformity to estimate the confidence in the predictions.

We start by defining the concept of a nonconformity measure which is a way of measuring how well an example fits to a set. A measure of fitness is introduced by a nonconformity measure A. For a sequence $z_1, z_2, ...z_n$ of examples, where the i_{th} example z_i is composed of objects and labels, $z_i = (x_i, y_i)$, x_i means objects and y_i means label, $x_i \subset X$ and $y_i \subset Y$. We can write $\wr z_1, z_2, ..., z_n \wr$ for the bag consisting of the examples $z_1, z_2, ..., z_n$, we can score a distance between z_i and the bag $\wr z_1, z_2, ..., z_n \wr / z_i$, expressed by $\alpha_i = A(\wr z_1, z_2, ..., z_n \wr / z_i, z_i)$, called the nonconformity score.

Non-conformity score can be based on a classical algorithm of prediction, in this paper we use the nearest neighbors algorithm [6]. The idea of using the nearest neighbors algorithm to measure the nonconformity of example z, (x, y), from the other examples is comparing x's distance to other examples' objects with the same label to its distance to others with different label.

$$\alpha = \frac{distance\ to\ z's\ nearest\ neighbor\ to\ other\ examples\ with\ the\ same\ label}{distance\ to\ z's\ nearest\ neighbor\ to\ other\ examples\ with\ a\ different\ label}$$

2.1 Prediction and Confidence

Assume that an i.i.d (independent and identically distributed) data are given: $z_1, z_2, ...z_{n-1}$. Now, we have an new example x_n and want to predict its label y_n. First of all, we give y_n a value which belongs to Y, then calculate the non-conformity score for each example, finally, compare α_n to the other α by using p-value: $\frac{|\{j=1,...n:\alpha_j \geq \alpha_n\}|}{n}$. If the p-value is small, then z_n is nonconforming, if it is large, then z_n is very conforming. After we tried every value in Y, each of the possible labels will get a p-value $p(y)$, then the label with the largest p-value will be our prediction and its corresponding confidence will be equal to $1-$ the second largest p-value.

2.2 Validity

In conformal predictor, we also could set a level of significance ϵ to find the prediction region which contains all labels $y \subset Y$ with the corresponding p-value larger than ϵ, that means we will have confidence $1 - \epsilon$ in our prediction about y, and the probability of the event that true label is not contained by prediction region should be ϵ. And we define the situation when the true label is not contained in prediction region as error. Thus, the prediction in on-line mode is under the guarantee and valid.

$$Prob\{pvalue \leq \epsilon\} \leq \epsilon$$

According to this, size of prediction region could be another criterion, for a specific significant level, smaller region sizes provide us more efficient predictions. In this paper, we will use number of uncertain prediction to express this property, and uncertain prediction means the situation when prediction region contains multi-classes.

The confidence corresponds to the minimal level, at which we can guarantee that the prediction is certain (consists of the only label) under i.i.d. assumption. For example, if $Y = \{1, 2, 3\}$, and $p(1) = 0.23$, $p(2) = 0.07$, $p(3) = 0.01$ is the results for one of the examples in on-line mode, the prediction is: 1 with confidence 0.93. We can say that true label is 1 or 2 if we wish to be right with probability 0.95, on this level we are not sure that it is 1. On the other hand, if it enough for us to be right with probability 0.90, then we can claim that it is 1. See [5] for details of conformal (confident) prediction.

We can find that if the prediction is very conforming, its confidence will be very high, the prediction is made depends on the objects of examples, which means the more useful features we use, the higher confidence we will get. So, we could use confidence to justify how useful the features are.

3 Data Description

We use abdominal pain dataset, it has 9 kinds of diagnosis as labels and 33 types of symptoms as object [7,10], which are sex, age, pain-site onset, pain-site present, aggravating factors, relieving factors, progress of pain, duration of pain, type of pain, severity, nausea, vomiting, anorexia, indigestion, jaundice, bowel habit, micturition, previous pain, previous surgery, drugs, mood, calor, abdominal movements, abdominal scar, abdominal distension, site of tenderness, rebound, guarding, rigidity, abdominal masses, murphy's test, bowel sounds and rectal examination. Each of symptoms contains different numbers of values, we can give two options for each value, 1 and 0, which means the patient has this value of one symptom or not, separately. After this step, we get 135 features in total. There are around 6000 examples in the original dataset where some of them have missing values, so we use 1153 of them which do not have missing values.

List of diagnostic groups is below:

Diagnostic Groups		
Group	Diagnosis	Number of Examples
D=1	Appendicitis (APP)	126
D=2	Diverticulitis (DIV)	28
D=3	Perforates Peptic Ulcer (PPU)	9
D=4	Non-Specific Abdominal Pain (NAP)	585
D=5	Cholecystitis (CHO)	53
D=6	Intestinal Obstruction (INO)	68
D=7	Pancreatitis (PAN)	11
D=8	Renal Colic (RCO)	60
D=9	Dyspepsia (DYS)	173

4 Methodology

Our goal is to find the most useful features' set for separating two kinds of diagnosis by using conformal predictors.

Firstly, we try to separate the two classes just by one feature. We take all examples from the dataset that belong to one of these two classes and do not contain missing values. Then we process it in on-line mode: prediction for a next example is based using all preceding ones as the training set, and then get the corresponding confidence of the single prediction. To assess performance we calculate average confidence of the examples, and Due to on-line processing, it is averaged over different sizes of the training set.

This was done for each feature, so the feature with the largest average confidence will be the first important feature of useful features' set.

Then we solve in the same way the question: what the second feature can be added to this one in order to maximize average confidence? After deciding this we will have list of two features, then we look for the third feature and so on. On each step a feature is added to the list of ones being used for prediction, and average confidence grows until adding more features appears not to be useful anymore, the confidence does not change much. The speed of the method is depending on the size of examples and how many m suitable features we want from the full feature set.

5 Results and Discussion

We choose a typical result for illustration, and you can find more results at appendix. Table 1 shows the order we get when we separate APP (D1) from DYS (D9).

The order of features listed in tables corresponds to stage of including features into the selected set of features.

Table 1. Separate APP (D1)from DYS (D9)

Order	Value	Symptom	Average Confidence
1	4/2	Pain-site present: right lower quadrant	0.16
2	24/0 or 24/1	Abdominal scar: present or absent	0.32
3	5/3	Aggravating factors: food	0.44
4	6/5	Relieving factors: nil	0.48
5	14/0 or 14/1	Indigestion: history of indigestion or no history	0.64
6	27/0 or 27/1	Rebound: present or absent	0.73
7	9/0	Type of pain: steady	0.81
8	26/2	Site of tenderness: right lower quadrant	0.86
9	33/4	Rectal examination: normal	0.88
10	2/1	Age: 10-19	0.91

Figure 1 shows us the tendence of confidence average while separate APP(D1) from DYS(D9) when we extend features size till take every feature into account. As we can seen from figure, there are just 102 features, because when we calculate for useful features, some features show same values as equal useful features and we just choose one of them for further steps. The confidences in this picture grows fast at the beginning till meet the peak, 0.97503, and then keep steady for a while, near the end part, it get a slight fall to 0.958. This kind of fall also happens in other separation cases we mentioned above and may cause by the influence from features which are not relevant with this separation.

And, the programme finish its learning process when confidence reached 0.95 with the useful feature size is around 16 out of 135 features and then reach a plateau.

Figure 1 shows that the confidence level is greater than 0.95 when useful features'size is 20, so we will compare the results of using these 20 features with using whole features for separation D1 from D9 by on-line conformal prediction, and significance level ϵ is 0.05.

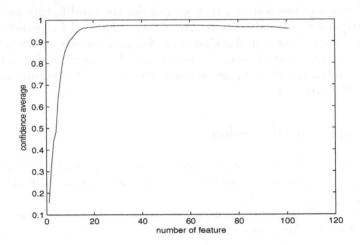

Fig. 1. Separate APP(D1) from DYS(D9)

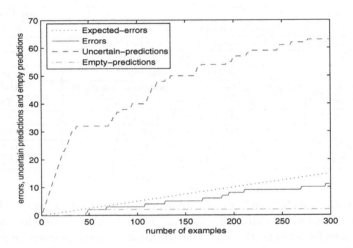

Fig. 2. Use the whole feature set to separate APP(D1) from DYS(D9)

Figure 2 presents the results of using all the features - we have 11 errors, 63 uncertain predictions and 2 empty predictions. Figure 3 shows the prediction results using only 20 selected features. It is clear that the number of prediction errors is reduced to 9 and we just get 42 uncertain predictions and 0 empty prediction. Thus, the selected 20 features give us better prediction results with less errors and more efficient size.

The following Table 2 shows us the comparison of results between full features and selected features in other binary classification subproblems by on-line conformal prediction. Compare with using all features, selected features could give us the same level of accuracy by small size. Because the predictions are under

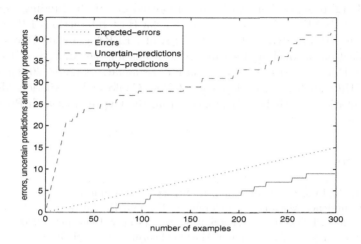

Fig. 3. Use 20 useful features to separate APP(D1) from DYS(D9)

Table 2. Results Comparison

subproblem	size	accuracy	full features uncertain predictions	size	selected features accuracy	size	the best results uncertain predictions
D1–D9	135	0.96	63	17	0.96	30	31
D2–D9	135	0.98	76	22	0.98	26	30
D3–D9	135	0.97	41	20	0.97	21	26
D5–D9	135	0.96	173	17	0.96	40	92
D6–D9	135	0.97	101	14	0.97	33	40
D8–D9	135	0.96	69	18	0.96	30	44

the guarantee, for a specific significant level, the best result is which has the most efficient prediction region. We will find how many features could provide us the best results which have the least uncertain predictions when significance level ϵ is 0.05.

6   Conclusion

As we can seen from above tables, at beginning, confidences are always low because few features are not enough for accurate predictions. As the number of features growing, the corresponding average confidence increases, and we could get desirable confidences by small number of features . One can then use a significance level in order to decide where to stop adding them. If a plateau is not reached, one can stop when confidence stops to grow, but this would probably mean that the underlying method of computing non-conformity measure was not very appropriate. So, conformal predictor could be an useful way of feature selection.

Acknowledgements. This work was supported in part by funding from BB-SRC for ERASySBio+ Programme: Salmonella Host Interactions PRoject European Consortium (SHIPREC) grant; VLA of DEFRA grant on Development and Application of Machine Learning Algorithms for the Analysis of Complex Veterinary Data Sets; EU FP7 grant O-PTM-Biomarkers (2008–2011); and by grant PLHRO/0506/22 (Development of New Conformal Prediction Methods with Applications in Medical Diagnosis) from the Cyprus Research Promotion Foundation.

References

1. Bellotti, T., Luo, Z., Gammerman, A.: Strangeness Minimisation Feature Selection with Confidence Machines. In: Corchado, E., Yin, H., Botti, V., Fyfe, C. (eds.) IDEAL 2006. LNCS, vol. 4224, pp. 978–985. Springer, Heidelberg (2006)
2. Guyon, I., Elisseeff, A.: Journal of Machine Learning Research 3, 1157–1182 (2003)
3. Quinlan, J.R.: Induction of decision trees. Machine Learning 1, 81–106 (1986)
4. Rosenblatt, F.: The perceptron: a probabilistic model for information storage and organization in the brain. Psychological Review 65, 386–408 (1959)
5. Vovk, V., Gammerman, A., Shafer, G.: Algorithmic Learning in a Random World. Springer, Heidelberg (2005)
6. Shafer, G., Vovk, V.: A Tutorial on Conformal Prediction. Journal of Machine Learning Research 9, 371–421 (2008)
7. Gammerman, A., Thatcher, A.R.: Bayesian Diagnostic Probabilisties without Assuming Independence of Symptoms. Methods Inf. Med. 30(1), 15–22 (1991)
8. Gammerman, A., Vovk, V.: Hedging Predictions in Machine Learning. The Computer Journal 50(2), 151–163 (2007)
9. Proedrou, K., Nouretdinov, I., Vovk, V., Gammerman, A.: Transductive Confidence Machines for Pattern Recognition. In: Elomaa, T., Mannila, H., Toivonen, H. (eds.) ECML 2002. K. Proedrou, I. Nouretdinov, V. Vovk, and A. Gammerman, vol. 2430, pp. 381–390. Springer, Heidelberg (2002)
10. Papadopoulos, H., Gammerman, A., Vovk, V.: Reliable Diagnosis of Acute Abdominal Pain with Conformal Prediction. Engineering Intelligent Systems 17(2-3), 127–137 (2009)

Appendix

Table 3 is for separating of classes DIV(D2) and DYS(D9), Table 4 is for CHO(D5)and DYS(D9) and Table 5 is for PPU(D3) and NAP(D4).

Table 3. separate DIV (D2)from DYS (D9)

order	value	symptom	average confidence
1	4/12	Pain-site present: epigastric	0.27
2	2/2	Age: 20-29	0.43
3	2/3	Age: 30-39	0.54
4	20/0 or 20/1	Drugs: being taken or not being taken	0.54
5	26/13	Site of tenderness: none	0.67
6	16/1	bowel habit: constipated	0.72
7	10/0 or 10/1	Severity of pain: moderate or severe	0.76
8	6/0	relieving factors: lying still	0.82
9	21/0	mood: normal	0.83
10	22/1	color: pale	0.86
11	3/5	Pain-site onset: lower half	0.88
12	3/3	Pain-site onset: left lower quadrant	0.90
13	11/0 or 11/1	Nausea: nausea present or no nausea	0.91
14	8/0	Duration of pain: under 12 hours	0.92
15	14/0 or 14/1	Indigestion: history of indigestion or no history	0.93
16	3/4	Pain-site onset: upper half	0.932
17	8/1	Duration of pain: 12-24 hours	0.938
18	4/5	Pain-site present: lower half	0.941

Table 4. separate CHO (D5)from DYS (D9)

order	value	symptom	average confidence
1	22/3	Color: jaundiced	0.06
2	31/0 or 31/1	Murphy's test: positive or negative	0.10
3	24/0 or 24/1	Abdominal scar: present or absent	0.18
4	6/0	Relieving factors: lying still	0.25
5	18/0 or 18/1	Previous pain: similar pain before or no pain before	0.32
6	1/0 or 1/1	sex: male or female	0.45
7	4/12	Pain-site present: epigastric	0.59
8	5/5	Aggravating factors: nil	0.64
9	26/0	Site of tenderness: right upper quadrant	0.69
10	10/0 or 10/1	Severity of pain: moderate or severe	0.76

Table 5. Separate PPU(D3) from NAP(D4)

order	value	symptom	average confidence
1	2/1	Age: 10-19	0.25
2	10/0 or 10/1	Severity of pain: moderate or severe	0.54
3	26/2	Site of tenderness	0.67
4	2/2	Age: 20-29	0.76
5	26/13	Site of tenderness: none	0.83
6	18/0 or 18/1	Previous pain: similar pain before or no pain before	0.87
7	22/0	Color: normal	0.89
8	3/8	Pain-site onset: central	0.92
9	21/2	Mood: anxious	0.93
10	8/2	Duration of pain: 24-48 hours	0.94
11	2/5	Age: 50-59	0.945
12	4/2	Pain-site present: right half or left half	0.946

Applying Conformal Prediction to the Bovine TB Diagnosing

Dmitry Adamskiy[2], Ilia Nouretdinov[2], Andy Mitchell[1],
Nick Coldham[1], and Alex Gammerman[2]

[1] Veterinary Laboratories Agency
{a.p.mitchell,n.g.coldham}@vla.defra.gsi.gov.uk
[2] Royal Holloway, University of London
{adamskiy,ilia,alex}@cs.rhul.ac.uk

Abstract. Conformal prediction is a recently developed flexible method which allows making valid predictions based on almost any underlying classification or regression algorithm. In this paper, conformal prediction technique is applied to the problem of diagnosing Bovine Tuberculosis. Specifically, we apply Nearest-Neighbours Conformal Predictor to the VETNET database in an attempt to allow the increase of the positive prediction rate of the existing Skin Test. Conformal prediction framework allows us to do so while controlling the risk of misclassifying true positives.

Keywords: conformal predition, bovine TB, online learning.

1 Introduction

Bovine Tuberculosis (bTB) is an infectious disease of cattle, caused by the bacterium *Mycobacterium bovis (M.bovis)*. The disease is widespread in certain areas of the UK (particularily South West England) and of major economic importance, costing the UK Government millions of pounds each year, since positive animals are slaughtered and compensation paid to the cattle owners. The main testing tool for diagnosing TB in cows is the Single Intradermal Cervical Tuberculin (SICCT) skin test. The procedure involves administering intradermally both Bovine and Avian Tuberculin PPDs and measuring the thickening of the skin.

Avian tuberculin is used to exclude unspecific reactions so the actual value of the skin test is a difference of thickenings: $(B_2 - B_1) - (A_2 - A_1)$, where A_1 and B_1 are the initial values of skin thickness measured in millimetres, A_2 and B_2 are skin thickness after the injection of avian and bovine tuberculin respectively.

If the cow is a reactor, in the sense that the test is positive,

$$(B_2 - B_1) - (A_2 - A_1) > T$$

where T is a threshold (usually 3mm), it is slaughtered and the post-mortem examination is performed, which may result in the detection of visible lesions

L. Iliadis et al. (Eds.): EANN/AIAI 2011, Part II, IFIP AICT 364, pp. 449–454, 2011.

typical of M.bovis. Furthermore, samples for some of the slaughtered cattle are sent for the bacteriological culture analysis.

The data on all the reactors is stored in the VETNET database. The data stored there includes (per reactor) the numeric test results A_1, A_2, B_1, B_2, and such features as age, herd identifier, date of the test, post-mortem and culture results (if any) and others([5]).

As there are two tests that can confirm the diagnosis after the cow is slaughtered, the definition of a truly positive cow could be different and here we use logical OR as such (the cow is positive if there are visible lesions or if the culture test was positive). The data in VETNET alone is not enough to judge about the efficiency of the skin test, as the post-mortem tests are not performed for the negative animals. However, it is believed that the positive prediction rate could be improved by taking into account some other factors apart from just the binary result of the skin test.

In this paper we state it as an online learning problem: given the history of the animals tested prior to the current reactor, we try to dislodge, at a given significance level, the hypothesis of this reactor being a real one. In what follows we introduce conformal predictors and describe how conformal predictors could be used for this task.

2 Conformal Prediction

Conformal prediction [1] is a way of making valid hedged predictions which does not require any assumption other than i.i.d.; the only assumption made is the i.i.d. assumption: the examples are generated from the same probability distribution independently of each other.

Also, it is possible to estimate confidence in the prediction of the given individual example. Detailed explanation of conformal prediction could be found in [1], here we outline the intuition behind it and the way it is applied to the problem stated.

Online setting implies that the sequence of examples $z_i = (x_i, y_i)$, $z_i \in Z = (X, Y)$ is revealed one by one and at each step after the object x_i is revealed the prediction is made. Original conformal prediction algorithm takes as a parameter the function called nonconformity measure and outputs the prediction set Γ. Thus the performance of the conformal predictor is measured in terms of validity and efficiency: how many errors the predictor makes ($y_i \notin \Gamma_i$) and how big is the set Γ_i. In case of the classification task it is also possible to output forced point prediction along with the measure of confidence in it.

A nonconformity measure formally is any measurable function taking bag of examples and a new example and returning a number specifying "nonconformity" (strangeness) of a given example to the set. The resulting conformal predictor will be valid no matter what function we choose, however in order to obtain an efficient predictor one should carefully select the reasonable one. Specifically there is a general scheme of defining nonconformity from any given point predictor.

Algorithm 1. Conformal Predictor for classification

Input: data examples $(x_1, y_1), (x_2, y_2), \ldots, (x_l, y_l) \in X \times Y$
Input: a new object $x_{l+1} \in X$
Input: a non-conformity measure $A : (z_i, \{z_1, \ldots, z_{l+1}\}) \to \alpha_i$ on pairs $z_i \in X \times Y$
Input(optional): a significance level γ
$z_1 = (x_1, y_1), \ldots, z_l = (x_l, y_l)$
for $y \in Y$ **do**
$\quad z_{l+1} = (x_{l+1}, y)$
$\quad$ **for** j in $1, 2, \ldots, l+1$ **do**
$\quad\quad \alpha_j = A(z_j, \{z_1, \ldots, z_l, z_{l+1}\})$
$\quad$ **end for**
$\quad p(y) = \frac{\#\{j=1,\ldots,l+1 : \alpha_j \geq \alpha_{l+1}\}}{l+1}$
end for
Output(optional): prediction set $R_{l+1}^\gamma = \{y : p(y) \geq 1 - \gamma\}$
Output: forced prediction $\hat{y}_{l+1} = \arg\max_y \{p(y)\}$
Output: confidence
$conf(\hat{y}_{l+1}) = 1 - \max_{y \neq \hat{y}_{l+1}} \{p(y)\}$

2.1 Mondrian Conformal Predictors

The algorithm 1 is valid in a sense that under the i.i.d. assuption it makes errors independently on each trial with probability less then $1 - \gamma$. However, sometimes we want to define the categories of the examples to have the category-wise validity. For instance, suppose that the examples fall into "easy to predict" and "hard to predict" categories, then the overall validity will be reached by conformal predictor, but the individual error rate for "hard to predict" objects could be worse.

In order to overcome this, Mondrian conformal predictor (first presented in [2]) is used. Here we are interested in label-wise validity, thus we predict it in a most simple form, see Algorithm 2 (where $|y = y_j|$ means quantity of j such that $y = y_j$).

3 Applying Conformal Prediction to the VETNET Database

We used the positively test cows from VEBUS subset of the original VETNET database. It includes 12873 false positives and 18673 true positives. In what follows the words "true positives" and "false positives" will refer to the skin test results.

After the preliminary study, it was discovered that the most relevant atrributes for classification are numeric value of skin test result, age and either the ID of the given test which is a herd identifier combined with a test date, or just indentifier of a herd (that may cover several tests performed at different time).

The extract from the VETNET database showing those features is shown in Table 1.

Algorithm 2. Mondrian Conformal Predictor for classification

Input: data examples $(x_1, y_1), (x_2, y_2), \ldots, (x_l, y_l) \in X \times Y$
Input: a new object $x_{l+1} \in X$
Input: a non-conformity measure $A : (z_i, \{z_1, \ldots, z_{l+1}\}) \to \alpha_i$ on pairs $z_i \in X \times Y$
Input(optional): a significance level γ
$z_1 = (x_1, y_1), \ldots, z_l = (x_l, y_l)$
for $y \in Y$ **do**
$\quad z_{l+1} = (x_{l+1}, y)$
$\quad$ **for** j in $1, 2, \ldots, l+1$ **do**
$\quad\quad \alpha_j = A(z_j, \{z_1, \ldots, z_l, z_{l+1}\})$
$\quad$ **end for**
$\quad p(y) = \frac{\#\{j=1,\ldots,l+1 : y_j = y, \alpha_j \geq \alpha_{l+1}\}}{|y = y_j|}$
end for
Output(optional): prediction set $R_{l+1}^\gamma = \{y : p(y) \geq 1 - \gamma\}$
Output: forced prediction $\hat{y}_{l+1} = \arg\max_y\{p(y)\}$
Output: confidence
$conf(\hat{y}_{l+1}) = 1 - \max_{y \neq \hat{y}_{l+1}}\{p(y)\}$

Table 1. Two entries in VETNET database

Test time (seconds)	CPHH (herd ID)	Lesions	Culture	Age	AvRes	BovRes
1188860400	35021001701	1	1	61	4	22
1218495600	37079003702	0	0	68	1	9

The first of these examples is a True Positive (Lesions or Culture test is positive) and the second is a False Positive (both Lesions and Culture tests are negative).

The task is to distinguish between these two classes.

Remind that the goal is to decrease number of cows being slaughtered. This means that we wish to discover as many cows as possible to be False Positives. On the other hand, the number of True Positives misclassified as False Positives should be strictly limited. So unlike standard conformal prediction, the role of two classes is different.

Thus we present a one-sided version of conformal predictor. Each new example is assigned only one p-value, that corresponds to True Positive hypothesis. Then for a selected significance level γ we mark a cow as a False Positive if $p < \gamma$. This allows us to set the level at which we tolerate the marking of true positive and we aim to mark as many false positives as possible.

The property of validity is interpreted in the following way: if a cow is True Positive, it is mismarked as a False Positive with probability at most γ. A trivial way to achieve this is to mark any cow with probability γ. So the result of conformal prediction can be considered as efficient only if the percentage of marked False Positives is essentially larger.

In our experiments we used the nonconformity measure presented in algorithm 3 based on k-Nearest-Neighbour algorithm.

Algorithm 3. kNN Nonconromity Measure for VETNET database

Input: a bag of data examples $z_1 = (x_1, y_1), z_2 = (x_2, y_2), \ldots, z_{l+1} = (x_{l+1}, y_{l+1}) \in$
$X \times Y$
Input: an example $z_i = (x_i, y_i)$ from this bag;
Input: a distance function $d(x_1, x_2) : X \times X \rightarrow \mathbb{R}^+$
$A(z_i, \{z_1, z_2, \ldots, z_{l+1}\}) = |\{j : x_j$ is amongst k nearest neighbours of x_i in
$x_1, \ldots, x_{i-1}, x_{i+1}, \ldots, x_{l+1}$ according to the distance d, and $y_j \neq y_{l+1}\}|$

A possible version of efficient predictor can be done setting $k = 50$ and using the following distance, which assigns the highest importance to comparison of herd IDs, second priority is given to the numerical value $(B_2 - B_1) - (A_2 - A_1)$ of skin test, and age is used as an additional source of information.

$$dist(x_1, x_2) = 100 S(x_1, x_2) + T(x_1, x_2) + |log_{10}(age(x_1)) - log_{10}(age(x_2))|$$

where $S(x_1, x_2) = 1$ if x_1 and x_2 belong to the same herd and 0 otherwise, $T(x_1, x_2)$ is the difference between numerical values of test results on x_1 and x_2. Thus, first all the animals within given test are considered as neighbours and then all the others. Experiments showed that this distance resulted in the efficient predictor though the validity property holds for other parameters as well. The results on the subset of VETNET database are summarized in the Table 2. The experiment was performed in an online mode with the data sorted by test date (as in real life).

Table 2. VETNET results

Significance level	Marked reactors within FP	Marked reactors within TP
1%	1267/12873	138/18673
5%	4880/12873	919/18673
10%	7971/12873	1904/18673

The i.i.d. assumption is clearly a simplification here, but as in some other conformal predictor applications(see [4]) we can see that it is not essentially broken and we can see that the validity property holds: the number of marked reactors within true positives is indeed the level that was set. The efficiency could be judged by the number of marked reactors within false positives: at a cost of misclassifying 10% of true positives it is possible to identify almost two thirds of test mistakes.

4 Conclusions and Future Work

We can see from the table above that the resulting predictions are valid and efficient. The disadvantage is not taking into account the delay (normally several weeks), needed to perform post mortem analysis. This actually might lead to overestimation of the importance of herd ID as a factor: when a skin test is performed on many cows from same farm same day, it is likely to be either correct or wrong on the most of them.

To perform the experiment more fairly, the online protocol can be replaced with "slow learning" one (described in [3]) where the labels are revealed not immediately, but with the delay. Preliminaty investigation show that herd ID in such case should be replaced with more specific attributes related to the illness history of a herd.

Acknowledgements. This work was supported in part by funding from the ERASySBio+ Salmonella Host Interactions PRoject European Consortium (SHIPREC) grant; VLA of DEFRA grant on Development and Application of Machine Learning Algorithms for the Analysis of Complex Veterinary Data Sets; EU FP7 grant O-PTM-Biomarkers (2008–2011); and by Cyprus Government grant: Development of New Venn Prediction Methods for Osteoporosis Risk Assessment.

References

1. Vovk, V., Gammerman, A., Shafer, G.: Algorithmic Learning in a Random World. Springer, Heidelberg (2005)
2. Vovk, V., Lindsay, D., Nouretdinov, I., Gammerman, A.: Modrian Confidence Machine. Working Paper #4 (2003)
3. Ryabko, D., Vovk, V., Gammerman, A.: Online Region Prediction with Real Teachers. Working Paper #7 (2003)
4. Nouretdinov, I., Burford, B., Gammerman, A.: Application of Inductive Confidence Machine to ICMLA Competition Data. In: Proc. ICMLA, pp. 435–438 (2009)
5. Mitchell, A., Johnson, R.: Private communications

Classifying Ductal Tree Structures Using Topological Descriptors of Branching

Angeliki Skoura[1], Vasileios Megalooikonomou[1],
Predrag R. Bakic[2], and Andrew D.A. Maidment[2]

[1] Department of Computer Engineering and Informatics,
University of Patras, 26504, Rio, Greece
[2] Department of Radiology, Hospital of the University of Pennsylvania,
3400 Spruce St, Philadelphia, Pennsylvania 19104, USA
{skoura,vasilis}@ceid.upatras.gr,
{predrag.bakic,andrew.maidment}@uphs.upenn.edu

Abstract. We propose a methodological framework for the classification of the tree-like structures of the ductal network of human breast regarding radiological findings related to breast cancer. Initially we perform the necessary preprocessing steps such as image segmentation in order to isolate the ductal tree structure from the background of x-ray galactograms. Afterwards, we employ tree characterization approaches to obtain a symbolic representation of the distribution of trees' branching points. Our methodology is based on Sholl analysis, a technique which uses concentric circles that radiate from the center of the region of interest. Finally, we apply the k-nearest neighbor classification scheme to characterize the tree-like ductal structures in galactograms in order to distinguish among different radiological findings. The experimental results are quite promising as the classification accuracy reaches up to 82% indicating that our methods may assist radiologists to identify image biomarkers in galactograms.

Keywords: medical image classification, galactography, tree-like structures.

1 Introduction

Several structures in human body follow a tree-like topology. Characteristic examples include the bronchial tree, the blood vessel network, the nervous system and the breast ductal network (Fig. 1). Nowadays, medical imaging modalities, such as Magnetic Resonance Imaging (MRI), Computed Tomography (CT) and radiographs have made available large series of images that visualize the above mentioned structures. Properties of the tree topology including spatial distribution of branching, tortuosity and asymmetry have been analyzed in literature so far and have been associated with altered function and pathology [1]. This type of information can be used to facilitate more accurate medical diagnoses.

For example regional changes in vessel tortuosity were utilized to identify early tumor development in human brain [2]. Moreover, the three dimensional analysis of the airway tree inside the lungs was used to distinguish pathologic formations from

L. Iliadis et al. (Eds.): EANN/AIAI 2011, Part II, IFIP AICT 364, pp. 455–463, 2011.

normal lung structures [3]. Similarly, careful examination of the morphology of the ductal network visualized in galactograms provided valuable information related to the risk of breast cancer [4].

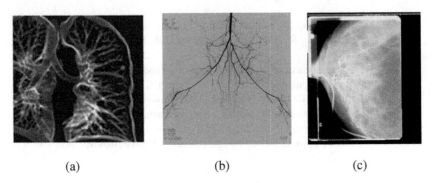

(a)   (b)   (c)

Fig. 1. Characteristic examples of tree shaped structures in medical images: (a) the human airway tree, (b) an angiography visualizing the hindlimb arterial network in a rabbit, (c) the network of human ducts depicted in a galactogram

The motivation of this study is to provide insight into the discriminative characteristics of ductal networks in normal galactograms and in galactograms with reported radiological findings. Here, we propose descriptors for trees' topological analysis: the Sholl Analysis, the Sectoring Analysis and the Combined Method. These descriptors focus on the spatial distribution of tree branching points and we employ this information to characterize normal or disease states of the breast ductal tree. In order to evaluate the new descriptors, we use the classification scheme of k-nearest neighbors. Our approach has the potential benefit of advancing our understanding in breast anatomy and physiology and could assist early cancer detection and cancer risk estimation. Moreover, the proposed methodology could also be applied to other tree like structures of the human body such us the blood vessel network in order to detect ischemic or thrombotic patterns, or the bronchial tree to provide a primary detection of lung cancer or other malformations.

2 Background

Several studies in the literature have demonstrated that examining the morphology of the ductal network can provide valuable insight into the development of breast cancer and assist in diagnosing pathological breast tissue. For example, Bakic et al. [5] proposed a quantitative method based on Ramification matrices (R-matrices) to classify galactograms regarding radiological findings. An R-matrix represents a description of branching structures at the topological level and the elements of such a matrix represent the probabilities of branching at various levels.

More recently, Megalooikonomou et al. [6] proposed a multi-step approach for representing and classifying trees in medical images including galactograms. The authors employed tree encoding techniques such as the depth-first string encoding and the Prüfer encoding to obtain a symbolic string representation of the tree's branching topology. Based on these encodings, the problem of comparing tree structures was reduced to the problem of string comparison. Moreover, they used the tf-idf text mining technique to assign a weight of significance to each string term. The classification was performed using the k-nearest neighbor scheme and the cosine similarity metric.

In the field of mammography, fractal analysis was also employed for characterizing the parenchymal pattern, distinguishing architectural distortion and detecting microcalcifications [7]. Fractal analysis can be used to describe properties that are not interpretable by the traditional Euclidean geometry and most of the above approaches utilize fractal analysis to characterize texture properties. In another approach, the tree asymmetry index was employed to characterize the topology of the ductal network in order to classify galactograms [8]. Finally, a combination of texture and branching descriptors was investigated to deepen the understanding of the relationships among the morphology, function and pathology of the ductal tree [9].

In this paper, we compare the proposed descriptors to the state-of-the-art Prüfer encoding approach and show that our methods compare favorably to it. A detailed description of the Prüfer encoding technique applied in galactograms can be found in [6].

3 Methodology

The goal of this paper is to investigate whether the distribution of branching points of ductal trees consists a discriminative factor concerning radiological findings in galactograms. Towards this direction, we present three methods that capture topological information regarding the tree branching points. The proposed methodology for analysis begins with the preprocessing of the images to segment the tree-like structures from the rest of the tissue depicted in the medical images. Then, we perform characterization of the extracted tree topologies applying the proposed techniques. Given a collection of such structures and the query tree, we perform similarity searches in order to find the ductal tree structures that are most similar to a query. Finally, we use the k-nearest neighbor classification scheme to evaluate the accuracy of the methods.

3.1 Image Preprocessing

Image preprocessing includes several steps needed to extract the tree structures from the background of the original medical images (Fig. 2(a)) and transform them into a suitable form for analysis. Initially, the Region Of Interest (ROI) of the galactograms, which is the part of the medical image where the ductal tree appears, is recognized (Fig. 2(b)). After focusing on the ROI, the segmentation of tree-like structures follows. Image segmentation is the boundary identification of objects or regions of special interest from the rest of the displayed objects or the background. In the application presented here, the task of segmentation has been done manually by medical experts (Fig. 2(c)). Then, the tree-like structures are reconstructed by identifying true positive branching points often resolving potential ambiguities such

as anastomoses occurring mostly as a result of two-dimensional acquisition artifacts
[10]. Although more sophisticated and fully-automated methods of reconstruction
could potentially be applied, such an approach is beyond the scope of this work as our
main objective is the feature analysis and the classification of the tree-like structures.
The final preprocessing step is the skeletonization of the ductal structures. This
thinning process that reduces most of the original foreground pixels in a binary image
to obtain a skeletal remnant that preserves the extent and the connectivity of the
original structure is necessary in order to detect the exact positions of the branching
points.

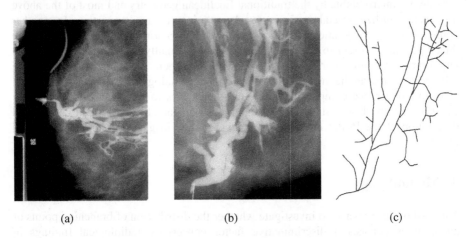

(a)   (b)   (c)

Fig. 2. Segmentation of a ductal tree: (a) an original galactogram, (b) the Region Of Interest of
the medical image showing enlarged the ductal network, (c) the segmented tree structure
manually traced by medical experts

3.2   Spatial Distribution of Tree Branching Points

3.2.1 Sholl Analysis. The first method that we propose for capturing information
about the spatial distribution of branching points is based on Sholl analysis [11].
According to Sholl's methodology, a number n of concentric circles having as center
the mass center of the ROI is applied. As the common center of all circles, we
consider the root node of the tree structure. More specifically if the first circle has a
radius b, the radii of the concentric $2^{nd}, 3^{rd}, 4^{th}, ..., n^{th}$ circle is $2 \cdot b, 3 \cdot b, 4 \cdot b, ..., n \cdot b$
correspondingly. Based on these circles, a vector that counts the number of tree
branching points between any two successive concentric circles is computed. The
vector's length is equal to the number n of circles of Sholl's analysis. The element i of
the Sholl's vector is computed as follows:

Sholl (i) = the number of branching points existing between the ith circle and $(i+1)^{th}$
circl (1)

For example, the application of Sholl Analysis Fig. 3(b) with radius b = 50 pixels to
the image Fig. 3(a) with dimensions 1150x790 pixels, produces the vector *Sholl* =
$[0, 0, 1, 1, 0, 1, 0, 1, 2, 1, 2, 0, 2, 3, 1, 1, 1, 6, 1, 3, 0, 2, 4, 3, 2, 3, 0, 0]$.

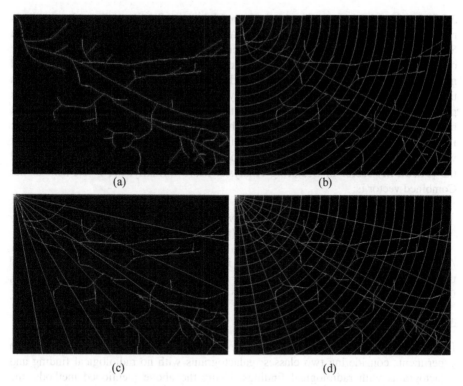

Fig. 3. Applying three methods to characterize the spatial distribution of tree's branching points: (a) the initial delineated ductal structure, (b) application of Sholl Analysis, (c) employment of Sectoring Analysis, (d) application of Combined Method

3.2.2 Sectoring Analysis. While Sholl's method captures the spatial distribution of branching points radiating out of the tree root, the second method we propose, the Sectoring Analysis, reflects the distribution of branching points based on consecutive clockwise sectors. As a typical ductal tree covers at most one out of the four quadrants of the plane, we focus only on the quadrant that contains all tree nodes. We omit the remaining three quadrant of the plane to avoid zero elements in the final vector. The quadrant, where the tree structure is depicted, is divided into a number m of consecutive sectors. The m sectors have as common starting point the root of the ductal tree and each triangular sector covers a part of the plane of a central angle equal to $90°/m$. According to the above partition of the plane, the Sectoring vector counts the number of tree branching points that belong to each sector. The vector's length is equal to the number m of Sectoring analysis. The ith element of the Sectoring vector is computed according to following formula:

$$Sectoring\ (i) = the\ number\ of\ branching\ points\ existing\ inside\ the\ i^{th}\ sector. \quad (2)$$

An example of the application of Sectoring method using $m = 8$ sectors is presented in Fig. 3(c). The resulted vector is $Sectoring = [0, 0, 3, 5, 13, 17, 3, 1]$.

3.2.3 Combined Method. The idea of combining the Sholl and Sectoring methods resulted in the third method we propose. According to this approach, we apply both Sholl and Sectoring Analysis to a tree structure. The center of Sholl as well the center of Sectoring Analysis is the root of the ductal tree. The application of the Combined method to an image divides one plane quadrant to a number of $n \cdot m$ cells, where n is the total number of circles of Sholl Analysis and m is the total number of sectors produced by Sectoring Analysis. The vector that reflects the spatial distribution of branching points using the Combined Method is constructed as Fig. 3(d) shows. The ith element of the Combined vector counts the branching points inside the ith cell where the ith cell is the cell produced by the $[(i \ div \ n)]$ circle of Sholl Analysis and the $(i \ mod \ m)$ sector of Sectoring Analysis. More formally, the ith element of the Combined vector is:

Combined (i) = the number of branching points existing inside the cell generated by the $[i \ div \ n]^{th}$ circle and the $(i \ mod \ m)^{th}$ sector.   (3)

Appling the Combined Method to the initial image Fig. 3(a) is presented in Fig. 3(d) and the Combined vector becomes $Combined = [0,0,0,0,0,0,0,0,0,0,0,0,0,0,0,0,0,0,0,1,0,$ $0,0,0,0,0,0,1,0,....]$ containing a total of $m \cdot n = 8 \cdot 28 = 224$ elements.

3.3 Galactogram Characterization

In order to characterize a clinical x-ray galactogram, we perform classification experiments considering two classes: galactograms with no radiological finding and galactograms with radiological findings. Using the above mentioned methods, the problem of galactogram classification is reduced to the problem of classification of the visualized tree structure. We employ the k-nearest neighbor classification scheme, which assigns the test tree to the class that appears most frequently among its neighbors. As the three methods presented above represent the spatial distribution of branching points with a numeric vector, we utilize the cosine similarity metric to retrieve the k closest neighboring tree structures. Given the vectors a and b of length t that represent the branching point's distribution of two trees correspondingly, the cosine similarity between them is computed by the following formula:

$$CosSim(a,b) = \frac{\vec{a}\cdot\vec{b}}{|\vec{a}|\cdot|\vec{b}|} = \frac{\sum_{i=1}^{t} a_i\cdot b_i}{\sqrt{\sum_{i=1}^{t} a_i^2\cdot\sum_{i=1}^{t} b_i^2}}.$$ (4)

4 Results

We applied the proposed methodology to a clinical galactogram dataset and we performed similarity search experiments and classification experiments to evaluate the accuracy of our methods. A detailed description of the dataset and the experimental results are presented below.

4.1 Dataset

Our dataset consisted of 50 x-ray galactograms performed at Thomas Jefferson University Hospital and the Hospital of the University of Pennsylvania. From these images, 28 corresponded to women with no reported galactographic findings (class

NF) and 22 to women with reported galactographic findings (class RF). The ductal trees visualized in the original galactograms were manually extracted and delineated by medical experts.

4.2 Similarity Experiments

We applied the proposed methods to the dataset and performed similarity search experiments. In similarity experiments, we considered each tree and its corresponding vector of spatial distribution of branching points as a query and retrieved the k-most similar trees based on the cosine similarity metric. The parameter k ranged from 1 to 5. To evaluate the methodology we reported the precision i.e., the percentage of neighboring images belonging to the same class as the query averaged over the entire dataset. As the two classes in our dataset are unbalanced, we randomly under-sampled the NF class to the size of the RF class and averaged the results over 100 sampling iterations. The following tables illustrate the precision obtained using the Sholl Analysis (Table1), the Sectoring Analysis (Table1), the Combined Method (Table2) and the tree characterization technique presented in [6] (Table2). We mention that in the experiments the parameter b of Sholl Analysis was equal to 10 pixels, the parameter m of Sectoring Analysis was equal to 16 sectors and the parameters b and m of Combined Method were 10 pixels and 16 sectors correspondingly.

Table 1. The obtained precision of similarity searches for Sholl and Sectoring Analysis.

	Precision of Similarity Searches					
k	**Sholl Analysis**			**Sectoring Analysis**		
	NF	RF	Total	NF	RF	Total
1	100%	100%	100%	100%	100%	100%
2	98.44%	74.06%	86.25%	99.06%	78.44%	88.75%
3	86.04%	65.62%	75.83%	87.29%	57.50%	72.40%
4	82.34%	65.31%	73.83%	83.13%	55.31%	69.22%
5	74.00%	60.75%	67.38%	74.00%	55.63%	64.81%

Table 2. The obtained precision of similarity searches for Combined and Prüfer/tf-idf methods

	Precision of Similarity Searches					
k	**Combined Method**			**Prüfer/tf-idf method**		
	NF	RF	Total	NF	RF	Total
1	100%	100%	100%	100%	100%	100%
2	96.25%	86.56%	91.41%	92.94%	74.53%	83.73%
3	85.21%	68.13%	76.67%	77.67%	66.92%	72.29%
4	78.91%	65.00%	71.95%	67.72%	66.70%	67.21%
5	70.13%	61.88%	66.00%	71.16%	60.16%	65.66%

4.3 Classification Experiments

Here, we performed leave-one-out k-nearest neighbor classification experiments using the cosine similarity metric. Also in this case, the values of k ranged from 1 to 5 and the NF class was under-sampled to the size of the smaller RF class. To evaluate the classification, we compared the accuracy obtained using the Sholl Analysis (Table3), the Sectoring Analysis (Table3), the Combined Method (Table4) and the state-of-the-art Prüfer/tf-idf method presented in [6] (Table4). The values of parameter b and m were equal to the values in similarity search experiments.

Table 3. The obtained classification accuracy using Sholl and Sectoring technique

	Classification Accuracy					
	Sholl Analysis			Sectoring Analysis		
k	NF	RF	Total	NF	RF	Total
1	95.00%	61.25%	78.13%	93.75%	46.88%	70.31%
2	96.88%	48.13%	72.50%	98.12%	56.87%	77.50%
3	69.58%	41.04%	55.31%	73.54%	39.37%	56.46%
4	73.39%	53.70%	63.54%	74.27%	40.36%	57.32%
5	66.63%	52.25%	59.44%	65.50%	42.13%	53.81%

Table 4. The obtained classification accuracy using Combined and Prüfer/tf-idf methods

	Classification Accuracy					
	Combined Method			Prüfer/tf-idf method		
k	NF	RF	Total	NF	RF	Total
1	91.87%	70.63%	81.25%	73.61%	53.61%	63.61%
2	92.50%	73.12%	82.81%	85.88%	49.06%	67.47%
3	68.75%	49.38%	59.06%	58.73%	50.92%	54.82%
4	73.59%	52.86%	63.23%	55.56%	56.60%	56.08%
5	62.87%	51.00%	56.94%	63.56%	50.20%	56.88%

5 Conclusion

We presented a new methodology for capturing branching properties of tree-like structures in medical images. Based on Sholl Analysis, we proposed Sectoring Analysis and we combined the two methods to obtain a more detailed analysis of the spatial distribution of branching points. We applied our approach to a dataset of clinical galactograms attempting to correlate the topological features of branching points with the underlying pathology of breast ductal trees. Our experimental results suggest that it is possible to deduce valuable information about the radiological findings from the topology of the ductal trees. Within our future research plans is the fine-tuning of the values of parameters b and m, as well as the application of statistical tests in order to check the statistical significance of the results. In addition, the study of other topological descriptors and the combination of all of them is a promising perspective to achieve even higher classification accuracy.

Acknowledgments. This research has been co-financed by the European Union (European Social Fund - ESF) and Greek national funds through the Operational Program "Education and Lifelong Learning" of the National Strategic Reference Framework (NSRF) - Research Funding Program: Heracleitus II. Investing in knowledge society through the European Social Fund.

References

1. Bullit, E., Muller, M.E., Jung, I., Lin, W., Aylward, S.: Analyzing attributes of vessel populations. Medical Image Analysis 24, 39–49 (2005)
2. Bullit, E., Zeng, D., Gerig, G., Aylward, S., Joshi, S., Smith, J.K., Lin, W., Ewend, M.G.: Vessel Tortuosity and Brain Malignancy: A Blinded Study. Academic Radiology 12, 1232–1240 (2005)
3. Tschirren, J., McLennan, G., Palagyi, K., Hoffman, E.A., Sonka, M.: Matching and anatomical labeling of human airway tree. IEEE Transactions in Medical Imaging 24, 701–716 (1996)
4. Dinkel, H.P., Trusen, A., Gassel, A.M., Rominger, M., Lourens, S., Muller, T., Tschammler, A.: Predictive value of galactographic patterns for benign and malignant neoplasms of the breast in patients with nipple discharge. The British Journal of Radiology 73, 706–714 (2000)
5. Bakic, P.R., Albert, M., Maidment, A.D.: Classification of galactograms with ramification matrices: preliminary results. Academic Radiology 10, 198–204 (2003)
6. Megalooikonomou, V., Barnathan, M., Kontos, D., Bakic, P.R., Maidment, A.D.: A Representation and Classification Scheme for Tree-like Structures in Medical Images: Analyzing the Branching Pattern of Ductal Trees in X-ray Galactograms. IEEE Transactions in Medical Imaging 28 (2009)
7. Kontos, D., Megalooikonomou, V., Javadi, A., Bakic, P.R., Maidment, A.D.: Classification of galactograms using fractal properties of the breast ductal network. In: Proc. IEEE International Symposium on Biomedical Imaging, pp. 1324–1327 (2006)
8. Skoura, A., Barnathan, M., Megalooikonomou, V.: Classification of ductal tree structures in galactograms. In: Proc. 6th IEEE International Symposium on Biomedical Imaging (2009)
9. Barnathan, M., Zhang, J., Kontos, D., Bakic, P.R., Maidment, A.D.A., Megalooikonomou, V.: Analyzing Tree-Like Structures in Biomedical Images Based on Texture and Branching: An Application to Breast Imaging. In: Krupinski, E.A. (ed.) IWDM 2008. LNCS, vol. 5116, pp. 25–32. Springer, Heidelberg (2008)
10. Moffat, D., Going, J.: Three dimensional anatomy of complete duct systems in human breast: pathological and developmental implications. Journal of Clinical Pathology 49, 48–52 (1996)
11. Sholl, D.: Dendritic organization in the neurons of the visual and motor cortices of the cat. J. Anat. 87, 387–406 (1953)

Intelligent Selection of Human miRNAs and Mouse mRNAs Related to Obstructive Nephropathy

Ioannis Valavanis[1,*], P. Moulos[1], Ilias Maglogiannis[2], Julie Klein[3], Joost Schanstra[3], and Aristotelis Chatziioannou[1,*]

[1] Institute of Biological Research and Biotechnology,
National Hellenic Research Foundation, Athens, Greece
[2] Department of Biomedical Informatics,
University of Central Greece, Lamia, Greece
[3] Institut National de la Santé et de la Recherche Médicale (INSERM),
U858, Toulouse, France
{ivalavan,pmoulos,achatzist}@eie.gr,
imaglo@ucg.gr,
{joost-peter.schanstra,julie.klein}@inserm.fr

Abstract. Obstructive Nephropathy (ON) is a renal disease and its pathology is believed to be magnified by various molecular processes. In the current study, we apply an intelligent workflow implemented in Rapidminer data mining platform to two different ON datasets. Our scope is to select the most important actors in two corresponding molecular information levels: human miRNA and mouse mRNA. A forward selection method with an embedded nearest neighbor classifier is initially applied to select the most important features in each level. The resulting features are next fed to classifiers appropriately tested utilizing a leave-one-out resampling technique in order to evaluate the relevance of the selected input features when used to classify subjects into output classes defined by ON severity. Preliminary results show that high classification accuracies are obtained, and are supported by the fact that the selected miRNAs or mRNAs have been found significant within differential expression analysis using the same datasets.

Keywords: obstructed nephropathy, miRNA, mRNA, feature selection, forward selection, k-nn classifier, classification tree.

1 Introduction

Obstructive nephropathy (ON) is a renal disease caused by impaired flow of urine or tubular fluid [1] and is the most frequent nephropathy observed in newborns and children. The improper flow of urine may be caused by the presence of an obstacle on the urinary tract, e.g. stenosis or abnormal implantation of the ureter in the kidney, and the resulting accumulation of urine within the kidney can lead to progressive alterations of the renal parenchyma, development of renal fibrosis and loss of renal

---

* Corresponding authors.

L. Iliadis et al. (Eds.): EANN/AIAI 2011, Part II, IFIP AICT 364, pp. 464–471, 2011.

function. However, ON is treatable and often reversible [2]. Common met in infants due to congenital abnormalities of the urinary tract, it represents 16.1% of all pediatric transplantations in North America [3].

In addition to the classical mechanical view of ON, there is evidence that the pathophysiological process of nephron destruction is magnified by cellular processes which can be classified into three broad categories: tubulointerstitial inflammation, tubular cell death and fibrosis [4-5]. The cellular interactions that regulate development of interstitial inflammation, tubular apoptosis and interstitial fibrosis are complex. Renal gene expression and protein production result to several biomarkers that include signaling molecules and receptors involved in macrophage recruitment and proliferation, tubular death signals and survival factors [6]. Given the afore described molecular aspect of ON pathology, it is important to advance its in depth knowledge using well established –omics profiling techniques combined with intelligent data mining methods that could identify the most important molecular actors in various levels, e.g. transcriptomics, proteomics and metabolomics .

Microarray experiments are one of the well knows examples of -omics profiling techniques that allow the systemic analysis and characterization of alterations in genes, RNA, proteins and metabolites, and offer the possibility of discovering novel biomarkers and pathways activated in disease or associated with disease conditions. In particular, they have become a major tool in medical knowledge discovery in order to: i) identify and categorize diagnostic or prognostic biomarkers ii) classify diseases iii) monitor the response to therapy, and iv) understand the mechanisms involved in the genesis of disease processes [7]. Microarray data pose a great challenge for computational techniques, because of their large dimensionality (up to several tens of thousands of genes) and their small sample sizes [8]. Furthermore, additional experimental complications like noise and variability make the analysis of microarray data an exciting domain that employs bioinformatics-driven methods to deal with these particular characteristics. These methods range from statistics or heuristics that identify differentially expressed genes between two different disease status to sophisticated data mining techniques that employ biologically inspired machine learning methodologies. In particular, the obvious need for dimension reduction was realized as early as the field of microarray analysis emerged [9-10] and the application of feature selection methodologies became a standard in the field [11].

In the current study, we employ a data mining framework towards the analysis of expression data related to ON that were derived through microarray techniques in two data levels: human miRNA and mice mRNA. Both available data sets are divided into subject groups based on ON severity. Data is analyzed within Rapidminer, a freely available open-source data mining platform that integrates fully the machine learning WEKA library, and permits easy data mining algorithms integration, process and usage of data and metadata [12-13]. Our final scope is to identify the most critical players in the two levels of molecular information. Towards this end, we apply a forward selection module with an embedded k-nearest neighbor (k-nn) classifier in order to select feature subsets of the greatest relevance, which are then tested for their generalization ability using other classifiers and the resampling technique of leave-one-out. Preliminary results here reported are further commented in relation to the ones obtained by applying statistical selection within the analysis of differential expression values in pairwise comparisons of disease status.

2 Dataset

The two –omics datasets analyzed within the current ON study correspond to i) human miRNA data and ii) mice mRNA data. Human miRNA dataset included children aged between two weeks and six months. Based on a set of clinical parameters, available samples belonged to three subsets 1) **Control** including children without any renal damage (8 subjects), 2) **NoOp** comprising children with mild obstruction who do not need to undergo surgery to repair the ureteropelvic junction (8 subjects), and 3) **Op**: children with severe obstruction who need surgery to repair and reconstruct the junction (10 subjects). The Agilent Human miRNA Microarray platform was used to measure expression values for a total number of 790 miRNAs. Three mice mRNA data groups were extracted after partial unilateral ureteral obstruction on neonatal mice in order to mimic the obstructive nephropathy syndrome as it occurs in children. These groups comprised non-operated **Control** mice (9 subjects), operated mice with **Mild** obstruction (5 subjects) and operated mice with **Severe** obstruction (4 subjects). Agilent's mice oligonucleotide microarrays were used to analyze the expression of about 41000 mouse transcripts, corresponding to 20873 genes. Prior to the analysis of both datasets within Rapidminer, expression values were normalized to (divided by) the average expression value in control subjects.

3 Methods

For both datasets, human miRNA and mouse mRNA, the same select and test protocol was followed in order to select the most important features (miRNAs and genes) and test their relevance to the classification of all subjects into corresponding ON related classes: Control, NoOp and Op for human miRNA, and Control, Mild and Severe for mouse genes. The protocol was implemented within the stand-alone Rapidminer platform [12-13] which includes in a single workflow all data mining steps (feature selection, classifier construction and evaluation) in appropriate operators (Fig. 1).

Forward Selection: This operator starts with an empty selection of features and, in each round, adds each unused feature of the given set of examples. For each added feature, the performance is estimated using an embedded operator (Fig. 2), which measures on a 10-fold cross-validation basis the average accuracy that the feature yields using a 6-nn classifier. The k-nn classifier[1] was used here due to the rather low computational cost it raises, compared to other alternatives e.g. artificial neural network or support vector machine, and the need for executing and evaluating the classifier for a large number of rounds within forward selection. An additional speculative number of rounds equal to three was used to ensure that the stopping

---

[1] A predifined number of nearest neighbors $k=6$ was set within forward selection. Other values for k, a classification tree module and a different validation technique (leave-one-out) were later used to evaluate the selected subsets of features.

criterion is fulfilled and that the algorithm does not get stuck in local optima. For both datasets, the maximum number of features was set to 30.

Classification: Following the forward selection, classification is performed to construct classifiers fed by the selected features and measure their performance, thus validating the relevance of the selected features. Here, three nearest neighbor based classifiers (k=1,6,12) and a classification tree (the gini index was used a split criterion) were used, while their average performance was measured using the

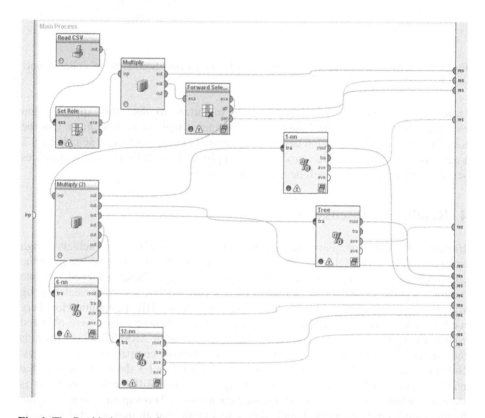

Fig. 1. The Rapidminer workflow: Read CSV operator reads the comma seperated value dataset file that contains all subject examples: features – genes or miRNAs – and the class they belong to. Set Role operator defines the corresponding entry as the target class. Forward Selection operator contains a recursive validation operator that uses an embedded 6-nn classifier (Fig. 2) and 10-fold cross validation. This operator outputs an examples set containing the selected features and the target class, as well as the average accuracy achieved by the selected features subset and the 6-nn classifier. The modified examples set is forwarded to 1-nn, 6-nn, 12-nn and tree classifiers all evaluated using leave-one-resampling. The Multiply operator provides copies of an examples set to be used by other operators. Numerical results (e.g. accuracies, confusion matrices) and classification models themselves can be forwarded to results section, where can be studied/used asynchronously.

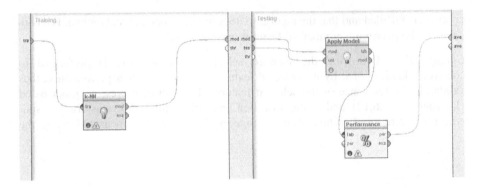

Fig. 2. The embedded training and testing process for a *k*-nn classification model in Rapidminer workflow

leave- *one-out resampling technique*. It is important to note that applying and evaluating a classifier within Rapidminer is a rather simple procedure, given that the user has set the implementation details of the classifier (e.g. number of neighbor for a k-nn classifier), the validation procedure to apply, and the attribute of the examples set fed to the classifier that will be used as a target class.

4   Results and Discussion

Human miRNA dataset: Forward Selection provided 7 miRNAs that achieved high accuracies both within selection process and, most importantly, when evaluated by the constructed classifiers using the leave-one-out strategy. These achieved average accuracies in the range 82%-93% (Table 1). Selected miRNA ids are presented along with *p*-values (t-test was applied), false discovery rates (FDR, Benjamini-Hochberg method was applied) and fold change natural values. All these values were obtained within a complementary analysis of miRNA dataset within ARMADA [14] towards the analysis of differential expression for two pairwise comparisons: Control vs. NoOp (Table 2) and Control vs. Op (Table 3). Tables 2 and 3 show that most of the miRNAs are found statistically significant (*p*<0.1) when differential expression is measured within one at least of the pairwise comparison. This supports their selection by the feature selection method applied here and imply that all may affect the disease outcome.

Mouse mRNA dataset: Forward Selection provided here three important features corresponding to mouse genes that provided well performing classifiers either within the feature selection process or following this process (Table 4). Pairwise comparisons, Control vs. Mild and Control vs. Severe, support the selection of the resulted genes which are presented with differential expression measurements (Tables 5,6) extracted similarly to miRNA data.

   Given the small size of both datasets, a more detailed analysis on both levels of information comprises our future work aiming finally at providing molecular actors that could act as ON biomarkers. Such an analysis should be undertaken towards feature selection by sophisticated and computationally costly data mining methods, e.g. artificial neural networks and evolutionary methods, along with a thorough exploration of the differential expression of miRNAs and mRNAs. Furthermore, relating resulted molecular players between the two levels of information, e.g. identify important mRNAs regulated by homologous mice miRNAs to the ones selected in humans, and exploiting established functional information on their role by controlled biological vocabularies (Gene Ontology Terms, KEGG Pathways) could further enlighten the molecular mechanisms beneath ON.

Table 1. Forward selection of human miRNAs: Mean accuracy obtained by the embedded 6-nn classifier within forward selection (10-fold cross validation) and mean accuracies obtained when the selected subset is fed to 1-nn, 6-nn, 12-nn classifiers and classification tree (leave-one-out resampling)

Forward Selection (embedded 6-nn)	1-nn	6-nn	12-nn	tree
96.67%	82.14%	92.86%	78.57%	82.14%

Table 2. miRNAs selected by forward feature selection: p-values, false discovery rates (FDR) and fold changes (NoOp/Control) in natural values for the pairwise comparison Control versus NoOp are presented

miRNA_ID	p-value	FDR	Fold Change (natural)
kshv-miR-K12-9*	0.000558	0.044043	1.19228
hsa-miR-125b	0.95032	0.978817	1.00416
hsa-miR-367*	0.433949	0.715699	1.0256
hsa-miR-199b-5p	0.140661	0.456205	1.0597
hsa-miR-377*	0.03673	0.268672	1.08498
ebv-miR-BART14*	0.031001	0.236668	1.10049
hsa-miR-509-3-5p	0.529799	0.775738	0.973679

Table 3. miRNAs selected by forward feature selection: p-values, false discovery rates (FDR) and fold changes (Op/Control) in natural values for the pairwise comparison Control versus Op are presented

miRNA_ID	p-value	FDR	Fold Change (natural)
kshv-miR-K12-9*	9.98E-05	0.024237	1.259912
hsa-miR-125b	0.030252	0.169496	0.847489
hsa-miR-367*	0.115856	0.347684	0.930161
hsa-miR-199b-5p	0.730141	0.868795	0.991497
hsa-miR-377*	0.550648	0.784874	1.023243
ebv-miR-BART14*	0.081845	0.297936	1.075465
hsa-miR-509-3-5p	0.007946	0.090974	1.123825

Table 4. Forward selection of mouse mRNAs: Mean accuracy obtained by the embedded 6-nn classifier within forward selection (10-fold cross validation) and mean accuracies obtained when the selected subset is fed to 1-nn, 6-nn, 12-nn classifiers and classification tree (leave-one-out resampling)

Forward Selection (embedded 6-nn)	1-nn	6-nn	12-nn	tree
100%	100%	100%	55.56%	94.44%

Table 5. mRNAs selected by forward feature selection (gene symbol, description and Genbank id is presented for each mRNA): p-values, false discovery rates (FDR) and fold changes (Mild/Control) in natural values for the pairwise comparison Control versus Mild are presented

Gene Symbol	Description	Genbank id	p-value	FDR	Fold Change (natural)
Tpm1	tropomyosin 1, alpha	NM_024427	0.002114	0.15	1.593333
Svs1	seminal vesicle secretory protein 1	NM_172888	0.06	0.53	0.77
Wnt4	wingless-related MMTV integration site 4	NM_009523	0.000338	0.09	1.42

Table 6. mRNAs selected by forward feature selection (gene symbol, description and Genbank id is presented for each mRNA): p-values, false discovery rates (FDR) and fold changes (Severe/Control) in natural values for the pairwise comparison Control versus Severe are presented

Gene Symbol	Description	Genbank id	p-value	FDR	Fold Change (natural)
Tpm1	tropomyosin 1, alpha	NM_024427	8.68E-06	0.008794	1.406667
Svs1	seminal vesicle secretory protein 1	NM_172888	0.18	0.77	1.2
Wnt4	wingless-related MMTV integration site 4	NM_009523	1.64E-06	0.005406	2.02

4 Conclusions

In the current study, an intelligent workflow aiming to the analysis of two –omics datasets, i.e. human miRNA and mouse mRNA of samples classified by ON severity derived using clinical parameters, was presented. The forward feature selection, first step within the workflow, selected 7 miRNAs and 3 mRNAs found as the most relevant molecular players towards the classification of samples into the corresponding classes in the two levels of molecular information. Preliminary results showed that the selected features yielded high accuracy measurements when fed to k-nn and tree based classifiers within the next step of the workflow, thus comprising candidate ON biomarkers to be studied within future work.

References

1. Klahr, S.: The geriatric patient with obstructive uropathy. Geriatr. Nephrol. Urol. 9, 101–107 (1999)
2. Klahr, S.: Obstructed Nephropathy. Internal Medicine 39(5), 355–361 (2000)
3. Bascands, J.L., Schanstra, J.P.: Obstructive nephropathy: Insights from genetically engineered animals. Kidney Int. 68, 925–937 (2005)
4. Ucero, A.C., Concalvesm, S., Benito-Martin, A., et al.: Obstructive renal injury: from fluid mechanics to molecular cell biology. Open Access Journal of Urology 2, 41–55 (2010)
5. Wen, J., Frøkiaer, J., Jørgensen, T., Djurhuus, J.: Obstructive nephropathy: an update of the experimental research. Urol. Res. 27(1), 29–39 (1999)
6. Chevalier, R.L.: Obstructive nephropathy: towards biomarker discovery and gene therapy. Nat. Clin. Pract. Nephrol. 2(3), 157–168 (2006)
7. Tarca, A.L., Romero, R., Draghici, S.: Analysis of microarray experiments of gene expression profiling. Am. J. Obstet. Gynecol. 195(2), 373–388 (2006)
8. Somorjai, R., et al.: Class prediction and discovery using gene microarray and proteomics mass spectroscopy data: curses, caveats, cautions. Bioinformatics 19, 1484–1491 (2003)
9. Alon, U., et al.: Broad patterns of gene expression revealed by clustering analysis of tumor and normal colon tissues probed by oligonucleotide arrays. Proc. Nat. Acad. Sci. USA 96, 6745–6750 (1999)
10. Ben-Dor, A., et al.: Tissue classification with gene expression profiles. J. Comput. Biol. 7, 559–584 (2000)
11. Saeys, Y., Inza, I., Larrañaga, P.: A review of feature selection techniques in bioinformatics. Bioinformatics 23(19), 2507–2517 (2007)
12. Mierswa, I., Wurst, M., Klinkenberg, R., Scholz, M., Euler, T.: YALE: Rapid Prototyping for Complex Data Mining Tasks. In: Proceedings of the 12th ACM SIGKDD International Conference on Knowledge Discovery and Data Mining, KDD 2006 (2006)
13. http://rapid-i.com/
14. Chatziioannou, A., Moulos, P., Kolisis, F.N.: Gene ARMADA: an integrated multi-analysis platform for microarray data implemented in MATLAB. BMC Bioinformatics 10, 354 (2009), doi:10.1186/1471-2105-10-354

Independent Component Clustering
for Skin Lesions Characterization

S.K. Tasoulis[1], C.N. Doukas[2], I. Maglogiannis[1], and V.P. Plagianakos[1]

[1] Department of Computer Science and Biomedical Informatics,
University of Central Greece, Papassiopoulou 2–4, Lamia, 35100, Greece
{stas,imaglo,vpp}@ucg.gr
[2] Department of Information and Communication Systems Engineering,
University of the Aegean, Karlovassi, 83200, Samos, Greece
doukas@aegean.gr

Abstract. In this paper, we propose a clustering technique for the recognition of pigmented skin lesions in dermatological images. It is known that computer vision-based diagnosis systems have been used aiming mostly at the early detection of skin cancer and more specifically the recognition of malignant melanoma tumor. The feature extraction is performed utilizing digital image processing methods, i.e. segmentation, border detection, color and texture processing. The proposed method combines an already successful clustering technique from the field of projection based clustering with a projection pursuit method. Experimental results show great performance on detecting the skin cancer.

Keywords: Pigmented Skin Lesion, Image Analysis, Feature Extraction, Unsupervised clustering, Cluster analysis, Independent Component Analysis, Projection Pursuit, Kernel density estimation.

1 Introduction

Several studies found in literature have proven that the analysis of dermatological images and the quantification of tissue lesion features may be of essential importance in dermatology [1,3]. The main goal is the early detection of malignant melanoma tumor, which is among the most frequent types of skin cancer, versus other types of non-malignant cutaneous diseases. The interest in melanoma is due to the fact that its incidence has increased faster than that of almost all other cancers and the annual incidence rates have increased on the order of $3 - 7\%$ in fair-skinned populations in recent decades [2].

The advanced cutaneous melanoma is still incurable, but when diagnosed at early stages it can be cured without complications. However, the differentiation of early melanoma from other non-malignant pigmented skin lesions is not trivial even for experienced dermatologists. In several cases, primary care physicians underestimate melanoma in its early stage [3]. To deal with this problem in several cases we utilize data mining methods. In particular, using clustering could be the key step to understand the differences between the types and subtypes of skin lesions.

L. Iliadis et al. (Eds.): EANN/AIAI 2011, Part II, IFIP AICT 364, pp. 472–482, 2011.

In this paper, based on an already saucerful clustering technique, we propose a new algorithmic framework for the skin lesion characterization. The paper is organized as follows: in Section 2 we present the image dataset, as well as the preprocessing and segmentation, and feature extraction techniques applied. Next, in Section 4 we present the ICA model. Section 5 is devoted to the proposed method and in Section 6 we investigate the efficiency of the proposed technique. The paper ends with concluding remarks.

2   Skin Lesions Image Analysis

The image data set used in this study is an extraction of the skin database that exists at the Vienna and the Athens General Hospital, kindly provided by Dr. Ganster. The whole data set consists of 3631 images, 972 of them are displaying nevus (dysplastic skin lesions), 2590 featuring non-dysplastic lesions and the rest 69 images contain malignant melanoma cases. The number of the melanoma images set is not small considering the fact that malignant melanoma cases in a primordial state are very rare. It is very common that many patients arrive at specialized hospitals with partially removed lesions.

The first step in an image analysis workflow is image segmentation, which in this case concerns the separation of the skin lesion from the healthy skin. For the special problem of skin lesion segmentation, mainly region-based segmentation methods are applied [5,9]. A simple approach is thresholding, which is based on the fact that the values of pixels that belong to a skin lesion differ from the values of the background.

In this study, a more sophisticated approach of a local/adaptive thresholding technique was adopted, where the window size, the threshold value and degree of overlap between successive moving windows were the procedure parameters. The details of this method may be found in [8]. Image analysis and feature extraction is performed by measurements on the pixels that represent a segmented object allowing non-visible features to be computed. Several studies have also proven the efficiency of border shape descriptors for the detection of malignant melanoma on both clinical and computer based evaluation methods [7,11]. Three types of features are utilized in this study: Border Features which cover the A and B parts of the ABCD-rule of dermatology, Color Features which correspond to the C rules and Textural Features, which are based on D rules. More specifically the extracted features are as follows:

Border features

- Thinness Ratio measures the circularity of the skin lesion defined as: $TR = 4\pi\text{Area}/(\text{perimeter})^2$.
- Border Asymmetry is computed as the percent of non-overlapping area after a hypothetical folding of the border around the greatest diameter or the maximum symmetry diameters.
- The variance of the distance of the border lesion points from the centroid location.

- Minimum, maximum, average and variance responses of the gradient operator, applied on the intensity image along the lesion border.

Color Features

- Plain RGB color plane average and variance responses for pixels within the
  lesion.
- Intensity, Hue, Saturation Color Space average and variance responses for
  pixels within the lesion: $I = \frac{R+G+B}{3}$, $\quad S = 1 - \frac{3}{R+G+B}[\min(R,G,B)]$, and

$$H = \begin{cases} W & , G > B, \\ 2\pi - W, & G < B, \\ 0 & , G = B, \end{cases} \text{ and } W = \arccos[\frac{R(1 - \frac{1}{2}(G+B))}{(R-G)^2 + (R-B)(G-B)^{\frac{1}{2}}}].$$

- Spherical coordinates LAB average and variance responses for pixels within
  the lesion:
  $L = \sqrt{R^2 + G^2 + B^2}$, $AngleA = \cos^{-1}[\frac{B}{L}]$,
  and $AngleB = \cos^{-1}[\frac{R}{L \sin(AngleA)}]$.

Texture features

- Dissimilarity, d, which is a measure related to contrast using linear increase
  of weights as one moves away from the GLCM (gray level co-occurrence
  matrix) diagonal: $d = \sum_{i,j=0}^{N-1} P_{i,j}\|i - j\|$, where i and j denote the rows
  and columns, respectively, N is the total number of rows and columns, and
  $P_{i,j} = \frac{V_{i,j}}{\sum_{i,j=0}^{N-1} V_{i,j}}$ is the normalization equation in which $V_{i,j}$ is the DN value
  of the cell i, j in the image window.
- Angular Second Moment, ASM, which is a measure related to orderliness,
  where $P_{i,j}$ is used as a weight to itself: $ASM = \sum_{i,j=0}^{N-1} iP_{i,j}^2$.
- GLCM Mean, μ_i , which differs from the familiar mean equation in the
  sense that it denotes the frequency of the occurrence of one pixel value
  in combination with a certain neighbor pixel value and is given by $\mu_i = \sum_{i,j=0}^{N-1} i(P_{i,j})$. For the symmetrical GLCM, holds that $\mu_i = \mu_j$.
- GLCM Standard Deviation, σ_i, which gives a measure of the dispersion of
  the values around the mean: $\sigma_i = \sqrt{\sum_{i,j=0}^{N-1} P_{i,j}(i - \mu_i)^2}$.

3   Clustering Background

The "divisive" hierarchical clustering techniques produce a nested sequence of
partitions, with a single, all-inclusive cluster at the top. Starting from this all-
inclusive cluster the nested sequence of partitions is constructed by iteratively
splitting clusters, until a termination criterion is satisfied. Any divisive clustering
algorithm can be characterized by the way it chooses to provide answers to the
following three questions:

Q_1: Which cluster to split further?
Q_2: How to split the selected cluster?
Q_3: When should the iteration terminate?

The projection based divisive clustering algorithms in particular, projects the high dimensional data onto a lower dimensional subspace to provide answers to the questions above, in a computationally efficient manner. Note also that certain answers to one of these questions may render obsolete one of the others. However, this is not always the case.

To formally describe the manner in which projection based divisive clustering algorithms operate, let us assume the data is represented by an $n \times a$ matrix D, whose each row represents a data sample d_i, for $i = 1, \ldots, n$. Finally, if A is the matrix with columns the vectors that denote the targeted subspace, then

$$D^P_{n \times k} = D_{n \times a} A_{a \times k},$$

is the projection of the data onto the lower k-dimensional subspace defined by the matrix A. The most studied technique for such data analysis is the Principal Component Analysis (PCA) [21]. PCA can be viewed as one of many possible procedures for projection pursuit [17] and is able to compute meaningful projections of high dimensional data [4,13,15,26].

4 Independent Component Analysis

Independent component analysis (ICA) [12,14] is a technique that finds underlying factors or independent components from multivariate (multidimensional) statistical data by maximizing the statistical independence of the estimated components. ICA defines a generative model for the observed multivariate data, which is typically given as a large database of samples. In the model, the data variables are assumed to be linear or nonlinear mixtures of some unknown latent variables, and the mixing system is also unknown. The latent variables are assumed non-gaussian and mutually independent, and they are called the independent components of the observed data. These independent components can be found by ICA. We can define ICA as follows. Let $x = (x_1, \ldots, x_n)$ be the random vector that represents the data and $s = (s_1, \ldots, s_n)$ be random vector that represents the components. The task is to transform the data x, using a linear static transformation W, as $s = Wx$, into maximally independent components s measured by some function of independence. The definition of independence for ICA that we utilize in this work is the maximization non-gaussianity. The most used measure of non-gaussianity is kurtosis and the second measure is given by negentropy. Kurtosis is zero for a gaussian random variable and nonzero for most non-gaussian random variables. Negentropy is based on the information-theoretic quantity of (differential) entropy. The entropy of a random variable can be interpreted as the degree of information that the observation of the variable gives. The more random and unstructured the variable is, the larger its entropy.

4.1   Relation to Projection Pursuit

The critical attribute of the ICA model is that we can use it to find directions for which the 1-dimensional projected data onto these directions show the least Gaussian distribution. It has been argued by Huber [18] and by Jones and Sibson [23] that the Gaussian distribution is the least interesting one, and that the most interesting directions are those that show the least Gaussian distribution. Interesting distribution can be a consider a distribution that captures the structure of the data. As such ICA can be considered as an 1-dimensional projection pursuit technique for finding directions of maximum non-gaussianity.

4.2   The FastICA Algorithm

To find the direction for maximum non-gaussianity, we utilize a well known fixed point algorithm. The FastICA algorithm [20] is a very efficient method for maximizing the objective function with respect to the selected measure of non-gaussianity. For this task it is assumed that the data is preprocessed by centering and whitening. Whitening can be achieved with principal component analysis or singular value decomposition. Whitening ensures that all dimensions are treated equally, before the algorithm is run. In this work, we only make use of the FastICA algorithm for one unit. The FastICA for one unit finds a direction w such that the projection $D_{n \times k}^{P} = D_{n \times a} w_{a \times 1}$, maximizes nongaussianity. Nongaussianity is here measured by the approximation of negentropy [19].

5   The Proposed Framework

A new clustering algorithm that was proposed recently [27], incorporates information about the true clusters in the data from the density of their projections on the principal components. Based on that principle, we introduce an new algorithmic scheme that utilize the ICA model to find optimal directions to project the data and then splits the data based on the density of their projections on these directions. In contrast to the previous approach, in this work, we choose as a splitting point the maximum of all the local minima of the projections density denoted by x^*. Finally, to estimate the ICA model, the FastICA algorithm is utilized. The proposed algorithmic scheme mICDC (refer to Algorithm 1) utilizes the following criteria:

- (Stopping Criterion) ST: Let $\Pi = \{\{C_i, P_i\}, \ i = 1, \ldots, k\}$ a partition of the data set $\mathcal{D}$ into k sets C_i, and the assorted projections P_i of them onto the direction of maximum nongaussianity. Let $\mathcal{X}$, be the set x_i^* of the density estimates $\hat{f}(x_i^*; h)$ of the projection P_i of the data of each $C_i \in \Pi$, $i = 1, \ldots, k$. Stop the procedure when the set $\mathcal{X}$ is empty.
- (Cluster Selection Criterion) CS: Let $\Pi = \{\{C_i, P_i\}, \ i = 1, \ldots, k\}$ a partition of the data set $\mathcal{D}$ into k sets C_i, and the assorted projections P_i of them onto the direction of maximum nongaussianity. Let $\mathcal{F}$ be the set of the

density estimates $f_i = \hat{f}(x_i^*; h)$ of x_i^* for the projection P_i of the data of each $C_i \in \Pi$, $i = 1, \ldots, k$. The next set to split is C_j, with $j = \arg\max_i\{f_i : f_i \in \mathcal{F}\}$.

- (Splitting Criterion) SPC: Let $\hat{f}'(x; h')$ be the kernel density estimation of the density of the projections $p_i \in \mathcal{P}$, and x^* the maximum of the minima. Then construct $P_1 = \{d_i \in \mathcal{D} : p_i \leqslant x^*\}$ and $P_2 = \{d_i \in \mathcal{D} : p_i > x^*\}$.

```
 1 Function mICDC (D)
 2 Get uᵖ the direction of maximum nongaussianity of D
 3 Calculate P = Duᵖ the projection of D to uᵖ
 4 Set Π = {{D, P}}
 5 repeat
 6     Select an element {C, Pᶜ} ∈ Π using Cluster Selection Criterion CS
 7     Split C into two sub-sets C₁, C₂, using Splitting Criterion SPC
 8     Remove {C, Pᶜ} from Π and set Π → Π ∪ {{C₁, Pᶜ¹}, {C₂, Pᶜ²}}, where
       Pᶜ¹, Pᶜ² are the projections of C₁, C₂ on the direction of maximum
       nongaussianity uᵖ¹, uᵖ² of C₁ and C₂, respectively
 9 until Stopping Criterion ST is not satisfied;
10 return Π the partition of D into |Π| clusters
```

Algorithm 1. The mICDC algorithm

The computational complexity of this approach, using a brute force technique, would be quadratic in the number of samples. However, it has been shown [16,28] that using techniques like the Fast Gauss Transform linear running time can be achieved for the Kernel Density Estimation, especially for the one dimensional case. To find the maximum of the minima, we only need to evaluate the density at n positions, in between the projected data points, since those are the only valid splitting points.

6 Experimental Analysis

As defined in Section 2, the image dataset used in our experiments consists of 3631 images; 972 of them are displaying nevus (dysplastic skin lesions), 2590 featuring non-dysplastic lesions, and the rest 69 images contain malignant melanoma cases. The number of the melanoma images is not so small considering the fact that malignant melanoma cases in a primordial state are very rare.

In our first experiment, in an attempt to effectively retrieve the malignant melanoma class, we consider the problem as a two class situation. Since we know beforehand the actual cluster number, we can control the number of clusters that the algorithm retrieves by setting a proper value of the bandwidth parameter for the density function. In [27] the bandwidth parameter was set by choosing a multiple of the h_{opt} bandwidth ("normal reference rule"), which is the bandwidth that minimizes the Mean Integrated Squared Error (MISE). This is given by:

$$h_{opt} = \sigma \left(\frac{4}{3n}\right)^{1/5}, \tag{1}$$

where σ is the standard deviation of the data. The starting multiplier value is set to 2 and then we set greater values in an attempt to get good clustering results with fewer clusters. Table 1 reports the clustering results with respect to the multiplier values. 100 experiments have been made for each case and the mean values and the respective standard deviation are presented.

Table 1. Results with respect to the mean clustering purity and V-measure (with the observed standard deviation in parenthesis) for several multiplier values for the full dataset

Multi.	mICDC		
	Purity	V-measure	Clusters
2	0.9959(0.00)	0.2788(0.07)	9.40(3.06)
4	0.9974(0.00)	0.6670(0.19)	5.28(2.81)
8	0.9986(0.00)	0.9005(0.06)	3(0.92)
	K-means		
	0.9979(0.00)	0.0673(0.00)	13(0)
	0.9978(0.00)	0.1013(0.00)	7(0)
	0.9878(0.00)	0.0913(0.06)	4(0)

To assess the quality of a data partition, additional external information not available to the algorithm, such as class labels, are used [22,24]. Consequently, the degree of correspondence between the resulting clusters and the classes assigned a priori to each object can be measured. For a dataset $\mathcal{D}$, let $\mathcal{L}$ be a set of labels $l_i \in \mathcal{L}$, for each point $d_i \in \mathcal{D}$, $i = 1, \ldots, n$, with l_i taking values in $\{1, \ldots, L\}$. Let a k-cluster partitioning $\Pi = \{\mathcal{C}_1, \ldots, \ldots, \mathcal{C}_k\}$. The purity of Π is defined as:

$$p(\Pi) = \frac{\sum_{j=1}^{k} \max\{|\{p_i \in \mathcal{C}_j : l_i = 1, \ldots, L\}|\}}{n}, \tag{2}$$

so that $0 \leq p(\Pi) \leq 1$. High values indicate that the majority of vectors in each cluster come from the same class, so in essence the partitioning is "pure" with respect to class labels.

However, cluster purity does not address the question of whether all members of a given class are included in a single cluster and therefore is expected to increase monotonically with the number of clusters in the result. For this reason, criteria like the V-measure [25] have been proposed. The V-measure tries to capture cluster homogeneity and completeness, which summarizes a clustering solution's success in including every point of a single class and no others. Again, high values corresponds to better performance. For details on how these are calculated, the interested reader should refer to [25].

As shown even for the case of a high multiplier value where the algorithm finds very few clusters, the performance remains at high levels. For comparison purposes, we also report the performance of well known k-means algorithm. For the results to be comparable, the number of clusters for the k-means algorithm is set to the values found by the mICDC algorithm. For the computation of k-means, we employ the Matlab function "kmeans".

To better understand the clustering results, firstly we employ the confusion matrices for the 4 cluster case (Table 2). Class 1 refers to the malignant melanoma class and class 2 to the rest of the dataset.

Table 2. Confusion Matrices of the mICDC and k-means algorithms for 4 clusters

	mICDC			k-means	
	Class 1	Class 2		Class 1	Class 2
cluster1	0	3558	cluster1	6	2436
cluster2	62	0	cluster2	62	0
cluster3	4	0	cluster3	1	884
cluster4	3	0	cluster4	0	238

As shown in this case, the mICDC algorithm does not split the much bigger class 2; so the rest of the clusters are considered to belong to class 1 (melanoma samples). This is a very important result, since one can easily conclude that the clusters with the much fewer samples constitute the malignant melanoma class. As shown in Table 3, the malignant melanoma class can also be found even when the algorithms retrieves more clusters.

Table 3. Confusion Matrices of the mICDC algorithm for 7 clusters

	mICDC				
	Class 1	Class 2		Class 1	Class 2
cluster1	5	0	cluster5	5	0
cluster2	46	0	cluster6	0	3558
cluster3	6	0	cluster7	3	0
cluster4	4	0			

Table 4. Results with respect to the mean clustering purity and V-measure (with the observed standard deviation in parenthesis) for different multiplier values for the dataset containing only nevus and the malignant melanoma classes

Multi.	mICDC		
	Purity	V-measure	Clusters
2	0.9809 (0.02)	0.3710 (0.23)	7.30 (2.98)
4	0.9520 (0.02)	0.2490 (0.40)	2.20 (0.63)
	K-means		
	0.9908 (0.00)	0.1718 (0.01)	10 (0)
	0.9452 (0.02)	0.1420 (0.02)	3 (0)

Table 5. Confusion Matrices of the mICDC algorithm for 7 clusters

mICDC					
	Class 1	Class 2		Class 1	Class 2
cluster1	55	0	cluster4	6	0
cluster2	4	0	cluster5	0	52
cluster3	0	920	cluster6	4	0

In our next experiment, we will perform clustering on the dataset containing only the displaying nevus and the malignant melanoma classes. The lack of samples of the non-dysplastic class makes the projection pursuit problem a bit more difficult, due loss of information. The results are presented in Table 4. In Table 5, we exhibit a high performance case of the mICDC algorithm. In this case the two classes have been perfectly separated and the majority of the displaying nevus samples are at the same cluster. However, there also exists a small cluster that does not belong to the malignant melanoma class.

7   Conclusions

In this paper, a clustering technique for the recognition of pigmented skin lesions in dermatological images is proposed. The images are preprocessed and feature extraction is performed utilizing digital image processing methods, i.e. segmentation, border detection, color and texture processing. The proposed clustering methodology combines an already successful clustering technique from the field of projection based clustering with a projection pursuit method. The new framework utilizes the ICA model to find optimal projections, and then it incorporates information from the density of the projected data to effectively retrieve the malignant melanoma class. Experimental results show great performance on detecting the skin cancer.

Acknowledgments. The authors thank the European Social Fund (ESF), Operational Program for EPEDVM and particularly the Program Herakleitos II, for financially supporting this work.

References

1. Maglogiannis, I., Doukas, C.: Overview of advanced computer vision systems for skin lesions characterization. IEEE Transactions on Information Technology in Biomedicine 13(5), 721–733 (2009)
2. Marks, R.: Epidemiology of melanoma. Clin. Exp. Dermatol. 25, 459–463 (2000)
3. Pariser, R., Pariser, D.: Primary care physicians errors in handling cutaneous disorders. J. Am. Acad. Dermatol. 17(3), 239–245 (1987)
4. Nilsson, M.: Hierarchical Clustering using non-greedy principal direction divisive partitioning. Information Retrieval 5(4), 311–321 (2002)
5. Zhang, Z., Stoecker, W., Moss, R.: Border detection on digitized skin tumor image. IEEE Transactions on Medical Imaging 19(11), 1128–1143 (2000)

6. Maglogiannis, I.: Automated segmentation and registration of dermatological images. Springer 2, 277–294 (2003)
7. Maglogiannis, I., Pavlopoulos, S., Koutsouris, D.: An integrated computer supported acquisition, handling and characterization system for pigmented skin lesions in dermatological images. IEEE Transactions on Information Technology in Biomedicine 9(1), 86–98 (2005)
8. Maglogiannis, I., Zafiropoulos, E., Kyranoudis, C.: Intelligent Segmentation and Classification of Pigmented Skin Lesions in Dermatological Images. In: Antoniou, G., Potamias, G., Spyropoulos, C., Plexousakis, D. (eds.) SETN 2006. LNCS (LNAI), vol. 3955, pp. 214–223. Springer, Heidelberg (2006)
9. Chung, D.H., Sapiro, G.: Segmenting skin lesions with partial-differential-equations-based image processing algorithms. IEEE Transactions on Medical Imaging 19(7), 763–767 (2000)
10. Nachbar, F., Stolz, W., Merkle, T., Cognetta, A.B., Vogt, T., Landthaler, M., Bilek, P., Braun-Falco, O., Plewig, G.: The abcd rule of dermatoscopy: High prospective value in the diagnosis of doubtful melanocytic skin lesions. J. Amer. Acad. Dermatol. 30(4), 551–559 (1994)
11. Stoecker, W.V., Li, W.W., Moss, R.H.: Automatic detection of asymmetry in skin tumors. Computerized Med. Imag. Graph 16(3), 191–197 (1992)
12. Hyvärinen, A., Karhunen, J., Oja, E.: Independent Component Analysis. John Wiley & Sons, Chichester (2001)
13. Boley, D.: Principal direction divisive partitioning. Data Mining and Knowledge Discovery 2(4), 325–344 (1998)
14. Comon, P.: Independent component analysis, a new concept? Signal Process 36, 287–314 (1994)
15. Dhillon, I.S.: Co-clustering documents and words using bipartite spectral graph partitioning. In: Proceedings of the seventh ACM SIGKDD International Conference on Knowledge Discovery and Data Mining, pp. 269–274. ACM, New York (2001)
16. Greengard, L., Strain, J.: The fast gauss transform. SIAM J. Sci. Stat. Comput. 12(1), 79–94 (1991)
17. Huber, P.: Projection Pursuit. Annals of Statistics 13, 435–475 (1985)
18. Huber, P.J.: Projection pursuit. Annals of Statistics 13(2), 435–475 (1985)
19. Hyvärinen, A., Oja, E.: Independent component analysis: algorithms and applications. Neural Networks 13(4-5), 411–430 (2000)
20. Hyvärinen, A.: Fast and robust fixed-point algorithms for independent component analysis. IEEE Transactions on Neural Networks 10, 626–634 (1999)
21. Jain, A.K., Dubes, R.C.: Algorithms for clustering data (1988)
22. Jain, A., Dubes, R.: Algorithms for Clustering Data. Prentice Hall, Englewood Cliffs (1988)
23. Jones, M.C., Sibson, R.: What is projection pursuit? Journal of the Royal Statistical Society. Series A (General) 150(1), 1–37 (1987)
24. Kogan, J.: Introduction to Clustering Large and High-Dimensional Data (2007)
25. Rosenberg, A., Hirschberg, J.: V-measure: A conditional entropy-based external cluster evaluation measure. In: 2007 Joint Conference on Empirical Methods in Natural Language Processing and Computational Natural Language Learning (EMNLP-CoNLL), pp. 410–420 (2007)

26. Tasoulis, S., Tasoulis, D.: Improving principal direction divisive clustering. In: 14th ACM SIGKDD International Conference on Knowledge Discovery and Data Mining (KDD 2008), Workshop on Data Mining using Matrices and Tensors, Las Vegas, USA (2008)

27. Tasoulis, S., Tasoulis, D., Plagianakos, V.: Enhancing Principal Direction Divisive Clustering. Pattern Recognition 43, 3391–3411 (2010)

28. Yang, C., Duraiswami, R., Gumerov, N.A., Davis, L.: Improved fast gauss transform and efficient kernel density estimation. In: Proceedings of Ninth IEEE International Conference on Computer Vision 2003, pp. 664–671 (2003)

A Comparison of Venn Machine with Platt's Method in Probabilistic Outputs

Chenzhe Zhou[1], Ilia Nouretdinov[1], Zhiyuan Luo[1], Dmitry Adamskiy[1], Luke Randell[2], Nick Coldham[2], and Alex Gammerman[1]

[1] Computer Learning Research Centre,
Royal Holloway, University of London, Egham, Surrey, TW20 0EX, UK
[2] Veterinary Laboratories Agency, Weybridge, Surrey, KT15 3NB, UK

Abstract. The main aim of this paper is to compare the results of several methods of prediction with confidence. In particular we compare the results of Venn Machine with Platt's Method of estimating confidence. The results are presented and discussed.

Keywords: Venn Machine, Support Vector Machine, Probabilistic Estimation.

1 Introduction

There are many machine learning algorithms that allow to make classification and regression estimation. However, many of them suffer from the absence of a confidence measure to assess the risk of error made by an individual prediction.

Sometimes, however, the confidence measure is introduced but very often it is an ad hoc measure. An example of this is a Platt's algorithm developed to estimate confidence for SVM[1]. We recently developed a set of new machine learning algorithms [2,3] that allow not just to make prediction but also to supply this prediction with a measure of confidence. What's more important is that this measure is valid and based on a well-developed algorithmic randomness theory.

The algorithm introduced in this paper is Venn Machine[3], a method that outputs the prediction with an interval of probability that prediction is correct. What follows is an introduction to Venn Machine and Platt's Method, then description of used data and results of experiments.

1.1 Venn Machine

Let us consider a training set consisting of object, x_i, and label, y_i, as pairs: $(x_1, y_1), \ldots, (x_{n-1}, y_{n-1})$. The possible labels are finite, that is, $y \in \mathbf{Y}$. Our task is to predict the label y_n for the new object x_n and give the estimation of the likelihood that our prediction is correct.

In brief, Venn Machine operates as follows. First, we define a *taxonomy* that can divide all examples into categories. Then, we try all the possible labels of the

L. Iliadis et al. (Eds.): EANN/AIAI 2011, Part II, IFIP AICT 364, pp. 483–490, 2011.

new object. In each attempt, we can calculate the frequencies of the labels in the category which the new object falls into. The minimum frequency is called the *quality* of this column. At last, we output the assumed label with the highest *quality* among all the columns as our prediction and output the minimum and the maximum frequencies of this column as the interval of the probability that this prediction is correct.

Taxonomy (or, more fully, *Venn taxonomy*) is a function A_n, $n \in \mathbf{N}$ of the space $\mathbf{Z}^{(n-1)} \times \mathbf{Z}$ that divide every example into one of the finite categories τ_i, $\tau_i \in \mathbf{T}$. Then we consider z_i as the pair (x_i, y_i),

$$\tau_i = A_n(\{z_1, \ldots, z_{i-1}, z_{i+1}, \ldots, z_n\}, z_i) \tag{1}$$

We assign z_i and z_j to the same category if and only if

$$A_n(\{z_1, \ldots, z_{i-1}, z_{i+1}, \ldots, z_n\}, z_i) = A_n(\{z_1, \ldots, z_{j-1}, z_{j+1}, \ldots, z_n\}, z_j) \tag{2}$$

Here is an example of a simplest *taxonomy* based on 1-nearest neighbour (1NN).

We assign the category of an example the same to the label of its nearest neighbour based on the distance between two objects (e.g. Euclidean distance).

$$A_n(\{z_1, \ldots, z_{i-1}, z_{i+1}, \ldots, z_n\}, z_i) = \tau_i = y_j \tag{3}$$

where

$$j = \arg \min_{j=1,\ldots,i-1,i+1,\ldots,n} ||x_i - x_j|| \tag{4}$$

For every attempt (x_n, y), of which the category is τ, let p_y be the empirical probability distribution of the labels in category τ.

$$p_y\{y'\} := \frac{|\{(x^*, y^*) \in \tau : y^* = y'\}|}{|\tau|} \tag{5}$$

this is a probability distribution on $\mathbf{Y}$. The set $P_n := \{p_y : y \in \mathbf{Y}\}$ is the multiprobability predictor consists of K probabilities, where $K = |Y|$.

After all attempts, we get a $K \times K$ matrix P. Let the *best* column with the highest quality, which is the minimum entry of a column, be j_{best}. j_{best} is our prediction and the interval of the probability that the prediction is correct is

$$[\min_{i=1,\ldots,K} P_{i,j_{best}}, \max_{i=1,\ldots,K} P_{i,j_{best}}] \tag{6}$$

1.2   Platt's Method

Standard Support Vector Machines (SVM) [4] only output the value of *sign* $(f(x_i))$, where f is the decision function. So we can say that SVM is a non-probabilistic binary linear classifier. But in many cases we are more interested in the belief that the label should be +1, that is, the probability $P(y = 1|x)$. Platt introduced a method to estimate posterior probabilities based on the decision function f by fitting a sigmoid for SVM.

$$P(y = 1|f) = \frac{1}{1 + exp(Af + B)} \tag{7}$$

The best parameter A and B are determined by using maximum likelihood estimation from a training set (f_i, y_i). Let us use regularized target probabilities t_i as the new training set (f_i, t_i) defined as:

$$t_i = \begin{cases} \frac{N_+ + 1}{N_+ + 2}, & \text{if } y_i = +1 \\ \frac{1}{N_- + 2}, & \text{if } y_i = -1 \end{cases} \tag{8}$$

where N_+ is the number of positive examples, while N_- is the number of negative examples. Then, the parameters A and B are found by minimizing the negative log likelihood of the training data, which is a cross-entropy error function.

$$-\sum_i (t_i \log (p_i) + (1 - t_i) \log(1 - p_i)) \longrightarrow \min_{p_i} \tag{9}$$

where the solution is

$$p_i = \frac{1}{1 + exp(Af_i + B)} \tag{10}$$

With parameters A and B we can calculate the posterior probability that the label should be +1 of every example using (10). But in many cases, probability that the prediction is correct is more useful and easy to compare with Venn Machine. In this binary classification problem, one example with the probability p_i means its label should be +1 with the likelihood of p_i, that is to say, its label should be −1 with the likelihood of $1 - p_i$. So we use the complementary probability when the probability is less than the optimal threshold (in this paper we set it to 0.5 as explained later).

2   Data Sets

The data sets we used in this paper is Salmonella mass spectrometry data provided by VLA[1] and Wisconsin Diagnostic Breast Cancer (WDBC) data from UCI.

The aim of the study of Salmonella data is to discriminate Salmonella vaccine strains from wild type field strains of the same serotype. We analysed the set of 50 vaccine strains (Gallivav vaccine strain) and 43 wild type strains. Both vaccine and wild type strains belong to the same serotype Salmonella enteritidis.

Each strain was represented by three spots; each spot produced 3 spot replicates. Therefore, there are 9 replicates per strain. Pre-processing was applied to each replicate and resulted in representation of each mass spectra as a vector of 25 features corresponding to the intensity of most common peaks. The median was later taken for each feature across replicates of the same strain. In the data set, label +1 corresponds to vaccine strains, label −1 to wild type strains. Table 1 shows some quantitive properties of the data set.

---

[1] Veterinary Laboratories Agency.

Table 1. Salmonella Data Set Features

Number of Instances	Number of Attributes	Number of Positive Examples	Number of Negative Examples
93	25	50	43

In Figure 1, there is a plot of the class-conditional densities $p(f|y = \pm 1)$ of Salmonella data. The plot shows histograms of the densities of the data set with bins 0.1 wide, derived from Leave-One-Out Cross-Validation. The solid line is $p(f|y = +1)$, while the dot line is $p(f|y = -1)$. What we observed from the plot is that this a linearly non-separable data set.

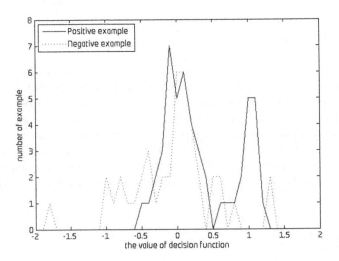

Fig. 1. The histograms for $p(f|y = \pm 1)$ for a linear SVM trained on the Salmonella Data Set

The second data set is Wisconsin Diagnostic Breast Cancer data. There are ten real-valued features computed for each cell nucleus, resulting in 30 features in the data set. These features are from a digitized image of a fine needle aspirate (FNA) of a breast mass. They describe characteristics of the cell nuclei present in the image. And the diagnosis includes two predicting fields, label +1 corresponding to Benign and label −1 corresponding to Malignant. Data set is linearly separable using all 30 input features. Table 2 shows some quantitive properties of the data set.

Table 2. Wisconsin Diagnostic Breast Cancer Data Set Features

Number of Instances	Number of Attributes	Number of Positive Examples	Number of Negative Examples
569	30	357	212

3   Empirical Result

There are two experiments in this paper to compare the performance of Venn Machine with the SVM+sigmoid combination in Platt's Method.

3.1   Taxonomy Design

The taxonomy used in both experiments is newly designed and it is based on the decision function the same as Platt's Method.

Let the number of categories $K_T = |T|$ and the taxonomy is further referred to as K_T-SVM. Then we train an SVM for the whole data $\{(x_1, y_1), \ldots, (x_n, y_n)\}$ and calculate the decision values for all examples. We put the examples into the same category if the decision values of them are in the same interval which is generated depending on K_T.

For an instance, if $K_T = 8$, the intervals can be $(-\infty, -1.5]$, $(-1.5, -1.0]$, $(-1.0, -0.5]$, $(-0.5, 0]$, $(0, 0.5]$, $(0.5, 1.0]$, $(1.0, 1.5]$, $(1.5, \infty)$.

3.2   Experiments

The first experiment dealing with Salmonella data set is using a radial basis function (i.e. RBF) kernel in SVM and a Venn Machine with 8-SVM taxonomy since Salmonella data is a linearly non-separable data set . And the second experiment dealing with WDBC data set is using a linear kernel in SVM (i.e. Standard SVM) and a Venn Machine with 6-SVM taxonomy. The Venn Machine can be compared to Platt's Method and the raw SVM for accuracies and estimated probabilities. Assuming equal loss for Type I and Type II errors, the optimal threshold for the Platt's Method is $P(y = 1|f) = 0.5$. And all of the results in this paper are presented using Leave-One-Out Cross-Validation (LOOCV).

Table 3 shows the parameters setting for experiments. The C value is the *cost* for the SVM. And the Underlying Algorithm is the algorithm used in the taxonomy for Venn Machine and the kernel used in SVM. The Kernal Parameter is σ, the parameter of RBF.

Table 4 is the results of experiments. The table lists the accuracies and the probabilistic outputs for raw SVM, Platt's Method, and Venn Machine using both data sets. For Platt's Method, the probabilistic output is the average estimated probability that the prediction is correct. And for Venn Machine, the probabilistic output is the average estimated interval of probability that the prediction is correct.

Table 3. Experimental Parameters

Data Set	Task	C	Underlying Algorithm	Kernal Parameter
Salmonella	SVM	1	RBF	0.05
	Platt's Method	1	RBF	0.05
	Venn Machine	1	8-SVM RBF	0.05
WDBC	SVM	1	Linear	
	Platt's Method	1	Linear	
	Venn Machine	1	6-SVM Linear	

Table 4. Experimental Results

Data Set	Task	Accuracy	Probabilistic Outputs
Salmonella	SVM	81.72%	
	Platt's Method	82.80%	84.77%
	Venn Machine	90.32%	[83.49%, 91.03%]
WDBC	SVM	97.72%	
	Platt's Method	98.07%	96.20%
	Venn Machine	98.24%	[97.22%, 98.27%]

3.3   Results

Table 5 lists some comparisons between two methods. As shown in the table, Venn Machine got better results in both two data sets. For Salmonella data set, Venn Machine got a significant improvement (7.52%) comparing with Platt's Method in accuracy when it used a 8-SVM RBF taxonomy. In the aspect of probabilistic outputs, Venn Machine output an interval of probability with the accuracy included while the probabilistic output of Platt's Method is 1.93% higher than the accuracy. For WDBC data set, Venn Machine increased by 0.52% in accuracy while Platt's Method got 0.35%. In the aspect of probabilistic outputs, Venn Machine output an interval of probability with the accuracy included while the probabilistic output of Platt's Method is 1.87% lower than the accuracy.

Sensitivity and specificity are also calculated and shown in Table 5. For Salmonella Data Set, Venn Machine got a outstanding result in sensitivity, 16.00% better than Platt's Method. It is obvious that Venn Machine got a better ability of identity salmonella vaccine. And for WDBC Data Set, they got approximate results in both sensitivity and specificity. It is hard to tell which method is better, but we can still find Venn Machine has made a slight improvement in both aspects.

Another interest thing we observed is that Platt's Method performs better on linearly separable data set (that is WDBC in this paper) than linearly non-separable data set (that is Salmonella data set), while Venn Machine can achieve

Table 5. Comparisons Between Two Methods

Data Set	Task	Accuracy	Probabilistic Outputs	Sensitivity	Specificity
Salmonella	Platt's Method	82.80%	84.77%	76.00%	67.44%
	Venn Machine	90.32%	[83.49%, 91.03%]	92.00%	67.44%
WDBC	Platt's Method	98.07%	96.20%	97.52%	99.02%
	Venn Machine	98.24%	[97.22%, 98.27%]	97.53%	99.50%

good results on both data sets. But it needs conducting experiments on more data sets to prove this.

Table 6 shows several examples in Salmonella data set predicted by Venn Machine and Platt's Method. For each example, the table contains the true label, prediction of Venn Machine and intervals of probability that the prediction is correct, the prediction of Platt's Method and the probabilistic outputs. The table indicates that both methods can be proper or erroneous. For instance, wild type strain 2, 4, 5 and vaccine strain 44 are both wrong for the two methods.

Table 6. Prediction for Individual Examples in Salmonella Data Set

No.	True Label	Prediction of VM	Probabilistic Outputs of VM	Prediction of PM	Probabilistic Outputs of PM
1	−1	−1	[88.89%, 100.00%]	−1	95.65%
2	−1	+1	[60.00%, 63.33%]	+1	77.72%
3	−1	−1	[88.89%, 100.00%]	−1	98.49%
4	−1	+1	[60.00%, 63.33%]	+1	63.98%
5	−1	+1	[60.00%, 63.33%]	+1	56.57%
6	−1	−1	[76.19%, 80.95%]	−1	78.57%
7	−1	−1	[76.19%, 80.95%]	−1	71.69%
...	...	...	...	...	...
44	+1	−1	[80.95%, 85.71%]	−1	71.92%
45	+1	+1	[90.00%, 93.33%]	+1	96.91%
46	+1	+1	[90.00%, 93.33%]	+1	77.96%
47	+1	+1	[56.67%, 60.00%]	−1	61.81%
48	+1	+1	[56.67%, 60.00%]	−1	58.43%
49	+1	+1	[90.00%, 93.33%]	+1	96.15%
50	+1	+1	[90.00%, 93.33%]	+1	94.12%
...	...	...	...	...	...

4   Conclusion

From our experience on these data sets we see the following. The Platt's estimation for the accuracy of prediction can be too optimistic or too pessimistic, while Venn's bounds estimate it more correctly: two-sided estimation is safer than single one. As for the accuracy itself, we see that if Platt's and Venn Machines are based on the same kind of SVM, accuracy of Venn Machine is also a bit better. This may be because Venn Machine do not rely on a fixed transformation of the SVM output, but makes its own transformation for each taxonomy, based on the actual data set.

We applied different probabilistic approaches to the dataset of Salmonella strains. As it can be seen from Figure 1 and Table 6, this data set is hard to separate: there are few errors in the class +1, but large part of examples from the class −1 seems to be hardly distinguishable from the class +1. This is why in this case we need to have individual assessment of prediction quality: being unable to make a confident prediction on any example, we still can select some of them where our prediction has higher chance to be correct.

The results have been observed on two particular data sets. We plan to conduct experiments on bigger data sets. Another possible direction is to compare Venn Machine and Platt's Method theoretically.

Acknowledgements. This work was supported in part by funding from BB-SRC for ERASySBio+ Programme: Salmonella Host Interactions PRoject European Consortium (SHIPREC) grant; VLA of DEFRA grant on Development and Application of Machine Learning Algorithms for the Analysis of Complex Veterinary Data Sets; EU FP7 grant O-PTM-Biomarkers (2008–2011); and by grant PLHRO/0506/22 (Development of New Conformal Prediction Methods with Applications in Medical Diagnosis) from the Cyprus Research Promotion Foundation.

References

1. Platt, J.C.: Probabilistic Outputs for Support Vector Machines and Comparisons to Regularized Likelihood Methods. In: Advances in Large Margin Classifiers, pp. 61–74. MIT Press, Cambridge (1999)
2. Gammerman, A., Vovk, V.: Hedging Predictions in Machine Learning (with discussion). The Computer Journal 50(2), 151–163 (2007)
3. Vovk, V., Gammerman, A., Shafer, G.: Algorithmic Learning in a Random World. Springer, Heidelberg (2005)
4. Vapnik, V.: The Nature of Statistical Learning Theory. Springer, Heidelberg (1995)

Author Index